Birkhäuser

Trends in Mathematics

Trends in Mathematics is a series devoted to the publication of volumes arising from conferences and lecture series focusing on a particular topic from any area of mathematics. Its aim is to make current developments available to the community as rapidly as possible without compromise to quality and to archive these for reference.

Proposals for volumes can be submitted using the Online Book Project Submission Form at our website www.birkhauser-science.com.

Material submitted for publication must be screened and prepared as follows:

All contributions should undergo a reviewing process similar to that carried out by journals and be checked for correct use of language which, as a rule, is English. Articles without proofs, or which do not contain any significantly new results, should be rejected. High quality survey papers, however, are welcome.

We expect the organizers to deliver manuscripts in a form that is essentially ready for direct reproduction. Any version of TEX is acceptable, but the entire collection of files must be in one particular dialect of TEX and unified according to simple instructions available from Birkhäuser.

Furthermore, in order to guarantee the timely appearance of the proceedings it is essential that the final version of the entire material be submitted no later than one year after the conference.

Ho-Hon Leung • R. Sivaraj • Firuz Kamalov
Editors

Recent Developments in Algebra and Analysis

International Conference on Recent Developments in Mathematics, Dubai, 2022 – Volume 1

Birkhäuser

Editors

Ho-Hon Leung
Department of Mathematical Sciences
United Arab Emirates University
Al Ain, United Arab Emirates

R. Sivaraj
Department of Mathematics
Dr. B. R. Ambedkar National Institute of Technology
Jalandhar, Punjab, India

Firuz Kamalov
Department of Electrical Engineering
Canadian University Dubai
Dubai, United Arab Emirates

ISSN 2297-0215 ISSN 2297-024X (electronic)
Trends in Mathematics
ISBN 978-3-031-37537-8 ISBN 978-3-031-37538-5 (eBook)
https://doi.org/10.1007/978-3-031-37538-5

This book is published under the imprint Birkhäuser, www.birkhauser-science.com by the registered company Springer Nature Switzerland AG
The registered company address is: Gewerbestrasse 11, 6330 Cham, Switzerland

Paper in this product is recyclable.

Preface

The Canadian University Dubai, UAE, and United Arab Emirates University, UAE, jointly organized the International Conference on Recent Developments in Mathematics (ICRDM 2022) during August 24–26, 2022, in Hybrid mode at Canadian University Dubai, UAE. The major objective of ICRDM 2022 is to promote scientific and educational activities toward the advancement of common man's life by improving the theory and practice of various disciplines of Mathematics. The conference was a grand success, and more than 500 participants (professors/scholars/students) enriched their knowledge in the wings of mathematics through ICRDM 2022. Over 200 leading researchers worldwide served in various capacities to organize ICRDM 2022. Thirty-one eminent speakers worldwide delivered the keynote address and invited talks in this conference. Three hundred seventy-six researchers submitted their quality research articles to ICRDM 2022 through EasyChair. We shortlisted more than 300 research articles for oral presentations authored by dynamic researchers around the world. After peer review, 119 manuscripts were shortlisted for publication in the Springer book series: *Trends in Mathematics*. We hope that ICRDM 2022 inspired several researchers in mathematics; shared research interest and information; and created a forum of collaboration to build a trust relationship. We feel honored and privileged to serve the best recent developments in the field of mathematics to the readers in two volumes. Volume I: *Recent Developments in Algebra and Analysis* and Volume II: *Advances in Mathematical Modeling and Scientific Computing*.

This book comprises the recent developments in Algebra and Analysis. A basic premise of this book is that the quality assurance is effectively achieved through the selection of quality research articles by the scientific committee that consists of several potential reviewers worldwide. This book comprises the contribution of several dynamic researchers in 37 chapters. Each chapter identifies the existing challenges in the areas of Algebra and Analysis and emphasizes the importance of establishing new theorems and algorithms to addresses the challenges. Each chapter presents a selection of research problem, furnishes theorems and algorithms suitable for solving the problem with sufficient mathematical background, and summarizes the obtained results to understand the domain of applicability. This book also provides a comprehensive literature survey which reveals the challenges, outcomes, and developments of higher-level mathematics in this decade. The theoretical coverage of this book is relatively at a higher level to meet the global orientation of Algebra and Analysis.

The target audience of this book is postgraduate students and researchers. This book promotes a vision of Algebra and Analysis as integral to modern science. Each chapter contains important information emphasizing Algebra and Analysis, intended for the professionals who already possesses a basic understanding. In this book, theoretically oriented readers will find an overview of Algebra and Analysis and applications. The readers can make use of the literature survey of this book to identify the current trends in Algebra and Analysis. It is our hope and expectation that this book will provide an effective learning experience and referenced resource for all young mathematicians in the areas of Algebra and Analysis.

As editors, we would like to express our sincere thanks to all the administrative authorities of Canadian University Dubai, UAE, and United Arab Emirates University, UAE, for their motivation and support. We also extend our profound thanks to all faculty members and staff members of the institutes. We especially thank all the members of the organizing committee of ICRDM 2022 who worked as a team by investing their time to make the conference as a grand success. We express our sincere gratitude to all the referees for spending their valuable time to review the manuscripts which led to substantial improvements and sort out the quality research papers for publication. We thank EasyChair platform for providing the manuscript submission and review service. We are thankful to the project coordinator and team members from Springer Nature for their commitment and dedication toward the publication of this book. The organizing committee is grateful to Dr. Chris Eder, Associate Editor, Mathematics, Birkhäuser, Springer Nature for his continuous encouragement and support toward the publication of this book.

Al Ain, United Arab Emirates — Ho-Hon Leung
Jalandhar, Punjab, India — R. Sivaraj
Dubai, United Arab Emirates — Firuz Kamalov

Contents

Part I
Algebra

Fuzzy Quotient BM-Algebra with Respect to a Fuzzy BM-Ideal

Julie Thomas and K. Indhira

Keywords BM-algebra · Fuzzy BM-subalgebra · Fuzzy BM-ideal

1 Introduction

We take into consideration a class of abstract algebra known as BM-algebra which was introduced by Kim and Kim [1]. This algebraic structure was given the new name TM-algebra by Megalai and Tamilarasi [2]. Several authors (see [3–6], and [7]) looked at the different characterizations of this structure and the relationship between other algebras and them. Zadeh et al. [8] and Ameri et al. [9] stated several elementary properties of finite BM-algebras. Application of the fuzzy set concept to group theory by Rosenfeld [10] led to the fuzzification of different algebraic structures including BM/TM-algebra. Saeid [11] and Megalai and Tamilarasi [12] explored the characteristics of the newly created algebraic structure known as fuzzy BM/TM-algebra after applying the fuzzy set theory to BM/TM-algebra. After that, several fuzzy structures in BM/TM-algebras were considered by many researchers (see [13–17], and [18]). Handam [19] considered the quotient structure of TM-algebra via an ideal. Thus there arises a gap of defining quotient fuzzy structure via a fuzzy ideal in BM-algebra, and we found that further research is needed in this regard. In this paper, we are trying to generalize the concept of quotient BM-algebra in the crisp case defined by Handam [19]. We define a compatible equivalence relation using a fuzzy BM-ideal and the constant θ in a BM-algebra X and study the quotient structure obtained using this.

J. Thomas · K. Indhira (✉)
Department of Mathematics, Vellore Institute of Technology, Vellore, Tamil Nadu, India
e-mail: kindhira@vit.ac.in

H.-H. Leung et al. (eds.), *Recent Developments in Algebra and Analysis*,
Trends in Mathematics, https://doi.org/10.1007/978-3-031-37538-5_1

2 Preliminaries

We are reminded of a few definitions and findings that are necessary for the sections that follow.

Definition 1 A BM-algebra is a triple with the notation $(X, *, \theta)$ in which X is a non-empty set with a fixed member θ and $*$ is a binary operation so that the two criterias

1. $x * \theta = x$
2. $(x * y) * (x * z) = z * y$

are met for all $x, y, z \in X$.

Definition 2 A subset $S \neq \phi$ of X is called a BM-subalgebra of X if $x * y \in S$ for all $x, y \in S$. A subset I of X is known as a BM-ideal of X if it holds the following two criterias:

1. $\theta \in I$.
2. $x * y \in I$ and $y \in I \implies x \in I$ for all $x, y \in X$.

Definition 3 Consider two BM-algebras $(X, *, \theta)$ and $(Y, *', \theta')$. If a mapping $\phi : X \to Y$ satisfies $\phi(x * y) = \phi(x) *' \phi(y)$ for all $x, y \in X$, then it is called a BM-homomorphism.

Definition 4 The pair (X, μ_A) in a set X is considered as a fuzzy set A where the function is $\mu_A : X \to [0, 1]$ and is named as the membership function of A. The set $U(\mu_A; \alpha) := \{x \in X | \mu_A(x) \geq \alpha\}$ for $\alpha \in [0, 1]$ is called an upper level set of A.

Definition 5 Let f be a mapping defined from X into Y, where $A = (X, \mu_A)$ and $B = (Y, \eta_B)$ be fuzzy sets in X and Y. Consequently $f(A)$ is a fuzzy set in $f(X)$ called the image of A under f, where $\mu_{f(A)}$ defined as

$$f(\mu_A)(y) = \begin{cases} sup\{\mu_A(x) | x \in f^{-1}(y) \neq \phi\} \\ 0 \text{ if } f^{-1}(y) = \phi \end{cases}$$

for all $y \in f(X)$.

Let $\eta_{f^{-1}(B)}$ is defined by $f^{-1}(\eta_B)(x) = \eta_B(f(x))$ for all $x \in X$, then the inverse image $f^{-1}(B)$ in X is again a fuzzy set in X.

Definition 6 Let f be any function from X to Y. Membership function μ_A of X is called f-invariant if $f(x) = f(y)$ implies $\mu_A(x) = \mu_A(y)\, \forall x, y \in X$.

Definition 7 A fuzzy subset $A = (X, \mu_A)$ in a BM-algebra X is known as a fuzzy BM-ideal of X, if

(i) $\mu_A(\theta) \geq \mu_A(x)$
(ii) $\mu_A(x) \geq min\{\mu_A(x * y), \mu_A(y)\}$

for all $x, y \in X$.

Theorem 1 *A is a fuzzy BM-ideal iff its level subset* μ_α *is a BM-ideal where* $\alpha \in Im(\mu_A)$.

Theorem 2 *If* $\varphi : X \to Y$ *is an epimorphism from BM-algebra* $(X, *, \theta)$ *onto another BM-algebra* $(Y, *', \theta')$, *then* $X/ker(\varphi) \cong Y$.

3 Construction of a Quotient BM-Algebra Using a Fuzzy BM-Ideal

Let $A = (X, \mu_A)$ be a fuzzy BM-ideal of a BM-algebra $(X, *, \theta)$.

Lemma 1 *Let* $x \sim_\theta y$ *if and only if* $\mu_A(x * y) = \mu_A(y * x) = \mu_A(\theta)$ *for any* $x, y \in X$. *Then* $\sim_\theta$ *is an equivalence relation on* X.

Proof Let $x, y, z \in X$. Since $\mu_A(x * x) = \mu_A(\theta)$ we get $x \sim_\theta x$. Hence $\sim_\theta$ is reflexive. Assume that $x \sim_\theta y$. Then we have $\mu_A(x * y) = \mu_A(y * x) = \mu_A(\theta)$ and hence $y \sim_\theta x$ which implies $\sim_\theta$ is symmetric. Now, suppose that $x \sim_\theta y$ and $y \sim_\theta z$. We have $\mu_A(x * y) = \mu_A(y * x) = \mu_A(\theta) = \mu_A(y * z) = \mu_A(z * y)$. Then

$$\begin{aligned}\mu_A(x * z) &\geq \min\{\mu_A((x * z) * (x * y)), \mu_A(x * y)\}\\ &= \min\{\mu_A(y * z), \mu_A(x * y)\}\\ &= \min\{\mu_A(\theta), \mu_A(\theta)\}\\ &= \mu_A(\theta)\end{aligned}$$

Since A is a fuzzy BM-ideal, we have $\mu_A(\theta) \geq \mu_A(z * x)$. Combining both we get $\mu_A(x * z) = \mu_A(\theta)$. Similarly, $\mu_A(z * x) = \mu_A(\theta)$. Thus, $x \sim_\theta z$, and hence $\sim_\theta$ is transitive.

Lemma 2 $\sim_\theta$ *is a congruence relation.*

Proof Assume that $x \sim_\theta y$ and $u \sim_\theta v$, for $x, y \in X$, which implies $\mu_A(x * y) = \mu_A(y * x) = \mu_A(\theta) = \mu_A(u * v) = \mu_A(v * u)$. Then

$$\begin{aligned}\mu_A((x * u) * (y * v)) &\geq \min\{\mu_A(((x * u) * (y * v)) * (x * y)), \mu_A(x * y)\}\\ &= \min\{\mu_A(((x * u) * (x * y)) * (y * v)), \mu_A(x * y)\}\\ &= \min\{\mu_A((y * u) * (y * v)), \mu_A(x * y)\}\\ &= \min\{\mu_A(v * u), \mu_A(x * y)\}\\ &= \mu_A(\theta)\end{aligned}$$

Then by (i) of Definition 7, we get $\mu_A((x*u)*(y*v)) = \mu_A(\theta)$.
Similarly, it can be shown that $\mu_A((y*v)*(x*u)) = \mu_A(\theta)$.
Hence, $x*u \sim_\theta y*v$, proving that $\sim_\theta$ is compatible.

Remark 1 We denote $A_x = \{y \in X \mid y \sim_\theta x\}$ for the corresponding equivalence class containing the element x and $X/A = \{A_x \mid x \in X\}$ for the set of equivalence classes of X.

Theorem 3 *Let A be a fuzzy BM-ideal of a BM-algebra X. Define a binary operation* $\circledast$ *on* X/A *by* $A_x \circledast A_y = A_{x*y}$ *for all* $x, y \in X$. *Then* $(X/A, \circledast, A_\theta)$ *is a BM-algebra called the fuzzy quotient BM-algebra.*

Proof Consider any $A_x, A_y \in X/A$.
Then $\circledast(A_x, A_y) = A_{x*y} \in X/A$.
Also, if $A_x = A_u$ and $A_y = A_v$, then $A_{x*y} = A_{u*v}$. Hence, $\circledast$ is well defined.
Now it is enough to show that X/A is a BM-algebra.
For, consider any $A_x \in X/A$.
We get $A_x \circledast A_\theta = A_{x*\theta} = A_x$.
Also, for any $A_x, A_y, A_z \in X/A$, we have
$(A_x \circledast A_y) \circledast (A_x \circledast A_z) = (A_{x*y} \circledast A_{x*z}) = A_{(x*y)*(x*z)} = A_{z*y}$.
Hence the proof.

Example 1 Consider a BM-algebra $(X, *, \theta)$ defined by the following table.

$*$	θ	a	b	c
θ	θ	a	c	b
a	a	θ	b	c
b	b	c	θ	a
c	c	b	a	θ

Define $\mu_A : X \to [0,1]$ by $\mu_A(\theta) = \mu_A(a) = 0.7$, $\mu_A(b) = \mu_A(c) = 0.4$. Then μ_A is a fuzzy BM-ideal. Then $a \sim_\theta \theta$ and $b \sim_\theta c$.
Thus, $A_\theta = A_a = \{\theta, a\}$ and $A_b = A_c = \{b, c\}$. Take $X/A = \{A_\theta, A_b\}$.
Then, $(X/A, \circledast, A_\theta)$ is a BM-algebra.

Lemma 3 *Let* $\varphi : X \to Y$ *be a homomorphism of BM-algebras. If* $B = (Y, \mu_B)$ *is a fuzzy BM-ideal of Y, then the pre-image* $\varphi^{-1}(B) = (X, \varphi^{-1}(\mu_B))$ *of B under* φ *is a fuzzy BM-ideal of X.*

Proof For any $x, y \in X$,

$$\begin{aligned}\varphi^{-1}(\mu_B)(x) &= \mu_B(\varphi(x)) \\ &\geq \min\{\mu_B((\varphi(x) *' \varphi(y))), \mu_B(\varphi(y))\} \\ &= \min\{\mu_B(\varphi(x*y)), \mu_B(\varphi(y))\} \\ &= \min\{\varphi^{-1}(\mu_B)(x*y), \varphi^{-1}(\mu_B)(y)\}\end{aligned}$$

Hence $\varphi^{-1}(B)$ is a fuzzy BM-ideal.

Theorem 4 *Let $\varphi : X \to Y$ be an epimorphism and $B = (Y, \mu_B)$ be a fuzzy BM-ideal of Y. Then $X/\varphi^{-1}(B) \cong Y/B$.*

Proof Let $A = \varphi^{-1}(B) = \left(X, \varphi^{-1}(\mu_B)\right)$. By Theorem 3 and previous Lemma 3, X/A and Y/B are (fuzzy) BM-algebras. Define $\eta : X/A \to Y/B$ by $\eta(A_x) = B_{\varphi(x)}$. Assume

$$A_x = A_y \implies \varphi^{-1}(\mu_B)(x * y) = \varphi^{-1}(\mu_B)(y * x) = \varphi^{-1}(\mu_B)(\theta)$$

$$\implies \mu_B(\varphi(x * y)) = \mu_B(\varphi(y * x)) = \mu_B(\varphi(\theta))$$

$$\implies \mu_B\left(\varphi(x) *' \varphi(y)\right) = \mu_B\left(\varphi(y) *' \varphi(x)\right) = \mu_B\left(\theta'\right)$$

$$\implies B_{\varphi(x)} = B_{\varphi(y)}$$

$$\implies \eta(A_x) = \eta(A_y)$$

Thus the map η is well defined. Consider

$$\eta\left(A_x \circledast A_y\right) = \eta\left(A_{x*y}\right)$$

$$= B_{\varphi(x*y)}$$

$$= B_{\varphi(x)*'\varphi(y)}$$

$$= B_{\varphi(x)} \circledast' B_{\varphi(y)}$$

$$= \eta(A_x) \circledast' \eta(A_y)$$

Thus, η is found to be a homomorphism.

Now, let $B_z \in Y/B$ for $z \in Y$.

Since φ is an onto homomorphism, $\exists x \in X$ so that $\varphi(x) = z$, which implies that

$$\eta(A_x) = B_{\varphi(x)} = B_z$$

proving that η is onto.

Now, assume $\eta(A_x) = \eta\left(A_y\right)$

$$\implies B_{\varphi(x)} = B_{\varphi(y)}$$

$$\implies \mu_B\left(\varphi(x) *' \varphi(y)\right) = \mu_B\left(\varphi(y) *' \varphi(x)\right) = \mu_B\left(\theta'\right)$$

$$\implies \mu_B(\varphi(x * y)) = \mu_B(\varphi(y * x)) = \mu_B(\varphi(\theta))$$

$$\Longrightarrow \varphi^{-1}(\mu_B)(x * y) = \varphi^{-1}(\mu_B)(y * x) = \varphi^{-1}(\mu_B)(\theta)$$

$$\Longrightarrow A_x = A_y$$

Hence the proof.

Lemma 4 *Let A be a fuzzy BM-ideal of X. If the upper level subset $\mu_\alpha = \{x \in X | \mu_A(x) \geq \alpha\} \neq \phi$ for all $\alpha \in [0, 1]$, then μ_α is found to be a BM-ideal of X and $\mu_A(\theta) = 1$.*

Proof Given $\mu_1 \neq \phi$, thus $\exists x \in \mu_1$ such that $\mu_A(x) = 1$.
Since $\mu_A(\theta) \geq \mu_A(x)$, we get $\mu_A(\theta) = 1$ and hence $\theta \in \mu_\alpha$ for any $\alpha \in [0, 1]$.
Now, assume $x * y \in \mu_\alpha$ and $y \in \mu_\alpha$.
Thus, we have $\mu_A(x * y) \geq \alpha$ and $\mu_A(y) \geq \alpha$. Then

$$\mu_A(x) \geq \min\{\mu_A(x * y), \mu_A(y)\} \geq \alpha$$

implying that $x \in \mu_\alpha$. Hence μ_α is a BM-ideal.

Theorem 5 *Suppose that $\mu_\alpha \neq \phi$ for a fuzzy BM-ideal A of X, for all $\alpha \in [0, 1]$. Then there exists a BM-ideal K of X/A with the property $\frac{(X/A)}{K} \cong X/\mu_\alpha$.*

Proof From Lemma 4, μ_α is a BM-ideal of X and $\mu_A(\theta) = 1$.
Define $\varphi : X/A \to X/\mu_\alpha$ by $\varphi(A_x) = \mu_{\alpha_x}$ for all $x \in X$.
If $A_x = A_y$, then
$\mu_A(x * y) = \mu_A(y * x) = \mu_A(\theta) = 1$
$\geq \alpha$ for all $\alpha \in [0, 1]$.
Thus, $x * y \in \mu_\alpha$, $y * x \in \mu_\alpha$, which implies $\mu_{\alpha_x} = \mu_{\alpha_y}$. Hence, φ is well defined.
Clearly, φ is onto. Now let $K = \ker(\varphi)$.
Then the proof follows by Theorem 2.

Theorem 6 *If A is a fuzzy BM-ideal of X with $\mu_1 \neq \phi$, then $X/A \cong X/\mu_1$.*

Proof It is enough to show that the epimorphism φ defined in Theorem 5 is one to one.
For example, let $\varphi(A_x) = \varphi(A_y)$.

$$\mu_{1_x} = \mu_{1_y} \Longrightarrow x * y \in \mu_1 \text{ and } y * x \in \mu_1$$

$$\Longrightarrow \mu_A(x * y) = \mu_A(y * x) = 1 = \mu_A(\theta)$$

$$\Longrightarrow A_x = A_y$$

Hence $\varphi : X/A \to X/\mu_1$ is an isomorphism.

Theorem 7 *Let φ be an epimorphism of BM-algebras from X onto Y and $\mu_1 = ker(\varphi)$. Then $X/A \cong Y$ where A is a fuzzy BM-ideal of X.*

Proof By Theorem 2, $X/\ker(\varphi) \cong Y$.
i.e., $X/\mu_1 \cong Y$.
But we have $X/A \cong X/\mu_1$ by Theorem 6.
Hence $X/A \cong Y$.

Let us try to prove the above theorem in a more general case.

Theorem 8 *Let A be a φ-invariant fuzzy BM-ideal of X such that $\phi \neq \mu_1 \subseteq$ $ker(\varphi)$, where φ is an epimorphism of BM-algebras from X onto Y. Then $X/A \cong Y$.*

Proof We define $\eta : X/A \to Y$ by $\eta(A_x) = \varphi(x)$.
Let $A_x = A_y$, then $\mu_A(x * y) = \mu_A(y * x) = \mu_A(\theta) = 1$.

$$\Longrightarrow x * y, y * x \in \mu_1 \subseteq \ker(\varphi)$$

$$\Longrightarrow \varphi(x * y) = \varphi(y * x) = \varphi(\theta) = \theta'$$

$$\Longrightarrow \varphi(x) *' \varphi(y) = \varphi(y) *' \varphi(x) = \theta'$$

$$\Longrightarrow \varphi(x) = \varphi(y)$$

Hence η is well defined.
Consider $\eta\left(A_x \circledast A_y\right) = \eta\left(A_{x*y}\right)$

$$= \varphi(x * y) = \varphi(x) *' \varphi(y)$$

$$= \eta(A_x) *' \eta(A_y)$$

Thus η is a homomorphism.
Clearly η is onto, since φ is onto.
Suppose $\eta(A_x) = \eta\left(A_y\right)$

$$\Longrightarrow \varphi(x) = \varphi(y)$$

$$\Longrightarrow \varphi(x) *' \varphi(y) = \theta' = \varphi(y) *' \varphi(x)$$

$$\Longrightarrow \varphi(x * y) = \varphi(y * x) = \theta' = \varphi(\theta)$$

$\Longrightarrow \mu_A(x * y) = \mu_A(y * x) = \mu_A(\theta)$ since A is φ-invariant.
$\Longrightarrow A_x = A_y$.
Hence η is an isomorphism.

Lemma 5 *Let A be a fuzzy BM-ideal of X. Then the natural homomorphism π_A : $X \to X/A$ defined by $\pi_A(x) = A_x$ is always an onto map. Analogous to this, if μ_A is the characteristic function $\chi_{\{\theta\}}$, then π_A is an isomorphism.*

Proof Clearly, if A is a fuzzy BM-ideal of X, then π_A is an epimorphism.

Now, suppose $\mu_A = \chi_{\{\theta\}}$. If $\pi_A(x) = \pi_A(y)$, then $A_x = A_y$ for $x, y \in X$.
$\Longrightarrow \chi_{\{\theta\}}(x * y) = \chi_{\{\theta\}}(y * x) = \chi_{\{\theta\}}(\theta) = 1$
$\Longrightarrow x * y = \theta$ and $y * x = \theta$
$\Longrightarrow x = y$,
proving that π_A is one to one.

Theorem 9 *Let $\varphi : X \to Y$ be a homomorphism of BM-algebras, and consider two fuzzy BM-ideals of X and Y, say $A = (X, \mu_A)$ and $B = (Y, \mu_B)$, such that $\varphi(\mu_A) \subseteq \mu_B$ and $\mu_A(\theta) \geq \mu_B(\theta')$. Then, there exists a homomorphism $\varphi_\sim : X/A \to Y/B$ with $\varphi_\sim \circ \pi_A = \pi_B \circ \varphi$. In other words, the diagram below is commutative.*

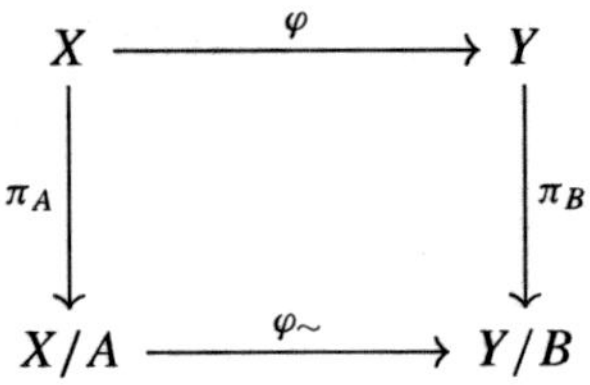

Proof Since $\varphi(\mu_A) \subseteq \mu_B$, we have
$\mu_B(\theta') \geq \varphi(\mu_A)(\theta') = \sup\{\mu_A(\varphi^{-1}(\theta'))\} \geq \mu_A(\theta)$.
But we assumed that $\mu_B(\theta') \leq \mu_A(\theta)$.
Hence $\mu_A(\theta) = \mu_B(\theta')$.
Now, define $\varphi_\sim : X/A \to Y/B$ by $\varphi_\sim(A_x) = B_{\varphi(x)}$.
Let $A_x = A_y$. Then
$\mu_A(x * y) = \mu_A(y * x) = \mu_A(\theta)$.
Now,
$\mu_B(\varphi(x) *' \varphi(y)) = \mu_B(\varphi(x * y))$
$\geq \varphi(\mu_A)(\varphi(x * y)) = \sup\{\mu_A(\varphi^{-1}(\varphi(x * y)))\}$
$\geq \mu_A(x * y)$ since $x * y \in f^{-1}(f(x * y))$
$= \mu_A(\theta)$
$= \mu_B(\theta')$
Similarly, $\mu_B(\varphi(y) *' \varphi(x)) = \mu_B(\theta')$.
$\Longrightarrow A_{\varphi(x)} = A_{\varphi(y)}$
$\Longrightarrow \varphi_\sim(A_x) = \varphi_\sim(A_y)$. Hence $\varphi_\sim$ is well defined.
Consider $\varphi_\sim(A_x) \circledast' \varphi(A_y) = B_{\varphi(x)} \circledast' B_{\varphi(y)}$

$$= B_{\varphi(x)*'\varphi(y)}$$

$$= B_{\varphi(x*y)}$$

$$= \varphi_\sim(A_{x*y})$$

$$= \varphi_\sim(A_x \circledast A_y)$$

Hence $\varphi_{\sim}$ is a homomorphism. Also for all $x \in X$,

$$
\begin{aligned}
(\varphi_{\sim} \circ \pi_A)(x) &= \varphi_{\sim}(\pi_A(x)) \\
&= \varphi_{\sim}(A_x) \\
&= B_{\varphi(x)} \\
&= \pi_B(\varphi(x)) \\
&= (\pi_B \circ \varphi)(x)
\end{aligned}
$$

4 Conclusion

We generalized the concept of quotient BM-algebra in the crisp case defined by Handam [19], to fuzzy case. We defined a compatible equivalence relation using a fuzzy BM-ideal and the constant θ in a BM-algebra X and studied the quotient structure obtained using this. This research can be further extended to see how the quotients and products behaves in the quotient fuzzy BM-algebra when equipped with a fuzzy topology.

References

1. Kim, C.B., Kim, H.S.: On BM-algebras. Scientiae Math. Japonicae **63**(3), 421–428 (2006)
2. Megalai, K., Tamilarasi, A.: TM-algebra - an introduction. IJCA Special Issue on CASCT (1), 17–23 (2010)
3. Saeid, A.B., Zarandi, A.: Vague set theory applied to BM-algebras. Int. J. Algebra **5**(5), 207–222 (2011)
4. Saeid, A.B.: BM-algebras defined by bipolar-valued sets. Indian J. Sci. Technol. **5**(2), 2071–2078 (2012)
5. Walendziak, A.: BM-algebras and related topics. Math Slovaca **64**(5), 1075–1082 (2014)
6. Prabpayak, C.: Some homomorphism properties of TM-algebras. Appl. Mech. Mat. **866**, 406–409 (2017)
7. Thomas, J., Indhira, K., Chandrasekaran, V.M.: BM-algebra with quasi-valuation map. AIP Conf. Proc. **2177**(1), 020029-1–020029-5 (2019)
8. Zadeh, S.A.N., Saeid, A.B., Borzooei, R.A., Radfar, A., Armai, R.: On finite BM-algebra. Kochi J. Math. **9**, 43–50 (2014)
9. Ameri, R., Saeid, A.B., Zadeh, S.A.N., Radfar, A., Borzooei, R.A.: On finite B-algebra. Afrika Matematika **26**, 825–847 (2015)
10. Rosenfeld, A.: Fuzzy groups. J. Math. Anal. Appl. **35**, 512–517 (1971)
11. Saeid, A.B.: Fuzzy BM-algebras. Indian J. Sci. Technol. **3**(5), 523–529 (2010)

12. Megalai, K., Tamilarasi, A.: Fuzzy subalgebras and fuzzy T-ideals in TM-algebra. J. Math. Stat. **7**(2), 107–111 (2011)
13. Saeid, A.B.: Redefined (anti) fuzzy BM-algebras. Math. Sci. J. **6**(2), 7–21 (2010)
14. Thomas, J., Indhira, K., Chandrasekaran, V.M.: T-normed fuzzy TM-subalgebra of TM-algebras. Int. J. Comput. Intel. Syst. **12**(2), 706–712 (2019)
15. Megalai, K., Tamilarasi,A.: On intutionistic fuzzy T-ideals in TM-algebra. In: Signal Processing, Image Processing and Pattern Recognition, pp. 328–335. Springer, Berlin (2011)
16. Mostafa, S.M., Naby, M.A., Elgendy, O.R.: Fuzzy TM-ideals of TM-algebras. J. Am. Sci. **7**(9), 17–21 (2011)
17. Ramachandran, T., Priya, T., Parimala, M.: Anti-fuzzy T-ideals of TM-algebras and its lower level cuts. Int. J. Comput. Appl. **43**(22), 17–22 (2012)
18. Kaviyarasu, M., Indhira, K., Chandrasekaran, V.M.: Direct product of intuitionistic fuzzy P-ideals of TM-algebras. Global J. Pure Appl. Math. **12**(3), 168–173 (2016)
19. Handam, A.H.: Quotient TM-algebras. Int. Math. Forum **8**(17–20), 949–956 (2013)

New Classes of the Quotient Permutation BN-Algebras in Permutation BN-Algebras

Enoch Suleiman, Abu Firas Al-Musawi, and Shuker Khalil

Keywords Cycles · Permutation · BN-algebras · Homomorphism · Subalgebra · Quotient

1 Introduction

BCK-algebras and BCI-algebras are two types of abstract algebras that were introduced by Y. Imai and K. Is'eki [1, 2]. It is well known that the BCK-algebra class is a proper subclass of the BCI-algebra class. Next, J. Neggers and H. S. Kim [3] suggested and examined a few B-algebra aspects that they believed would be of interest. Chang Bum Kim [4] introduced the notion of BN-algebras, which is a generalization of B-algebras. Permutation sets are given by Alsalem [5]. Some properties and applications for the permutations in symmetric and alternating groups are studied [6–13]. A permutation set is as a nonclassical set like fuzzy sets [14–20], soft sets [21–26], and nano sets [27]. After that, the notations of permutation B-algebra [28], permutation BF-algebra [29], and permutation BH-algebra [30] are shown, and some results are studied with their applications using permutation sets. We proposed permutation BN-algebras as a new class of BN-algebras and listed some of their key characteristics in this study: {1}-commutative, condition (D), permutation BN-subalgebra, permutation BN-normal, and permutation BN_1-algebra. Also, the relationships between permutation BN-algebra and other classes like permutation BH/B/BF-algebras are given. Moreover, we explored some new notions in permutation theory for the first time. We also examined BN-algebra

E. Suleiman
Department of Mathematics, Federal University Gashua, Gashua, Yobe State, Nigeria

A. F. Al-Musawi
Department of Mathematics, College of Education for Pure Sciences, University of Basrah, Basra, Iraq

S. Khalil (✉)
Department of Mathematics, College of Science, University of Basrah, Basrah, Iraq

H.-H. Leung et al. (eds.), *Recent Developments in Algebra and Analysis*,
Trends in Mathematics, https://doi.org/10.1007/978-3-031-37538-5_2

homomorphism, equivalence relations, quotient permutation BN-algebras, and BN-algebra isomorphism theorems.

2 Preliminaries

The fundamental ideas and facts necessary for this subject will be reviewed in this section.

Definition 2.1: [3] Let $X \neq \varnothing$ and 0 be a constant with a binary operation $*$. We say that $(X, *, \varnothing)$ is a *B-algebra* if it satisfies the following conditions:

(a) $x * x = 0$,
(b) $x * 0 = x$,
(c) $(x * y) * z = x * (z * (0 * y)), \forall\, x, y, z \in X$.

Definition 2.2: [4] Let $K(\tau)$ be the class of all algebras of Type $\tau = (2, 0)$. By a *BN-algebra*, we mean a system $(X; *, 0)$ in which the following axioms are satisfied:

1. $x * x = 0, \forall x \in X$.
2. $x * 0 = x, \forall x \in X$.
3. $(x * y) * z = (0 * z) * (y * x), \forall x, y, z \in X$.

We say that 0 is the unit in X.

Definition 2.3: [31] Let $K(\tau)$ be the class of all algebras of Type $\tau = (2, 0)$. By a *BF-algebra*, we mean a system $(X; *, 0)$ in which the following axioms are satisfied:

4. $x * x = 0, \forall x \in X$.
5. $x * 0 = x, \forall x \in X$.
6. $0 * (x * y) = y * x, \forall x, y \in X$.

We say that 0 is the unit in X.

Definition 2.4: [5]
For any permutation $\beta = \prod_{i=1}^{c(\beta)} \lambda_i$ in a symmetric group S_n, where $\{\lambda_i\}_{i=1}^{c(\beta)}$ is a composite of pairwise disjoint cycles $\{\lambda_i\}_{i=1}^{c(\beta)}$ where $\lambda_i = \left(t_1^i, t_2^i, \ldots, t_{\alpha_i}^i\right), 1 \leq i \leq c(\beta)$, for some $1 \leq \alpha_i, c(\beta) \leq n$. If $\lambda = (t_1, t_2, \ldots, t_k)$ is $k-$cycle in S_n, we define $\beta-$set as $\lambda^\beta = \{t_1, t_2, \ldots, t_k\}$ and is called $\beta-$set of cycle λ. So the $\beta-$sets of $\{\lambda_i\}_{i=1}^{c(\beta)}$ are defined by $\left\{\lambda_i^\beta = \left\{t_1^i, t_2^i, \ldots, t_{\alpha_i}^i\right\} | 1 \leq i \leq c(\beta)\right\}$.

Definition 2.5: [28] Let $X = \left\{\lambda_i^\beta\right\}_{i=1}^{c(\beta)} \cup \{1\}$ be a collection of β-sets $\left\{\lambda_i^\beta\right\}_{i=1}^{c(\beta)}$ with $\{1\}$, where β is a permutation in the symmetric group $G = S_n$. We say $(X, \#, \{1\})$ is a *permutation B-algebra* (PB-A), if the binary operation $\# : X \times X \longrightarrow X$ satisfies the following conditions:

1. $\lambda_i^\beta \# \lambda_i^\beta = \{1\}$,
2. $\lambda_i^\beta \# \{1\} = \lambda_i^\beta$,
3. $\left(\lambda_i^\beta \# \lambda_j^\beta\right) \# \lambda_k^\beta = \lambda_i^\beta \# \left(\lambda_k^\beta \# \left(\{1\} \# \lambda_j^\beta\right)\right), \forall \lambda_i^\beta, \lambda_j^\beta, \lambda_k^\beta \in X$.

Definition 2.6: [29] We say $(X, \#, \{1\})$ is a *permutation BF-algebra* (PBF-A), where β is a permutation in the symmetric group $G = S_n$ and $X = \left\{\lambda_i^\beta\right\}_{i=1}^{c(\beta)} \cup \{1\}$, if # satisfies the following conditions:

1. $\lambda_i^\beta \# \lambda_i^\beta = \{1\}$,
2. $\lambda_i^\beta \# \{1\} = \lambda_i^\beta, \forall \lambda_i^\beta, \lambda_j^\beta \in X$.
3. $\lambda_i^\beta \# \lambda_j^\beta = \{1\}$ *and* $\lambda_j^\beta \# \lambda_i^\beta = \{1\} \Longrightarrow \lambda_i^\beta = \lambda_j^\beta, \forall \lambda_i^\beta, \lambda_j^\beta \in X$.

Definition 2.7: [30] Let X be a collection of β-sets $\left\{\lambda_i^\beta\right\}_{i=1}^{c(\beta)}$, where β is a permutation in the symmetric group $G = S_n$ with {1}. Then $X = \left\{\lambda_i^\beta\right\}_{i=1}^{c(\beta)} \cup \{1\}$ with a binary operation $\# : X \times X \longrightarrow X$ is said to be a permutation BH-algebra (PBH − A) if # satisfies the following condition:

1. $\lambda_i^\beta \# \lambda_i^\beta = \{1\}$,
2. $\lambda_i^\beta \# \lambda_j^\beta = \{1\}$ and $\lambda_j^\beta \# \lambda_i^\beta = \{1\} \Longrightarrow \lambda_i^\beta = \lambda_j^\beta$.
3. $\lambda_i^\beta \# \{1\} = \lambda_i^\beta, \quad \forall \lambda_i^\beta, \lambda_j^\beta \in X$.

Also, we say that {1} is the fixed element in X. It is denoted by $(X, \#, \{1\})$.

3 Permutation BN-Algebras

In this section, we will look at some of the fundamental characteristics of permutation BN-algebras (PBN-As) and study some novel implications in the field.

Definition 3.1: Let $X = \left\{\lambda_i^\beta\right\}_{i=1}^{c(\beta)}$ be a collection of β-sets, where β is a permutation in the symmetric group $G = S_n$. Then X is said to be a *permutation BN-algebra* (PBG-A) if there exists a mapping $\# : X \times X \longrightarrow X$ such that

1. $\lambda_i^\beta \# \lambda_i^\beta = \{1\}, \quad \forall \lambda_i^\beta \in X$
2. $\lambda_i^\beta \# \{1\} = \lambda_i^\beta, \quad \forall \lambda_i^\beta \in X$
3. $\left(\lambda_i^\beta \# \lambda_j^\beta\right) \# \lambda_k^\beta = \left(\{1\} \# \lambda_k^\beta\right) \# \left(\lambda_j^\beta \# \lambda_i^\beta\right), \quad \forall \lambda_i^\beta, \lambda_j^\beta, \lambda_k^\beta \in X$.

We say that {1} is the fixed element in X.

Example 3.2:

Let (S_{10},o) be a symmetric group and $\beta = \begin{pmatrix} 1 & 2 & 3 & 4 & 5 & 6 & 7 & 8 & 9 & 10 & 11 \\ 4 & 8 & 1 & 9 & 5 & 2 & 7 & 3 & 6 & 11 & 10 \end{pmatrix}$ be a permutation in S_{11}. Since $\beta = \begin{pmatrix} 1 & 2 & 3 & 4 & 5 & 6 & 7 & 8 & 9 & 10 & 11 \\ 4 & 8 & 1 & 9 & 5 & 2 & 7 & 3 & 6 & 11 & 10 \end{pmatrix} = (1496283)(7)\,(10\ 11)$.

Therefore, we have $X = \left\{\lambda_i^{\beta}\right\}_{i=1}^{3} \cup \{1\} = \{\{1,4,9,6,2,8,3\},\{7\},\{10,11\}\}$. Define $\# : X \times X \longrightarrow X$ by $\#\left(\lambda_i^{\beta},\lambda_j^{\beta}\right) = \lambda_i^{\beta} \# \lambda_j^{\beta} = \lambda_k^{\beta}$, where λ_k^{β} its cycle λ_k such that $\lambda_k{=}\lambda_i o \lambda_j^{-1}$, where λ_i and λ_j are cycles for λ_i^{β} and λ_i^{β}, respectively. Here we have (i) $\lambda_i o \lambda_j^{-1} = (1) \rightarrow \lambda_i^{\beta} \# \lambda_j^{\beta} = \{1\}$, and (ii) $\lambda_i o (1)^{-1} = \lambda_i \Rightarrow \lambda_i^{\beta} \# \{1\} = \lambda_i^{\beta}$,

(iii) For any $\lambda_i^{\beta}, \lambda_j^{\beta} \in X$, we have

$$\left(\lambda_i o\, \lambda_j^{-1}\right) o\, \lambda_k^{-1} = \begin{cases} \lambda_i o\, \lambda_j^{-1} o\, \lambda_k^{-1}, if\ j \neq k \neq i \neq j, \\ \lambda_k^{-1}, if\ j \neq k \neq i = j, \\ \lambda_j^{-1}, if\ i = k \neq j, \\ \lambda_i o\, \lambda_k^{-1} o\, \lambda_k^{-1}, if\ i \neq k = j, \end{cases} \quad \ldots\ldots(*).$$

In other side, we consider that $\left((1) o \lambda_k^{-1}\right) o \left(\left(\lambda_j o\, \lambda_i^{-1}\right)\right)^{-1} = \lambda_k{}^{-1} o \left(\lambda_i o \lambda_j^{-1}\right) =$

$$\begin{cases} \lambda_i o\, \lambda_j^{-1} o\, \lambda_k^{-1}, if\ j \neq k \neq i \neq j, \\ \lambda_k^{-1}, if\ j \neq k \neq i = j, \\ \lambda_j^{-1}, if\ i = k \neq j, \\ \lambda_i o\, \lambda_k^{-1} o\, \lambda_k^{-1}, if\ i \neq k = j, \end{cases} \quad \ldots.(**)$$

By (*) and (**), we have $\left(\lambda_i o\, \lambda_j^{-1}\right) o\ \lambda_k^{-1} = \left((1) o \lambda_k^{-1}\right) o \left(\left(\lambda_j o\, \lambda_i^{-1}\right)\right)^{-1}$. Hence, $\left(\lambda_i^{\beta} \# \lambda_j^{\beta}\right) \# \left(\{1\} \# \lambda_j^{\beta}\right) = \lambda_i^{\beta}, \forall \lambda_i^{\beta}, \lambda_j^{\beta} \in X$. Then, X is a (PBN-A).

Proposition 3.3: If $(X, \#, \{1\})$ is a (PBN-A), then $(X, \#, \{1\})$ is a (PBF-A).

Proof: Put $\lambda_k^{\beta} = \{1\}$ in (3) of Definition 3.1; we have that $\lambda_i^{\beta} \# \lambda_j^{\beta} = \{1\} \# \left(\lambda_j^{\beta} \# \lambda_i^{\beta}\right)$. Hence $(X, \#, \{1\})$ is a (PBF-A). Note that the opposite is not true in all cases.

Proposition 3.4: If $(X, \#, \{1\})$ is a (PBN-A), then **(1)** $\{1\} \# \left(\{1\} \# \lambda_i^{\beta}\right) = \lambda_i^{\beta}$, **(2)** $\lambda_j^{\beta} \# \lambda_i^{\beta} = \left(\{1\} \# \lambda_i^{\beta}\right) \# \left(\{1\} \# \lambda_j^{\beta}\right)$, **(3)** $\left(\{1\} \# \lambda_i^{\beta}\right) \# \lambda_j^{\beta} = \left(\{1\} \# \lambda_j^{\beta}\right) \# \lambda_i^{\beta}$, **(4)** $\lambda_i^{\beta} \# \lambda_j^{\beta} = \{1\} \Longrightarrow \lambda_j^{\beta} \# \lambda_i^{\beta} = \{1\}$, **(5)** $\{1\} \# \lambda_i^{\beta} = \{1\} \# \lambda_j^{\beta} \Longrightarrow \lambda_i^{\beta} = \lambda_j^{\beta}$, **(6)** $\left(\lambda_i^{\beta} \# \lambda_k^{\beta}\right) \# \left(\lambda_j^{\beta} \# \lambda_k^{\beta}\right) = \left(\lambda_k^{\beta} \# \lambda_j^{\beta}\right) \# \left(\lambda_k^{\beta} \# \lambda_i^{\beta}\right), \forall \lambda_i^{\beta}, \lambda_j^{\beta}, \lambda_k^{\beta} \in X$.

Proof: **(1)** Put $\lambda_j^{\beta} = \{1\}, \lambda_k^{\beta} = \{1\}$ in (3) of Definition 3.1, and then $\left(\lambda_i^{\beta} \# \{1\}\right) \# \{1\} = (\{1\} \# \{1\}) \# \left(\{1\} \# \lambda_i^{\beta}\right)$. By (1) and (2) of Definition 3.1, we have that

$\{1\} \# \left(\{1\} \# \lambda_i^\beta\right) = \lambda_i^\beta$. **(2)** By (2) and (3) of Definition 3.1, we have that $\lambda_j^\beta \# \lambda_i^\beta = \left(\lambda_j^\beta \# \{1\}\right) \# \lambda_i^\beta = \left(\{1\} \# \lambda_i^\beta\right) \# \left(\{1\} \# \lambda_j^\beta\right)$. **(3)** By (2) and (3) of Definition 3.1, we have that $\left(\{1\} \# \lambda_i^\beta\right) \# \lambda_j^\beta = \left(\{1\} \# \lambda_j^\beta\right) \# \lambda_i^\beta$. **(4)** If $\lambda_i^\beta \# \lambda_j^\beta = \{1\}$, then $\{1\} = \{1\} \# \{1\} = \{1\} \# \left(\lambda_i^\beta \# \lambda_j^\beta\right) = \lambda_j^\beta \# \lambda_i^\beta$ [by Proposition (3.3)]. **(5)** If $\{1\} \# \lambda_i^\beta = \{1\} \# \lambda_j^\beta$, then by (1), we have that $\lambda_i^\beta = \{1\} \# \left(\{1\} \# \lambda_i^\beta\right) = \{1\} \# \left(\{1\} \# \lambda_j^\beta\right) = \lambda_j^\beta$. **(6)** By (3) of Definition 3.1 and Proposition 3.3, we have that $\left(\lambda_i^\beta \# \lambda_k^\beta\right) \# \left(\lambda_j^\beta \# \lambda_k^\beta\right) = \left(\{1\} \# \left(\lambda_j^\beta \# \lambda_k^\beta\right)\right) \# \left(\lambda_k^\beta \# \lambda_i^\beta\right) = \left(\lambda_k^\beta \# \lambda_j^\beta\right) \# \left(\lambda_k^\beta \# \lambda_i^\beta\right)$.

Definition 3.5: Let $(X, \#, \{1\})$ be a (PBN-A) which is said to be permutation $\{1\}$-commutative BN-algebra (P1CBN – A) if $\lambda_i^\beta \# \left(\{1\} \# \lambda_j^\beta\right) = \lambda_j^\beta \# \left(\{1\} \# \lambda_i^\beta\right) \forall \lambda_i^\beta, \lambda_j^\beta \in X$.

Proposition 3.6: If $(X, \#, \{1\})$ is a (PBN-A), then it is a (P1CBN – A).

Proof: Let $\lambda_i^\beta, \lambda_j^\beta \in X$, and then $\lambda_i^\beta \# \left(\{1\} \# \lambda_j^\beta\right) = \left(\{1\} \# \left(\{1\} \# \lambda_i^\beta\right)\right) \# \left(\{1\} \# \lambda_j^\beta\right)$ (By Proposition 3.4).

$= \left(\{1\} \# \left(\{1\} \# \lambda_j^\beta\right)\right) \# \left(\left(\lambda_i^\beta \# \lambda_j^\beta\right) \# \{1\}\right)$ (By (3) of Definition 3.1)

$= \lambda_j^\beta \# \left(\{1\} \# \lambda_i^\beta\right)$ (By Proposition 3.4 (1) and (2) of Definition 3.1).

Proposition 3.7: If $(X, \#, \{1\})$ is a (P1CBN – A), then it is a (PBN-A).

Proof: Let $\lambda_i^\beta, \lambda_j^\beta, \lambda_k^\beta \in X$, and then $\left(\{1\} \# \lambda_k^\beta\right) \# \left(\lambda_j^\beta \# \lambda_i^\beta\right) = \left(\{1\} \# \lambda_k^\beta\right) \# \left(\{1\} \# \left(\lambda_i^\beta \# \lambda_j^\beta\right)\right)$

(By (3) of Definition 2.5) $= \left(\lambda_i^\beta \# \lambda_j^\beta\right) \# \left(\{1\} \# \left(\{1\} \# \lambda_k^\beta\right)\right)$ (By Definition 3.5)

$= \left(\lambda_i^\beta \# \lambda_j^\beta\right) \# \lambda_k^\beta$ (By (3) of Definition 2.5 and (3) of Definition 3.1). Hence, $(X, \#, \{1\})$ is a (PBN-A).

Proposition 3.8: $(X, \#, \{1\})$ is a (P1CBN – A) if and only if it is a (PBN-A).

Proof: If $(X, \#, \{1\})$ is a (P1CBN – A), by Proposition 3.7, it is a (PBN-A).

Conversely, if $(X, \#, \{1\})$ is a permutation BN-algebra, then from Proposition 3.6, it is $\{1\}$-commutative. Now it follows by Proposition 3.10 that $(X, \#, \{1\})$ is a permutation BF-algebra.

Proposition 3.9: If $(X, \#, \{1\})$ is a (P1CBN − A), then $\left(\{1\} \# \lambda_i^{\beta}\right) \# \left(\{1\} \# \lambda_j^{\beta}\right) = \lambda_j^{\beta} \# \lambda_i^{\beta} \quad \forall \lambda_i^{\beta}, \lambda_j^{\beta} \in X$.

Proposition 3.10: If $(X, \#, \{1\})$ is a permutation B-algebra, then $\{1\} \# \left(\lambda_i^{\beta} \# \lambda_j^{\beta}\right) = \lambda_j^{\beta} \# \lambda_i^{\beta} \forall \lambda_i^{\beta}, \lambda_j^{\beta} \in X$.

Corollary 3.11: Every {1}-commutative permutation B-algebra $(X, \#, \{1\})$ is a (PBN-A).

Proof: By Proposition 3.8 and Proposition 3.10, we get the proof is hold.

Note that the converse is not always true.

Proposition 3.12: Let $(X, \#, \{1\})$ be an abelian group. Then $(X, \#, \{1\})$ is a (PBN-A), if

$$\lambda_i^{\beta} \# \lambda_j^{\beta} = \lambda_i^{\beta} o \lambda_j^{\beta-1}, \quad \forall \lambda_i^{\beta}, \lambda_j^{\beta} \in X.$$

Proof: We have $\lambda_i^{\beta} \# \lambda_i^{\beta} = \lambda_i^{\beta} o \lambda_i^{\beta-1} = \{1\}$ and $\lambda_i^{\beta} \# \{1\} = \lambda_i^{\beta} o \{1\}^{-1} = \lambda_i^{\beta} o \{1\} = \lambda_i^{\beta}$.

Now for all $\lambda_i^{\beta}, \lambda_j^{\beta}, \lambda_k^{\beta} \in X$, $\left(\lambda_i^{\beta} \# \lambda_j^{\beta}\right) \# \lambda_k^{\beta} = \left(\lambda_i^{\beta} o \lambda_j^{\beta-1}\right) o \lambda_k^{\beta-1} = \lambda_k^{\beta-1} o \left(\lambda_i^{\beta} o \lambda_j^{\beta-1}\right)$

$= \lambda_k^{\beta-1} o \left(\lambda_j^{\beta-1} o \lambda_i^{\beta}\right) = \lambda_k^{\beta-1} o \left(\lambda_i^{\beta-1} o \lambda_j^{\beta}\right)^{-1} = \lambda_k^{\beta-1} \# \left(\lambda_i^{\beta-1} o \lambda_j^{\beta}\right) = \left(\{1\} \# \lambda_k^{\beta}\right) \# \left(\lambda_j^{\beta} o \lambda_i^{\beta-1}\right) = \left(\{1\} \# \lambda_k^{\beta}\right) \# \left(\lambda_j^{\beta} \# \lambda_i^{\beta}\right)$. Hence, $(X, \#, \{1\})$ is a (PBN-A).

Proposition 3.13: Let $(X, \#, \{1\})$ be a (PBN-A) with $\{1\} \# \left(\{1\} \# \lambda_i^{\beta}\right) = \lambda_i^{\beta}, \forall \lambda_i^{\beta} \in X$. Then $(X, \#, \{1\})$ is (P1CBN − A) if and only if $\left(\{1\} \# \lambda_i^{\beta}\right) \# \left(\{1\} \# \lambda_j^{\beta}\right) = \lambda_j^{\beta} \# \lambda_i^{\beta} \forall \lambda_i^{\beta}, \lambda_j^{\beta} \in X$.

Proof: If $(X, \#, \{1\})$ is (P1CBN − A), then $\left(\{1\} \# \lambda_i^{\beta}\right) \# \left(\{1\} \# \lambda_j^{\beta}\right) = \lambda_j^{\beta} \# \left(\{1\} \# \left(\{1\} \# \lambda_i^{\beta}\right)\right) = \lambda_j^{\beta} \# \lambda_i^{\beta} \forall \lambda_i^{\beta}, \lambda_j^{\beta} \in X$. Conversely, if $\left(\{1\} \# \lambda_i^{\beta}\right) \# \left(\{1\} \# \lambda_j^{\beta}\right) = \lambda_j^{\beta} \# \lambda_i^{\beta} \forall \lambda_i^{\beta}, \lambda_j^{\beta} \in X$. So $\lambda_i^{\beta} \# \left(\{1\} \# \lambda_j^{\beta}\right) = \left(\{1\} \# \left(\{1\} \# \lambda_i^{\beta}\right)\right) \# \left(\{1\} \# \lambda_j^{\beta}\right) = \lambda_j^{\beta} \# \left(\{1\} \# \lambda_i^{\beta}\right)$.

Definition 3.14: A permutation algebra $(X, \#, \{1\})$ is said to have *condition (D)* if $\left(\lambda_i^{\beta} \# \lambda_j^{\beta}\right) \# \lambda_k^{\beta} = \lambda_i^{\beta} \# \left(\lambda_k^{\beta} \# \lambda_j^{\beta}\right) \forall \lambda_i^{\beta}, \lambda_j^{\beta}, \lambda_k^{\beta} \in X$.

Proposition 3.15: If $(X, \#, \{1\})$ is a (PBN-A) and satisfies condition (D), then **(1)** $\{1\} \# \lambda_i^{\beta} = \lambda_i^{\beta}$, and **(2)** $\lambda_i^{\beta} \# \lambda_j^{\beta} = \lambda_j^{\beta} \# \lambda_i^{\beta}, \quad \forall \lambda_i^{\beta}, \lambda_j^{\beta} \in X$.

Proof: **(1)** Put $\lambda_i^\beta = \{1\}$, $\lambda_k^\beta = \{1\}$ in condition (D), and then we have that $\{1\} \# \lambda_j^\beta = \{1\} \# \left(\{1\} \# \lambda_j^\beta\right) = \lambda_j^\beta$.By Proposition 3.4 (1). **(2)** $\lambda_i^\beta \# \lambda_j^\beta = \lambda_i^\beta \# \left(\{1\} \# \lambda_j^\beta\right) = \lambda_j^\beta \# \left(\{1\} \# \lambda_i^\beta\right) = \lambda_j^\beta \# \lambda_i^\beta$. By Proposition 3.6 and (1).

Proposition 3.16: If $(X, \#, \{1\})$ is a (PBN-A) and satisfies condition (D), then it is a (PB-A).

Proof: Let $\lambda_i^\beta, \lambda_j^\beta, \lambda_k^\beta \in X$, and then $\lambda_i^\beta \# \left(\lambda_k^\beta \# \left(\{1\} \# \lambda_j^\beta\right)\right) = \lambda_i^\beta \# \left(\lambda_k^\beta \# \lambda_j^\beta\right)$

(By Proposition 3.15 (1))

$= \left(\lambda_i^\beta \# \lambda_j^\beta\right) \# \lambda_k^\beta$ (By condition (D)). Hence, $(X, \#, \{1\})$ is (PB-A).

Proposition 3.17: If $(X, \#, \{1\})$ is a (PBN-A) and satisfies condition (D), then $(X, \#, \{1\})$ is an abelian group.

Proof: Since $(X, \#, \{1\})$ is a (PBN-A), $\lambda_i^\beta \# \lambda_i^\beta = \{1\} \forall \lambda_i^\beta \in X$. We considered λ_i^β as its own inverse, i.e., $\lambda_i^{\beta-1} = \lambda_i^\beta$. Now by (2) of Definition 3.1 and Proposition 3.15 (1), we have that

$\lambda_i^\beta \# \{1\} = \{1\} \# \lambda_i^\beta = \lambda_i^\beta$. That is, $\{1\}$ is the identity element of X. Since also that

$\left(\lambda_i^\beta \# \lambda_j^\beta\right) \# \lambda_k^\beta = \lambda_i^\beta \# \left(\lambda_k^\beta \# \lambda_j^\beta\right) = \lambda_i^\beta \# \left(\lambda_j^\beta \# \lambda_k^\beta\right)$.

By Proposition 3.15 (2), the associative law holds. Proposition 3.15 (2) also shows that $(X, \#, \{1\})$ is an abelian group.

Proposition 3.18: If $(X, \#, \{1\})$ is a (PBN-A) and satisfies condition (D), then it is a (PBH-A).

Proof: Let $\lambda_i^\beta \# \lambda_j^\beta = \{1\}$ and $\lambda_j^\beta \# \lambda_i^\beta = \{1\}$. Then, by Proposition 3.15, we have that $\lambda_i^\beta = \lambda_i^\beta \# \{1\} = \lambda_i^\beta \# \left(\lambda_j^\beta \# \lambda_i^\beta\right) = \left(\lambda_i^\beta \# \lambda_i^\beta\right) \# \lambda_j^\beta = \{1\} \# \lambda_j^\beta = \lambda_j^\beta$. Hence, $(X, \#, \{1\})$ is a (PBH-A).

Definition 3.19: A *permutation coxeter algebra* (PCA) is a set $X \neq \varnothing$ with a constant $\{1\}$ and a binary operation " $\#$ " such that **(1)** $\lambda_i^\beta \# \lambda_i^\beta = \{1\}$, **(2)** $\lambda_i^\beta \# \{1\} = \lambda_i^\beta$, and **(3)** $\left(\lambda_i^\beta \# \lambda_j^\beta\right) \# \lambda_k^\beta = \lambda_i^\beta \# \left(\lambda_j^\beta \# \lambda_k^\beta\right), \forall \lambda_i^\beta, \lambda_j^\beta, \lambda_k^\beta \in X$.

Proposition 3.20: If $(X, \#, \{1\})$ is a (PCA), then **(1)** $\{1\} \# \lambda_i^\beta = \lambda_i^\beta$, and **(2)** $\lambda_i^\beta \# \lambda_j^\beta = \lambda_j^\beta \# \lambda_i^\beta \forall \lambda_i^\beta, \lambda_j^\beta \in X$.

Proof: **(1)** For all $\lambda_i^\beta \in X$, we have that $\lambda_i^\beta = \lambda_i^\beta \# \{1\} = \lambda_i^\beta \# \left(\lambda_i^\beta \# \lambda_i^\beta\right) = \left(\lambda_i^\beta \# \lambda_i^\beta\right) \# \lambda_i^\beta = \{1\} \# \lambda_i^\beta$, **(2)** $\lambda_j^\beta = \{1\} \# \lambda_j^\beta$ (By (1))

$= \left[\left(\lambda_i^\beta \# \lambda_j^\beta\right) \# \left(\lambda_i^\beta \# \lambda_j^\beta\right)\right] \# \lambda_j^\beta = \left(\lambda_i^\beta \# \lambda_j^\beta\right) \# \left[\left(\lambda_i^\beta \# \lambda_j^\beta\right) \# \lambda_j^\beta\right] = \left(\lambda_i^\beta \# \lambda_j^\beta\right) \# \left[\lambda_i^\beta \# \left(\lambda_j^\beta \# \lambda_j^\beta\right)\right] = \left(\lambda_i^\beta \# \lambda_j^\beta\right) \# \left(\lambda_i^\beta \# \{1\}\right) = \left(\lambda_i^\beta \# \lambda_j^\beta\right) \# \lambda_i^\beta$. Now multiplying λ_i^β to the right side, we have that $\lambda_j^\beta \# \lambda_i^\beta = \left[\left(\lambda_i^\beta \# \lambda_j^\beta\right) \# \lambda_i^\beta\right] \# \lambda_i^\beta = \left(\lambda_i^\beta \# \lambda_j^\beta\right) \# \left(\lambda_i^\beta \# \lambda_i^\beta\right) = \left(\lambda_i^\beta \# \lambda_j^\beta\right) \# \{1\} = \lambda_i^\beta \# \lambda_j^\beta$

Proposition 3.21: Every (PCA) is a (PBN-A).

Proof: Let $(X, \#, \{1\})$ be a (PCA). For all $\lambda_i^\beta, \lambda_j^\beta, \lambda_k^\beta \in X$, we have that $\left(\{1\} \# \lambda_k^\beta\right) \# \left(\lambda_j^\beta \# \lambda_i^\beta\right) = \left(\lambda_j^\beta \# \lambda_i^\beta\right) \# \left(\{1\} \# \lambda_k^\beta\right)$ (By Proposition 3.20 (2))

$= \left(\lambda_i^\beta \# \lambda_j^\beta\right) \# \lambda_k^\beta$ (By (3) of Definition 3.1). Hence, $(X, \#, \{1\})$ is a (PBN-A). Note that the opposite of above proposition is not necessarily always true.

Proposition 3.22: $(X, \#, \{1\})$ is a (PBN-A) and satisfies condition (D) if and only if it is a (PCA).

Proof: For all $\lambda_i^\beta, \lambda_j^\beta, \lambda_k^\beta \in X$, $\left(\lambda_i^\beta \# \lambda_j^\beta\right) \# \lambda_k^\beta = \lambda_i^\beta \# \left(\lambda_k^\beta \# \lambda_j^\beta\right)$ (Condition) (D) $= \lambda_i^\beta \# \left(\lambda_j^\beta \# \lambda_k^\beta\right)$ (From Proposition 3.20 (2)). Hence, $(X, \#, \{1\})$ is a (PCA).

Conversely, assume that $(X, \#, \{1\})$ is a (PCA). By Proposition 3.12, we only need to prove condition (D). For all $\lambda_i^\beta, \lambda_j^\beta, \lambda_k^\beta \in X$, $\left(\lambda_i^\beta \# \lambda_j^\beta\right) \# \lambda_k^\beta = \lambda_i^\beta \# \left(\lambda_j^\beta \# \lambda_k^\beta\right)$ (X being a (PCA))

$= \lambda_i^\beta \# \left(\lambda_k^\beta \# \lambda_j^\beta\right)$ (By Proposition 3.20 (2))

Definition 3.23: Let $(X, \#, \{1\})$ be a (PBN-A) and let $\varnothing \neq S \subseteq X$. S is said to be a *permutation BN-subalgebra* (PBN-SA) of X if $\lambda_i^\beta \# \lambda_j^\beta \in S$ for all $\lambda_i^\beta, \lambda_j^\beta \in S$. S, and it is said to be *normal* of X if $\left(\lambda_i^\beta \# \lambda_m^\beta\right) \# \left(\lambda_j^\beta \# \lambda_n^\beta\right) \in S$ whenever $\lambda_i^\beta \# \lambda_j^\beta, \lambda_m^\beta \# \lambda_n^\beta \in S$.

Proposition 3.24: Every normal subset S of a (PBN-A) $(X, \#, \{1\})$ is a (PBN-SA) of X.

Proof: If $\lambda_i^\beta, \lambda_j^\beta \in S$, then $\lambda_i^\beta \# \{1\}, \lambda_j^\beta \# \{1\} \in S$. Since S is normal, $\lambda_i^\beta \# \lambda_j^\beta = \left(\lambda_i^\beta \# \lambda_j^\beta\right) \# (\{1\} \# \{1\}) \in S$. Thus, S is a (PBN-SA) of X.

Lemma 3.25: Let S be a normal (PBN-SA) $(X, \#, \{1\})$. If $\lambda_i^\beta \# \lambda_j^\beta \in S$, then $\lambda_j^\beta \# \lambda_i^\beta \in S$.

Proof: Let $\lambda_i^\beta \# \lambda_j^\beta \in S$. Since $\lambda_j^\beta \# \lambda_j^\beta = \{1\} \in S$ and S normal, then $\lambda_j^\beta \# \lambda_i^\beta = \left(\lambda_j^\beta \# \lambda_i^\beta\right) \# \left(\lambda_j^\beta \# \lambda_j^\beta\right) \in S$.

Definition 3.26: Let $(X, \#, \{1\})$ be a (PBN-A) and let S be a normal (PBN-SA) of X. Defining a relation $\sim_S$ on X by $\lambda_i^\beta \sim_S \lambda_j^\beta$ if and only if $\lambda_i^\beta \# \lambda_j^\beta \in S$, where $\lambda_i^\beta, \lambda_j^\beta \in X$. Then, $\sim_S$ is an equivalence relation on X. We denote the equivalence class containing λ_i^β by $\left[\lambda_i^\beta\right]_S$. i.e., $\left[\lambda_i^\beta\right]_S := \left\{\lambda_j^\beta \in X | \lambda_i^\beta \sim_S \lambda_j^\beta\right\}$, and let $X/S := \left\{\left[\lambda_i^\beta\right]_S | \lambda_i^\beta \in X\right\}$ be defined by $\left[\lambda_i^\beta\right]_S \# \left[\lambda_j^\beta\right]_S = \left[\lambda_i^\beta \# \lambda_j^\beta\right]_S$, and then X/S is said to be the *quotient permutation BN-algebra* (QPBN-A) of X by S.

Proposition 3.27: Let S be normal (PBN-SA) of a (PBN-A) $(X, \#, \{1\})$. Then X/S is a (QPBN-A).

Proof: If we define $\left[\lambda_i^\beta\right]_S \# \left[\lambda_j^\beta\right]_S = \left[\lambda_i^\beta \# \lambda_j^\beta\right]_S$, then the operation "$\#$" is well-defined, since if $\lambda_i^\beta \sim_S \lambda_m^\beta$ and $\lambda_j^\beta \sim_S \lambda_n^\beta$, then $\lambda_i^\beta \# \lambda_m^\beta \in S$ and $\lambda_j^\beta \# \lambda_n^\beta \in S$ imply that $\left(\lambda_i^\beta \# \lambda_j^\beta\right) \# \left(\lambda_m^\beta \# \lambda_n^\beta\right) \in S$. By normality of S, so $\lambda_i^\beta \# \lambda_j^\beta \sim_S \lambda_m^\beta \# \lambda_n^\beta$ and so $\left[\lambda_i^\beta \# \lambda_j^\beta\right]_S = \left[\lambda_m^\beta \# \lambda_n^\beta\right]_S$.

Consider $[\{1\}]_S = \left\{\lambda_i^\beta \in X \mid \lambda_i^\beta \sim_S \{1\}\right\} = \left\{\lambda_i^\beta \in X \mid \lambda_i^\beta \# \{1\} \in S\right\} = \left\{\lambda_i^\beta \in X \mid \lambda_i^\beta \in S\right\} = S$.

Definition 3.28: Let each of $(X, \#, \{1\}_X)$ and $(Y, \#, \{1\}_Y)$ be (PBN-A). A mapping $\theta : X \longrightarrow Y$ is called a *homomorphism* from X to Y if $\theta\left(\lambda_i^\beta \# \lambda_j^\beta\right) = \theta\left(\lambda_i^\beta\right) \# \theta\left(\lambda_j^\beta\right)$, $\forall \lambda_i^\beta, \lambda_j^\beta \in X$. Note that $\theta(\{1\}_X) = \{1\}_Y$. Indeed, $\theta(\{1\}_X) = \theta\left(\lambda_i^\beta \# \lambda_i^\beta\right) = \theta\left(\lambda_i^\beta\right) \# \theta\left(\lambda_i^\beta\right) = \{1\}_Y$. The *kernel* of the homomorphism denoted by $Ker\theta = \left\{\lambda_i^\beta \in X | \theta\left(\lambda_i^\beta\right) = \{1\}_Y\right\}$. Note that $Ker\theta$ is a subset of X.

Remark 3.29: Let S be a normal (PBN-SA) of a (PBN-A) X. Then the mapping $\gamma : X \longrightarrow X/S$ given by $\gamma\left(\lambda_i^\beta\right) := \left[\lambda_i^\beta\right]_S$ is an epimorphism of permutation BN-algebras and $Ker\gamma = S$.

Definition 3.30: A (PBN-A) $(X, \#, \{1\})$ is called a *permutation BN_1-algebra* (PBN_1-A) if it satisfies the following condition: $\lambda_i^\beta = \left(\lambda_i^\beta \# \lambda_j^\beta\right) \# \lambda_j^\beta \forall \lambda_i^\beta, \lambda_j^\beta \in X$.

Corollary 3.31: Let $(X, \#, \{1\})$ be a (PBN_1-A). If $\lambda_i^\beta \# \lambda_j^\beta = \{1\}$, then $\lambda_i^\beta = \lambda_j^\beta$.

Proof: Substituting $\lambda_i^\beta := \lambda_j^\beta$ for all $\lambda_i^\beta, \lambda_j^\beta \in X$ in $\lambda_i^\beta = \left(\lambda_i^\beta \# \lambda_j^\beta\right) \# \lambda_j^\beta$, we have that $\lambda_j^\beta = \left(\lambda_j^\beta \# \lambda_j^\beta\right) \# \lambda_j^\beta = \{1\} \# \lambda_j^\beta$. Now taking $\lambda_i^\beta \# \lambda_j^\beta = \{1\}$, we have that $\lambda_i^\beta = \left(\lambda_i^\beta \# \lambda_j^\beta\right) \# \lambda_j^\beta = \{1\} \# \lambda_j^\beta = \lambda_j^\beta$.

Proposition 3.32: Let $\theta : X \longrightarrow Y$ be a homomorphism from a (PBN-A) $(X, \#, \{1\}_X)$ into a (PBN_1-A) $(Y, \#, \{1\}_Y)$. Then the kernel of θ, $Ker\theta$is a normal (PBN-SA) of X.

Proof: Since $\{1\}_X \in Ker\theta$, then $Ker\theta \neq \varnothing$. If $\lambda_i^\beta, \lambda_j^\beta \in Ker\theta$, then $\theta\left(\lambda_i^\beta \# \lambda_j^\beta\right) = \theta\left(\lambda_i^\beta\right) \# \theta\left(\lambda_j^\beta\right) = \{1\}_Y \# \{1\}_Y = \{1\}_Y$, i.e., $\lambda_i^\beta \# \lambda_j^\beta \in Ker\theta$. Hence, $Ker\theta$ is a (PBN-SA) of X. Let $\lambda_i^\beta \# \lambda_j^\beta, \lambda_m^\beta \# \lambda_n^\beta \in Ker\theta$. Then $\theta\left(\lambda_i^\beta \# \lambda_j^\beta\right) = \theta\left(\lambda_i^\beta\right) \# \theta\left(\lambda_j^\beta\right) = \{1\}_Y$ and $\theta\left(\lambda_m^\beta \# \lambda_n^\beta\right) = \theta\left(\lambda_m^\beta\right) \# \theta\left(\lambda_n^\beta\right) = \{1\}_Y$. Since Y is a (PBN_1-A), by Corollary 3.31, $\theta\left(\lambda_i^\beta\right) = \theta\left(\lambda_j^\beta\right)$ and $\theta\left(\lambda_m^\beta\right) = \theta\left(\lambda_n^\beta\right)$. Hence, $\theta\left(\left(\lambda_i^\beta \# \lambda_m^\beta\right) \# \left(\lambda_j^\beta \# \lambda_n^\beta\right)\right) = \theta\left(\lambda_i^\beta \# \lambda_m^\beta\right) \# \theta\left(\lambda_j^\beta \# \lambda_n^\beta\right) = \left(\theta\left(\lambda_i^\beta\right) \# \theta\left(\lambda_m^\beta\right)\right) \# \left(\theta\left(\lambda_j^\beta\right) \# \theta\left(\lambda_n^\beta\right)\right) = \left(\theta\left(\lambda_i^\beta\right) \# \theta\left(\lambda_m^\beta\right)\right) \# \left(\theta\left(\lambda_i^\beta\right) \# \theta\left(\lambda_m^\beta\right)\right) = \{1\}_Y$. Thus, $\left(\lambda_i^\beta \# \lambda_m^\beta\right) \# \left(\lambda_j^\beta \# \lambda_n^\beta\right) \in Ker\theta$. Hence, $Ker\theta$ is a normal (PBN-SA) of X.

Corollary 3.33: Let $\theta : X \longrightarrow Y$ be a homomorphism from a (PBN-A) $(X, \#, \{1\}_X)$ into a (PBN_1-A) $(Y, \#, \{1\}_Y)$. Then, $X/Ker\theta \cong Im\theta$. In particular, if θ is surjective, then $X/Ker\theta \cong Y$.

4 Conclusion

In this research, some new notions and results are investigated and proven using the composition of BN-algebra with permutation sets such as permutation BN-algebras, a form of permutation group-derived, permutation BN-subalgebra, and permutation BN-normal. In upcoming work, we'll use the theory of neutrosophic sets to compose neutrosophic sets and BN-algebra, after which we'll think about a few ideas and research the outcomes.

References

1. Imai, Y., Iseki, K.: On axiom system of propositional calculi, XIV. Proc. Japan Acad. **42**(1), 19–22 (1966)
2. Iseki, K.: An algebra related with a propositional calculus. Proc. Japan Acad. **42**(1), 26–29 (1966)
3. Neggers, J., Kim, H.S.: On B-algebras. Mate. Vesnik. **54**, 21–29 (2002)
4. Chang Bum Kim: On BN-algebras, Kyungpook. Math. J. **53**, 175–184 (2013)
5. Alsalem, S.: The permutation topological spaces and their bases. Basrah J. Sci. **32**(1), 28–42 (2014)
6. Khalil, S.M., Abbas, N.M.A.: Characteristics of the number of conjugacy classes and P-regular classes in finite symmetric groups. IOP Conf. Ser. Mater. Sci. Eng. **571**, 012007 (2019). https://doi.org/10.1088/1757-899X/571/1/012007

7. Torki, M.M., Khalil, S.M.: New types of finite groups and generated algorithm to determine the integer factorization by excel. AIP Conf. Proc. **2290**, 040020 (2020)
8. Khalil, S.M., Hameed, F.: An algorithm for generating permutations in symmetric groups using soft spaces with general study and basic properties of permutations spaces. J. Theor. Appl. Inf. Technol. **96**, 2445–2457 (2018)
9. Khalil, S.M., Rajah, A.: Solving Class Equation $x^d = \beta$ in an Alternating Group for all $n \in \theta$ & $\beta \in H_n \cap C^\alpha$. Arab J. Basic Appl. Sci. **16**, 38–45 (2014)
10. Khalil, S.M., Suleiman, E., Torki, M.M.: Generated new classes of permutation I/B-algebras. J. Discrete Math. Sci. Cryptogr. **25**(1), 31–40 (2022)
11. Khalil, S.M., Abbas, N.M.: Applications on new category of the symmetric groups. AIP Conf. Proc. **2290**, 040004 (2020)
12. Khalil, S.M., Suleiman, E., Ali Abbas, N.M.: New technical to generate permutation measurable spaces. In: 2021 1st Babylon International Conference on Information Technology and Science (BICITS), pp. 160–163 (2021). https://doi.org/10.1109/BICITS51482.2021.9509892
13. Khalil, S.M., Rajah, A.: Solving the class equation $x^d = \beta$ in an alternating group for each $\beta \in H \cap C^\alpha$ and $n \notin \theta$. Arab J. Basic Appl. Sci. **10**, 42–50 (2011)
14. Khalil, S.M., Ulrazaq, M., Abdul-Ghani, S.: Abu Firas Al-Musawi, $\sigma-$Algebra and $\sigma-$Baire in Fuzzy Soft Setting. Adv. Fuzzy Syst., 10 (2018)
15. Khalil, S.M., Hassan, A.: Applications of fuzzy soft $\rho-$ideals in $\rho-$algebras. Fuzzy Inf. Eng. **10**, 467–475 (2018)
16. Abdul-Ghani, S.A., Khalil, S.M., Abd Ulrazaq, M., Al-Musawi, A.F.: New branch of intuitionistic fuzzification in algebras with their applications. Int. J. Math. Math. Sci., 6 (2018)
17. Khalil, S.M., Hasab, M.H.: Decision making using new distances of intuitionistic fuzzy sets and study their application in the universities, INFUS. Adv Intell. Syst. Comput. **1197**, 390–396 (2020)
18. Khalil, S., A. Hassan A, H. Alaskar, W. Khan, Hussain A.: Fuzzy Logical Algebra and Study of the Effectiveness of Medications for COVID-19. Mathematics. **9**(22), 28–38 (2021)
19. Khalil, S.M., Hassan, A.N.: New class of algebraic fuzzy systems using cubic soft sets with their applications. IOP Conf. Ser. Mater. Sci. Eng. **928** (2020). https://doi.org/10.1088/1757-899X/928/4/042019
20. Khalil, S.M.: Dissimilarity fuzzy soft points and their applications. Fuzzy Inf. Eng. **8**, 281–294 (2016)
21. Khalil, S.M., Abdul-Ghani, S.A.: Soft M-ideals and soft S-ideals in soft S-algebras. IOP Conf. Series: J. Phys. **1234**, 012100 (2019)
22. Hasan, M.A., Khalil, S.M., Abbas, N.M.A.: Characteristics of the soft-(1, 2)-gprw–closed sets in soft bi-topological spaces, Conf. IT-ELA, 9253110, 103–108 (2020).
23. Abbas, N.M.A., Khalil, S.M., Hamza, A.A.: On α^*-continuous and contra α^*-continuous mappings in topological spaces with soft setting. Int. J. Nonlinear Anal. Appl. **12**, 1107–1113 (2021)
24. Khalil, S.M., Hameed, F.: Applications on cyclic soft symmetric. IOP Conf. Series: J. Phys. **1530**, 012046 (2020)
25. Hasan, M.A., Abbas, N.M.A., Khalil, S.M.: On soft α^*-open sets and soft contra α^*-continuous mappings in soft topological spaces. J. Interdiscip. Math. **24**, 729–734 (2021)
26. Al-Musawi, A.M., Khalil, S.M., Ulrazaq, M.A.: Soft (1,2)-strongly open maps in bi-topological spaces. IOP Conf. Ser. Mater. Sci. Eng. **571**, 012002 (2019). https://doi.org/10.1088/1757-899X/571/1/012002
27. Khalil, S.M., Abbas, N.M.A.: On nano with their applications in medical field. AIP Conf. Proc. **2290**, 040002 (2020)
28. Alsalem, S., Al Musawi, A., Suleiman, E.: On Permutation B-center and Derived Permutation B-algebras. IEEE (2022) To appear
29. Suleiman, E., Hasan, M., Yerima, N.: Normal Permutation BF-Subalgebras in Permutation BF-Algebras. IEEE (2022) To appear
30. Alsalem, S., Al Musawi, A., Suleiman, E.: On maximal permutation $BH-$Ideals of Permutation $BH-$Algebras. IEEE (2022) To appear
31. Walendziak, A.: On BF-algebras. Math. Slovaca. **57**(2), 119–128 (2007)

Interpretation of Skew Ideals with Relators in Join Skew Semilattice

Sri Rama Ravi Kumar E., Siva Ram Prasad J., Baby Rani Ch., Nageswar Rao T., and Venkateswara Rao M.

Keywords Skew lattice · Distributive semilattice · Ideal · Skew ideal

1 Introduction

In the research of algebra, the order structure is crucial. Skew lattice is the most common type of lattice. A Hasse diagram can represent the skew lattice with order structure, but it cannot represent its algebraic structure. Leech [1] has started researching in the coset structure of skew lattices. Characterization for more types of skew lattices uses only coset laws [2–4]. Skew ideals of skew lattices were created to characterize the skewness of skew lattices, which are inextricably linked to the concept of normality as well the natural preorder structure. In the literature, skew lattices have received little attention. As a result, they are still a relatively new area of study. Semilattices are a branch of nonclassical logic. In review of algebra, semilattices, which are related to nonclassical logic, are consistently present. In semilattices, there are several notions of distributivity, one of which is the 0-distributivity concept introduced and investigated by Cvetko-vah. K [5]. Hickman R.C. [6] introduced and investigated mildly distributive semilattices, which are another interesting class of distributivity. The distributive semilattice is defined as a join semilattice $(\mathbb{S}, \vee)$ only if $x_1 \vee x_2 \geq y$, for $x_1, x_2, y \in \mathbb{S}$, the elements z_1 and z_2 exists in $\mathbb{S}$ such as $x_1 \geq z_1$, $x_2 \geq z_2$ and $y = z_1 \vee z_2$ [7]. Even skew ideals can be obtained from a partial order, but their behavior is more akin to that of skew lattices. Costa Pita J [8] investigated ideals in skew lattices. This manuscript

Sri Rama Ravi Kumar E. (✉) · Siva Ram Prasad J. · Baby Rani Ch.
V R Siddhartha Engineering College, Vijayawada, India
e-mail: esrravikumar@vrsiddhartha.ac.in

Nageswar Rao T.
Koneru Lakshmaiah Education Foundation, Vaddeswaram, India
e-mail: tnraothota@kluniversity.in

Venkateswara Rao M.
Government Degree College, Avanigadda, India

H.-H. Leung et al. (eds.), *Recent Developments in Algebra and Analysis*,
Trends in Mathematics, https://doi.org/10.1007/978-3-031-37538-5_3

introduces and investigates the skew semilattice, as well as its characterization. We have the distributive semilattice necessary and sufficient theorem and are working on the concept of normal skew semilattice. Skew ideals derived from partial order, on the other hand, appear to be more closely related to the unique properties of skew lattices. Furthermore, we design an order skew ideal in skew semilattice and prove the skew semilattice characterization theorem by defining new through relators.

2 Distributive Skew Semilattice

2.1 *Skew Lattice*

We define a skew lattice as a non-empty set $\mathbb{S}$ supplied by exactly two binary operations $\vee$ (join) and $\wedge$ (meet) and satisfying the following properties:

Associative:- $(\mathbf{g} \vee \mathbf{h}) \vee \mathbf{z} = \mathbf{g} \vee (\mathbf{h} \vee \mathbf{z})$ **and** $(\mathbf{g} \wedge \mathbf{h}) \wedge \mathbf{z} = \mathbf{g} \wedge (\mathbf{h} \wedge \mathbf{z})$.
Idempotent:- $\mathbf{g} \vee \mathbf{g} = \mathbf{g}$ **and** $\mathbf{g} \wedge \mathbf{g} = \mathbf{g}$.
Absorption:- $(\mathbf{h} \wedge \mathbf{g}) \vee \mathbf{g} = \mathbf{g} = \mathbf{g} \vee (\mathbf{g} \wedge \mathbf{h})$ **and** $(\mathbf{h} \vee \mathbf{g}) \wedge \mathbf{g} = \mathbf{g} = \mathbf{g} \wedge (\mathbf{g} \vee \mathbf{h})$.

2.2 *Note*

The skew lattices are defined by absorption dualities because the binary operations $\vee$ (join) and $\wedge$ (meet) are associative and idempotent:

$s \vee t = s$ assuming and only assuming $s \wedge t = t$
$s \vee t = t$ assuming and only assuming $s \wedge t = s$

Skew lattice $\mathbb{S}$ has a natural partial ordering $\geq$, similar to order in lattice, and it is defined by $f_1 \geq f_2$ whenever $f_1 \wedge f_2 = f_2 = f_2 \wedge f_1$ or dually $f_1 \vee f_2 = f_1 = f_2 \vee f_1$ for $f_1, f_2 \in \mathbb{S}$. Leech J [8]'s Lemma (2.3) on skew lattice ideals is a useful observation for skew lattices characterized by regular partial order.

2.3 *Lemma*

In the skew lattice and if $p_1, q_1 \in \mathbb{S}$, then $p_1 \geq q_1$ whenever $q_1 = p_1 \wedge q_1 \wedge p_1$ or dually $p_1 = q_1 \vee p_1 \vee q_1$.

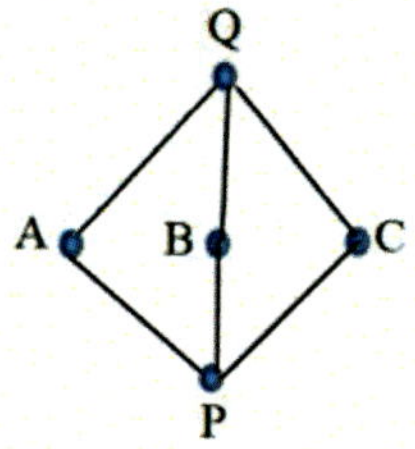

Fig. 1 Example of skew semilattice

2.4 Join Skew Semilattice

A skew semilattice is a set $\mathbb{S}$ that is not empty, with $(\mathbb{S}, \vee)$ if it is both right join skew semilattice and left join skew semilattice.

2.5 Right Join Skew Semilattice

Right join skew semilattice is the semilattice $\mathbb{S}$ satisfying the identity $p_1 \vee q_1 \vee p_1 = q_1 \vee p_1$ for $p_1, q_1 \in \mathbb{S}$.

2.6 Left Join Skew Semilattice

Left skew semilattice is the semilattice $\mathbb{S}$ satisfying identity $p_1 \vee q_1 \vee p_1 = p_1 \vee q_1$, for $p_1, q_1 \in \mathbb{S}$.

2.7 Example of Skew Semilattice

In Fig. 1, we observe $A \vee B \vee C = A \vee B$, and $A \vee B \vee C = B \vee C$, $A \vee C \vee A = A \vee C$, and $A \vee C \vee A = C \vee A$; $P \vee A \vee P = P \vee A$ and $P \vee A \vee P = A \vee P$; $B \vee C \vee B = B \vee C$ and $B \vee C \vee B = C \vee B$; $A \vee Q \vee A = A \vee Q$ and $A \vee Q \vee A = Q \vee A$; $B \vee Q \vee B = B \vee Q$ and $B \vee Q \vee B = Q \vee B$, $C \vee Q \vee C = C \vee Q$ and $C \vee Q \vee C = Q \vee C$, etc. So the given semilattice $\mathbb{S}$ (Fig. 1) is both left skew semilattice and right skew semilattice. Therefore, $\mathbb{S}$ is skew semilattice.

The partial order $(D_{12}, /)$ given in Fig. 2 is not skew semilattice, since $1 \vee 2 \vee 4 = 4 \neq 1 \vee 2$ but $1 \vee 2 \vee 4 = 2 \vee 4 = 4$; hence, it is a right skew semilattice but not left skew semilattice.

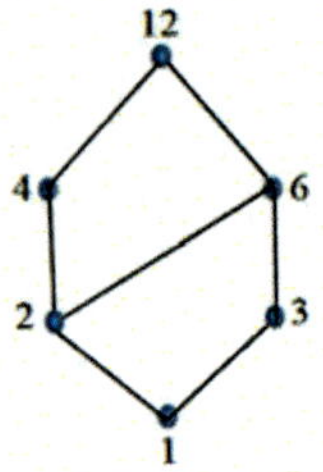

Fig. 2 Counterexample of skew semilattice

2.8 *Distributive Skew Semilattice*

A skew semilattice $\mathbb{S}$ is claimed to be distributive if for all **g, h, z** with $\mathbf{g} \leq \mathbf{h} \vee \mathbf{z}$, their exists $\mathbf{h}^1 \leq \mathbf{h}$ and $\mathbf{z}^1 \leq \mathbf{z}$ such that $\mathbf{g} = \mathbf{h}^1 \vee \mathbf{z}^1$; it follows an equivalent property that $\mathbf{g} \vee (\mathbf{h} \wedge \mathbf{z}) \vee \mathbf{g} = (\mathbf{g} \vee \mathbf{h} \vee \mathbf{g}) \wedge (\mathbf{g} \vee \mathbf{z} \vee \mathbf{g})\ \forall\ \mathbf{g}, \mathbf{h}, \mathbf{z}$ in $\mathbb{S}$.

2.9 *Example*

If $\mathbb{S}$ is skew semilattice, then for **d, e** in $\mathbb{S}$ we have $\mathbf{d} \vee \mathbf{e} \vee \mathbf{e} = \mathbf{d} \vee \mathbf{e}$ and $\mathbf{d} \vee \mathbf{e} \vee \mathbf{d} = \mathbf{e} \vee \mathbf{d}$, and then, similarly we may have $\mathbf{x} \vee (\mathbf{y} \wedge \mathbf{z}) \vee \mathbf{x} = (\mathbf{y} \wedge \mathbf{z}) \vee \mathbf{x} = (\mathbf{y} \vee \mathbf{x}) \wedge (\mathbf{z} \vee \mathbf{x})$ – (i) and also $(\mathbf{x} \vee \mathbf{y} \vee \mathbf{x}) \wedge (\mathbf{x} \vee \mathbf{z} \vee \mathbf{x}) = (\mathbf{y} \vee \mathbf{b}) \wedge (\mathbf{z} \vee \mathbf{x})$ – (ii); therefore, from (i) and (ii), we have $\mathbf{x} \vee (\mathbf{y} \wedge \mathbf{z}) \vee \mathbf{x} = (\mathbf{x} \vee \mathbf{y} \vee \mathbf{x}) \wedge (\mathbf{x} \vee \mathbf{z} \vee \mathbf{x})$; therefore, $\mathbb{S}$ is a distributive skew semilattice.

2.10 *Skew Normal Semilattice*

The skew semilattice $\mathbb{S}$ is said to be normal if it satisfies the property that $\mathbf{m} \vee \mathbf{n} \vee \mathrm{z} \vee \mathrm{w} = \mathbf{m} \vee \mathrm{z} \vee \mathbf{n} \vee \mathrm{w}$ for **m**, **n**, z, w $\in \mathbb{S}$.

2.11 *Skew Ideal*

If $\mathrm{p} \in \mathbb{S}$ and $\mathrm{q} \in \mathcal{I}$, such as $\mathrm{q} \geq \mathrm{p}$, imply $\mathrm{p} \in \mathcal{I}$, then $\mathcal{I}$ is called skew ideal of, where $\mathcal{I}$ is non-empty subset of a join skew semilattice $\mathbb{S}$.

Example: A subset $\mathcal{I} = \{\mathrm{a, c, p, q}\}$ of skew semilattice $\mathbb{S}$ of Fig. 1 is a skew ideal of $\mathbb{S}$.

2.12 *Distributive Ideal*

Each ideal $\mathcal{I}$ of a join skew semilattice $\mathbb{S}$ is known as distributive ideal for all $q \geq p$ and p in $\mathcal{I}$, such that $\mathcal{I} \vee (q \wedge z) \vee \mathcal{I} = (\mathcal{I} \vee q) \wedge (\mathcal{I} \vee z)$.

2.13 *Result*

If $\mathbf{k}_1$ and k_2 are distributive ideals of skew semilattice $\mathbb{S}$, then $k_1 \vee k_2$ is also distributive ideal of $\mathbb{S}$.

Proof: Assume k_1 and k_2 are distributive ideals of skew semilattice $\mathbb{S}$.

$$\begin{aligned}
\text{Now } [(k_1 \vee k_2) \vee (\mathbf{v} \wedge z) \vee (k_1 \vee k_2)] &= [(p_1 \vee p_2)) \vee (\mathbf{v} \wedge z) \vee (p_1 \vee p_2)], \\
&\quad \text{for } (p_1 \vee p_2) \in k_1 \vee k_2 \text{ where } p_1 \in k_1 \text{ and } p_2 \in k_2 \\
&= [p_1 \vee (p_2 \vee (\mathbf{v} \wedge z)) \vee (p_1 \vee p_2)] \\
&= [p_1 \vee \{ (p_2 \vee \mathbf{v} \vee p_2) \wedge (p_2 \vee z \vee p_2)\} \vee (p_1 \vee p_2)] \\
&= [(p_1 \vee \{ (p_2 \vee \mathbf{v} \vee p_2)\} \wedge \{ (p_2 \vee z \vee p_2)\} \vee p_1) \vee (p_1 \vee p_2)] \\
&= (p_1 \vee p_2 \vee \mathbf{v} \vee p_1 \vee p_2) \wedge (p_1 \vee p_2 \vee z \vee p_1 \vee p_2)] \\
&= [(k_1 \vee k_2) \vee \mathbf{v} \vee (k_1 \vee k_2)] \wedge [(k_1 \vee k_2) \vee z \vee (k_1 \vee k_2)]
\end{aligned}$$

Therefore, $k_1 \vee k_2$ is distributive ideal of $\mathbb{S}$.

2.14 *Theorem*

$(\mathbb{S}, \geq)$ is distributive skew semilattice if and only if $p_1 \vee q_1 \geq w$, for w, p_1, q_1, $\mathbf{p}$ as well as $\mathbf{q}$ are in $\mathbb{S}$ such as $p_1 \geq \mathbf{p}$, $q_1 \geq \mathbf{q}$ with $w = \mathbf{p} \vee \mathbf{q}$.

Proof: $(\mathbb{S}, \geq)$ is a distributive skew semilattice, by definition; it follows

$p \vee (q \wedge z) \vee p = (p \vee q \vee p) \wedge (p \vee z \vee p)$ for all p, q, z in $\mathbb{S}$. To prove that the condition $p_1 \vee q_1 \geq w$, such as $p_1 \geq \mathbf{p}$, $q_1 \geq \mathbf{q}$ with $w = \mathbf{p} \vee \mathbf{q}$, for w, p_1, q_1; $\mathbf{p}$ as well as **q are** in $\mathbb{S}$.

Consider the elements $\mathbf{p}$, $\mathbf{q}$ in $\mathbb{S}$ such as $p_1 \geq \mathbf{p}$, $q_1 \geq \mathbf{q}$.

$$\begin{aligned}
\text{Now consider } (\mathbf{p} \vee \mathbf{q}) \vee (\mathbf{q} \wedge z) \vee (\mathbf{p} \vee \mathbf{q}) &= [(\mathbf{p} \vee \mathbf{q} \vee \mathbf{q}) \wedge (\mathbf{p} \vee \mathbf{q} \vee z)] \vee (\mathbf{p} \vee \mathbf{q}) \\
&= [(\mathbf{p} \vee \mathbf{q} \vee \mathbf{q}) \vee (\mathbf{p} \vee \mathbf{q})] \wedge [(\mathbf{p} \vee \mathbf{q} \vee z) \vee (\mathbf{p} \vee \mathbf{q})] \\
&= (w \vee \mathbf{q} \vee w) \wedge (w \vee z \vee w) \text{ for } w = (\mathbf{p} \vee \mathbf{q}) \in \mathbb{S}
\end{aligned}$$

Therefore, $w \vee (\mathbf{q} \wedge z) \vee w = (w \vee \mathbf{q} \vee w) \wedge (w \vee z \vee w)$ for $w = (\mathbf{p} \vee \mathbf{q}) \in \mathbb{S}$.

Hence, for w, p_1, q_1 in $\mathbb{S}$, **p** and **q** in $\mathbb{S}$ such as $p_1 \geq \mathbf{p}$, $q_1 \geq \mathbf{q}$, then $p_1 \vee q_1 \geq \mathbf{p} \vee \mathbf{q}$, implies $p_1 \vee q_1 \geq w$, with $w = \mathbf{p} \vee \mathbf{q}$.

Conversely: Let the condition $p_1 \vee q_1 \geq w$, for w, p_1, q_1 in $\mathbb{S}$, **p** and **q** in $\mathbb{S}$ such as $p_1 \geq \mathbf{p}$, $q_1 \geq \mathbf{q}$ with $w = \mathbf{p} \vee \mathbf{q}$.

Consider $(g \vee h \vee g) \wedge (g \vee z \vee g) = (w \vee g) \wedge (g \vee z \vee g)$ for $w = g \vee h$

$$= g \vee (w \wedge z) = g \vee ((g \vee h) \wedge z)$$
$$= g \vee (g \wedge z \vee h \wedge z) = g \vee (g \wedge z) \vee (h \wedge z)$$
$$= g \vee (h \wedge z) \vee g.$$

Also $p_1 \vee (\mathbf{q} \wedge z) \vee p_1 \geq \mathbf{p} \vee (\mathbf{q} \wedge z) \vee \mathbf{p}$.

And also $p_1 \vee (\mathbf{q} \wedge z) \vee p_1 \geq p_1 \vee (q_1 \wedge z)$.

Thus,

$$p_1 \vee (q_1 \wedge z) \geq \boldsymbol{p} \vee (\boldsymbol{q} \wedge z) \vee \boldsymbol{p}. \tag{1}$$

Also $p_1 \vee (q_1 \wedge z) = (p_1 \vee q_1) \wedge (p_1 \vee z) = (p_1 \vee q_1 \vee p_1) \wedge (p_1 \vee z \vee p_1)$
$\geq (\mathbf{p} \vee \mathbf{q} \vee \mathbf{p}) \wedge (\mathbf{p} \vee z \vee \mathbf{p})$.

Thus,

$$p_1 \vee (q_1 \wedge z) \geq (\boldsymbol{p} \vee \boldsymbol{q} \vee \boldsymbol{p}) \wedge (\boldsymbol{p} \vee z \vee \boldsymbol{p}). \tag{2}$$

As a result from (1) and (2) $\mathbf{p} \vee (\mathbf{q} \wedge z) \vee \mathbf{p} = (\mathbf{p} \vee \mathbf{q} \vee \mathbf{p}) \wedge (\mathbf{p} \vee z \vee \mathbf{p})$ for all u, v, z in $\mathbb{S}$.

Hence, $\mathbb{S}$ is a distributive skew semilattice.

2.15 Result

All ideals of a skew semilattice form a skew semilattice.

Proof: Assuming $\mathbb{S}$ is a skew semilattice as well as I to be the skew Ideal of $\mathbb{S}$, then for all p in $\mathbb{S}$ and q in I such as $y \geq x$ implicit $x \in I$. Now for $y \geq x$, implicit $x \vee y = y$ and $x \vee y \vee x = (x \vee y) \vee x = y \vee x = y$. Therefore, $x \vee y \vee x = y \vee x$, with partial order "$\geq$." Therefore I is right skew semilattice. Similarly, $x \vee y \vee x = x \vee (y \vee x) = x \vee y$ as $y \vee x = y$; thus, the skew ideal I is left skew semilattice. Therefore, I is both left and right skew semilattices; hence, every ideal is a skew semilattice of a skew semilattice $\mathbb{S}$.

3 Characterization of Skew Semilattice

3.1 Increasing Subset

A subset $I \subseteq \mathbb{S}$ is called an increasing subset, if for each z in $\mathbb{S}$, such as z$\in I$ and z $\leq$ v, then v $\in I$.

3.2 Order Ideal

If I is an increasing subset and for each m, n $\in I$, k $\in \mathbb{S}$, such as m $\geq$ k and n $\geq$ k, the subset I of the skew semilattice S is then referred to as an order ideal.

3.3 Theorem

The skew semilattice $\mathbb{S}$ is a distributive only if (i) $\mathbb{S}$ is directed below (ii) Id($\mathbb{S}$) for all order ideals of $\mathbb{S}$, a distributive skew semilattice, where Id($\mathbb{S}$) is set of all proper order ideals of $\mathbb{S}$.

Proof: Let a skew semilattice $\mathbb{S}$ be distributive. Let m, n, w $\in \mathbb{S}$, then for m $\vee$ n $\geq$ w, e and f in $\mathbb{S}$ such as m $\geq$ e, n $\geq$ f and w = e $\vee$ f. Since m $\vee$ n $\in \mathbb{S}$ and m $\leq$ m $\vee$ n and also e $\leq$ m, f $\leq$ n, thus e $\vee$ f $\leq$ m $\vee$ n which implies m = e $\vee$ f. As f $\leq$ e $\vee$ f = m, then f in $\mathbb{S}$ such as f $\leq$ m and f $\leq$ n for all m, n in $\mathbb{S}$. Therefore, $\mathbb{S}$ is directed below semilattice. Let Id($\mathbb{S}$) be a set of all proper order ideals of $\mathbb{S}$.

Now to show that Id($\mathbb{S}$) is distributive, let $I_1 \vee I_2 \geq I_3$ for all $I_1, I_2, I_3 \in$ Id($\mathbb{S}$) $\subseteq \mathbb{S}$, for m $\in I_1$, n $\in I_2$, and w $\in I_3$, and we have m $\vee$ n $\geq$ w. However, $\mathbb{S}$ is distributive e and f in $\mathbb{S}$ such as m $\geq$ e, n $\geq$ f and w = e $\vee$ f. Id($\mathbb{S}$) denotes all order ideals of $\mathbb{S}$ that are increasing sets; thus in I, the subset of $\mathbb{S}$ is an order ideal, and the increasing subset I and m, n $\in I$, then probably i $\in I$ such as m $\geq$ i, n $\geq$ I, which implies m $\vee$ n $\geq$ i and as $I \in$ Id($\mathbb{S}$) are increasing set, and for e $\in I$ and f $\geq$ e, we have f $\in I$. Hence, for m $\vee$ n $\geq$ e $\vee$ f = f $\in I$, which implies e $\vee$ f $\in I$ and as m $\vee$ n = w $\in I_3 \subseteq$ Id($\mathbb{S}$). Hence, Id($\mathbb{S}$) is distributive. Let us suppose $\mathbb{S}$ is directed below and Id($\mathbb{S}$) of all order ideals of $\mathbb{S}$ is distributive; we prove that $\mathbb{S}$ is distributive. Let for w in $\mathbb{S}$ and m, n $\in I \subseteq \mathbb{S}$, and consider m $\vee$ n $\geq$ w and as $I \in$ Id($\mathbb{S}$) and I is increasing subset as i in I such as m $\geq$ i and n $\geq$ i, and then m $\vee$ n $\geq$ i. Thus, w = i which is i $\vee$ i $\in \mathbb{S}$. Hence, $\mathbb{S}$ is distributive.

3.4 Theorem

Assume $\mathbb{S}$ as a skew semilattice and $I \subseteq \mathbb{S}$ and then I as a skew ideal if and only if the following equivalence holds:

For all $p_1, q_1 \in \mathbb{S}$, $p_1 \vee q_1 \vee p_1 \in I$ if and only if $p_1, q_1 \in I$.

Proof: When $I \subseteq \mathbb{S}$, is a skew ideal of $\mathbb{S}$, then for every $p_1 \in \mathbb{S}$ & $q_1 \in I$, with $q_1 \geq p_1$, which gives $p_1 \in I$. Since $q_1 \geq p_1, q_1 \vee p_1 = q_1$. Now $p_1 \vee (q_1 \vee p_1) = p_1 \vee q_1 = q_1$, then $p_1 \vee q_1 \vee p_1 = q_1 \in I$, implies $p_1 \vee q_1 \vee p_1 \in I$ for $p_1, q_1 \in I$. Let $p_1, q \in \mathbb{S}$, such as $p_1 \vee q_1 \vee p_1 \in I$ and as $p_1 \vee q_1 \vee p_1 \geq p_1$ and $p_1 \vee q_1 \vee p_1 \geq q_1$, and I is an ideal, implies $p_1, q_1 \in I$, for $p_1 \vee q_1 \vee p_1 \in I$. Conversely, suppose for all p_1, $q_1 \in \mathbb{S}$, the equivalence condition holds, for $p_1, q_1 \in$ I if and only if $p_1 \vee q_1 \vee p_1 \in I$. To prove that I is skew ideal of $\mathbb{S}$, since $p_1 \vee q_1 \vee p_1 \in I$, for $p_1, q_1 \in I$ thus I is closed under the operation $\vee$. Let $q_1 \in I$ and $p_1 \in \mathbb{S}$ such as $q_1 \geq p_1$, also $p_1 \vee q_1 \vee p_1 = p_1 \vee q_1 = q_1 \in I$ and $p_1 \vee q_1 \vee p_1 = q_1 \vee p_1 = q_1 \in I$. Therefore, $p_1 \in I$. Hence, I is a skew ideal of $\mathbb{S}$.

3.5 Theorem

If $\mathbb{S}$ is skew normal semilattice, then $o \vee p \vee (o \vee p \vee o) = o \vee p \vee o$ for o, p in $\mathbb{S}$.

Proof: If $\mathbb{S}$ is normal skew semilattice, then, $o \vee (p \vee z) \vee w = o \vee (z \vee p) \vee w$, for o, p, z, w in $\mathbb{S}$.

Now consider
$$\begin{aligned} o \vee p \vee (o \vee p \vee o) &= o \vee p \vee (p \vee o) \vee o \\ &= o \vee (p \vee o) \vee (p \vee o) \\ &= (o \vee p \vee o) \vee (p \vee o) \end{aligned}$$

Thus,

$$o \vee p \vee (o \vee p \vee o) = (o \vee p \vee o) \vee (p \vee o)\,. \tag{3}$$

Similarly
$$\begin{aligned} o \vee p \vee o = o \vee p \vee o \vee o = o \vee p \vee p \vee o \vee o &= o \vee p \vee (o \vee p) \vee o \\ &= (o \vee p \vee o) \vee (p \vee o) \end{aligned}$$

Thus,

$$o \vee p \vee o = (o \vee p \vee o) \vee (p \vee o)\,. \tag{4}$$

From (3) and (4), we have $o \vee p \vee (o \vee p \vee o) = o \vee p \vee o$.

3.6 *Preorder Ideal*

If $J \subseteq \mathbb{S}$, then J is called a preorder skew ideal of skew semilattice $\mathbb{S}$, for $m \geq n$ and $n \in J$ and if $m, n \in J$, then $m \vee n \in J$.

3.7 *Relator*

Let $\mathbb{S}$ be skew semilattice, for $m, n \in \mathbb{S}$, and then the relator $<m, n>$ is that m related to n and is defined as $<o, p> = \{x \in \mathbb{S} / p \leq x \vee o\}$.

3.8 *Theorem*

The subsequent conditions are analogous for the skew semilattice $\mathbb{S}$.

(a) $\mathbb{S}$ is distributive skew semilattice.
(b) The set $<m, n> \in Id(\mathbb{S})$ for all $m, n \in \mathbb{S}$.
(c) $<J, I> \in Id(\mathbb{S})$ for all $J \in J_i(\mathbb{S})$ and $I \in Id(\mathbb{S})$, where $Id(\mathbb{S})$ is of all proper order ideals of $\mathbb{S}$, and $J_i(\mathbb{S})$ is of all preorder ideals of $\mathbb{S}$.

Proof: Let $\mathbb{S}$ be distributive skew semilattice.

To prove that **(a)** $\Rightarrow$ **(b)**, let $m, n \in \mathbb{S}$. For $p \in \mathbb{S}$, such as $p \in <m, n>$ and $p \leq q$, we have $n \leq p \vee m \leq q \vee m$, which implies $n \leq q \vee m$, which implies $q \in <m, n>$. Therefore, $<m, n>$ is increasing set.

Now to show that $<m, n> \in Id(\mathbb{S})$, i.e., $<m, n>$ is an order ideal of $\mathbb{S}$. Let $x, y \in <m, n>$ then $n \leq p \vee m$ and $n \leq q \vee m$. Since $\mathbb{S}$ is distributive, p_1, m_1 in $\mathbb{S}$ such as $p \geq p_1$, $m \geq m_1$ and $n = p_1 \vee m_1$ in particular, we have $n \geq p_1$, and then $p \vee m \geq n \geq p_1$ implies $p \vee m \geq p_1$ and again $\mathbb{S}$ is distributive and then their exist p_2 and m_2 in $\mathbb{S}$ such as $p \geq p_2$ and $m \geq m_2$ and $p_1 = p_2 \vee m_2$. Also $p \geq p_1 = p_2 \vee m_2$, which implies $p \geq m_2$ and $q \geq m_2$ and $p_2 \vee m \geq p_2 \vee (m_1 \vee m_2) = (p_2 \vee m_2) \vee m_1 = p_1 \vee m_1 = n$. Therefore, $n \leq p_2 \vee m \leq p \vee m$. Therefore, $<m, n >$ is an order ideal of $\mathbb{S}$. Hence, $<m, n> \in Id(\mathbb{S})$.

To prove that **(b)** $\Rightarrow$ **(c)**, let $<m, n> \in Id(\mathbb{S})$ for all $m, n \in \mathbb{S}$. To show that $<J, I> \in Id(\mathbb{S})$ for all J is an element of $J_i(\mathbb{S})$ (set of all un-order ideals) and $I \in Id(\mathbb{S})$. Let $p \in <J, I>$ and $p \leq q$ for any $p \in \mathbb{S}$. Then $I \leq p \vee J \leq q \vee J$, which implies $q \in <J, I>$. Therefore, $<J, I>$ is increasing. Now to show that $<J, I>$ is an order ideal, let $m, n \in <J, I>$ then (p_1, q_1) and $(p_2, q_2) \in <J, I>$ such as $m \in <p_1, q_1>$ and $n \in <p_2, q_2>$, and then $q_1 \leq m \vee p_1$ and $q_2 \leq n \vee p_2$. Since J is a preorder ideal, then $p = p_1 \vee p_2 \in F$; and I is order ideal, q in I such as $q_1 \geq q$, and $q_2 \geq q$. Then $p \geq p_1$ and $p \geq p_2$. Now $m \vee p \geq m \vee p_1$ and $n \vee p \geq n \vee p_2$, which implies $m \vee p \geq q_1 \geq v$ and $n \vee p \geq q_2 \geq v$; therefore, we have $q \leq m \vee p$ and $q \leq n \vee p$. Therefore, $m \in$

$<p, q>$ and $n \in <p, q>$, and therefore, by hypothesis, there exist $z \in <p, q>$ such as $m \geq z$ and $n \geq z$, where $<p, q> \subseteq <J, I>$. Hence $<J, I>$ is an order ideal.

To show that **(c)** $\Rightarrow$ **(a)**, let $<J, I>$ represent order ideal. Let m, n, $z \in \mathbb{S}$, such as $m \vee n \geq z$. Since $<m, z> = <(m], (z]>$ and n, $z \in <m, z>$, by hypothesis an element $n_1 \in <m, z>$, thus $z \leq n_1$, which implies $n_1 \leq z$, and $n_1 \leq z \leq m \vee n$ gives $n_1 \leq n$. Then, we have $n_1 \leq n$ and $n_1 \leq z$. Then, $n_1 \vee m \geq z$ and z, $m \in <n_1, z>$. As $<n_1, z>$ is representing order Ideal, exists

Thus,

$$m_1 \in < n_1, z > \textit{ such as } z \geq m_1 \textit{ and } m \geq m_1 \textit{ and } z \leq m_1 \vee n_1. \tag{5}$$

But

$$z \geq n_1 \textit{ and } z \geq m_1 \quad \textit{which gives } z \geq m_1 \vee n_1. \tag{6}$$

Therefore, from (5) and (6), we have $z = m_1 \vee n_1$. Hence, $\mathbb{S}$ is distributive semilattice.

4 Conclusion

In this paper, we present and investigate the skew semilattice and the distributive skew semilattice, as well as the necessary and sufficient theorem of the distributive semilattice. We also investigate the notion of normal skew semilattice. In addition, we define an order skew ideal in skew semilattice and provide the skew semilattice characterization theorem by defining some new through relators.

References

1. Leech, J.: The geometric structure of skew lattices. Trans. Amer. Math. Soc. **335**, 823–842 (1993)
2. Celani, S.A.: Topological representation of distributive semilattice. Scientiae Math. Japonicae Online. **8**, 41–51 (2003)
3. Pita Costa, J.: On the coset structure of a skew lattice. Demonstratio Mathematica. **44**(4), 1–19 (2011)
4. Pita Costa, J.: On ideals of a skew lattice. arXiv:1207.0231 v2 Demonstratio Mathematica (2012)
5. Cvetko-vah, K., Pita Costa, J.: On the coset laws for skew lattices. Semigroup Forum. **8**, 395–411 (2011)
6. Hickman, R.C.: Mildly distributive semilattices. J. Austral. Math. Soc. (Series A). **36**, 287–315 (1984)
7. Grätzer, G.: General Lattice Theory. Birkhäuser Verlag (1998)
8. Pita Costa, J.: Coset laws for categorical skew lattices. Algebra Univ. **68**, 75–89 (2012)

On Certain Semigroups of Order-Decreasing Full Contraction Mappings of a Finite Chain

Muhammad Mansur Zubairu, Abdullahi Umar, and Jaafar Abdulkadir Aliyu

Keywords Order decreasing and order preserving full contractions semigroup · Adequate and dense semigroups · Rank properties

1 Introduction

Let $[n]$ be a finite chain say $\{1, 2, \ldots, n\}$ and denote $\mathcal{P}_n$ (resp., $\mathcal{T}_n$) to be the partial transformation semigroup on the chain $[n]$ (resp., semigroup of full transformations on $[n]$). An element $\alpha \in \mathcal{T}_n$ is *order preserving* (resp., *order reversing*) if $(\forall a, b \in [n])$ $a \leq b$ implies $a\alpha \leq b\alpha$ (resp., $a\alpha \geq b\alpha$); *order increasing* (resp., *order decreasing*) if $(\forall a \in [n])$ $a \leq a\alpha$ (resp., $a\alpha \leq a$); and a *contraction* if $(\forall a, b \in [n])$ $|a\alpha - b\alpha| \leq |a - b|$. The collection of all contraction mappings on $[n]$ denoted by $\mathcal{CT}_n$ is known as the *semigroup of full contraction mappings*. The study of various semigroup of contractions was first proposed in 2013 by Umar and Alkharousi [12], where they give notations for the semigroups. We will also adopt these notations for the semigroups considered in this paper. Let

$$\mathcal{DCT}_n = \{\alpha \in \mathcal{CT}_n : \text{for all } a \in [n],\ a\alpha \leq a\} \tag{1}$$

denote order decreasing full contraction semigroup,

$$\mathcal{ODCT}_n = \{\alpha \in \mathcal{DCT}_n : (\text{for all } a, b \in [n])\ a \leq b \Rightarrow a\alpha \leq b\alpha\} \tag{2}$$

M. M. Zubairu (✉) · J. A. Aliyu
Department of Mathematical Sciences, Bayero University Kano, Kano, Nigeria
e-mail: mmzubairu.mth@buk.edu.ng

A. Umar
Khalifa University, Abu Dhabi, UAE
e-mail: abdullahi.umar@ku.ac.ae

H.-H. Leung et al. (eds.), *Recent Developments in Algebra and Analysis*, Trends in Mathematics, https://doi.org/10.1007/978-3-031-37538-5_4

denote order preserving and order decreasing full contraction semigroup,

$$\mathcal{OCT}_n = \{\alpha \in \mathcal{CT}_n : (\text{for all } a, b \in [n])\ a \leq y \ \Rightarrow\ a\alpha \leq b\alpha\}$$

be order-preserving full contractions semigroup, and

$$\mathcal{ORCT}_n = \mathcal{OCT}_n \cup \{\alpha \in \mathcal{CT}_n : (\forall a, b \in [n])\ a \leq b \ \Rightarrow\ a\alpha \geq b\alpha\ \}$$

be order preserving or order reversing full contraction semigroup. It is clear that $\mathcal{ORCT}_n$ is a subsemigroup of $\mathcal{CT}_n$, whereas $\mathcal{OCT}_n$ and $\mathcal{ODCT}_n$ are subsemigroups of $\mathcal{ORCT}_n$ and $\mathcal{DCT}_n$, respectively. A complete characterization of Green's equivalences for $\mathcal{CT}_n$ were obtained by Ali et al., [2]. The combinatorial results for $\mathcal{ORCT}_n$,, $\mathcal{OCT}_n$, and $\mathcal{ODCT}_n$ were investigated by Adeshola and Umar [1]. Furthermore, the ranks of $\mathcal{OCT}_n$ and $\mathcal{ORCT}_n$ and the rank of their two-sided ideals were obtained by Kemal [10] and Leyla [3], respectively. However, it appears that the rank and algebraic properties of the semigroup $\mathcal{DCT}_n$ and its subsemigroup $\mathcal{ODCT}_n$ have not been investigated. This chapter intends to study Green's equivalences, their starred analogue, and rank properties of these semigroups.

In the current section, we give a brief introduction and introduce some basic notations and definitions. Moreover, we characterize the elements of $\mathcal{DCT}_n$. Section 2 of this chapter constitute s characterization of all Green's equivalence and the regular elements in $\mathcal{DCT}_n$ and $\mathcal{ODCT}_n$, respectively. In Sect. 3, we characterize the starred analogue of Green's equivalences and show that if $S \in \{\mathcal{DCT}_n, \mathcal{ODCT}_n\}$, then S is left abundant for all n but not right abundant for $n \geq 3$. Moreover, we show that the semigroup $\mathcal{DCT}_n$ is a dense semigroup with a cover. In Sect. 4, we show that $\mathcal{ODCT}_n$ is left adequate and also investigate its rank.

For a contraction α in $\mathcal{CT}_n$, we shall denote Im α, rank α, and id_A to be the image of α, $|\text{Im}\,\alpha|$, and identity map on $A \subseteq [n]$, respectively. For two elements say $\alpha,\ \beta \in \mathcal{CT}_n$, their composition shall be as $a(\alpha \circ \beta) = ((a)\alpha)\beta$ for all a in $[n]$, the notation $\alpha\beta$ shall be adopted to denote $\alpha \circ \beta$ in our subsequent discussions. For a semigroup S, an element $d \in S$ is called an *idempotent* if $d^2 = d$. It is known that the condition Im $\epsilon = F(\epsilon)$ (where $F(\epsilon) = \{a \in [n] : a\epsilon = a\}$) is a necessary and sufficient condition for $\epsilon \in \mathcal{T}_n$ to be an idempotent. If S is a commutative semigroup and all its elements are idempotents (i.e., $S = E(S)$), then S is said to be a *semilattice*. In this case, for all $v, u \in S$, $u^2 = u$, and $vu = uv$. For basic concepts in semigroup theory, the reader may refer to Howie [7].

Next, given any transformation α in $\mathcal{CT}_n$, the domain of α is partitioned into *blocks* by the relation ker $\alpha = \{(a, b) \in [n] \times [n] : a\alpha = b\alpha\}$, so that by Adeshola and Umar [[1], Lemma 1.2] α can be expressed as

$$\alpha = \begin{pmatrix} D_1 & D_2 & \cdots & D_r \\ t+1 & t+2 & \cdots & t+r \end{pmatrix} \ (1 \leq r \leq n \text{ and for some integer } t), \tag{3}$$

where D_i for all $1 \leq i \leq r$ are equivalence classes of the relation ker α, i.e., $D_i = (t+i)\alpha^{-1}$ for all $i \in \{1, \ldots, r\}$. We shall denote the partition of $[n]$ (by the relation ker α) by **Ker** $\alpha = \{D_1, D_2, \ldots, D_r\}$ so that $[n] = D_1 \cup D_2 \cup \ldots \cup D_r$ where $(1 \leq r \leq n)$. Now if $\alpha \in \mathcal{OCT}_n$, then we see that $D_i < D_j$ if and only if $i < j$.

Elements of $\mathcal{DCT}_n$ ($\mathcal{ODCT}_n$) can be expressed as in the following lemma:

Lemma 1 *Every element $\alpha \in \mathcal{DCT}_n$ ($\mathcal{ODCT}_n$) of rank $r \in [n]$ can be expressed as*

$$\alpha = \begin{pmatrix} D_1 & D_2 & \cdots & D_r \\ 1 & 2 & \cdots & r \end{pmatrix}.$$

Proof Let $\alpha \in \mathcal{DCT}_n$ ($\mathcal{ODCT}_n$) be as expressed in Eq. (3), i.e.,

$$\alpha = \begin{pmatrix} D_1 & D_2 & \cdots & D_r \\ t+1 & t+2 & \cdots & t+r \end{pmatrix}.$$

Notice that $t+1 < t+2 < \ldots < t+r$; moreover, since α is order-decreasing, then $t+1 \leq a$ for all $a \in D_1 \cup D_2 \cup \ldots \cup D_r$. In particular, $t+1 \leq 1$, which implies $t+1 = 1$, and so $t = 0$. Thus $D_i\alpha = i$ for $1 \leq i \leq r$, as required. □

Let

$$\alpha = \begin{pmatrix} D_1 & D_2 & \cdots & D_r \\ 1 & 2 & \cdots & r \end{pmatrix} \text{ and } \beta = \begin{pmatrix} B_1 & B_2 & \cdots & B_r \\ 1 & 2 & \cdots & r \end{pmatrix} \in \mathcal{DCT}_n \quad (1 \leq r \leq n). \tag{4}$$

The corollary below follows directly as a consequence of Lemma 1.

Corollary 1 *Let $\gamma, \delta \in S \in \{\mathcal{DCT}_n, \mathcal{ODCT}_n\}$:*

(a) $|\text{Im } \gamma| = |\text{Im } \delta| \Leftrightarrow \text{Im } \gamma = \text{Im } \delta$;
(b) $\ker \gamma = \ker \delta \Leftrightarrow \gamma = \delta$.

2 Regularity and Green's Equivalences

For a semigroup S, an element $u \in S$ is *regular* if $u = uvu$ for some $v \in S$; S is *a regular semigroup* if all its elements are regular. For definitions of the five Green's equivalences: $\mathcal{L}$, $\mathcal{D}$, $\mathcal{R}$, $\mathcal{J}$, and $\mathcal{H}$, the reader may refer to Howie [7]. It is a known result that, if S is finite, the equivalence $\mathcal{J} = \mathcal{D}$. The characterizations of Green's relations on various semigroups of transformations were examined and studied by many authors (see, e.g., [2, 15]). It is also known that $\mathcal{C}_n$ (where $\mathcal{C}_n$ denotes the order-preserving and order-decreasing full transformation semigroup) is $\mathcal{J}$−trivial [9]. Thus, we state and prove the theorem below.

Theorem 1 *Let $\mathcal{DCT}_n$ be as in Eq. (1) and let $\alpha, \beta \in \mathcal{DCT}_n$ be as expressed in Eq. (4). Then*

(i) $(\alpha, \beta) \in \mathcal{R} \Leftrightarrow \beta = \alpha$;
(ii) $(\alpha, \beta) \in \mathcal{L} \Leftrightarrow \min i\alpha^{-1} = \min i\beta^{-1}$ *for all* $1 \leq i \leq r$.

Proof The proof is similar to that of Lemma 1.2 in [11] coupled with the fact that $\operatorname{Im}\alpha = \{1, 2, \ldots, r\} = \operatorname{Im}\beta$ by Lemma 1. □

Notice that in this case, $\mathcal{DCT}_n$ is said to be $\mathcal{R}-$trivial semigroup. Thus, we have the following corollaries:

Corollary 2 *On $\mathcal{DCT}_n$, $\mathcal{H} = \mathcal{R}$ and $\mathcal{L} = \mathcal{D} = \mathcal{J}$.*

Corollary 3 *An element $\alpha \in \mathcal{DCT}_n$ is regular if and only if α is an idempotent.*

As a consequence of the above corollary, we now have the following lemma.

Lemma 2 *Let $\alpha \in \mathcal{DCT}_n$ be as expressed in Eq. (4). Then α is an idempotent if and only if $\min i\alpha^{-1} = i$ for all $1 \leq i \leq r$.*

We now give the characterizations of the five Green's equivalences on $\mathcal{ODCT}_n$.

Theorem 2 *Let $\mathcal{ODCT}_n$ be as in Eq. (2). Then $\mathcal{ODCT}_n$ is $\mathcal{J}-$trivial.*

Proof Observe that $\mathcal{ODCT}_n$ is a subsemigroup of $\mathcal{C}_n$, then $\mathcal{ODCT}_n$ is $\mathcal{J}-$trivial. □

Consequently, the corollaries below follow.

Corollary 4 *On the semigroup $\mathcal{ODCT}_n$, $\mathcal{L} = \mathcal{R} = \mathcal{D} = \mathcal{H} = \mathcal{J}$.*

Now since $\mathcal{ODCT}_n$ is $\mathcal{R}$ trivial, then we have the following:

Corollary 5 *An element $\alpha \in \mathcal{ODCT}_n$ is regular if and only if α is an idempotent.*

3 Starred Green's Equivalences

The relation $\mathcal{L}^*$ on S is defined as: $(u, v) \in \mathcal{L}^* \Leftrightarrow (u, v) \in \mathcal{L}(P)$ (i.e., $(u, v) \in \mathcal{L}$ on a semigroup P) for some semigroup P, where S is a subsemigroup of P; $\mathcal{R}^*$ is defined in a similar way, and $\mathcal{D}^*$ is the join of the relations $\mathcal{R}^*$ and $\mathcal{L}^*$, where the intersection of the relations $\mathcal{R}^*$ and $\mathcal{L}^*$ is $\mathcal{H}^*$. A semigroup S is said to be *left abundant* (resp., *right abundant*) if every $\mathcal{L}^* - class$ (resp., every $\mathcal{R}^* - class$) contains an idempotent, and it is said to be *abundant* if it is both left and right abundant. If $E(S)$ is a subsemigroup of a left abundant(resp., right abundant) semigroup S, then S is called *left quasi adequate*(resp., *right quasi adequate*); if $E(S)$ is commutative, then it is said to be *left adequate* (resp., *right adequate*); and if S is both left and right adequate, it is called *adequate* (see [5] for more details on adequate semigroups). If a semigroup is not regular, then there is a need to investigate the class to which the semigroup belongs. To carry out such investigation, one would naturally characterize its starred Green's equivalences. We are going to investigate regularity, characterize the starred Green's equivalences, and show that if $S \in \{\mathcal{DCT}_n, \mathcal{ODCT}_n\}$, then S is left abundant; moreover, we show that $\mathcal{ODCT}_n$ is left adequate. As in [7], the relations $\mathcal{L}^*$ and $\mathcal{R}^*$ have the following characterizations:

$$\mathcal{L}^* = \{(\alpha,\ \beta) : (\text{for all } \mu, \lambda \in S^1)\ \ \alpha\mu = \alpha\lambda \Leftrightarrow \beta\mu = \beta\lambda\} \tag{5}$$

and

$$\mathcal{R}^* = \{(\alpha,\ \beta) : (\text{for all } \mu, \lambda \in S^1)\ \ \mu\alpha = \lambda\alpha \Leftrightarrow \mu\beta = \lambda\beta\}. \tag{6}$$

We now give characterizations of all the Starred Green's equivalence on $S \in \{\mathcal{DCT}_n, \mathcal{ODCT}_n\}$ in the theorem below. The proof of the theorem is a simplified version of the proof of Theorem 1 in [13].

Theorem 3 *Let $\alpha, \beta \in S \in \{\mathcal{DCT}_n, \mathcal{ODCT}_n\}$. Then*

(i) $(\alpha,\ \beta) \in \mathcal{L}^* \Leftrightarrow \operatorname{Im}\alpha = \operatorname{Im}\beta$;
(ii) $(\alpha,\ \beta) \in \mathcal{R}^* \Leftrightarrow \alpha = \beta$;
(iii) $\mathcal{H}^* = \mathcal{R}^*$;
(iv) $\mathcal{D}^* = \mathcal{L}^*$.

Proof

(i) Let $\alpha, \beta \in S$ be as expressed in Eq. (4) and suppose $(\alpha, \beta) \in \mathcal{L}^*$. Notice that $\operatorname{Im}\alpha = \{1, \ldots, r\}$. Now consider $\mu = \begin{pmatrix} 1 \cdots \{i, \ldots, n\} \\ 1 \cdots \quad i \end{pmatrix}$ $(1 \leq r \leq i \leq n)$. Then clearly $\mu \in S$ and $\alpha \circ \begin{pmatrix} 1 \cdots \{i, \ldots, n\} \\ 1 \cdots \quad i \end{pmatrix} = \alpha \circ id_{[n]} \Leftrightarrow \beta \circ \begin{pmatrix} 1\ 2 \cdots \{i, \ldots, n\} \\ 1\ 2 \cdots \quad i \end{pmatrix} = \beta \circ id_{[n]}$ (by Eq. (5)), which implies that $\operatorname{Im}\alpha \subseteq \operatorname{Im}\beta$. We can show in a similar way that $\operatorname{Im}\beta \subseteq \operatorname{Im}\alpha$, as such $\operatorname{Im}\alpha = \operatorname{Im}\beta$.
Conversely, suppose $\operatorname{Im}\alpha = \operatorname{Im}\beta$. Then by Howie ([7], Exercise 2.6.17) $\alpha\mathcal{L}^{\mathcal{P}_n}\beta$ and following from the definition that $\alpha\mathcal{L}^*\beta$, the result follows.

(ii) Let $\alpha, \beta \in S$ be as expressed in Eq. (4). Suppose $(\alpha,\ \beta) \in \mathcal{R}^*$. Then $(a,\ b) \in \ker\alpha \Leftrightarrow a\alpha = b\alpha \Longleftrightarrow \begin{pmatrix} [n] \\ a \end{pmatrix} \circ \alpha = \begin{pmatrix} [n] \\ b \end{pmatrix} \circ \alpha \Longleftrightarrow \begin{pmatrix} [n] \\ a \end{pmatrix} \circ \beta = \begin{pmatrix} [n] \\ b \end{pmatrix} \circ \beta$ (by Eq. (6)) $\Longleftrightarrow a\beta = b\beta \Longleftrightarrow (a,\ b) \in \ker\beta$.
Therefore $\ker\ \alpha = \ker\ \beta$. Thus by Corollary 1 (b) $\beta = \alpha$. The converse is clear. Thus $(\alpha,\ \beta) \in \mathcal{R}^*$.

(iii) The result follows from (i) and (ii).

(iv) Since S is $\mathcal{R}^*$ trivial then $\mathcal{L}^* = \mathcal{D}^*$.

□

We are now going to show that on $\mathcal{ODCT}_n$, $\mathcal{J}^* = \mathcal{D}^*$, but before then we note the lemma below.

Lemma 3 ([5], Lemma 1.7(3)) *If $x, y \in S$. Then $y \in J^*(x)$ if and only if there exist $x_0, x_1, \ldots, x_n \in S$, $a_1, a_2, \ldots, a_n$, $b_1, b_2, \ldots, b_n \in S^1$ such that $x = x_0$, $y = x_n$ and $(y_i, a_i x_{i-1} b_i) \in \mathcal{D}^*$ for $i = 1, 2, \ldots, n$.*

Now we prove an analogue of [[11], Lemma 2.13.].

Lemma 4 *Let $\beta, \alpha \in S \in \{\mathcal{DCT}_n, \mathcal{ODCT}_n\}$. If $\alpha \in J^*(\beta)$ then $\text{Im}\,\alpha \subseteq \text{Im}\,\beta$.*

Proof Suppose $\alpha \in \mathcal{J}^*(\beta)$, $(\alpha, \beta \in S)$. Thus by Lemma 3, there are $\beta_0, \beta_1, \ldots, \beta_n \in S$, $\lambda_1, \lambda_2, \ldots, \lambda_n$, $\mu_1, \mu_2, \ldots, \mu_n \in S^1$ such that $\beta = \beta_0$, $\alpha = \beta_n$ and $(\beta_i, \lambda_i \beta_{i-1} \mu_i) \in \mathcal{D}^*$ for $i \in [n]$. Thus using Theorem 3(iv), $\text{Im}\,\beta_i = \text{Im}\,\lambda_i \beta_{i-1} \mu_i \subseteq \text{Im}\,\beta_{i-1}$ for $i \in [n]$. This means that $\text{Im}\,\alpha \subseteq \text{Im}\,\beta$. □

Lemma 5 *For $S \in \{\mathcal{DCT}_n, \mathcal{ODCT}_n\}$, $\mathcal{J}^* = \mathcal{D}{*}^*$.*

Proof Let $\alpha, \beta \in S$. Clearly $\mathcal{D}^* \subseteq \mathcal{J}^*$. Now for $\mathcal{J}^* \subseteq \mathcal{D}^*$, let $(\alpha, \beta) \in \mathcal{J}^*$, i.e., $\alpha \in \mathcal{J}^*(\beta)$, and $\beta \in \mathcal{J}^*(\alpha)$. Thus, Lemma 4 ensures that $\text{Im}\,\beta = \text{Im}\,\alpha$. Thus, using Theorem 3(i) and (ii), we see that $(\alpha, \beta) \in \mathcal{D}^*$, as required. □

Now we show in the lemma below that $S \in \{\mathcal{DCT}_n, \mathcal{ODCT}_n\}$ is left abundant.

Lemma 6 *The semigroup $S \in \{\mathcal{DCT}_n, \mathcal{ODCT}_n\}$ is left abundant.*

Proof Let $\alpha \in S$ be as expressed in Eq. (4) and let L^*_α be an $\mathcal{L}^*$-class of α in S. Denote $\epsilon = \begin{pmatrix} 1\ 2 \cdots \{r, r+1, \ldots, n\} \\ 1\ 2 \cdots \quad\quad r \end{pmatrix} \in \mathcal{ODCT}_n$, $(1 \leq r \leq n)$. Clearly ϵ is an idempotent in S; moreover, $\text{Im}\,\alpha = \text{Im}\,\epsilon$, and so by Theorem 3(i), we see that $(\alpha, \epsilon) \in \mathcal{L}^*$, which means $\epsilon \in L^*_\alpha$. Since L^*_α is an arbitrary $\mathcal{L}^*$- class of α in S, then S is left abundant, as required. □

Remark 1 In contrast with [[14], Lemma 1.20], the semigroup $S \in \{\mathcal{DCT}_n, \mathcal{ODCT}_n\}$ is not right abundant for $n \geq 3$.

For a counterexample, consider $\alpha = \begin{pmatrix} \{1, 2\}\ 3 \\ 1 \quad 2 \end{pmatrix} \in S \in \{\mathcal{DCT}_3, \mathcal{ODCT}_3\}$. It is clear that $R^*_\alpha = \left\{\begin{pmatrix} \{1, 2\}\ 3 \\ 1 \quad 2 \end{pmatrix}\right\}$ has no idempotent.

However, the semigroup $S \in \{\mathcal{DCT}_n, \mathcal{ODCT}_n\}$ is right abundant for $1 \leq n \leq 2$, which is also in contrast with [[14], Remark 1.21].

3.1 The Cover for $\mathcal{DCT}_n$

A semigroup S is an *$E-$semigroup* if $E(S)$ is a subsemigroup. A subsemigroup K of S is said to be *dense* if for all $s \in S$, $ss' \in K$ and $s''s \in K$ for some s' and $s'' \in S$; it is said to be *unitary* if $\forall t, t' \in K$ and $a \in S$, $at' \in K$ and $ta \in K$ implies $a \in K$ [6, 8]. A semigroup S is *E unitary* if its set of idempotent (i.e., $E(S)$) is a unitary subsemigroup; it is said to be *$E-$dense* if $E(S)$ is dense; and it is called *$E-$unitary dense* if $E(S)$ is unitary dense subsemigroup. For the semigroups S and K, an epimorphism $\varphi : K \to S$, which is injective on the idempotents, is said to be a *covering*; as such, K is said to be a *cover* for S. Many classes of semigroups, regular and non-regular, are shown to be dense and/or unitary semigroups; in particular, the $E-$*semigroup* has been shown to have a cover in [8]. For basic concepts and structural theory of unitary and dense semigroups, we refer the reader to [6, 8]. We will show that $\mathcal{DCT}_n$ is a dense semigroup with cover with the help of some results from Jorge et al., [8].

Before we begin our investigation, we first note the following results from [8].

Proposition 1 *[[8], Proposition. 1.1] For an $E-$semigroup S where $E(S) = E$, the statements below correspond.*

(i) S is an $E-$dense semigroup;
(ii) $\forall a \in S$, $aa' \in E$ for some $a' \in S$;
(iii) $\forall a \in S$, $a''a \in E$ for some $a'' \in S$.

Theorem 4 *[[8], Theorem. 2.1] Let S be a semigroup. Then*

(i) If S is an $E-$semigroup, then S has an $E-$unitary cover.
(ii) If S is an $E-$dense semigroup, then S has an $E-$unitary dense cover.
(iii) If S is an orthodox semigroup, then S has an $E-$unitary orthodox cover.

We prove the following lemma.

Lemma 7 *$E(\mathcal{DCT}_n)$ is a subsemigroup of $\mathcal{DCT}_n$.*

Proof Let $\epsilon_1 = \begin{pmatrix} D_1 & D_2 & \cdots & D_r \\ 1 & 2 & \cdots & r \end{pmatrix}$ and $\epsilon_2 = \begin{pmatrix} B_1 & B_2 & \cdots & B_m \\ 1 & 2 & \cdots & m \end{pmatrix}$ be idempotents in $\mathcal{DCT}_n$. There are two cases to consider, i.e., either $r \leq m$ or $m < r$.

Now suppose $r \leq m$. Notice that $i\epsilon_2 = i$ for $i = 1, \ldots, r \leq m$. Thus,

$$\epsilon_1\epsilon_2 = \begin{pmatrix} D_1 & D_2 & \cdots & D_r \\ 1 & 2 & \cdots & r \end{pmatrix} = \epsilon_1 \in E(\mathcal{DCT}_n).$$

Now if $m < r$. Then $\epsilon_1\epsilon_2 = \begin{pmatrix} D'_1 & D'_2 & \cdots & D'_m \\ 1 & 2 & \cdots & m \end{pmatrix}$, where $D'_i = D_i \cup B$ $(1 \leq i \leq m)$ for some $B \subseteq \{m+1, m+2, \ldots, r, r+1, \ldots, n\}$, i.e., the elements in the set $\{m+1, m+2, \ldots, r, r+1, \ldots, n\}$ are distributed to the blocks D_i for some $1 \leq i \leq m$ in the following manner: given any $x \in \{m+1, m+2, \ldots, r, r+1, \ldots, n\}$, if $x \in B_i$ (for some $1 \leq i \leq m$) then $x\epsilon_2 = i \in D'_i$. As such x is placed in D'_i. Notice that $i \in D'_i$ for $1 \leq i \leq m$, as such $\epsilon_1\epsilon_2$ is an idempotent, as required. □

The lemma below is a direct consequence of Lemma 7.

Lemma 8 *$\mathcal{DCT}_n$ is an $E-$ semigroup.*

Now, we prove the lemma below.

Lemma 9 *For every $\alpha \in \mathcal{DCT}_n$, there exists $\beta \in \mathcal{DCT}_n$ such that $\alpha\beta \in E(\mathcal{DCT}_n)$.*

Proof Let $\alpha \in \mathcal{DCT}_n$ be as expressed in Eq. (7). Now let $\beta = \begin{pmatrix} [n] \\ 1 \end{pmatrix} \in \mathcal{DCT}_n$.

Then $\alpha\beta = \begin{pmatrix} [n] \\ 1 \end{pmatrix}$ which is obviously an idempotent. □

Consequently we have proved the following theorem by Lemma 8 and 9 and by Proposition (1) and Theorem (4), respectively.

Theorem 5 *Let $\mathcal{DCT}_n$ be as defined in Eq. (1). Then we have:*

(i) $\mathcal{DCT}_n$ is $E-$ dense.
(ii) $\mathcal{DCT}_n$ has an $E-$unitary and $E-$unitary dense cover.

4 Rank of $\mathcal{ODCT}_n$

Let $T \subseteq S$ $(T \neq \emptyset)$. The notation $\langle T \rangle$ denotes the subsemigroup generated by the subset T, which is defined as the intersection of all subsemigroups of S containing T. If T is finite and $\langle T \rangle = S$, then S is called a *finitely generated semigroup*. Moreover, the rank of S is defined and denoted by

$$\text{rank } S = \min\{|T| : \langle T \rangle = S\}.$$

Now let Reg($\mathcal{ORCT}_n$) be the collection of regular elements of $\mathcal{ORCT}_n$. Then, we first note the following result about idempotents in $\mathcal{ORCT}_n$ from [13].

Lemma 10 ([13], Lemma 13) *Let ϵ be an idempotent in $(\mathcal{ORCT}_n)$. Then ϵ can be expressed as*

$$\begin{pmatrix} \{1, \dots, d+1\} & d+2 \cdots d+r-1 & \{d+r, \dots, n\} \\ d+1 & d+2 \cdots d+r-1 & d+r \end{pmatrix}.$$

We now prove the lemma below, crucial to the main result.

Lemma 11 *Every $\epsilon \in E(\mathcal{ODCT}_n)$ can be expressed as*

$$\epsilon = \begin{pmatrix} 1\ 2 \cdots & \{r, r+1, \dots, n\} \\ 1\ 2 \cdots & r \end{pmatrix} \quad \text{where } r \in [n].$$

Proof The proof follows from Lemmas 10 and 1. □

We show in the next theorem that the collection of all idempotents in $\mathcal{ODCT}_n$ (i.e., $E(\mathcal{ODCT}_n)$) is a semilattice.

Theorem 6 *$E(\mathcal{ODCT}_n)$ is a semilattice.*

Proof Let $\epsilon, \eta \in E(\mathcal{OCDT}_n)$. Then by Lemma 11, we may denote ϵ and η by

$$\epsilon =, \begin{pmatrix} 1 \cdots \{m, m+1, \dots, n\} \\ 1 \cdots \quad m \end{pmatrix} \text{ and } \eta = \begin{pmatrix} 1 \cdots \{r, r+1, \dots, n\} \\ 1 \cdots \quad r \end{pmatrix}$$

for $m, r \in [n]$.

Thus, there are two cases to consider:

If $m \leq p$. Then $\epsilon\eta = \begin{pmatrix} 1 \cdots \{m, m+1, \dots, n\} \\ 1 \cdots \quad m \end{pmatrix} = \eta\epsilon = \epsilon \in E(\mathcal{ODCT}_n)$.

If $r < m$. Then $\epsilon\eta = \begin{pmatrix} 1 \cdots \{r, r+1, \dots, n\} \\ 1 \cdots \quad r \end{pmatrix} = \eta\epsilon = \eta \in E(\mathcal{ODCT}_n)$.

Thus $E(\mathcal{ODCT}_n)$ is a semilattice. □

Now by Theorem 6 and Lemma 6, we readily have the result below.

Theorem 7 *Let $\mathcal{ODCT}_n$ be as defined in Eq. (2). Then $\mathcal{ODCT}_n$ is left adequate.*

Next, we state the following well-known result from [4] as a lemma below.

Lemma 12 *In a finite $\mathcal{J}$ trivial semigroup S, every minimal generating set is (unique) minimum.*

Let $G_r = \{\alpha \in \mathcal{ODCT}_n : |\mathrm{Im}\,\alpha| = r\}$ and $K_r = \{\alpha \in \mathcal{ODCT}_n : |\mathrm{Im}\,\alpha| \leq r\}$. It is worth noting that $K_r = G_1 \cup G_2 \cup \dots \cup G_r$ $(1 \leq r \leq n)$. Now we have the following lemma.

Lemma 13 *For $1 \leq r \leq n-2$, $G_r \subseteq \langle G_{r+1} \rangle$.* □

Proof Let $\alpha \in G_r$, then by Lemma 1, we may let $\alpha = \begin{pmatrix} D_1 \cdots D_r \\ 1 \cdots r \end{pmatrix}$, where $1 \leq r \leq n-2$. Next now let $D_r' \cup D_p'' = D_r$ with $D_r' \neq \emptyset$, $D_r'' \neq \emptyset$, $D_r' \cap D_r'' = \emptyset$ and $D_r' < D_r''$. Now denote δ and ρ as:

$$\delta = \begin{pmatrix} D_1 \cdots D_{r-1} & D_r' & D_r'' \\ 1 \cdots r-1 & p & r+1 \end{pmatrix} \text{ and } \rho = \begin{pmatrix} 1 \cdots r-1 & \{r, r+1\} & \{r+2, \dots, n\} \\ 1 \cdots r-1 & r & r+1 \end{pmatrix}.$$

Notice that $\delta, \rho \in G_{r+1}$. It is easy to see that $\alpha = \delta\rho \in \langle G_{r+1} \rangle$. Hence $G_r \subseteq \langle G_{r+1} \rangle$. □

We now have the corollary below.

Corollary 6 *For $1 \leq k \leq n-1$, $G_k \subseteq \langle G_{n-1} \rangle$.*

Proof Suppose $1 \leq k \leq n-1$,; then, by Lemma 13, we see that $G_k \subseteq \langle G_{k+1} \rangle$ and similarly $G_{k+1} \subseteq \langle G_{k+2} \rangle$, which implies that $\langle G_k \rangle \subseteq \langle G_{k+2} \rangle$. Therefore, $G_k \subseteq \langle G_{k+1} \rangle \subseteq \langle G_{k+2} \rangle$. If we continue in this fashion, we see that $G_k \subseteq \langle G_{k+1} \rangle \subseteq \langle G_{k+2} \rangle \subseteq \dots \subseteq \langle G_{n-2} \rangle \subseteq \langle G_{n-1} \rangle$, as required. □

Lemma 14 *In $\mathcal{ODCT}_n$, $|G_{n-1}| = n-1$.*

Proof Notice that if $\alpha \in G_{n-1}$, then α is of the form $\alpha = \begin{pmatrix} D_1 & \cdots & D_{n-1} \\ 1 & \cdots & n-1 \end{pmatrix}$, where $D_i < D_j$ if and only if $i < j$. It is now clear that the order of G_{n-1} is equal to the number of subsets of the set $[n]$ of the form $\{i, i+1\}$ $(1 \leq i \leq n-1)$, which is $n-1$. □

The following lemma gives us the rank of $\mathcal{ODCT}_n \setminus \{id_n\}$.

Lemma 15 *In $\mathcal{ODCT}_n$, rank $(K_{n-1}) = n-1$.*

Proof To prove that the rank $(K_{n-1}) = n-1$, it is enough to show that G_{n-1} is a minimal generating set of K_{n-1}, i.e., $K_{n-1} = \langle G_{n-1} \rangle$ and $\langle G_{n-1} \backslash \{\tau\} \rangle \neq K_{n-1}$ for any $\tau \in G_{n-1}$. It is clear by Corollary 6 that $\langle G_{n-1} \rangle = K_{n-1}$.

Now observe that

$$G_{n-1} = \left\{ \begin{pmatrix} \{1,2\} & 3 & \cdots & n \\ 1 & 2 & \cdots & n-1 \end{pmatrix}, \begin{pmatrix} 1, & \{2,3\} & 4 & \cdots & n \\ 1 & 2 & 3 & \cdots & n-1 \end{pmatrix}, \ldots, \begin{pmatrix} 1 & \cdots & n-2 & \{n-1, n\} \\ 1 & \cdots & n-2 & n-1 \end{pmatrix} \right\}. \quad (7)$$

Take $\tau_i = \begin{pmatrix} 1 & 2 & \cdots & \{i, i+1\} & \cdots & n-2 & n-1 & n \\ 1 & 2 & \cdots & i & \cdots & n-3 & n-2 & n-1 \end{pmatrix} \in G_{n-1}$ for $i = 1, \ldots, n-1$. Then one can easily verify that for any $\alpha, \beta \in G_{n-1} \backslash \{\tau_i\}$, $h(\alpha\beta) < n-1$, $h(\tau_i \alpha) < n-1$, $h(\alpha \tau_i) < n-1$ and moreover, $\alpha \tau_{n-1} = \alpha$ for all $\alpha \in G_{n-1}$. Thus G_{n-1} is a minimal generating set for K_{n-1}. Thus since S is a finite $\mathcal{J}$ trivial semigroup then by Lemma 12, G_{n-1} is the (unique) minimum generating set for K_{n-1}. □

Finally the rank of $\mathcal{ODCT}_n$ is given in the theorem below.

Theorem 8 *Let $\mathcal{ODCT}_n$ be as defined in Eq. (2). Then rank $(\mathcal{ODCT}_n) = n$.*

Proof Notice that $K_{n-1} = \mathcal{ODCT}_n \backslash \{id_{[n]}\}$. Therefore the rank $(\mathcal{ODCT}_n) =$ rank $(K_{n-1}) + 1 = n$, as required. □

References

1. Adeshola, A.D., Umar, A.: Combinatorial results for certain semigroups of order-preserving full contraction mappings of a finite Chain. JCMCC 106, 37–49 (2018)
2. Ali, B., Umar, A., Zubairu, M.M.: Regularity and Green's relation for the semigroups of partial and full contractions of a finite chain. (Accepted, to appear, Scientific Africa)
3. Bugay, L.: On the ranks of certain ideal of monotone contractions. Hecet. J. Math. Stat. **49**(6), 1988–1996 (2020)
4. Doyen, J.: Equipotence et unicite de systemes generateurs minimaux dans certains monoides. Semigroup Forum **28**, 341–346 (1984)
5. Fountain, J.B.: Abundant semigroups. Proc. Lond. Math. Soc. **44**, 113–125 (1982)

6. Fountain, J.B.: E-unitary dense covers of E-dense monoids. Bull. London Math. Soc. **22**, 353–358 (1990)
7. Howie, J. M.: Fundamental of semigroup theory. London Mathematical Society, New series 12. The Clarendon Press, Oxford University Press, Oxford (1995)
8. Jorge, A., Jean-Eric, P. Pascal, W.: Semigroups whose idempotents form a subsemigroup. In: Mathematical Proceedings, vol. 111, pp. 241–253. Cambridge University Press, Cambridge (1992)
9. Laradji, A., Umar, A.: On certain finite semigroups of order-decreasing transformations I. Semigroup Forum 69(1), 184–200 (2004)
10. Toker, K.: Rank of some subsemigroups of full contraction mapping on a finite chain. J. BAUN Inst. Sci. Technol. **22**(2), 403–414 (2020)
11. Umar, A.: On the semigroups of order decreasing finite full transformations. Proc. R. Soc. Edinburgh Sect. A **120**, 129–142 (1992)
12. Umar, A., Al-Kharousi, F.: Studies in semigroup of contraction mappings of a finite chain. The Research Council of Oman Research grant proposal No. ORG/CBS/12/007, 6th March 2012
13. Umar, A., Zubairu, M. M.: On certain semigroups of contractions mappings of a finite chain. Algebra Descrete Maths. **32**(2), 299–320 (2021)
14. Umar, A., Zubairu, M.M.: On certain semigroups of full contractions of a finite chain. arXiv:1804.10057vl
15. Zubairu, M.M., Umar, A., Aliyu, J.A.: On certain Semigroup of Order-decreasing Full Contraction Mappings of a Finite Chain (2020). https://doi.org/10.48550/arXiv.2204.07427

$\hat{Z}$-ideals) of $\hat{Z}$-algebras, also given some excellent outcomes [15]. Following that, definitions of fuzzy sets, interval-valued fuzzy sets, and $\hat{Z}$-ideals in $\hat{Z}$-algebras have been provided. These characteristics of fuzzy $\hat{Z}$-subalgebras and fuzzy $\hat{Z}$-ideals also included the aspects of $\hat{Z}$-homomorphism and Cartesian product on fuzzy α-translations and fuzzy β-multiplications of $\hat{Z}$-algebras [16]. In 2012, Jun [6] highlighted and focused the characteristics of a cubic set using a fuzzy set and an interval_valued_fuzzy_set. Further, in 2010, Jun [7] explored and researched the subject of cubic_subalgebras with cubic_ideals in BCK/BCI-algebras. Jun [5] promoted the development of cubic subgroups in 2011 after applying this concept of cubic_sets to a group. In [10], the concept of neutrosophic algebra is introduced, along with the idea of an ideal within the context of neutrosophic algebra. Additionally, the concepts of the kernel and neutrosophic quotient algebra were provided. Several key properties of neutrosophic algebra are identified and explored. Moreover, demonstrated that every quotient neutrosophic algebra is indeed a quotient algebra. The focus of this research is to promote neutrosophic_fuzzy sets in $\hat{Z}$-algebra through the use of specific theorems and examples, as well as to evaluate some of their properties.

2 Preliminaries

The below section describes basic definitions of fuzzy sets and $\hat{Z}$-algebra, as well as their main properties.

Definition 2.1 [18] The fuzzy set ϱ from the universal set X is defined to be ϱ (x): X→ [0,1] for each elements x ∈ X, and ϱ (x) is known as the membership value of x.

Definition 2.2 [9] Let ϱ_1 and ϱ_2 be the fuzzy sets from the universal set X the union of ϱ_1 and ϱ_2; is represented as $\varrho_1 \cup \varrho_2$ is defined to be $(\varrho_1 \cup \varrho_2)(x) = \max\{\varrho_1(x), \varrho_2(x)\}\ \forall\, x \in X$.

Definition 2.3 [9] Let ϱ_1 and ϱ_2 be any two fuzzy sets from the universal set X, the intersection of ϱ_1 & ϱ_2, it is expressed as $\varrho_1 \cap \varrho_2$ is defined to be $(\varrho_1 \cap \varrho_2)(x) = \min\{\varrho_1(x), \varrho_2(x)\}\ \forall\, x \in X$.

Definition 2.4 [9] Let neutrosophic_set in X be in structure of the form $\vartheta = \{(x: \varrho_{\mathscr{T}}(x), \varrho_{\mathscr{I}}(x), \varrho_{\mathscr{F}}(x)/x \in X\}$ where $\varrho_{\mathscr{T}}, \varrho_{\mathscr{I}}, \varrho_{\mathscr{F}}$ are fuzzy sets in X; which is denoted by $\varrho_{\mathscr{T}}(x)$ *is a*true_membership function, $\varrho_{\mathscr{I}}(x)$ *is*an indeterminate_membership function & $\varrho_{\mathscr{F}}(x)$ *is* an false_membership function respectively.

Definition 2.5 [1] Suppose X be the non-empty subset with binary operation $*$ and constant; then, (X, $*$, 0) is $\hat{Z}$– algebra if

(i) $x * 0 = 0$ (ii) $0 * x = x$ (iii) $x * x = x$
(iv) $x * y = y * x$, when $x \neq 0$ and $y \neq 0\ \forall\, x, y \in X$

Table 1 Consider the example for Ẑ-algebra

*	0	Ω_1	Ω_2	Ω_3	Ω_4
0	0	Ω_1	Ω_2	Ω_3	Ω_4
Ω_1	0	Ω_1	Ω_3	Ω_4	Ω_2
Ω_2	0	Ω_3	Ω_2	Ω_1	Ω_1
Ω_3	0	Ω_4	Ω_1	Ω_3	Ω_3
Ω_4	0	Ω_2	Ω_1	Ω_3	Ω_4

Example 2.6 Let X = $\{0, \Omega_1, \Omega_2, \Omega_3, \Omega_4\}$ be the set with 0 as a constant, and a binary operation $*$ is to be defined on X by cayley's table (Table 1).

Definition 2.7 [1] If X is a non-empty subset of neutrosophic_Ẑ-algebra, it is called to be Ẑ-subalgebra of X:

$$x * y \in X \forall x, y \in X$$

Definition 2.8 [2] Let U be the subset in universe X; the sup property of a fuzzy set ϱ is referred to as $\varrho(x_0) = \sup_{x \in U} \varrho(x)$, if $\exists$ x, $x_0 \in U$.

Definition 2.9 [17] Let the intuitionistic fuzzy set be ϱ which assigns the membership function μ_ϱ as X $\to$ [0,1], and τ_ϱ as X $\to$ [0,1] is nonmembership function, and then it is known to be sup_inf property, and then for any subset U of X then $\exists$ $(x_0) \in U \ni \mu_\varrho(x_0) = \sup_{x \in U}(\mu_\varrho(x))$ and $\tau_\varrho(x_0) = \inf_{x \in U}(\tau_\varrho(x))$.

Definition 2.10 A neutrosophic_fuzzy set ϱ in a set X is referred to as an sup_sup_inf property if the subset U of X then $\exists$ $x_0 \in U \ni \varrho_{\mathscr{T}}(x_0) = \sup_{x \in U}(\varrho_{\mathscr{T}}(x))$, $\varrho_{\mathscr{I}}(x_0) = \sup_{x \in U}(\varrho_{\mathscr{I}}(x))$, $\varrho_{\mathscr{F}}(x_0) = \inf_{x \in U}(\varrho_{\mathscr{F}}(x))$.

Definition 2.11 Let $\vartheta = \{x, \varrho_{\mathscr{T},\mathscr{I},\mathscr{F}}(x) \ / \ x \in X\}$ be the neutrosophic_set in Ẑ and ∂ maps from X $\to$ Ỳ then the image of ϑ under ∂, $\partial(\vartheta)$ represented as $\{\partial_{sup}(\varrho_{\mathscr{T}}), \partial_{sup}(\varrho_{\mathscr{I}}), \partial_{inf}(\varrho_{\mathscr{F}}), x \in \grave{Y}$.

(i) $$\partial_{\sup}(\varrho_{\mathscr{T}})(y) = \begin{cases} \sup\limits_{x \in \partial^{-1}(y)} \varrho_{\mathscr{T}}(x) & \text{if } \partial^{-1}(y) \neq \phi \\ 1 & \text{otherwise} \end{cases}$$

(ii) $$\partial_{\sup}(\varrho_{\mathscr{I}})(y) = \begin{cases} \sup\limits_{x \in \partial^{-1}(y)} \varrho_{\mathscr{I}}(x) & \text{if } \partial^{-1}(y) \neq \phi \\ 1 & \text{otherwise} \end{cases}$$

(iii) $$\partial_{\inf}(\varrho_{\mathscr{F}})(y) = \begin{cases} \inf\limits_{x \in \partial^{-1}(y)} \varrho_{\mathscr{F}}(x) & \text{if } \partial^{-1}(y) \neq \phi \\ 0 & \text{otherwise} \end{cases}$$

Definition 2.12 If ∂ : X → Ỳ be a function. Let $\varrho_{\mathcal{T}_1,\mathcal{I}_1,\mathcal{F}_1}$ & $\varrho_{\mathcal{T}_2,\mathcal{I}_2,\mathcal{F}_2}$ be two neutrosophic_set in X and Ỳ, respectively, and then the inverse image of $\varrho_{\mathcal{T},\mathcal{I},\mathcal{F}}$ under ∂ is defined by ∂^{-1} $(\varrho_{\mathcal{T}_2,\mathcal{I}_2,\mathcal{F}_2})$ = {x, ∂^{-1} $(\varrho_{\mathcal{T}_2}(x))$, ∂^{-1} $(\varrho_{\mathcal{I}_2}(x))$, ∂^{-1} $(\varrho_{\mathcal{F}_2}(x))$, / x ∈ Ẑ} such that $\partial^{-1}(\varrho_{\mathcal{T}_2})\ \partial\ (x) = \varrho_{\mathcal{T}_2}(\partial\ (x))$, $\partial^{-1}(\varrho_{\mathcal{I}_2})\ \partial(x) = \varrho_{\mathcal{I}_2}(\partial\ (x))$, $\partial^{-1}(\varrho_{\mathcal{F}_2})\ \partial\ (x) = \varrho_{\mathcal{F}_2}(\partial\ (x))$.

Definition 2.13 [16] Let (Ẑ, ∗,0) and (Ẑ′, ∗′, 0′) the two Ẑ-algebra, and then the mapping from ɧ: (Ẑ, ∗,0) → (Ẑ′, ∗′,0′) is known as Ẑ-homomorphism of Ẑ-algebra if

$$\lambda\ (x * y) = \lambda\ (x) *' \lambda\ (y).$$

Definition 2.14 [16] Let h be an Ẑ-endomorphism of neutrosophic_Ẑ-algebras and $\vartheta = \{x, \varrho_{\mathcal{T},\mathcal{I},\mathcal{F}}\ (x)\ /x \in X\}$ be the neutrosophic_set in X, and then define a new fuzzy set ϱ^h in X, as $\vartheta_{\varrho^h}\ (x) = \vartheta_{\varrho}(h(x))\ \forall\ x \in X$.

3 Neutrosophic in Ẑ-Subalgebra

This section defines neutrosophic Ẑ-algebra and discussed some fascinating results.

Definition 3.1 Suppose (X, ∗,0) be the Ẑ-algebra with operation "∗" and constant 0, then the neutrosophic_set $\vartheta = \{x : \varrho_{\mathcal{T}}, \varrho_{\mathcal{I}}, \varrho_{\mathcal{F}} / x \in X\}$; it is defined to be neutrosophic_Ẑ- subalgebra of X.

(i) $\varrho_{\mathcal{T}}(x * y) \geq \min\{\varrho_{\mathcal{T}}(x), \varrho_{\mathcal{T}}\ (y)\}$
(ii) $\varrho_{\mathcal{I}}(x * y) \geq \min\{\varrho_{\mathcal{I}}(x), \varrho_{\mathcal{I}}\ (y)\}$
(iii) $\varrho_{\mathcal{F}}(x * y) \leq \max\{\varrho_{\mathcal{F}}\ (x), \varrho_{\mathcal{F}}\ (y)\}$

Example 3.2 Consider Ẑ– algebra defined the Example 2.6. and the following neutrosophic_fuzzy set defined on X is a neutrosophic_Ẑ- subalgebra of X.

$$\varrho_{\mathcal{T},\mathcal{I},\mathcal{F}} = \begin{cases} 0.5 & x = 0 \text{ when } (x = 0, y \neq 0) \text{ or } (x \neq 0, y = 0) \\ 0.8 & x = \Omega_1, \Omega_2 \\ 0.4 & x = \Omega_3, \Omega_4 \end{cases}$$

Proposition 3.3 If $\vartheta = \{x, \varrho_{\mathcal{T}}, \varrho_{\mathcal{I}}, \varrho_{\mathcal{F}} : x \in \hat{Z}\}$ is a neutrosophic_Ẑ– subalgebra of X.

Then 1. $\varrho_{\mathcal{T}}\ (0) \geq \varrho_{\mathcal{T}}\ (x)$, $\varrho_{\mathcal{I}}\ (0) \geq \varrho_{\mathcal{I}}\ (x)$, $\varrho_{\mathcal{F}}\ (0) \leq \varrho_{\mathcal{F}}\ (x)$, ∀x ∈ X.

2. $\varrho_{\mathcal{T}}(0) \geq \varrho_{\mathcal{T}}(x^*) \geq \varrho_{\mathcal{T}}(x)$, $\varrho_{\mathcal{I}}\ (0) \geq \varrho_{\mathcal{I}}\ (x^*) \geq \varrho_{\mathcal{I}}\ (x)$, $\varrho_{\mathcal{F}}\ (0) \leq \varrho_{\mathcal{F}}\ (x^*) \leq \varrho_{\mathcal{F}}\ (x)$, where $x^* = 0 * x\ \forall\ x \in X$.

Proof: For every x ∈X,

1. $$\varrho_{\mathscr{T}}(0) = \varrho_{\mathscr{T}}(x * x)$$
$$\geq \min\{\varrho_{\mathscr{T}}(x), \varrho_{\mathscr{T}}(x)\} = \varrho_{\mathscr{T}}(x)$$
Similarly, $\varrho_{\mathscr{I}}(0) = \varrho_{\mathscr{I}}(x)$
$$\varrho_{\mathscr{F}}(0) = \varrho_{\mathscr{F}}(x * x)$$
$$\leq \max\{\varrho_{\mathscr{F}}(x), \varrho_{\mathscr{F}}(x)\} = \varrho_{\mathscr{F}}(x)$$

2. $$\varrho_{\mathscr{T}}(x^*) = \varrho_{\mathscr{T}}(0 * x)$$
$$\geq \{\min(\varrho_{\mathscr{T}}(0), E\varrho_{\mathscr{T}}(x))\}$$
$$\varrho_{\mathscr{T}}(x^*) = \varrho_{\mathscr{T}}(x)$$
$$\therefore \varrho_{\mathscr{T}}(x^*) \geq \varrho_{\mathscr{T}}(x)$$

Similarly, $\varrho_{\mathscr{I}}(x*) \geq \varrho_{\mathscr{I}}(x)$
$$\varrho_{\mathscr{F}}(x^*) = \varrho_{\mathscr{F}}(0 * x)$$
$$\leq \max\{\varrho_{\mathscr{F}}(0) * \varrho_{\mathscr{F}}(x)\}$$
Hence, $\varrho_{\mathscr{F}}(x^*) \leq \varrho_{\mathscr{F}}(x)$

Theorem 3.4 If $\vartheta = \{x: \varrho_{\mathscr{T},\mathscr{I},\mathscr{F}}(x) \ / \ x \in \hat{Z}\}$ and $\varrho_{\mathscr{T}_1\mathscr{I}_1\mathscr{F}_1}, \varrho_{\mathscr{T}_2\mathscr{I}_2\mathscr{F}_2}$ be two neutrosophic_Ẑ-subalgebras of X, then the intersection is also a neutrosophic_Ẑ-subalgebras of X.

Proof: Let $\vartheta = \{x, y \in \hat{Z}, \varrho_{\mathscr{T},\mathscr{I},\mathscr{F}}(x) / x \in X\}$ be a neutrosophic_Ẑ-subalgebras of X for all x, y ∈X. Then

$$(\varrho_{\mathscr{T}_1} \cap \varrho_{\mathscr{T}_2})(x * y) = \min\{\varrho_{\mathscr{T}_1}(x * y), \varrho_{\mathscr{T}_2}(x * y)\}$$
$$\geq \{\min\{\varrho_{\mathscr{T}_1}(x), \varrho_{\mathscr{T}_1}(y)\}, \min\{\varrho_{\mathscr{T}_2}(x), \varrho_{\mathscr{T}_2}(y)\}\}$$
$$= \min\{\varrho_{\mathscr{T}_1}(x), \varrho_{\mathscr{T}_2}(x)\}, \min\{\varrho_{\mathscr{T}_1}(y), \varrho_{\mathscr{T}_2}(y)\}$$
$$= \min\{\varrho_{\mathscr{T}_1} \cap \varrho_{\mathscr{T}_2}(x)\}, \varrho_{\mathscr{T}_1} \cap \varrho_{\mathscr{T}_2}(y)$$
$$\therefore \varrho_{\mathscr{T}_1} \cap \varrho_{\mathscr{T}_2})(x * y) \geq \min\{(\varrho_{\mathscr{T}_1} \cap \varrho_{\mathscr{T}_2})(x), (\varrho_{\mathscr{T}_1} \cap (\varrho_{\mathscr{T}_2})(y)\}$$
Similarly, $(\varrho_{\mathscr{I}_1} \cap \varrho_{\mathscr{I}_2})(x * y) \geq \min\{(\varrho_{\mathscr{I}_1} \cap \varrho_{\mathscr{I}_2})(x), (\varrho_{\mathscr{I}_1} \cap \varrho_{\mathscr{I}_2})(y)\}$
$$(\varrho_{\mathscr{F}_1} \cap \varrho_{\mathscr{F}_2})(x * y) = \max\{\varrho_{\mathscr{F}_1}(x * y), \varrho_{\mathscr{F}_2}(x * y)\}$$
$$\leq \{\max\{\varrho_{\mathscr{F}_1}(x), \varrho_{\mathscr{F}_1}(y)\}, \max\{\eta_{\mathscr{F}_2}(x), \varrho_{\mathscr{F}_2}(y)\}\}$$
$$= \max\{\varrho_{\mathscr{F}_1}(x), \varrho_{\mathscr{F}_2}(x)\}, \max\{\varrho_{\mathscr{F}_1}(y), \varrho_{\mathscr{F}_2}(y)\}$$
$$= \max\{(\varrho_{\mathscr{F}_1} \cap \varrho_{\mathscr{F}_2})(x), (\varrho_{\mathscr{F}_1} \cap \varrho_{\mathscr{F}_2})(y)\}$$
$$\therefore (\varrho_{\mathscr{F}_1} \cap \eta_{\mathscr{F}_2})(x * y) \leq \max\{(\varrho_{\mathscr{F}_1} \cap \varrho_{\mathscr{F}_2})(x), (\varrho_{\mathscr{F}_1} \cap \varrho_{\mathscr{F}_2})(y)\}$$
Hence, $\varrho_{\mathscr{T}_1}$ and $\varrho_{\mathscr{T}_2}$ are a neutrosophic_Ẑ-subalgebras of X.

Theorem 3.5 If $\vartheta = \{x, \varrho_{\mathscr{T},\mathscr{I},\mathscr{F}}(x) : x \in X\}$ is the neutrosophic_Ẑ-subalgebra of X. If ∃ a sequence $\{x_n\}$ of X ∋ $\lim_{n\to\infty} \varrho_{\mathscr{T}}(x_n) = 1$, $\lim_{n\to\infty} \varrho_{\mathscr{I}}(x_n) = 1$, $\lim_{n\to\infty} \varrho_{\mathscr{F}}(x_n) = 0$ Then $\varrho_{\mathscr{T}}(x_n) = 1$, $\varrho_{\mathscr{I}}(x_n) = 1$, $\varrho_{\mathscr{F}}(x_n) = 0$.

Proof: Consider the Proposition 3.3, $\varrho_{\mathscr{T}}(0) \geq \varrho_{\mathscr{T}}(x)\ \forall x \in X$.

Then, we have $\varrho_{\mathscr{T}}(0) \geq \varrho_{\mathscr{T}}(x)\ \forall\ x \in X$
$$\varrho_{\mathscr{T}}(0) \geq \lim_{n\to\infty} \varrho_{\mathscr{T}}(x_n)$$
Hence, $\varrho_{\mathscr{T}}(0) = 1$.

Similarly, $\varrho_{\mathscr{I}}(0) = 1$

$\varrho_{\mathscr{F}}(0) \leq \varrho_{\mathscr{F}}(x)\ \forall x \in X$
$\varrho_{\mathscr{F}}(0) \leq \lim_{n\to\infty} \varrho_{\mathscr{F}}(x_n)$
Hence, $\eta_{\mathscr{F}}(0) = 0$

Theorem 3.6 If $\vartheta = \{x : \varrho_{\mathscr{T},\mathscr{I},\mathscr{F}} / x \in X\}$ is the neutrosophic_Ẑ-subalgebra of X, then the set $X_{\varrho_{\mathscr{T},\mathscr{I},\mathscr{F}}} = \{x \in X / \varrho_{\mathscr{T},\mathscr{I},\mathscr{F}}(x) = \varrho_{\mathscr{T},\mathscr{I},\mathscr{F}}(0)\ \}$ is also a subalgebra of X.

Proof: For any x, $y \in X_{\varrho_{\mathscr{T},\mathscr{I},\mathscr{F}}}$

$\varrho_{\mathscr{T}}(x) = \varrho_{\mathscr{T}}(0) = \varrho_{\mathscr{T}}(y)$
$\varrho_{\mathscr{T}}(x * y) \geq \{\min(\varrho_{\mathscr{T}}(x), \varrho_{\mathscr{T}}(y)\}$
$= \min\{\varrho_{\mathscr{T}}(0), \varrho_{\mathscr{T}}(0)\}$
$= \min\{\varrho_{\mathscr{T}}(0)\}$

Similarly, $\varrho_{\mathscr{I}}(x * y) = \min\{\varrho_{\mathscr{I}}(0)\}$
$\varrho_{\mathscr{F}}(x * y) \leq \{\max(\varrho_{\mathscr{F}}(x), \varrho_{\mathscr{F}}(y)\}$
$= \max\{\varrho_{\mathscr{F}}(0), \varrho_{\mathscr{F}}(0)\}$
$= \max\{\varrho_{\mathscr{F}}(0)\}$
$\therefore X_{\varrho_{\mathscr{T},\mathscr{I},\mathscr{F}}}$ is the subalgebra of X

Theorem 3.7 Let (X, ∗, 0) and $(X', *', 0')$ be two Ẑ-algebras and $\Psi: X \to X'$ be a homomorphism. If $\varrho_{\mathscr{T},\mathscr{I},\mathscr{F}}$ is a neutrosophic_Ẑ-subalgebra of X, which is defined as

$\Psi(\varrho_{\mathscr{T},\mathscr{I},\mathscr{F}}) = \{x, (\varrho_{\mathscr{T},\mathscr{I},\mathscr{F}}(x) = \varrho_{\mathscr{T},\mathscr{I},\mathscr{F}}(\Psi(x))\}$ then $\Psi(\varrho_{\mathscr{T},\mathscr{I},\mathscr{F}})$ is the neutrosophic_Ẑ– subalgebra of X.

Proof: Let x, $y \in X$

$(\varrho_{\mathscr{T}})_{\Psi}(x * y) = \varrho_{\mathscr{T}}(\Psi(x * y)$
$= \varrho_{\mathscr{T}}(\Psi(x) * \Psi(y)$
$\geq \min\{\varrho_{\mathscr{T}}(\Psi(x), \varrho_{\mathscr{T}}(\Psi(y)\}$
$(\varrho_{\mathscr{T}})_{\Psi} = \min\{(\varrho_{\mathscr{T}})_{\Psi}(x), (\varrho_{\mathscr{T}})_{\Psi}(y)\}$
Similarly, $(\varrho_{\mathscr{I}})_{\Psi} = \min\{(\varrho_{\mathscr{I}})_{\Psi}(x), (\varrho_{\mathscr{I}})_{\Psi}(y)\}$
$(\varrho_{\mathscr{F}})_{\Psi}(x * y) = \varrho_{\mathscr{F}}(\Psi(x * y))$
$= \varrho_{\mathscr{F}}(\Psi(x) * \Psi(y))$
$\leq \max\{\varrho_{\mathscr{F}}(\Psi(x)), \varrho_{\mathscr{F}}(\Psi(y))$
$(\varrho_{\mathscr{F}})_{\Psi} = \max\{(\varrho_{\mathscr{F}})_{\Psi}(x), (\varrho_{\mathscr{F}})_{\Psi}(y)\}$
Therefore, $\Psi(\varrho_{\mathscr{T},\mathscr{I},\mathscr{F}})$ is the neutrosophic_Ẑ-subalgebra of X'.

Remark 3.8 The set which is denoted by $I\varrho_{\mathscr{T},\mathscr{I},\mathscr{F}}$ is also the subalgebra of X which is defined to be $I\varrho_{\mathscr{T},\mathscr{I},\mathscr{F}} = \{x \in X / \varrho_{\mathscr{T},\mathscr{I},\mathscr{F}}(x) = \varrho_{\mathscr{T},\mathscr{I},\mathscr{F}}(0)\}$.

Theorem 3.9 Let $\vartheta = \{x, \varrho_{\mathscr{T},\mathscr{I},\mathscr{F}}(x) : x \in X\}$ be the neutrosophic_Ẑ-algebra of X. Then, ∃ a set $I\varrho_{\mathscr{T},\mathscr{I},\mathscr{F}}$ is also the subalgebra of X.

Proof: Let x, $y \in I\varrho_{\mathscr{T},\mathscr{I},\mathscr{F}}$

Then $\varrho_{\mathcal{T},\mathcal{I},\mathcal{F}}$ (x) = $\varrho_{\mathcal{T},\mathcal{I},\mathcal{F}}$ (0) = $\varrho_{\mathcal{T},\mathcal{I},\mathcal{F}}$ (y).
Consider

$$\begin{aligned}\varrho_{\mathcal{T}}(x * y) &\geq \min\{\varrho_{\mathcal{T}}(x), \varrho_{\mathcal{T}}(y)\}\\ &\geq \min\{\varrho_{\mathcal{T}}(0), \varrho_{\mathcal{T}}(0)\}\\ &= \varrho_{\mathcal{T}}(0)\end{aligned}$$

$$\therefore \varrho_{\mathcal{T}}(x * y) \geq \varrho_{\mathcal{T}}(0),$$

Consider Proposition 3.3, $\varrho_{\mathcal{T}}(x * y) \geq \varrho_{\mathcal{T}}(0)$
Then, there exist $\varrho_{\mathcal{T}}$ ((x ∗ y) ≥ $\varrho_{\mathcal{T}}$ (0) or equivalently, (x ∗ y) ∈ $I\varrho_{\mathcal{T},\mathcal{I},\mathcal{F}}$

Similarly, $\varrho_{\mathcal{I}}(x * y) \geq \varrho_{\mathcal{I}}(0)$

$$\begin{aligned}\varrho_{\mathcal{F}}(x * y) &\leq \max\{\varrho_{\mathcal{F}}(x), \varrho_{\mathcal{F}}(y)\}\\ &\leq \max\{\varrho_{\mathcal{F}}(0), \varrho_{\mathcal{F}}(0)\}\\ &= \varrho_{\mathcal{F}}(0)\end{aligned}$$

$\therefore \varrho_{\mathcal{F}}(x * y) \leq \varrho_{\mathcal{F}}(0)$ Using Proposition 3.3
Thus, $\varrho_{\mathcal{F}}$ ((x ∗ y) = $\varrho_{\mathcal{F}}$ (0) or x ∗ y ∈ $I\varrho_{\mathcal{T},\mathcal{I},\mathcal{F}}$
Hence, the set $I\varrho_{\mathcal{T},\mathcal{I},\mathcal{F}}$ are Ẑ-subalgebras of X.

4 Homomorphism of Neutrosophic_Ẑ-Subalgebra

Some interesting results on homomorphism of neutrosophic_Ẑ-subalgebra are studied in this section.

Theorem 4.1 If $\partial : X \rightarrow \grave{Y}$ is the homomorphism of Ẑ-subalgebra. If $\varrho_{\mathcal{T},\mathcal{I},\mathcal{F}}$ be the neutrosophic_Ẑ -subalgebra of Y, then $\partial^{-1}(\varrho_{\mathcal{T},\mathcal{I},\mathcal{F}}) = \{(\partial^{-1}(\varrho_{\mathcal{T}}), \partial^{-1}(\varrho_{\mathcal{I}}), \partial^{-1}(\varrho_{\mathcal{F}})\}$ is also the neutrosophic_Ẑ-subalgebra of Ỳ, where $\partial^{-1}(\varrho_{\mathcal{T}}(x)) = \varrho_{\mathcal{T}}\partial(x)$, $\partial^{-1}(\varrho_{\mathcal{I}}(x)) = \varrho_{\mathcal{I}}\partial(x)$, $\partial^{-1}(\varrho_{\mathcal{F}}(x)) = \varrho_{\mathcal{F}}\partial(x)$, for every x ∈ X.

Proof: Given $\varrho_{\mathcal{T},\mathcal{I},\mathcal{F}}$ be the neutrosophic_ Ẑ-subalgebra of Ỳ
Let, x, y ∈ X.

$$\begin{aligned}\text{Then, } \partial^{-1}(\varrho_{\mathcal{T}}(x*y) &= \varrho_{\mathcal{T}}\partial(x*y)\\ &= \varrho_{\mathcal{T}}\partial(x) * \varrho_{\mathcal{T}}\partial(y)\\ &\geq \min\{\partial(\varrho_{\mathcal{T}}(x)) * \partial(\varrho_{\mathcal{T}}(y))\}\\ &= \min\{\partial^{-1}(\varrho_{\mathcal{T}}(x)) * \partial^{-1}(\varrho_{\mathcal{T}}(y))\}\end{aligned}$$

$$\partial^{-1}(\varrho_{\mathcal{T}}(x*y) \geq \min\{\partial^{-1}(\varrho_{\mathcal{T}}(x)) * \partial^{-1}(\varrho_{\mathcal{T}}(y))\}$$

$$\begin{aligned}\partial^{-1}(\varrho_{\mathcal{I}})(x*y) &= \varrho_{\mathcal{I}}(\partial(x*y))\\ &= \varrho_{\mathcal{I}}(\partial(x) * \partial(y))\\ &\geq \min\{\varrho_{\mathcal{I}}(\partial(x)), \varrho_{\mathcal{I}}(\partial(x))\}\\ &= \min\{\partial^{-1}(\varrho_{\mathcal{I}}(x)), \partial^{-1}(\varrho_{\mathcal{I}}(y))\}\end{aligned}$$

$$\begin{aligned}\partial^{-1}(\varrho_{\mathcal{F}})(x*y) &= \varrho_{\mathcal{F}}(\partial(x* y))\\ &= \varrho_{\mathcal{F}}(\partial(x) * \partial(y))\\ &\leq \max\{\varrho_{\mathcal{F}}(\partial(x)), \varrho_{\mathcal{F}}(\partial(x))\}\\ &= \max\{\partial^{-1}(\varrho_{\mathcal{F}}(x)), \partial^{-1}(\varrho_{\mathcal{F}}(y))\}\end{aligned}$$

2. Hemavathi, P., Muralikrishna, P., Palanivel, P.: A note on interval valued fuzzy β -subalgebras. Global J. Pure Appl. Math. **11**(4), 2553–2560 (2015)
3. Imai, Y., Iseki, K.: On axioms system of propositional calculi, XIV. Proc. Jpn. Acad. **42**, 19–22 (1966)
4. Iseki, K.: An algebra related with a propositional calculus. Proc. Jpn. Acad. **142**, 26–29 (1966)
5. Jun, Y.B., Kim, C.S., Yang, K.O.: Cubic sets. Ann. Fuzzy Math. Inf. **4**(1), 83–98 (2012)
6. Jun, Y.B., Jung, T., Kim, M.S.: Cubic subgroups. Ann. Fuzzy Math. Inf. **2**(1), 9–15 (2011)
7. Jun, Y.B., Kim, C.S., Yang, K.O.: Cubic subalgebras and Ideals of BCK/BCI – algebras. Far East J. Math. Sci. **44**(2), 239–250 (2010)
8. Muralikrishna, P., Manokaran, S.: MBJ neutrosophic β-ideal of β-algebra. Neutrosophic Sets Syst. **35** (2020)
9. Muralikrishna, P., Vinodkumar, R., Palani, G.: Neutrosophic cubic β -subalgebra. Neutrosophic Sets Syst. **48**(1) (2022)
10. Nagaiah, T., Bhaskar, L., Narasimha Swamy, P., Broumi, S.: A study on neutrosophic algebra. Neutrosophic Sets Syst. **50** (2022)
11. Smarandache, F.: Neutrosophic set – a generalization of the intuitionistic fuzzy set. Int. J. Pure Appl. Math. (2004)
12. Smarandache, F.: A unifying field in logics. In: Neutrosophy: Neutrosophic Probability, Set and Logic. American Research Press, Rehoboth (1999)
13. Sowmiya, S., Jeyalakshmi, P.: Fuzzy algebraic structures in $\hat{Z}$– algebras. World J. Eng. Res. Technol. **5**(4), 74–88 (2019)
14. Sowmiya, S., Jeyalakshmi, P.: On fuzzy $\hat{Z}$– ideals in $\hat{Z}$– algebras. Global J. Pure Appl. Math. **15**(4), 505–516 (2019)
15. Sowmiya, S., Jeyalakshmi, P.: Fuzzy α – translations and Fuzzy β – Multiplication of $\hat{Z}$–algebras. Adv. Math. Sci. J. **9**(3), 1287–1292 (2020)
16. Sowmiya, S., Jeyalakshmi, P.: $\hat{Z}$– Homomorphism and cartesian product on fuzzy α – translations and Fuzzy β – multiplication of $\hat{Z}$– algebras. AIP Conf. Proc. **226** (2020)
17. Manokaran, S., Prakjasam, M.: On MBJ-neutrosophic β -subalgebra. Neutrosophic Sets Syst. **28** (2019)
18. Zadeh, L.A.: Fuzzy sets. Inf. Control. **8**, 338–353 (1965)
19. Zadeh, L.A.: The concept of a linguistic variable and its application to approximate reasoning. Inf. Sci. **8**, 199–249 (1975)

Controllable Fuzzy Neutrosophic Soft Matrices

M. Kavitha and P. Murugadas

Keywords Fuzzy Neutrosophic Soft Matrices (FuNeSoMas) · Controllable Fuzzy Neutrosophic Soft Matrices (CoFuNeSoMas) · Nilpotent Fuzzy Neutrosophic Soft Matrices (NiFuNeSoMas) · Fuzzy Neutrosophic Soft Relation (FuNeSoRe) · Transitivity canonical form

1 Introduction

The concept of fuzzy sets was founded by Zadeh [19]. Intuitionistic Fuzzy Sets (InFuSes) introduced by Atanassov [2] are appropriate for such a situation. But the intuitionistic fuzzy sets can only handle the incomplete information considering both the truth-membership (or simply membership) and falsity-membership (or non-membership) values. It does not handle the indeterminate and inconsistent information, which exists in belief system. Smarandache [16] introduced the concept of Neutrosophic Set (NeSe), which is a mathematical tool for handling problems involving imprecise, indeterminacy, and inconsistent data.

In our regular everyday life, we face situations that require procedures allowing certain flexibility in information processing capacity. Molodtsov [12] addressed soft set theory problems successfully. In their early work, soft set was described purely as a mathematical method to model uncertainties. The researchers can pick any kind of parameters of any nature they wish in order to facilitate the decision-making procedure as there is a varied way of picturing the objects.

Maji [10] have done further research on soft set theory. Presence of vagueness demanded Fuzzy Soft Set (FuSoSe) to come into picture. But satisfactory evaluation of membership values is not always possible because of the insufficiency in the

M. Kavitha
Department of Mathematics, Bharath Institute of Higher Education and Research, Chennai, India

P. Murugadas (✉)
Department of Mathematics, Government Arts and Science College Veerapandi, Theni, India
e-mail: bodi_muruga@yahoo.com

H.-H. Leung et al. (eds.), *Recent Developments in Algebra and Analysis*,
Trends in Mathematics, https://doi.org/10.1007/978-3-031-37538-5_6

available information (besides the presence of vagueness) situation. Evaluation of non-membership values is also not always possible for the same reason, and as a result, there exists an indeterministic part upon which hesitation survives. Certainly fuzzy soft set theory is not suitable to solve such problems. In those situations, Intuitionistic Fuzzy Soft Set theory (InFuSoSs) [11] may be more applicable. Now in the parlance of soft set theory, there is hardly any limitation to select the nature of the criteria, and as most of the parameters or criteria (which are words or sentences) are neutrosophic in nature, Maji [9] has been motivated to combine the concept of soft set and neutrosophic set to make the new mathematical model neutrosophic soft set and has given an algorithm to solve a decision-making problem.

The theory of a fuzzy matrix is very useful in the discussion of fuzzy relations. We can represent basic propositions of the theory of fuzzy relations in terms of matrix operations. Furthermore, we can deal with the fuzzy relations in the matrix form. In the study of the theory of fuzzy matrix, a canonical form of some fuzzy matrices has received increasing attention. For example, Kim and Roush [8] studied the Idempotent fuzzy matrices. Xin [18] introduced the idea for Convergence of powers of controllable fuzzy matrices. Padder and Murugadas [15] are presented the max-min opetarion on InFuMa. Broumi et al. [3] redefined the notion of neutrosophic set in a new way and put forward the concept of neutrosophic soft matrix and different types of matrices in neutrosophic soft theory. They have introduced some new operations and properties on these matrices. The minimal solution was done by Kavitha et al. [5], based on the notion of FuNeSoMa given by Arokiarani and Sumathi [1]. As time goes, some works on FuNeSoMa were done by Kavitha et al. [4–7]. The Monotone interval fuzzy neutrosophic soft eigenproblem, and Monotone fuzzy neutrosophic soft eigenspace structures in max-min algebra and Solvability of system of neutrosophic soft linear equations were investigated by Murugadas et al. [13, 14]. Also, two kinds of fuzzy neutrosophic soft matrices presented by Uma et al. [17].

In this chapter, we study and prove some properties of controllable and Idompotent FuNeSoMas. However, we have developed an algorithm for controllable and nilpotent FuNeSoMas and reduce a controllable FuNeSoMa to canonical form. One of these results enables us to construct an idempotent and controllable FuNeSoMa from a given one, and this is the main result in the chapter.

2 Preliminaries

For basic result refer [5, 16–18].

2.1 Main Results

Let $R = \langle r_{ij}^T, r_{ij}^I, r_{ij}^F \rangle$ and $S = \langle s_{ij}^T, s_{ij}^I, s_{ij}^F \rangle$ be square FuNeSoMa with elements in [0,1].

- $R \vee S = [\langle r_{ij}^T, r_{ij}^I, r_{ij}^F \rangle \vee \langle s_{ij}^T, s_{ij}^I, s_{ij}^F \rangle]$,
- $R \wedge S = [\langle r_{ij}^T, r_{ij}^I, r_{ij}^F \rangle \wedge \langle s_{ij}^T, s_{ij}^I, s_{ij}^F \rangle]$,
- $R \ominus S = [\langle r_{ij}^T, r_{ij}^I, r_{ij}^F \rangle \ominus \langle s_{ij}^T, s_{ij}^I, s_{ij}^F \rangle]$,

$$\langle r_{ij}^T, r_{ij}^I, r_{ij}^F \rangle \ominus \langle s_{ij}^T, s_{ij}^I, s_{ij}^F \rangle := \begin{Bmatrix} \langle r_{ij}^T, r_{ij}^I, r_{ij}^F \rangle & If \quad \langle r_{ij}^T, r_{ij}^I, r_{ij}^F \rangle > \langle s_{ij}^T, s_{ij}^I, s_{ij}^F \rangle \\ \langle 0, 0, 1 \rangle & If \quad \langle r_{ij}^T, r_{ij}^I, r_{ij}^F \rangle \leq \langle s_{ij}^T, s_{ij}^I, s_{ij}^F \rangle \end{Bmatrix}$$

- $R \times S = [(\langle r_{i1}^T, r_{i1}^I, r_{i1}^F \rangle \wedge \langle s_{1j}^T, s_{1j}^I, s_{1j}^F \rangle) \vee (\langle r_{i2}^T, r_{i2}^I, r_{i2}^F \rangle \wedge \langle s_{2j}^T, s_{2j}^I, s_{2j}^F \rangle) \vee \ldots \vee (\langle r_{in}^T, r_{in}^I, r_{in}^F \rangle \wedge \langle s_{nj}^T, s_{nj}^I, s_{nj}^F \rangle)]$,
- $R^{k+1} = R^k \times R, \quad k = \{0, 1, 2, \ldots\}$,
- $R^0 = I$,
- $R' = \langle r_{ji}^T, r_{ji}^I, r_{ji}^F \rangle$ the transpose of R,
- $\Delta R = R \ominus R' = \Delta \langle r_{ij}^T, r_{ij}^I, r_{ij}^F \rangle = \langle r_{ij}^T, r_{ij}^I, r_{ij}^F \rangle \ominus \langle r_{ji}^T, r_{ji}^I, r_{ji}^F \rangle$,
- $\nabla R = R \wedge R' = \nabla \langle r_{ij}^T, r_{ij}^I, r_{ij}^F \rangle = \langle r_{ij}^T, r_{ij}^I, r_{ij}^F \rangle \wedge \langle r_{ji}^T, r_{ji}^I, r_{ji}^F \rangle$,
- $R \leq S$ iff $(\langle r_{ij}^T, r_{ij}^I, r_{ij}^F \rangle \leq \langle s_{ij}^T, s_{ij}^I, s_{ij}^F \rangle \forall\, i, j \in \{1, 2, \ldots, n\})$,
- $R \Psi S$ iff $(\langle s_{ij}^T, s_{ij}^I, s_{ij}^F \rangle = \langle 0, 0, 1 \rangle \Rightarrow \langle r_{ij}^T, r_{ij}^I, r_{ij}^F \rangle = \langle 0, 0, 1 \rangle \ \forall i, j \in \{1, 2, \ldots, n\})$,

FuNeSoMa R is said to be

- Transitive if $\langle r_{ij}^T, r_{ij}^I, r_{ij}^F \rangle^2 \leq \langle r_{ij}^T, r_{ij}^I, r_{ij}^F \rangle$;
- Idempotent if $\langle r_{ij}^T, r_{ij}^I, r_{ij}^F \rangle^2 = \langle r_{ij}^T, r_{ij}^I, r_{ij}^F \rangle$;
- Nilpotent if $\langle r_{ij}^T, r_{ij}^I, r_{ij}^F \rangle^n = \langle 0, 0, 1 \rangle$;
- Symmetric if $\langle r_{ij}^T, r_{ij}^I, r_{ij}^F \rangle = \langle r_{ji}^T, r_{ji}^I, r_{ji}^F \rangle$;
- ST iff for any index $i, j, k \in \{1, 2, \ldots, n\}$, with $i \neq j, i \neq k, j \neq k$, such that $\langle r_{ik}^T, r_{ik}^I, r_{ik}^F \rangle > \langle r_{ki}^T, r_{ki}^I, r_{ki}^F \rangle$ and $\langle r_{kj}^T, r_{kj}^I, r_{kj}^F \rangle > \langle r_{jk}^T, r_{jk}^I, r_{jk}^F \rangle$, we have $\langle r_{ij}^T, r_{ij}^I, r_{ij}^F \rangle > \langle r_{ji}^T, r_{ji}^I, r_{ji}^F \rangle$;
- Strictly Lower (Upper) Triangular (SL(U)T) if $\langle r_{ij}^T, r_{ij}^I, r_{ij}^F \rangle = \langle 0, 0, 1 \rangle \forall i \leq j (i \leq j)$.

Theorem 1 *Consider a NiFuNeSoMa N and Symmetric FuNeSoMa (SyFuNeSoMa) S. For a FuNeSoMa R given by* $R = N \vee S \exists$ *a Pemutation FuNeSoMa (PeFuNeSoMa) P* $\ni T = \langle t_{ij}^T, t_{ij}^I, t_{ij}^F \rangle = \langle p_{ij}^T, p_{ij}^I, p_{ij}^F \rangle \times \langle r_{ij}^T, r_{ij}^I, r_{ij}^F \rangle \times \langle p_{ji}^T, p_{ji}^I, p_{ji}^F \rangle$ *satisfies* $\langle t_{ij}^T, t_{ij}^I, t_{ij}^F \rangle \geq \langle t_{ji}^T, t_{ji}^I, t_{ji}^F \rangle$ *for* $i > j$.

Proof $\langle t^T_{ij}, t^I_{ij}, t^F_{ij}\rangle = \langle p^T_{ij}, p^I_{ij}, p^F_{ij}\rangle \times \langle r^T_{ij}, r^I_{ij}, r^F_{ij}\rangle \times \langle p^T_{ji}, p^I_{ji}, p^F_{ji}\rangle$
$= \langle p^T_{ij}, p^I_{ij}, p^F_{ij}\rangle \times (\langle n^T_{ij}, n^I_{ij}, n^F_{ij}\rangle \vee \langle s^T_{ij}, s^I_{ij}, s^F_{ij}\rangle) \times \langle p^T_{ji}, p^I_{ji}, p^F_{ji}\rangle$
$= ((p^T_{ij}, p^I_{ij}, p^F_{ij}\rangle \times \langle n^T_{ij}, n^I_{ij}, n^F_{ij}\rangle \times \langle p^T_{ji}, p^I_{ji}, p^F_{ji}\rangle) \vee (\langle p^T_{ij}, p^I_{ij}, p^F_{ij}\rangle \times \langle s^T_{ij}, s^I_{ij}, s^F_{ij}\rangle \times \langle p^T_{ji}, p^I_{ji}, p^F_{ji}\rangle)$.
Since N is NiFuNeSoMa, $((p^T_{ij}, p^I_{ij}, p^F_{ij}\rangle \times \langle n^T_{ij}, n^I_{ij}, n^F_{ij}\rangle \times \langle p^T_{ji}, p^I_{ji}, p^F_{ji}\rangle)$ becomes strictly lower triangler for some PeFuNeSoMa P.

Thus since $((p^T_{ij}, p^I_{ij}, p^F_{ij}\rangle \times \langle n^T_{ij}, n^I_{ij}, n^F_{ij}\rangle \times \langle p^T_{ji}, p^I_{ji}, p^F_{ji}\rangle)$ is symmetric,

T satisfies $\langle t^T_{ij}, t^I_{ij}, t^F_{ij}\rangle \geq \langle t^T_{ji}, t^I_{ji}, t^F_{ji}\rangle$ for $i > j$ by choosing such a PeFuNeSoMa P.

Remark 1 If $N = \langle n^T_{ij}, n^I_{ij}, n^F_{ij}\rangle$ is NiFuNeSoMa, then there exists a PeFuNeSoMa P such that $\langle p^T_{ij}, p^I_{ij}, p^F_{ij}\rangle \times \langle n^T_{ij}, n^I_{ij}, n^F_{ij}\rangle \times \langle p^T_{ji}, p^I_{ji}, p^F_{ji}\rangle$ is SL(U)T.

Remark 2 The NiFuNeSoMa, R has not less than a null row and atleast one null column.

Remark 3 If R is NiFuNeSoMa iff $\langle r^T_{ii}, r^I_{ii}, r^F_{ii}\rangle^{(k)} = \langle 0, 0, 1\rangle$, to a little $i \in I_n$(Index) and little $k \in I_n$, for $R^k := [\langle r^T_{ij}, r^I_{ij}, r^F_{ij}\rangle^{(k)}]$.

Theorem 2 *For any FuNeSoMa R,* $\langle r^T_{ij}, r^I_{ij}, r^F_{ij}\rangle = \Delta\langle r^T_{ij}, r^I_{ij}, r^F_{ij}\rangle \vee \nabla\langle r^T_{ij}, r^I_{ij}, r^F_{ij}\rangle$.

3 Controllable Fuzzy Neutrosophic Soft Matrices

Here we establish some basic properties of FuNeSoMas. In the ensuing discussion, we pact only with SqFuNeSoMas.

Proposition 1 *For a FuNeSoMa N. If* $\exists$ *a PeFuNeSoMa* $P \ni \langle p^T_{ij}, p^I_{ij}, p^F_{ij}\rangle \times \langle n^T_{ij}, n^I_{ij}, n^F_{ij}\rangle \times \langle p^T_{ji}, p^I_{ji}, p^F_{ji}\rangle$ *is SL(U)T, then N is a NiFuNeSoMa.*

Proof Let

$$S = \langle p^T_{ij}, p^I_{ij}, p^F_{ij}\rangle \times \langle n^T_{ij}, n^I_{ij}, n^F_{ij}\rangle \times \langle p^T_{ji}, p^I_{ji}, p^F_{ji}\rangle = \begin{bmatrix} \langle 0,0,1\rangle & \langle 0,0,1\rangle \\ & \ddots & \\ * & & \langle 0,0,1\rangle \end{bmatrix}$$

we can prove the direct multiplication that S^2 is also SL(U)T, and consequently $S^3, S^4, \ldots$ all powers of S. All diagonals are zero in $S, S^2, S^3, \ldots,$ so by Remark 3, S is nilpotent. As $\langle p^T_{ij}, p^I_{ij}, p^F_{ij}\rangle \times \langle p^T_{ji}, p^I_{ji}, p^F_{ji}\rangle = \langle p^T_{ji}, p^I_{ji}, p^F_{ji}\rangle \times \langle p^T_{ij}, p^I_{ij}, p^F_{ij}\rangle = \langle 1, 1, 0\rangle$, then multiplying S on the left by $\langle p^T_{ij}, p^I_{ij}, p^F_{ij}\rangle$, we get $\langle p^T_{ji}, p^I_{ji}, p^F_{ji}\rangle \times \langle s^T_{ij}, s^I_{ij}, s^F_{ij}\rangle \times \langle p^T_{ij}, p^I_{ij}, p^F_{ij}\rangle = \langle n^T_{ij}, n^I_{ij}, n^F_{ij}\rangle$, so we find the N^n,

that is $\langle n^T_{ij}, n^I_{ij}, n^F_{ij}\rangle^n = \langle p^T_{ji}, p^I_{ji}, p^F_{ji}\rangle \times \langle s^T_{ij}, s^I_{ij}, s^F_{ij}\rangle \times \langle p^T_{ij}, p^I_{ij}, p^F_{ij}\rangle \times \langle p^T_{ji}, p^I_{ji}, p^F_{ji}\rangle \times \langle s^T_{ij}, s^I_{ij}, s^F_{ij}\rangle \times \langle p^T_{ij}, p^I_{ij}, p^F_{ij}\rangle \times \ldots \times \langle p^T_{ji}, p^I_{ji}, p^F_{ji}\rangle \times \langle s^T_{ij}, s^I_{ij}, s^F_{ij}\rangle \times \langle p^T_{ij}, p^I_{ij}, p^F_{ij}\rangle$
$= \langle p^T_{ji}, p^I_{ji}, p^F_{ji}\rangle \times \langle s^T_{ij}, s^I_{ij}, s^F_{ij}\rangle^n \times \langle p^T_{ij}, p^I_{ij}, p^F_{ij}\rangle$
$= \langle p^T_{ji}, p^I_{ji}, p^F_{ji}\rangle \times \langle 0, 0, 1\rangle \times \langle p^T_{ij}, p^I_{ij}, p^F_{ij}\rangle = \langle 0, 0, 1\rangle$.

Theorem 3 *A FuNeSoMa N is nilpotent iff* $\exists$ *a PeFuNeSoMa* $\langle p^T_{ij}, p^I_{ij}, p^F_{ij}\rangle \ni \langle p^T_{ij}, p^I_{ij}, p^F_{ij}\rangle \times \langle n^T_{ij}, n^I_{ij}, n^F_{ij}\rangle \times \langle p^T_{ji}, p^I_{ji}, p^F_{ji}\rangle$ *is SL(U)T.*

Note 1 Let $\langle r^T_{ij}, r^I_{ij}, r^F_{ij}\rangle$ be a FuNeSoMa, $\langle p^T_{ij}, p^I_{ij}, p^F_{ij}\rangle$ is PeFuNeSoMa. Let $T = \langle t^T_{ij}, t^I_{ij}, t^F_{ij}\rangle = \langle p^T_{ij}, p^I_{ij}, p^F_{ij}\rangle \times \langle r^T_{ij}, r^I_{ij}, r^F_{ij}\rangle \times \langle p^T_{ji}, p^I_{ji}, p^F_{ji}\rangle$. The element which lies in the $(i, j)^{th}$ entry of R lies next in the $(h, k)^{th}$ of T iff $\langle p^T_{hi}, p^I_{hi}, p^F_{hi}\rangle = \langle p^T_{kj}, p^I_{kj}, p^F_{kj}\rangle = \langle 1, 1, 0\rangle$.

Theorem 4 *Let* $R = \langle r^T_{ij}, r^I_{ij}, r^F_{ij}\rangle$ *be a FuNeSoMa,* $P = \langle p^T_{ij}, p^I_{ij}, p^F_{ij}\rangle$ *is PeFuNeSoMa. Then*

$$\langle p^T_{ij}, p^I_{ij}, p^F_{ij}\rangle \times (\Delta\langle r^T_{ij}, r^I_{ij}, r^F_{ij}\rangle) \times \langle p^T_{ji}, p^I_{ji}, p^F_{ji}\rangle = \Delta(\langle p^T_{ij}, p^I_{ij}, p^F_{ij}\rangle \times \langle r^T_{ij}, r^I_{ij}, r^F_{ij}\rangle) \times \langle p^T_{ji}, p^I_{ji}, p^F_{ji}\rangle, \quad (1)$$

$$\langle p^T_{ij}, p^I_{ij}, p^F_{ij}\rangle \times (\nabla\langle r^T_{ij}, r^I_{ij}, r^F_{ij}\rangle) \times \langle p^T_{ji}, p^I_{ji}, p^F_{ji}\rangle = \nabla(\langle p^T_{ij}, p^I_{ij}, p^F_{ij}\rangle \times \langle r^T_{ij}, r^I_{ij}, r^F_{ij}\rangle \times \langle p^T_{ji}, p^I_{ji}, p^F_{ji}\rangle). \quad (2)$$

Definition 1 We say a FuNeSoMa R is controllabel from belove (above), if $\exists$ a PeFuNeSoMa $P \ni \langle t^T_{ij}, t^I_{ij}, t^F_{ij}\rangle = \langle p^T_{ij}, p^I_{ij}, p^F_{ij}\rangle \times \langle r^T_{ij}, r^I_{ij}, r^F_{ij}\rangle \times \langle p^T_{ji}, p^I_{ji}, p^F_{ji}\rangle$ satisfies $\langle t^T_{ij}, t^I_{ij}, t^F_{ij}\rangle \geq \langle t^T_{ji}, t^I_{ji}, t^F_{ji}\rangle$ $(\langle t^T_{ij}, t^I_{ij}, t^F_{ij}\rangle \leq \langle t^T_{ji}, t^I_{ji}, t^F_{ji}\rangle)$ a long as $i > j$. A FuNeSoMa $R = \langle r^T_{ij}, r^I_{ij}, r^F_{ij}\rangle$ is said to be controlled from below (above), if $\langle r^T_{ij}, r^I_{ij}, r^F_{ij}\rangle \geq \langle r^T_{ji}, r^I_{ji}, r^F_{ji}\rangle(\langle r^T_{ij}, r^I_{ij}, r^F_{ij}\rangle \leq \langle r^T_{ji}, r^I_{ji}, r^F_{ji}\rangle)$as long $i > j$.

Theorem 5 *The next statements are analogous:*

- *(1)* $\langle r^T_{ji}, r^I_{ji}, r^F_{ji}\rangle$ *is CFB(A).*
- *(2) There exists a PeFuNeSoMa* $\langle p^T_{ij}, p^I_{ij}, p^F_{ij}\rangle$ *such that* $\Delta(\langle p^T_{ij}, p^I_{ij}, p^F_{ij}\rangle \times \langle r^T_{ij}, r^I_{ij}, r^F_{ij}\rangle \times \langle p^T_{ji}, p^I_{ji}, p^F_{ji}\rangle)$ *is SL(U)T.*
- *(3) There exists a PeFuNeSoMa* $\langle p^T_{ij}, p^I_{ij}, p^F_{ij}\rangle$ *such that* $\langle p^T_{ij}, p^I_{ij}, p^F_{ij}\rangle \times (\Delta\langle r^T_{ij}, r^I_{ij}, r^F_{ij}\rangle) \times \langle p^T_{ji}, p^I_{ji}, p^F_{ji}\rangle)$ *is SL(U)T.*
- *(4)*$\Delta\langle r^T_{ij}, r^I_{ij}, r^F_{ij}\rangle$

Corollary 1 $\langle r^T_{ij}, r^I_{ij}, r^F_{ij}\rangle$ *is CFB iff* $\langle r^T_{ij}, r^I_{ij}, r^F_{ij}\rangle$ *is CFA.*

Note 2 Let R and S be FuNeSoMas, and P is a PeFuNeSoM. Then $R\Psi S$ iff $\langle p_{ij}^T, p_{ij}^I, p_{ij}^F\rangle \times \langle r_{ij}^T, r_{ij}^I, r_{ij}^F\rangle \times \langle p_{ji}^T, p_{ji}^I, p_{ji}^F\rangle \Psi \langle p_{ij}^T, p_{ij}^I, p_{ij}^F\rangle \times \langle s_{ij}^T, s_{ij}^I, s_{ij}^F\rangle \times \langle p_{ji}^T, p_{ji}^I, p_{ji}^F\rangle$.

Proof: The proof is obvious.

Theorem 6 *Let R, S be a FuNeSoMas, and $\Delta R\Psi\Delta S$. If S is controllable, then R is controllable.*

4 Reduction of Controllable Matrix to Canonical Form

Lemma 1 *Let $R = (\langle r_{ij}^T, r_{ij}^I, r_{ij}^F\rangle)$ and $S = (\langle s_{ij}^T, s_{ij}^I, s_{ij}^F\rangle)$ be $n \times n$ FuNeSoMa of the form*

$$R = \begin{bmatrix} \langle 0,0,1\rangle \,\vdots\, \langle 0,0,1\rangle \cdots \langle 0,0,1\rangle \\ \cdots\cdots\cdots\cdots\cdots \\ \alpha \quad\vdots\quad R_1 \end{bmatrix},$$

$$S = \begin{bmatrix} \langle 0,0,1\rangle \,\vdots\, \langle 0,0,1\rangle \cdots \langle 0,0,1\rangle \\ \cdots\cdots\cdots\cdots\cdots \\ \beta \quad\vdots\quad S_1 \end{bmatrix},$$

where α and β are $(n-1) \times 1$ FuNeSoMa and R_1 and S_1 are FuNeSoMa of order (n-1). Then

(i)

$$R \times S = \begin{bmatrix} \langle 0,0,1\rangle \,\vdots\, \langle 0,0,1\rangle \cdots \langle 0,0,1\rangle \\ \cdots\cdots\cdots\cdots\cdots \\ R_1 \times \beta \,\vdots\, R_1 \times s_1 \end{bmatrix}$$

(ii)

$$R^n = \begin{bmatrix} \langle 0,0,1\rangle \,\vdots\, \langle 0,0,1\rangle \cdots \langle 0,0,1\rangle \\ \cdots\cdots\cdots\cdots\cdots \\ R_1^{n-1} \times \alpha \,\vdots\, R_1^n \end{bmatrix}$$

(iii) R is NiFuNeSoMa iff R_1 is nilpotent.

Remark 4 Let $R = (\langle r_{ij}^T, r_{ij}^I, r_{ij}^F \rangle) \in FuNeSoMa_n$, and R have no less than one $\langle 0, 0, 1\rangle$ row (say, the ith row). Let i^{th} row $\rightarrow$ I-row and vice versa and do the same for column; then we have

$$R^* = \begin{bmatrix} \langle 0,0,1\rangle & \vdots & \langle 0,0,1\rangle \cdots \langle 0,0,1\rangle \\ & \cdots\cdots\cdots & \\ * & \vdots & \\ \vdots & \vdots & \\ * & \vdots & R_1 \end{bmatrix}$$

Lemma 2 *R is NiFuNeSoMa iff R_1 is NiFuNeSoMa. By Lemma 1, we have the following for NiFuNeSoMa R.*

Algorithm 1 **Step 1.** Check R, for a null row and null column; if anyone is missing, then R fails to be nilpotent. End.
Check R for both zero row and zero column. If not, then R is not nilpotent. If R has both conditions, then do interchange as mentioned; then, we have

$$\langle p_{ij}^T, p_{ij}^I, p_{ij}^F\rangle_1 \times R \times \langle p_{ji}^T, p_{ji}^I, p_{ji}^F\rangle_1 = \begin{bmatrix} \langle 0,0,1\rangle & \vdots & \langle 0,0,1\rangle \cdots \langle 0,0,1\rangle \\ & \cdots\cdots\cdots & \\ * & \vdots & R_1 \end{bmatrix}$$

where $\langle p_{ij}^T, p_{ij}^I, p_{ij}^F\rangle_1 = \langle p_{ij}^T, p_{ij}^I, p_{ij}^F\rangle(1, i_1)$. Next step
Step 2. Check R for both zero row and zero column. If not then R_1 is not nilpotent, Stop.
If R_1 satisfies desired conditions, i.e., in R_1, the i_2^th row is a null row. The new form $\langle p_{ij}^T, p_{ij}^I, p_{ij}^F\rangle_2 \times \langle p_{ij}^T, p_{ij}^I, p_{ij}^F\rangle_1 \times R \times \langle p_{ji}^T, p_{ji}^I, p_{ji}^F\rangle_1 \times \langle p_{ji}^T, p_{ji}^I, p_{ji}^F\rangle_2$ from FuNeSoMa $\langle p_{ij}^T, p_{ij}^I, p_{ij}^F\rangle_1 \times R \times \langle p_{ji}^T, p_{ji}^I, p_{ji}^F\rangle_1$ by interchanging the $(i_2 + 1)$-th row with II row and $(i_2 + 1)$-th column with the II column such that

$$\langle p_{ij}^T, p_{ij}^I, p_{ij}^F\rangle_2 \times \langle p_{ij}^T, p_{ij}^I, p_{ij}^F\rangle_1 \times R \times \langle p_{ji}^T, p_{ji}^I, p_{ji}^F\rangle_1 \times \langle p_{ji}^T, p_{ji}^I, p_{ji}^F\rangle_2$$

$$= \begin{bmatrix} \langle 0,0,1\rangle & \langle 0,0,1\rangle & \vdots & \langle 0,0,1\rangle \ldots \langle 0,0,1\rangle \\ * & \langle 0,0,1\rangle & \vdots & \langle 0,0,1\rangle \ldots \langle 0,0,1\rangle \\ & & \cdots\cdots\cdots & \\ * & * & \vdots & \\ \vdots & \vdots & \vdots & R_2 \\ * & * & \vdots & \end{bmatrix}$$

where $\langle p_{ij}^T, p_{ij}^I, p_{ij}^F\rangle_2 = \langle p_{ij}^T, p_{ij}^I, p_{ij}^F\rangle(2, i_2 + 1)$. Next step.

Step 3. Check R_2, for a null row and null column. If not, R_2 is not nilpotent; thus, Lemma 2 implies R_1 and R are not nilpotent, stop.
Else if in R_1, the i_3^{th} row of R_2 is null, the new FuNeSoMa of the form $\langle p_{ij}^T, p_{ij}^I, p_{ij}^F\rangle_3 \times \langle p_{ij}^T, p_{ij}^I, p_{ij}^F\rangle_2 \times \langle p_{ij}^T, p_{ij}^I, p_{ij}^F\rangle_1 \times R \times \langle p_{ji}^T, p_{ji}^I, p_{ji}^F\rangle_1 \times \langle p_{ji}^T, p_{ji}^I, p_{ji}^F\rangle_2 \times \langle p_{ji}^T, p_{ji}^I, p_{ji}^F\rangle_3$ from matrix $\langle p_{ij}^T, p_{ij}^I, p_{ij}^F\rangle_2 \times \langle p_{ij}^T, p_{ij}^I, p_{ij}^F\rangle_1 \times R \times \langle p_{ji}^T, p_{ji}^I, p_{ji}^F\rangle_1 \times \langle p_{ji}^T, p_{ji}^I, p_{ji}^F\rangle_2$ then changes the (i_3+2)-th row with II row and (i_3+2)-th column with the II column $\ni$,

$$\langle p_{ij}^T, p_{ij}^I, p_{ij}^F\rangle_3 \times \langle p_{ij}^T, p_{ij}^I, p_{ij}^F\rangle_2 \times \langle p_{ij}^T, p_{ij}^I, p_{ij}^F\rangle_1 \times R \times \langle p_{ji}^T, p_{ji}^I, p_{ji}^F\rangle_1 \times \langle p_{ji}^T, p_{ji}^I, p_{ji}^F\rangle_2 \times \langle p_{ji}^T, p_{ji}^I, p_{ji}^F\rangle_3$$

$$= \left[\begin{array}{cccc:ccc} \langle 0,0,1\rangle & \langle 0,0,1\rangle & \langle 0,0,1\rangle & & \langle 0,0,1\rangle & \ldots & \langle 0,0,1\rangle \\ * & \langle 0,0,1\rangle & \langle 0,0,1\rangle & & \langle 0,0,1\rangle & \ldots & \langle 0,0,1\rangle \\ * & * & \langle 0,0,1\rangle & & \langle 0,0,1\rangle & \ldots & \langle 0,0,1\rangle \\ & \cdots & \cdots & \cdots & \cdots & \cdots & \\ * & * & * & & & & \\ \vdots & \vdots & \vdots & & & R_3 & \\ * & * & * & & & & \end{array}\right]$$

where $\langle p_{ij}^T, p_{ij}^I, p_{ij}^F\rangle_3 = \langle p_{ij}^T, p_{ij}^I, p_{ij}^F\rangle(3, i_3+2)$. Next step.

Continuing like this, finally we get

$$\langle p_{ij}^T, p_{ij}^I, p_{ij}^F\rangle_n \times \langle p_{ij}^T, p_{ij}^I, p_{ij}^F\rangle_{n-1} \times \ldots \times \langle p_{ij}^T, p_{ij}^I, p_{ij}^F\rangle_2 \times \langle p_{ij}^T, p_{ij}^I, p_{ij}^F\rangle_1 \times R \times \langle p_{ji}^T, p_{ji}^I, p_{ji}^F\rangle_1 \times \langle p_{ji}^T, p_{ji}^I, p_{ji}^F\rangle_2 \times \ldots \times P_{n-1} \times \langle p_{ji}^T, p_{ji}^I, p_{ji}^F\rangle_n =$$

$$\left[\begin{array}{cccc:ccc} \langle 0,0,1\rangle & \langle 0,0,1\rangle & \langle 0,0,1\rangle & & \langle 0,0,1\rangle & \ldots & \langle 0,0,1\rangle \\ * & \langle 0,0,1\rangle & \langle 0,0,1\rangle & & \langle 0,0,1\rangle & \ldots & \langle 0,0,1\rangle \\ * & * & \langle 0,0,1\rangle & & \langle 0,0,1\rangle & \ldots & \langle 0,0,1\rangle \\ & \cdots & \cdots & \cdots & \cdots & \cdots & \\ * & * & * & & & & \\ \vdots & \vdots & \vdots & & & R_n & \\ * & * & * & & & & \end{array}\right]$$

where $\langle p_{ij}^T, p_{ij}^I, p_{ij}^F\rangle_m = \langle p_{ij}^T, p_{ij}^I, p_{ij}^F\rangle(m, i_m+m-1)$, $m \in I_n$.

In the event if R_n fails to satisfy both the condition, then R is not nilpotent.
Else,

$$R_n = \begin{bmatrix} \langle 0,0,1\rangle & \langle 0,0,1\rangle \\ & \ddots & \\ * & & \langle 0,0,1\rangle \end{bmatrix}$$

then by Lemma 2, R is nilpotent by the sequence of actions. Then, $\langle p^T_{ij}, p^I_{ij}, p^F_{ij}\rangle = \langle p^T_{ij}, p^I_{ij}, p^F_{ij}\rangle_t \times \langle p^T_{ij}, p^I_{ij}, p^F_{ij}\rangle_{t-1} \times \ldots \times \langle p^T_{ij}, p^I_{ij}, p^F_{ij}\rangle_2 \times \langle p^T_{ij}, p^I_{ij}, p^F_{ij}\rangle_1$.
We obtain,

$$\langle p^T_{ij}, p^I_{ij}, p^F_{ij}\rangle \times R \times \langle p^T_{ji}, p^I_{ji}, p^F_{ji}\rangle = \begin{bmatrix} \langle 0,0,1\rangle & \langle 0,0,1\rangle \\ & \ddots & \\ * & & \langle 0,0,1\rangle \end{bmatrix}$$

which indeed SLT. Algorithm to curtail CoFuNeSoMa to canonical form.

Algorithm 2 Step 1. By Algorithm 1, we can check if ΔR is nilpotent or not. Thus R is CoFuNeSoMa or not by Theorem 5.
Step 2. If R is CoFuNeSoMa, then by Step I, we get a permutation matrix P, i.e., $\langle p^T_{ij}, p^I_{ij}, p^F_{ij}\rangle \times (\Delta R) \times \langle p^T_{ji}, p^I_{ji}, p^F_{ji}\rangle$, which is SLT. So $\langle p^T_{ij}, p^I_{ij}, p^F_{ij}\rangle \times R \times \langle p^T_{ji}, p^I_{ji}, p^F_{ji}\rangle$ is canonical form of R. Stop.

5 Conclusion

In this article the controllable fuzzy neutrosophic soft matrix is defined. Further, various properties of nilpotent and controllable fuzzy neutrosophic soft matrices are showed. We have developed an algorithm for controllable and nilpotent fuzzy neutrosophic soft matrices and reduced a controllable fuzzy neutrosophic soft matrix to canonical form.

References

1. Arockiarani, I., Sumathi, I.R.: A fuzzy neutrosophic soft matrix approach in decision making. JGRMA **2**(2), 14–23 (2014)
2. Atanassov, K.: Intuitionistic fuzzy fets. Fuzzy Sets and System, **20**, 87–96 (1986)
3. Broumi, S., Sahin, R., Smarandache, F.: Generalized interval neutrosophic soft set and its decision making problem. J. New Results Sci. **7**, 29–47 (2014). https://doi.org/10.5281/zenodo.49000
4. Kavitha, M., Murugadas, P.: Eigenspace of a circulant fuzzy neutrosophic soft matrix. Neutrosophic Sets Syst. **50**, 591–601 (2022)

5. Kavitha, M., Murugadas, P., Sriram. S.: Minimal solution of fuzzy neutrosophic soft matrix. J. Linear Topol. Algebra **6**, 171–189 (2017)
6. Kavitha, M., Murugadas, P., Sriram, S.: On the power of fuzzy neutrosophic soft matrix. J. Linear Topol. Algebra **7**, 133–147 (2018)
7. Kavitha, M., Murugadas, P., Sriram, S.: Priodicity of interval fuzzy neutrosophic soft matrices. Adv. Math. Sci. J. **9**, 1661–1670 (2020). https://doi.org/10.37418/amsj.9.4.22
8. Kim, K.H., Roush,F.W.: Idempotent fuzzy matrices. Seminar Report, Alabama State Univ., Montgomery, AL (1981)
9. Maji, P.K.: A neutrosophic soft set approach to a decision making problem. Ann. Fuzzy Math. Inf. **3**(2), 313–319 (2012)
10. Maji, P.K., Biswas. R., Ray. R.: Fuzzy soft set. J. Fuzzy Math. **9**, 589–602 (2001)
11. Maji, P.K., Roy, A.R., Biswas. R.: On intuitionistic fuzzy soft set. Int. J. Fuzzy Math. **12**(3), 669–684 (2004). https://doi.org/10.3233/IFS-2012-0601
12. Molodtsov, D.: Soft set theory first results. Comput. Math. Appl. **37**(4–5), 19–31 (1999). https://doi.org/10.1016/S0898-1221(99)00056-5
13. Murugadas, P., Kavitha, M., Sriram, S.: Monotone interval fuzzy neutrosophic soft eigenproblem. Malaya J. Matematik **1**, 342–350 (2019). https://doi.org/10.26637/MJM0S01/0062
14. Murugadas, P., Kavitha, M.: Solvability of system of neutrosophic soft linear equations. Neutrosophic Sets Syst. **40**, 254–269 (2021)
15. Padder, R.A., Murugadas, P.: Max-max operation on intuitionistic fuzzy matrix. Ann. Fuzzy Math. Inf. Article in Press
16. Smarandache, F.: Neutrosophic set, a generalization of the intuitionistic fuzzy set. Int. J. Pure Appl. Math. **24**(3), 287–297 (2005)
17. Uma, R., Sriram, S., Murugadas, P.: Fuzzy neutrosophic soft matrices of type-I and type-II. Fuzzy Inf. Eng. **13**(2) (2021). https://doi.org/10.1080/16168658.2021.1923621
18. Xin, L.J.: Convergence of powers of controllable fuzzy matrices. Fuzzy Sets Syst. **45**, 83–88 (1994). https://doi.org/10.21917/ijsc.2016.0184
19. Zadeh, L.A.: Fuzzy sets. J. Inf. Control **8**(3), 338–353 (1965)

An MCDM Based on Neutrosophic Fuzzy SAW Method for New Entrepreneurs in Organic Farming

T. Porchudar, A. Merceline Anita, M. Arul Jeya Shalini, and J. Jeba Jesintha

Keywords Neutrosophic Fuzzy SAW method · Linguistic variables · Neutrosophic fuzzy numbers · Organic farming

1 Introduction

Multi-criteria decision-making(MCDM) is a significant approach to make a decision in accordance with the preference on the people who decide things. At every step in our lives, we make decisions to perform all our actions. The theory of decision-making is just an attempt at codification of our decision-making process in a mathematically tractable form [6]. During decision-making in some cases, there is no method for assessing the data clearly; it can be easily evaluated in terms of linguistic variables. The idea of fuzzy sets, which can be used to improve these kind of situations [11], was first introduced by Prof. Lotfi A. Zadeh of University of California in 1965. MCDM is a concept that enables us to select the most appropriate alternative by evaluating them in terms of many criteria. A widely used and popularly known MCDM is simple additive weighting method, which calculates the weight values for ranking each alternatives over specific criteria.

Since India is an Agrarian country, the future of small-scale farming lies in the hands of youth as they are considered as potential future farmers. In the recent years, we see that the interest in organic farming is booming among youngsters. Many IT professionals and other young professionals are leaving their lucrative jobs to start organic farming. Though they are interested, most of them were drawn into the world of organic farming without knowing the ground reality, and this paves the way to face so many difficulties during the initial times. Hence, this problem is taken into account and implemented with NFSAW method. The paper is structured as follows:

T. Porchudar · A. Merceline Anita (✉)
Department of Mathematics, Sacred Heart College, Tirupattur, Tamil Nadu, India

M. Arul Jeya Shalini · J. Jeba Jesintha
Department of Mathematics, Women's Christian College, Chennai, Tamil Nadu, India

H.-H. Leung et al. (eds.), *Recent Developments in Algebra and Analysis*, Trends in Mathematics, https://doi.org/10.1007/978-3-031-37538-5_7

Introduction and historical background of NFSAW method are dealt first, followed by basic definitions. An algorithm is presented for our case study with comparative analysis, and finally, it's concluded with the arrived results.

2 Literature Review

Churchman and Ackoff introduced the method called simple additive weighting (SAW) in 1945, and they utilized SAW method to select the portfolio problem. In 2005, Modarres and Sadi Nezhad published a paper using preference ratio in fuzzy simple additive weighting (FSAW) method. In 2013, Prabhjot Kaur and Sanjay Kumar developed fuzzy SAW method into Intuitionistic fuzzy SAW method for vendor selection [5]. In 2013, Manish Sagar, Jayaswal, and Kushwah explored fuzzy SAW method for maintenance strategy selection [8]. In 2016, E. Roszkowskaa and D. Kacprzakb extended the linguistic FSAW and linguistic fuzzy analytical hierarchy process (FTOPSIS) techniques [7].

In 2016, Budi Praseliyo, Niswah Baroroh, and Dwi Rufiyanti utilized this fuzzy SAW Method for taking decisions in human resource recruitment [3]. In 2018, Wini Waziana, Rita Irviani, Oktaviani Satria, and Adino Kurniawan have utilized this fuzzy simple additive weighting for helping out farmers and to determine the recipients in their breeding farm [10]. Further SAW method has also been applied in interval valued neutrosophic set for the selection of insurance options [2] and in the selection of achieving students in faculty level [4]. In 2019, this fuzzy SAW method was extended to neutrosophic fuzzy SAW method by D. Ajay and J. Aldring, who proposed an application for this neutrosophic fuzzy SAW method [1]. In 2020, Nguyen Tho Thong has extended TOPSIS method, and it has evolved in dynamic neutrosophic environment [9]. Earlier, there was no study on using neutrosophic sets to solve MCDM problems. Most recently, multiple scholarly approaches are emerging in MCDM using neutrosophic sets, which mainly deal with neutrality.

3 Basic Concepts

This chapter elaborates some of the fundamental ideas behind the fuzzy set and the neutrosophic set.

Definition 3.1 (Fuzzy Membership [1]) Let χ be the universal set. The membership function μ_A by which a fuzzy set A is usually defined is in the form $\mu_A : \chi \to [0, 1]$; the values obtained are called the membership values.

Definition 3.2 (Linguistic Variables [1]) In fuzzy logic, a linguistic variable is a variable whose values are phrases in either natural or in artificial language.

Definition 3.3 (Triangular Fuzzy Number [1]) Triangular fuzzy number is a triplet $A = \{x : (u, v, w)\}$, where the smallest likely value is "u," the most probable value is "v," and the largest possible value is "w" of any fuzzy event.

$$\mu_A(x) = \begin{cases} 0, & x \le u \\ \frac{x-u}{v-u}, & u < x \le v \\ \frac{w-x}{w-v}, & v < x < w \\ 0, & x \ge w \end{cases}$$

Definition 3.4 (Neutrosophic Set [1]) Let U be the universe of discourse and C be a subset of U. Each element $b \in U$ has degree of true indeterminacy and false membership in C. The neutrosophic set is $C_{NS} = \{< b, T_C(b), F_C(b), I_C(b) > : b \in \mathrm{X}\}$ where $T_C(b)$, $I_C(b)$, and $F_C(b)$ represent the degree of truth, indeterminacy, and falsity membership functions, respectively, which take their values in the unit closed interval. We have no restriction on the sum of $T_C(b)$, $F_C(b)$, and $I_C(b)$. It satisfies the following relation: $0 \le T_C(b) + F_C(b) + I_C(b) \le 3$.

4 Algorithm

Step 1 Determine the criteria C_j from a group of experts X_k for the decision-making problem.

Step 2 Select the relevant truth, false, and indeterminacy membership rating values of each criterion in terms of the linguistic variables by the experts.

Step 3 Fuzzify the linguistic variable of each criterion in terms of fuzzy triangular number.

Step 4 Find the average fuzzy scores L^i_j of triangular fuzzy numbers $(l^1_1, m^2_1, n^3_1), (l^1_2, m^2_2, n^3_2) \ldots, (l^1_j, m^2_j, n^3_j)$ defuzzified values, and normalized weight w_j for each criterion.

1. Average fuzzy scores $L^i_j = \frac{(l^i_1 + l^i_2 + \ldots + l^i_i)}{j}$, where $i = 1, 2, 3$.
2. Defuzzified value $(e) = \frac{(l+m+n)}{3}$, where $l = L^1_j, m = L^2_j, n = L^3_j$
3. Normalized values (w) = Defuzzified value of the criteria/sum of all defuzzified values.

Step 5: Find the centroid weight value $W_j = (\alpha + 2\beta + \gamma)/4$, where α, β, γ are normalized weighted values of truth, indeterminacy, and false membership function, respectively.

Step 6 Assign the applicable neutrosophic rating values (truth, false, and indeterminacy membership values) for each alternative A_i over a criteria C_j as linguistic variables by experts' opinion.

Step 7 Repeat step 4, and find the average fuzzy score and the defuzzified score of each alternative on criteria.

Step 8 Form normalized decision matrix for truth, indeterminacy, and false membership function, corresponding to each alternative over all criteria.

Step 9 Evaluate $N_{ij} = \frac{p_{ij}+\lambda q_{ij}+(1-\lambda)r_{ij}}{2}$, where p is the normalized truth membership value, q is the normalized indeterminacy membership function value, and r is the normalized false membership function value.

Step 10 Find the combined normalized neutrosophic decision matrix.

Step 11 Calculate the total scores of each alternative using $TS = N_{ij} * W_j$. Finally, the highest score is chosen as the ideal alternative.

5 Case Study

This chapter speaks about a real-world issue in the agricultural sector, which is applied to the NFSAW method. We choose five experts $(X_1, X_2, X_3, X_4, X_5)$ to analyze the best alternative for youngsters who urge to be successful entrepreneurs in organic farming. Five entrepreneur jobs in organic farming are chosen as alternatives. The following alternatives are A_1, beekeeping; A_2, community farming; A_3, integrated farming; A_4, organic store; and A_5, millet mill. Seven criteria are classified and taken as basic characteristics to be evaluated and checked before an entrepreneur starts with his/her idea. The list of criteria taken by an entrepreneur in organic farming includes C_1, initial investment rate; C_2, labor/manual support; C_3, awareness/training; C_4, youngster's interest rate; C_5, market demand rate; C_6, profit rate; and C_7, additional income rate. The truth membership (T_m) rating values to each criteria assigned by the experts (Table 1) in terms of the linguistic variables.

Transforming the linguistic variables of truth membership rating values in terms of fuzzy triangular number is shown in Table 2, similarly, transforming the linguistic variables of false and indeterminacy membership [1] rating values in terms of fuzzy triangular number.

Table 1 Truth membership rating values

Criteria/experts	X_1	X_2	X_3	X_4	X_5
Initial investment C_1	H	H	H	MH	ML
Manual support C_2	H	M	H	ML	L
Awareness and training C_3	H	VH	VH	H	VL
Youngsters' interest rate C_4	MH	VH	VH	VH	MH
Market demand rate C_5	M	VH	M	VH	H
Profit rate C_6	H	ML	MH	H	H
Additional income rate C_7	H	VH	H	VH	M

Table 2 Truth membership T_m with triangular fuzzy number

	X_1	X_2	X_3	X_4	X_5
C_1	(0.7,0.9,1)	(0.7,0.9,1)	(0.7,0.9,1)	(0.5,0.7,0.9)	(0.1,0.3,0.5)
C_2	(0.7,0.9,1)	(0.3,0.5,0.7)	(0.7,0.9,1)	(0.1,0.3,0.5)	(0,0.1,0.3)
C_3	(0.7,0.9,1.0)	(0.9,1.0,1.0)	(0.9,1.0,1.0)	(0.7,0.9,1.0)	(0.5,0.7,0.9)
C_4	(0.5,0.7,0.9)	(0.9,1.0,1.0)	(0.9,1.0,1.0)	(0.9,1.0,1.0)	(0.5,0.7,0.9)
C_5	(0.7,0.9,1.0)	(0.9,1.0,1.0)	(0.3,0.5,0.7)	(0.9,1.0,1.0)	(0.7,0.9,1.0)
C_6	(0.7,0.9,1.0)	(0.1,0.3,0.5)	(0.5,0.7,0.9)	(0.7,0.9,1.0)	(0.7,0.9,1.0)
C_7	(0.7,0.9,1.0)	(0.9,1.0,1.0)	(0.7,0.9,1.0)	(0.9,1.0,1.0)	(0.3,0.5,0.7)

Table 3 Normalized weight value for T_m

Criteria	Average fuzzy score	Defuzzified value	Normalized weight T_m
C_1	(0.54,0.74,0.88)	0.72	0.1433
C_2	(0.36,0.54,0.7)	0.533	0.1061
C_3	(0.74,0.9,0.98)	0.524	0.1043
C_4	(0.74,0.88,0.96)	0.86	0.1712
C_5	(0.7,0.86,0.94)	0.833	0.1658
C_6	(0.54,0.74,0.88)	0.72	0.1433
C_7	(0.7,0.86,0.94)	0.833	0.1658

Table 4 Normalized weight value for I_M

Criteria	Average fuzzy score	Defuzzified value	Normalized weight I_M
C_1	(0.02,0.08,0.22)	0.1066	0.0981
C_2	(0.12,0.2,0.34)	0.22	0.2025
C_3	(0.02,0.08,0.22)	0.1066	0.0981
C_4	(0.02,0.08,0.22)	0.1066	0.0981
C_5	(0.02,0.08,0.22)	0.1066	0.0981
C_6	(0.16,0.24,0.38)	0.26	0.2393
C_7	(0.04,0.16,0.34)	0.18	0.1685

We calculate the average fuzzy score L^i_j, defuzzified values (e), and normalized weighted values (w) for truth membership function of criteria in Table 3 using step 4.

Similarly, continuing the same process for indeterminacy (I_M) in Table 4 and false membership (F_M) rating values in Table 5 to get normalized weight values (w).

Now, the centroid weighted value W_j for all the criteria is calculated.

$$W_j = \frac{(\alpha + 2\beta + \gamma)}{4}, \ W_1 = \frac{0.1433 + 2(0.0981) + 0.1455}{4} = 0.121.$$

Similarly, the remaining values are calculated. Next, assign the applicable neutrosophic rating values by experts (truth T_M, false F_M, and indeterminacy I_M

Table 5 Normalized weight value for F_M

Criteria	Average fuzzy score	Defuzzified value	Normalized weight (I_M)
C_1	(0.08,0.18,0.34)	0.2	0.1455
C_2	(0.16,0.26,0.4)	0.2733	0.1988
C_3	(0.14,0.26,0.084)	0.1613	0.1173
C_4	(0.06,0.1,0.22)	0.1266	0.0921
C_5	(0.04,0.14,0.3)	0.16	0.1164
C_6	(0.2,0.3,0.44)	0.3133	0.2279
C_7	(0.04,0.12,0.26)	0.14	0.1018

Table 6 Normalized decision matrix T_m

T_m	C_1	C_2	C_3	C_4	C_5	C_6	C_7
A_1	0.8400	0.8319	0.9918	1	1	0.8906	0.6093
A_2	0.9199	0.7999	0.9674	0.8434	0.8162	1	0.9375
A_3	0.9360	0.9519	0.8129	0.7912	0.8529	0.9218	1
A_4	1	1	1	0.9392	0.8235	0.8828	0.9140
A_5	0.8479	0.6879	0.8129	0.6347	0.7500	0.8983	0.7421

Table 7 Normalized decision matrix I_m

I_m	C_1	C_2	C_3	C_4	C_5	C_6	C_7
A_1	0.9399	0.5745	0.2727	0.6153	0.4893	1	0.7885
A_2	0.7200	0.8723	0.5818	1	0.6594	0.6001	0.5383
A_3	0.6001	0.6808	0.5272	0.8461	0.4679	0.2907	1
A_4	0.6001	1	1	0.6407	1	0.5455	0.4229
A_5	1	0.8298	0.6001	0.4869	0.5745	0.7092	0.9422

Table 8 Normalized decision matrix F_m

F_m	C_1	C_2	C_3	C_4	C_5	C_6	C_7
A_1	0.7092	0.4679	0.5	0.3876	0.8937	0.6001	0.6094
A_2	0.6363	0.8084	0.5756	1	0.5955	1	0.6563
A_3	1	0.8084	1	0.6123	0.7660	0.9798	0.8593
A_4	0.8363	0.8723	0.5756	0.7755	1	0.8199	0.3436
A_5	0.9537	1	0.5454	0.7550	0.9361	0.8799	1

membership function) to each alternative A_j on criteria C_j as linguistic variables and then transforming [1] linguistic variables to triangular fuzzy number. Using step 4, average fuzzy score and defuzzified values are calculated, and we derive normalized decision matrices for truth membership, indeterminacy, and false membership, which are provided in Tables 6, 7, and 8, respectively.

Calculate the values of N_{ij} for $\lambda = 0.5$. We arrive N_{11} and N_{12} as follows.

$$N_{11} = \frac{0.8400 + (0.5)(0.9399) + (1 - 0.5)(0.7092)}{2} = 0.8325$$

Table 9 Combined normalized decision matrix

N_{ij}	C_1	C_2	C_3	C_4	C_5	C_6	C_7
A_1	0.8325	0.6765	0.6890	0.7507	0.8457	0.8453	0.6541
A_2	0.7990	0.8201	0.7730	0.9217	0.7963	0.90002	0.7674
A_3	0.8680	0.8482	0.7882	0.7602	0.7349	0.7785	0.9648
A_4	0.8591	0.9680	0.8939	0.8236	0.9117	0.7827	0.6486
A_5	0.9148	0.8014	0.6928	0.6278	0.7526	0.8464	0.8566

Table 10 Comparative ranking

Methods	Ranking order
NFSAW	$A_2 > A_4 > A_3 > A_5 > A_1$
TOPSIS	$A_2 > A_3 > A_1 > A_5 > A_4$
WASPAS	$A_2 > A_1 > A_3 > A_5 > A_4$
WSM	$A_2 > A_1 > A_5 > A_3 > A_4$
WPM	$A_2 > A_1 > A_3 > A_5 > A_4$

$$N_{12} = \frac{0.8319 + (0.5)(0.5745) + (1 - 0.5)(0.4679)}{2} = 0.6765$$

Similarly, the remaining values are calculated and used to form the Combined Normalized Neutrosophic Decision Matrix, which is tabulated in Table 9.

Finally, the total score of each alternative is obtained by $N_{ij} * W_j$.

$$A_1 = (0.8325 \times 0.121) + (0.6765 \times 0.177) + (0.6890 \times 0.104) +$$
$$(0.7507 \times 0.114) + (0.8457 \times 0.119) + (0.8453 \times 0.212) + (0.6541 \times 0.151)$$
$$= 0.7563$$
$$A_2 = (0.7990 \times 0.121) + (0.8201 \times 0.177) + (0.7730 \times 0.104) +$$
$$(0.9217 \times 0.114) + (0.7963 \times 0.119) + (0.9000 \times 0.212) + (0.7674 \times 0.151)$$
$$= 0.9446$$

Therefore, the total score of alternatives $A_1 = 0.7563$ and $A_2 = 0.9446$ is obtained. Similarly, the remaining values for each alternative is calculated.

6 Comparative Study

In this section, the results obtained through the proposed (NFSAW) method is compared with the existing fuzzy methods such as TOPSIS, weighted product model (WPM), weighted sum model (WSM), and weighted aggregated sum product assessment (WASPAS). From Table 10, it can be observed that the selection of alternative (A_2) as the preferred choice by NFSAW is validated by the existing

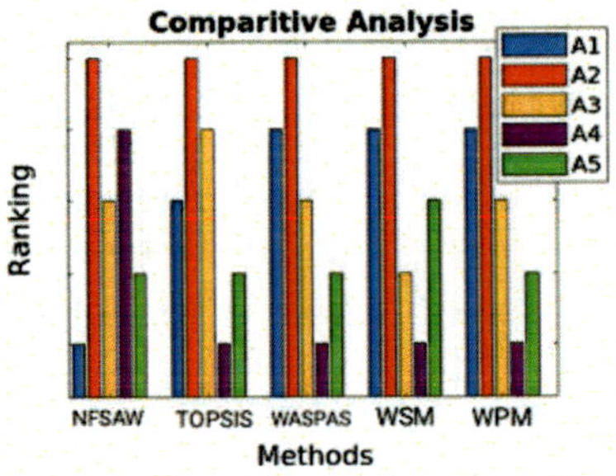

Fig. 1 Graphical representation of different MCDM methods

methods. The slight variation in the ranking order of other alternatives can be attributed to the sensitive nature of NFSAW method. The results are illustrated with graphical representation in Figure 1.

7 Conclusion

By applying NFSAW method, rankings of the alternatives are obtained with accuracy. The rankings for the alternative are of the order $A_2 > A_4 > A_3 > A_5 > A_1$, i.e., community farming > organic store > integrated farming > millet mills > beekeeping. It is evident that community farming(A_2) is ranked first in NFSAW method, and the same has been the best alternative in all the other existing methods. This shows that the best choice for an entrepreneur during their initial days is to start with community farming. Further, many such real-life oriented research work can be extended using neutrosophic sets.

Alternative	Total score	Rank
A_1	0.7563	V
A_2	0.9446	I
A_3	0.8219	III
A_4	0.8345	II
A_5	0.7845	IV

References

1. Ajay, D., Manivel, M., Aldring, J.: Neutrosophic fuzzy SAW method and its application. Int. J. Anal. Exp. Modal Anal., **XI**, 881–887 (2019)
2. Bolturk, E., Karasan, A., Kahraman, C.: Simple additive weighting and weighted produced methods using neutrosophic sets. In: Fuzzy Multi-Criteria Decision Making Using Neutrosophic Sets, pp. 647–676. Springer, Cham (2019)

3. Budi, P., Niswah, B., Dwi, R.: Fuzzy simple additive weighting method in the decision making of human resource recruitment. Lontar Komputer **7**, 174–181 (2016)
4. Deni, W., Sudana, O., Ssanitha, A.: Analysis and implementation fuzzy multi attributes decision making SAW method for selection of achieving students in faculty level. Int. J. Pure Appl. Math., **10**, 674–680 (2013)
5. Kaw, P., Kumar, S.: An intuitionistic fuzzy simple additive weighting method for selection of vendors. ISRO J. Bus. Manag. **15**, 78–81 (2013). ISSN 22 77 4106
6. Klir, G., Folger, T.A.: Fuzzy Sets, Uncertainty and Information. PHI Learning Private Limited, New Delhi (2010)
7. Roszkowskar, E., Kacprza, D.: The fuzzy saw and fuzzy TOPSIS procedures based on ordered fuzzy number. Inf. Sci. **369**, 564–584 (2016). INS 12370
8. Sugar, M.K., Jayaswal, P., Kushwah, K.: Exploring fuzzy SAW method for maintenance strategy selection problem of material handling equipment. Int. J. Curr. Eng. Technol., **3**, 600–605 (2013). ISSN 22 77 4106
9. Thong, N.T., Lan, L.T.H., Son, L.H.: Extended TOPSIS method with unknown weight information in dynamic neutrosophic environment. MDPI J. Math. **8**, 2–15 (2020)
10. Waziana, W., Irviani, R., Oktaviani, I., Satria, F., Kurniawan, D., Maseleno, A.: Fuzzy simple additive weighting for determination of receipients breeding farm program. Int. J. Pure Appl. Math., **118**, 93–100 (2018)
11. Zadeh, L.A.: Fuzzy sets. Inf. Control **8**, 338–353 (1965)

Decision-Making Problem Based on Complex Picture Fuzzy Soft Set Using ELECTRE I Method

S. Anita Shanthi and T. Umamakeswari

Keywords Complex picture fuzzy soft aggregation matrix · Concordance index · Discordance index · Fiberglass

1 Introduction

Zadeh [1] introduced fuzzy set. Atanassav [2] developed IF sets. Hatami-Marbini et al. [3] proposed ELECTRE I method in fuzzy environment. Aytac et al. [4] developed fuzzy ELECTRE I to select suitable catering firm. Wu et al. [5] proposed IF ELECTRE method for solving MCDM problems. Ramot et al. [6, 7] defined CPS and some basic operations on CPS. Liu et al. [8] dealt with TODIM and ELECTRE II method based on decision-making framework. Rouyendegh [9] used ELECTRE method to solve MCDM problems using IF data. Cuong et al. [10, 11] introduced PFS and defined some operations on PFS. Garg et al. [12, 13] developed MCDM problems on CIFS. Akram et al. [14, 15] extended ELECTRE I method to Pythagorean fuzzy environment. Further, they [16] extended it to hesitant Pythagorean fuzzy sets. Akram et al. [17] developed BN TOPSIS and ECLECTRE I method to solve MCDM problems. Further, in [18], they accomplished BF TOPSIS and ELECTRE I method and CSF ELECTRE I method in [19, 20]. Seenivasan et al. [21] designed a robust fuzzy ranking approach.

In Sect. 2, basic definitions needed for the development of the method are provided. Section 3 deals with the procedure of ELECTRE I method on CPFSS. In Sect. 4, results and discussions are specified.

S. Anita Shanthi (✉) · T. Umamakeswari
Department of Mathematics, Annamalai University, Annamalainagar, Tamilnadu, India

H.-H. Leung et al. (eds.), *Recent Developments in Algebra and Analysis*,
Trends in Mathematics, https://doi.org/10.1007/978-3-031-37538-5_8

2 ELECTRE I Method for CPFSS

Definition of CPFS set is given in [22].

Definition 1 CPFSSs are represented as $p \times q$ matrix denoted by CPFSM.

$$CPFSM =$$

$$\begin{array}{c} \\ C_1 \\ C_2 \\ \vdots \\ C_p \end{array} \begin{pmatrix} e_1 & e_2 & \cdots & e_q \\ (\mu_{11}e^{i\alpha_{11}\pi}, \xi_{11}e^{i\gamma_{11}\pi}, \nu_{11}e^{i\beta_{11}\pi}) & (\mu_{12}e^{i\alpha_{12}\pi}, \xi_{12}e^{i\gamma_{12}\pi}, \nu_{12}e^{i\beta_{12}\pi}) & \cdots & (\mu_{1s}e^{i\alpha_{1s}\pi}, \xi_{1s}e^{i\gamma_{1s}\pi}, \nu_{1s}e^{i\beta_{1s}\pi}) \\ (\mu_{21}e^{i\alpha_{21}\pi}, \xi_{21}e^{i\gamma_{21}\pi}, \nu_{21}e^{i\beta_{21}\pi}) & (\mu_{22}e^{i\alpha_{22}\pi}, \xi_{22}e^{i\gamma_{22}\pi}, \nu_{22}e^{i\beta_{22}\pi}) & \cdots & (\mu_{2s}e^{i\alpha_{2s}\pi}, \xi_{2s}e^{i\gamma_{2s}\pi}, \nu_{2s}e^{i\beta_{2s}\pi}) \\ \vdots & \vdots & \cdots & \vdots \\ (\mu_{r1}e^{i\alpha_{r1}\pi}, \xi_{r1}e^{i\gamma_{r1}\pi}, \nu_{r1}e^{i\beta_{r1}\pi}) & (\mu_{r2}e^{i\alpha_{r2}\pi}, \xi_{r2}e^{i\gamma_{r2}\pi}, \nu_{r2}e^{i\beta_{r2}\pi}) & \cdots & (\mu_{rs}e^{i\alpha_{rs}\pi}, \xi_{rs}e^{i\gamma_{rs}\pi}, \nu_{rs}e^{i\beta_{rs}\pi}) \end{pmatrix}$$

Definition 2 Given an CPFSS,
$(CPF, E) = \{\varphi, ((\mu_{ij}e^{i\alpha_{ij}\pi}(\varphi), \xi_{ij}e^{i\gamma_{ij}\pi}(\varphi), \nu_{ij}e^{i\beta_{ij}\pi}(\varphi))) : \varphi \in U\}$,
$\eta_{ij} = 1 - |\mu_{ij}e^{i\alpha_{ij}\pi} - \xi_{ij}e^{i\gamma_{ij}\pi} - \nu_{ij}e^{i\beta a_{ij}\pi}|$ is the degree of fuzziness. The CPFS entropy measure CPE_j is,

$$CPE_j = \frac{1}{r}\sum_{i=1}^{p} \eta_{ij},\ j = 1, 2, \ldots, q.$$

$$\text{Weights } w_i = \frac{1-CPE_j}{\sum_{j=1}^{q}(1-CPE_j)}, i = 1, 2, \cdots, p.$$

Weight value $w = (w_1, w_2, \ldots, w_p)$ satisfies $\sum_{i=1}^{p} w_i = 1$.

Definition 3 CPFS concordance (C) set is defined as

$$CPFSC_{pq} = \{j/\mu_{CPF_{pj}(\epsilon)}(\varphi) < \mu_{CPF_{qj}(\epsilon)}(\varphi), \xi_{CPF_{pj}(\epsilon)}(\varphi) < \xi_{CPF_{qj}(\epsilon)}(\varphi), \nu_{CPF_{pj}(\epsilon)}(\varphi)$$
$$\alpha_{CPF_{pj}(\epsilon)}(\varphi) < \alpha_{CPF_{qj}(\epsilon)}(\varphi), \gamma_{CPF_{pj}(\epsilon)}(\varphi) < \gamma_{CPF_{qj}(\epsilon)}(\varphi), > \nu_{CPF_{qj}(\epsilon)}(\varphi)$$
$$\beta_{CPF_{pj}(\epsilon)}(\varphi) > \beta_{CPF_{qj}(\epsilon)}(\varphi)\}$$

for the terms on amplitude and phase $\forall\, \varphi \in U, p \neq q$ and $p, q = 1, 2, \cdots, r$.

Definition 4 CPFS discordance (D) set is defined as
$CPFSD_{pq} =$
$\{j/\mu_{CPF_{F_{pj}(\epsilon)}}(\varphi) > \mu_{CPF_{F_{qj}(\epsilon)}}(\varphi), \xi_{CPF_{F_{pj}(\epsilon)}}(\varphi) < \xi_{CPF_{F_{qj}(\epsilon)}}(\varphi), \nu_{CPF_{F_{pj}(\epsilon)}}(\varphi) < \nu_{CPF_{F_{qj}(\epsilon)}}(\varphi)$
$\alpha_{CPF_{F_{pj}(\epsilon)}}(\varphi) > \alpha_{CPF_{F_{qj}(\epsilon)}}(\varphi), \gamma_{F_{pj}(\epsilon)}(\varphi) < \gamma_{CPF_{F_{qj}(\epsilon)}}(\varphi),$
$\beta_{CPF_{F_{pj}(\epsilon)}}(\varphi) < \beta_{CPF_{F_{qj}(\epsilon)}}(\varphi)\}$
for the terms on amplitude and phase $\forall\ \varphi \in U, p \neq q$ and $p, q = 1, 2, \cdots, r$.

Definition 5 CPFS C matrix is CM_{pq}.

$$CM_{pq} = \begin{array}{c} \\ C_1 \\ C_2 \\ \\ C_p \end{array} \begin{array}{c} \begin{array}{cccc} e_1 & e_2 & \cdots & e_q \end{array} \\ \begin{pmatrix} - & am_{12} & \cdots & am_{1q} \\ am_{21} & - & \cdots & am_{2q} \\ \cdots & \cdots & \cdots & \cdots \\ am_{p1} & am_{p2} & \cdots & - \end{pmatrix} \end{array}.$$

CPFS C index am_{pq}s are determined as $am_{pq} = \sum_{j \in CM_{pq}} w_j,\ j = 1, 2, \ldots, q.$

Definition 6 CPFS D matrix is DM_{pq}.

$$DM_{pq} = \begin{array}{c} \\ C_1 \\ C_2 \\ \\ C_p \end{array} \begin{array}{c} \begin{array}{cccc} e_1 & e_2 & \cdots & e_q \end{array} \\ \begin{pmatrix} - & bm_{12} & \cdots & bm_{1q} \\ bm_{21} & - & \cdots & bm_{2q} \\ \cdots & \cdots & \cdots & \cdots \\ bm_{p1} & bm_{p2} & \cdots & - \end{pmatrix} \end{array}$$

CPFS D index bm_{pq} are determined as

$$bm_{pq} = \frac{\max\limits_{j \in DM_{pq}} \sqrt{\frac{1}{6} \sum\limits_{\delta=\mu,\xi,\nu,\omega=\alpha,\gamma,\beta} [\delta_{pj} e^{i\omega_{pj}\pi} - \delta_{qj} e^{i\omega_{qj}\pi}]^2}}{\max\limits_{j} \sqrt{\frac{1}{6} \sum\limits_{\delta=\mu,\xi,\nu,\omega=\alpha,\gamma,\beta} [\delta_{pj} e^{i\omega_{pj}\pi} - \delta_{qj} e^{i\omega_{qj}\pi}]^2}}.$$

Definition 7 To rank the alternatives, threshold values such as C and D levels are to be computed. CPFS C level $\overline{\varphi}$ and CPFS D level $\overline{z}$ are bounded by means of CPFS C and D index.

CPFS C level $\overline{\varphi} = \frac{1}{r(r-1)} \sum\limits_{p=1,q\neq 1}^{r} \sum\limits_{p\neq 1,q=1}^{r} am_{pq}$.

CPFS D level $\overline{z} = \frac{1}{r(r-1)} \sum\limits_{p=1,q\neq 1}^{r} \sum\limits_{p\neq 1,q=1}^{r} bm_{pq}$.

Definition 8 By the CPFS C level $\overline{\varphi}$, the CPFS C dominance matrix (dom) K is computed as:

$$K = \begin{matrix} & \begin{matrix} e_1 & e_2 & \cdots & e_q \end{matrix} \\ \begin{matrix} C_1 \\ C_2 \\ \\ C_p \end{matrix} & \begin{pmatrix} - & k_{12} & \cdots & k_{1q} \\ k_{21} & - & \cdots & k_{2q} \\ \cdots & \cdots & \cdots & \cdots \\ k_{p1} & k_{p2} & \cdots & - \end{pmatrix} \end{matrix}$$

The values of k_{pq} are evaluated as $k_{pq} = 1 \quad if \quad am_{pq} \geq \overline{\varphi}, \quad 0 \quad if \quad am_{pq} < \varphi.$

Definition 9 By the CPFS D level $\overline{z}$, the CPFS D dom L is computed as:

$$L = \begin{matrix} & \begin{matrix} e_1 & e_2 & \cdots & e_q \end{matrix} \\ \begin{matrix} C_1 \\ C_2 \\ \\ C_p \end{matrix} & \begin{pmatrix} - & l_{12} & \cdots & l_{1q} \\ l_{21} & - & \cdots & l_{2q} \\ \cdots & \cdots & \cdots & \cdots \\ l_{p1} & l_{p2} & \cdots & - \end{pmatrix} \end{matrix}$$

The values of l_{pq} are evaluated as $l_{pq} = 1 \quad if \quad bm_{pq} \leq \overline{z}, \quad 0 \quad if \quad bm_{pq} > \overline{z}$

Definition 10 CPFS aggregated dom M is determined by peer-to peer multiplication of the elements of K and L.

$$M = \begin{matrix} & \begin{matrix} e_1 & e_2 & \cdots & e_q \end{matrix} \\ \begin{matrix} C_1 \\ C_2 \\ \\ C_p \end{matrix} & \begin{pmatrix} - & m_{12} & \cdots & m_{1q} \\ m_{21} & - & \cdots & m_{2q} \\ \cdots & \cdots & \cdots & \cdots \\ m_{p1} & m_{p2} & \cdots & - \end{pmatrix} \end{matrix}$$

where $m_{pq} = k_{pq}.l_{pq}$. A simple directed graph can be drawn using the values of m_{pq}, which connects the alternatives specifically.

3 Procedure

Step 1: Compute CPFSM.
Step 2: Determine weights w_j by Definition 2.
Step 3: Compute CPFS C set by Definition 3.
Step 4: Compute CPFS D set by Definition 4.
Step 5: Determine CPFS C Matrix by Definition 5.
Step 6: Determine CPFS D Matrix by Definition 6 and CPFS C and D level by Definition 7.
Step 7: Calculate CPFS C dom by Definition 8.
Step 8: Calculate CPFS D dom by Definition 9.

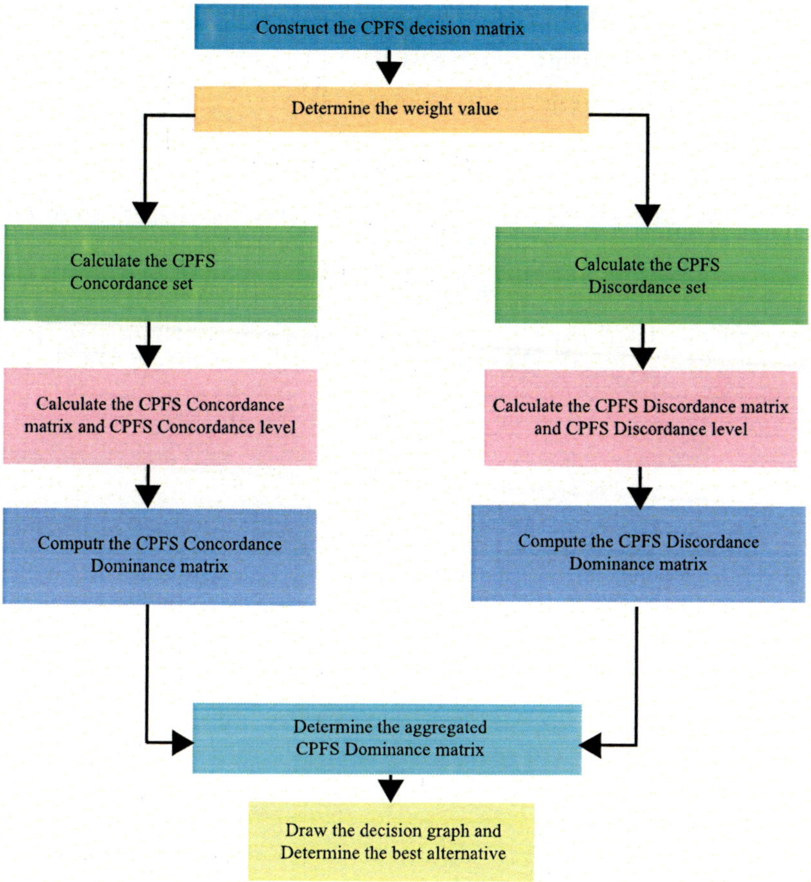

Fig. 1 CPFS ELECTRE I method

Step 9: Determine CPFS aggregated dom by Definition 10, and draw the decision graph. Determine the best alternative. Flowchart of CPFS ELECTRE I method is given in Fig. 1.

4 Results and Discussions

Fiberglass is durable and versatile. Hence, it has a wide range of uses. Fiberglass components are as follows:

A-type is acknowledged as alkali glass. It is opposing to C-type and has certain resemblance to window glass.

C-type is designated as a chemical glass.

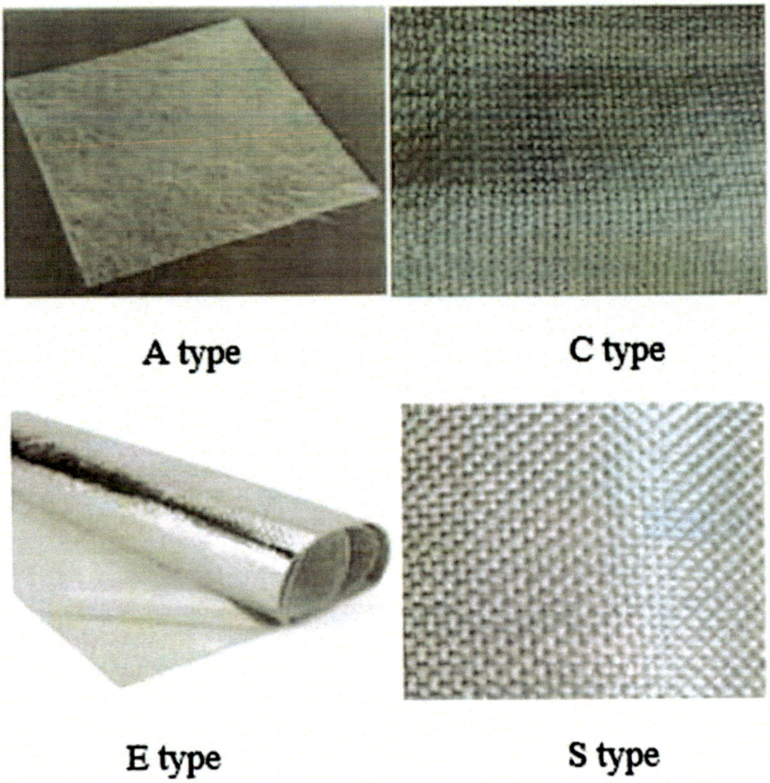

Fig. 2 Types of fiberglass

E-type is electrical glass.
S-type is identified as structural glass.

Types of fiberglass are given in Fig. 2.
Four types of fiberglasses, C_1, C_2, C_3, C_4, are chosen as alternatives. To determine the best alternative, fiberglass materials are evaluated based on the following parameters: e_1 = density, e_2 = tensile strength, e_3 = modulus, and e_4 = elongation at break. The best type of fiberglass based on this concept is found.

Step 1 CPFS sets C_1, C_2, C_3, C_4 are represented in Table 1.

Step 2 The weight values w_j are,
$w_1 = 0.3596$, $w_2 = 0.1815$, $w_3 = 0.0960$, $w_4 = 0.3627$.

Step 3 CPFS C set:

Table 1 Decision matrix in CPFS environment

U	e_1	e_2
C_1	$0.2e^{0.5\pi}$, $0.3e^{0.3\pi}$, $0.45e^{0.6\pi}$	$0.3e^{0.5\pi}$, $0.2e^{0.3\pi}$, $0.41e^{0.51\pi}$
C_2	$0.3e^{0.6\pi}$, $0.4e^{0.5\pi}$, $0.2e^{0.3\pi}$	$0.3e^{0.6\pi}$, $0.3e^{0.3\pi}$, $0.4e^{0.6\pi}$
C_3	$0.4e^{0.6\pi}$, $0.4e^{0.4\pi}$, $0.3e^{0.5\pi}$	$0.4e^{0.4\pi}$, $0.3e^{0.1\pi}$, $0.3e^{0.53\pi}$
C_4	$0.1e^{0.4\pi}$, $0.2e^{0.2\pi}$, $0.4e^{0.61\pi}$	$0.41e^{0.61\pi}$, $0.31e^{0.7\pi}$, $0.1e^{0.5\pi}$
U	e_3	e_4
C_1	$0.1e^{0.2\pi}$, $0.3e^{0.3\pi}$, $0.4e^{0.6\pi}$	$0.3e^{0.5\pi}$, $0.5e^{0.7\pi}$, $0.1e^{0.3\pi}$
C_2	$0.3e^{0.4\pi}$, $0.4e^{0.6\pi}$, $0.2e^{0.5\pi}$	$0.2e^{0.4\pi}$, $0.3e^{0.5\pi}$, $0.4e^{0.6\pi}$
C_3	$0.4e^{0.7\pi}$, $0.35e^{0.4\pi}$, $0.2e^{0.59\pi}$	$0.12e^{0.3\pi}$, $0.1e^{0.4\pi}$, $0.6e^{0.61\pi}$
C_4	$0.3e^{0.5\pi}$, $0.2e^{0.3\pi}$, $0.1e^{0.2\pi}$	$0.5e^{0.5\pi}$, $0.3e^{0.6\pi}$, $0.2e^{0.2\pi}$

$$CPFSC_{pq} = \begin{array}{c} \\ C_1 \\ C_2 \\ C_3 \\ C_4 \end{array} \begin{array}{c} \begin{array}{cccc} 1 & 2 & 3 & 4 \end{array} \\ \begin{pmatrix} - & \{1,3\} & \{1,3\} & \{2\} \\ \{4\} & - & \{3\} & \{2\} \\ \{4\} & \{4\} & - & \{2,4\} \\ \{1\} & \{1\} & \{1\} & - \end{pmatrix} \end{array}$$

Step 4 CPFS D set:

$$CPFSD_{pq} = \begin{array}{c} \\ C_1 \\ C_2 \\ C_3 \\ C_4 \end{array} \begin{array}{c} \begin{array}{cccc} 1 & 2 & 3 & 4 \end{array} \\ \begin{pmatrix} - & \{4\} & \{4\} & \{1\} \\ \{1,3\} & - & \{4\} & \{1\} \\ \{1,3\} & \{3\} & - & \{1\} \\ \{2\} & \{2\} & \{2,4\} & - \end{pmatrix} \end{array}$$

Step 5 CPFS C Matrix is:

$$CM_{pq} = \begin{pmatrix} - & 0.4556 & 0.4556 & 0.1815 \\ 0.3627 & - & 0.0960 & 0.1815 \\ 0.3627 & 0.3627 & - & 0.5442 \\ 0.3596 & 0.3596 & 0.3596 & - \end{pmatrix}$$

CPFS C level $\overline{\varphi} = 0.3$.

Step 6 CPFS D Matrix is:

$$DM_{pq} = \begin{pmatrix} - & 0.1904 & 0.3009 & 0.0530 \\ 0.1598 & - & 0.1105 & 0.2042 \\ 0.2155 & 0.0813 & - & 0.1751 \\ 0.1543 & 0.1782 & 0.2646 & - \end{pmatrix}$$

Fig. 3 Decision graph

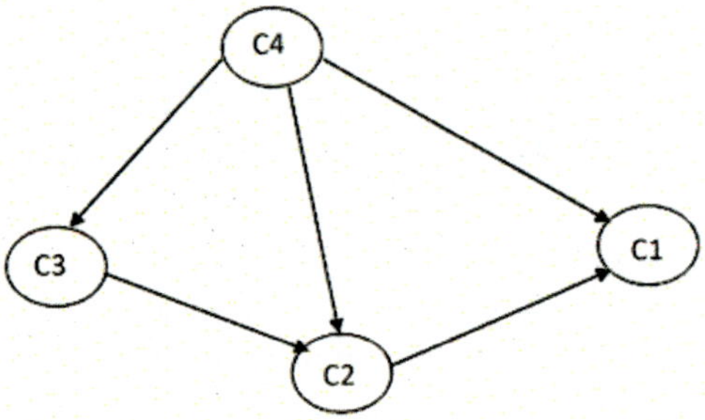

CPFS D level $\overline{z} = 0.17$.

Step 7 CPFS C dom is:

$$K = \begin{pmatrix} - & 1 & 1 & 0 \\ 1 & - & 0 & 0 \\ 1 & 1 & - & 1 \\ 1 & 1 & 1 & - \end{pmatrix}$$

Step 8 CPFS D dom is:

$$L = \begin{pmatrix} - & 0 & 0 & 1 \\ 1 & - & 1 & 0 \\ 0 & 1 & - & 0 \\ 1 & 1 & 1 & - \end{pmatrix}$$

Step 9 Compute CPFS aggregate dom, and decision graph (Fig. 3) is constructed based on these values.

$$M = \begin{pmatrix} - & 0 & 0 & 0 \\ 1 & - & 0 & 0 \\ 0 & 1 & - & 0 \\ 1 & 1 & 1 & - \end{pmatrix}$$

From matrix M for alternative C_2, the value 1 exists at column C_1. So an arrow is drawn from C_2 to C_1. The maximum number of arrows are from C_4 to C_1, C_2, C_3. Hence, C_4 (S-type) is the best fiberglass.

5 Conclusion

In this article, ELECTRE I method on CPFSS is developed. CPFS entropy, weights, C index, and D index are evaluated. The aggregated CPFS D matrix is computed, and decision graph is drawn. Taking four types of fiberglass, A-type, C-type, E-type, and S-type, and four properties present in these fiberglass as parameters, it is determined that S-type is the best as it has higher values of tensile strength, modulus, and elongation when compared to the other three types of fiberglass.

Conflict of Interest The authors declare that they have no conflict of interest.

Acknowledgments The authors are thankful to the reviewers for their valuable comments, which have greatly helped in improving the quality of the paper.

References

1. Zadeh, L.A.: Fuzzy sets. Inf. Control **8**(3), 338–353 (1965). https://doi.org/10.2307/2272014
2. Atanassov, K.: Intuitionistic fuzzy sets. Fuzzy Sets Syst. **20**(1), 87–96 (1986). https://doi.org/10.1016/S0165-0114(86)80034-3
3. Hatami-Marbini, A., Tavana, M.: An extension of the ELECTRE I method for group decision-making under a fuzzy environment. Omega **39**(4), 373–386 (2011). https://doi.org/10.1016/j.omega.2010.09.001
4. Aytac, E., Tus Isik, A., Kundakci, N.: Fuzzy ELECTRE-I method for evaluating catering firm alternatives. Ege Acad. Rev. **11**, 125–134 (2011)
5. Wu, M.C., Chen, T.Y.: The ELECTRE multicriteria analysis approach based on Atanassov's intuitionistic fuzzy sets. Exp. Syst. Appl. **38**(10), 12318–12327 (2011). https://doi.org/10.1016/j.eswa.2011.04.010
6. Ramot, D., Milo, R., Friedman, M., Kandel, A.: Complex Fuzzy sets. IEEE Trans. Fuzzy Syst. **10**(2), 171–186 (2002). https://doi.org/10.1109/91.995119
7. Ramot, D., Friedan, M., Langholz, G., Kandel, A.: Complex Information fuzzy logic. IEEE Trans. Fuzzy Syst. **11**(4), 450–461 (2003)
8. Liu, H.W., Wang, G.J.: Multi criteria decision making methods based on intuitionistic fuzzy sets. Europ. J. Oper. Res. **179**(1), 220–233 (2007). https://doi.org/10.1016/j.ejor.2006.04.009
9. Rouyendegh, B.D.: The intuitionistic fuzzy ELECTRE model. Int. J. Manage. Sci. Eng. Manage. **13**(2), 139–145 (2018). https://doi.org/10.1080/17509653.2017.1349625
10. Cuong, B.C.: Picture Fuzzy sets. J. Comput. Sci. Cybern. **30**(4), 409–420 (2014). https://doi.org/10.156225/1813-9663/30/4/5032
11. Cuong, B.C., Kreinovich, V.: Picture Fuzzy Sets – a new concept for computational intelligence problems. In: IEEE Third World Congress on Information and Communication Technologies WICT, pp. 1–6 (2013). https://doi.org/10.1109/WICT.2013.7113099
12. Grag, H., Rani, D.: Some generalized complex intuitionistic fuzzy aggregation operators and their application to multi criteria decision-making process. Arabian J. Sci. Eng. **44**(3), 2679–2698 (2019). https://doi.org/10.1007/s13369-018-3413-x
13. Rani, D., Garg, H.: Distance measures between the complex intuitionistic fuzzy sets and its applications to the decision-making process. Int. J. Comput. Intel. Syst. **7**(5), 423–439 (2017). https://doi.org/10.1615/Int.J.UncertaintyQuantification.2017020356
14. Akram, M., Garg, H., Ilyas, F.: Multi-criteria group decision making based on ELECTRE I method in Pythagorean fuzzy information. Soft Comput. **24**, 3425–3453 (2020). https://doi.org/10.1007/s00500-019-04105-0

15. Akram, M., Luqman, A., Alcantud, J.R.: Risk evaluation in failure modes and effects analysis: hybrid TOPSIS and ECLECTRE I solutions with Pythagorean fuzzy information. Neural Comput. Appl. **33**(11), 5675–5703 (2021). https://doi.org/10.1007/s00521-020-05350-3
16. Akram, M., Luqman, A., Kahraman, C.: Hesitant pythagorean fuzzy ELECTRE-II method for multi-criteria decision-making problems. Appl. Soft Comput. **108**, 1–25 (2021). https://doi.org/10.1016/j.asoc.2021.107479
17. Akram, M., Shumaiza, Smarandache, F.: Decision making with bipolar neutrosophic TOPSIS and bipolar neutrosophic ELECTRE-I. Axioms **7**(2), 1–25 (2018). https://doi.org/10.3390/axioms7020033
18. Akram, M., Shumaiza, Arshad, M.: Bipolar fuzzy TOPSIS and bipolar fuzzy ELECTRE-I method to diagnosis. Comput. Appl. Math. **39**(7), 1–21 (2020). https://doi.org/10.1007/s40314-019-0980-8
19. Akram, M., Al-Kennani, A.N., Shabir, M.: Enhancing ELECTRE I method with complex spherical fuzzy information. Int. J. Comput. Intel. Syst. **14**(190), 1–31 (2021). https://doi.org/10.1007/s44196-021-00038-5
20. Akram, M., Garg, H., Zahid, K.: Extension of ELECTRE-I and TOPSIS method for group decision-making under complex Pythagorean fuzzy environment. Iran. J. Fuzzy Syst. **17**(5), 147–164 (2020). https://doi.org/10.22111/IJFS.2020.5522
21. Seenivasan, M., Ramesh, R.: Heterogeneous queueing system with cost and profit appraises by robust fuzzy ranking approach. In: 4th International Congress on Life, Social and Health Sciences in a Changing World Congress proceedings Book, pp. 215–223 (2022)
22. Anita Shanthi, S., Umamakeswari, T., Saranya, M.: MCDM method on complex picture fuzzy soft environment. Mat. Today Proc. **51**, 2375–2379 (2022). https://doi.org/10.1016/j.matpr.2021.11.583

Resultant of an Equivariant Polynomial System with Respect to Direct Product of Symmetric Groups

Sonagnon Julien Owolabi, Ibrahim Nonkané, and Joel Tossa

Keywords Commutative algebra · Symbolic computation · Resultant · Discriminant · Divided difference · Direct product of symmetric groups

1 Motivation and Introduction

Solving algebraic systems of polynomial equations $f_1, f_2, \ldots, f_n$ in several variable is a fundamental problem with in computational algebra with many applications (cryptology, robotics, biology, physic, coding theory, etc.. . .). The analysis of such systems is based on the study of the resultant [4]. System which are invariant under the action of a group may be of great importance since symmetry is very relevant in physical sciences as it has to with energy. Thus, Laurent Busé and Anna Karasoulou have studied the resultant of an equivariant polynomial system with respect to $\mathcal{S}_n$ group of permutations on a set of variables $\{x_1, \ldots, x_n\}$ [2]. They developed a nice decomposition of that resultant, which leads to the decomposition of the discriminant of a symmetric polynomial. In some situations, the permutations among the set $x_1, \ldots, x_n$ may not be effective in the sense that some action may hindered or neglected, and in this case, the symmetric group would not be of the best description of the symmetry. For example, the coordinates $x_1, \ldots, x_n$ of the particles of a given molecule may be separated into two subsets, $\{x_1, \ldots, x_p\}$ and $\{x_{p+1}, \ldots, x_n\}$, which do not interact. The symmetry is therefore described by the direct product of symmetric groups $\mathcal{S}_p \times \mathcal{S}_{n-p}$,, where $\mathcal{S}_p$ and $\mathcal{S}_{n-p}$ are groups of permutations on $\{x_1, \ldots, x_p\}$ and $\{x_{p+1}, \ldots, x_n\}$, respectively. A similar situation may occur when the coordinates separated into three or more subsets, leading to a

S. J. Owolabi · J. Tossa
Institut de Mathématiques et de Sciences Physiques, IMSP, Université d' Abomey-Calavi, Abomey-Calavi, Benin
e-mail: julien.owolabi@imsp-auc.org; joel.tossa@imsp-uac.org

I. Nonkané (✉)
Départément d'économie et de mathématiques appliquées, IUFIC, Université Thomas Sankara, Saaba, Burkina Faso

H.-H. Leung et al. (eds.), *Recent Developments in Algebra and Analysis*,
Trends in Mathematics, https://doi.org/10.1007/978-3-031-37538-5_9

product of three or more symmetric groups. Therefore, we think that some results of [2] may be generalizes to systems that are equivariant with respect to the product of symmetric groups, even to other groups. In this chapter, we attempt to study the resultant of an equivariant system with respect to the direct product of two subgroups of $\mathcal{S}_n$. We realize that the techniques that have been used is the case of the symmetric groups [2] work for the case for the direct product of symmetric groups, then we make great use of them in this paper. This chapter is somehow a variant of [2], and we mainly refer to it for the proofs.

A polynomial system $\mathcal{A} = \{f_1, f_2, \ldots, f_n\}$ is said to be *equivariant with respect* to a finite group G if for all $g \in G$, $f_i \in \mathcal{A}$, $g(f_i) \in \mathcal{A}$, $i = 1, \ldots, n$. In other words $\mathcal{A}$ is globally stable under the finite group G see [3]. Let a system of n homogeneous polynomials $f^{\{1\}}, \ldots, f^{\{p\}}, f^{\{p+1\}}, \ldots, f^{\{n\}}$ of same degree d equivariant to the direct product $\mathcal{S}_{\{1,\ldots,p\}} \times \mathcal{S}_{\{p+1,\ldots,n\}}$ of two symmetric subgroups of $\mathcal{S}_n$ with $1 \leqslant p < n$. the action of $\mathcal{S}_{\{1,\ldots,p\}} \times \mathcal{S}_{\{p+1,\ldots,n\}}$ on $f^{\{1\}}, \ldots, f^{\{p\}}, f^{\{p+1\}}, \ldots, f^{\{n\}}$ is described as follows. Let $\sigma_1 \in \mathcal{S}_{\{1,\ldots,p\}}$, and $\sigma_2 \in \mathcal{S}_{\{p+1,\ldots,n\}}$ we have for all $i = 1, \ldots, n$

$$\begin{aligned}(\sigma_1, \sigma_2)\left(f^{\{i\}}\right)&(x_1, \ldots, x_p, x_{p+1}, \ldots, x_n)\\ &= f^{\{i\}}(x_{\sigma_1(1)}, \ldots x_{\sigma_1(p)}, x_{\sigma_2(p+1)}, \ldots, x_{\sigma_2(n)}).\end{aligned}$$

We assume that for all $k \in \{1, \ldots, p\}\sigma = (\sigma_1, \sigma_2) \in \mathcal{S}_{\{1,\ldots,p\}} \times \mathcal{S}_{\{p+1,\ldots,n\}}$,

$$\sigma\left(f^{\{k\}}\right) = \begin{cases}\sigma_1\left(f^{\{k\}}\right) = f^{\{\sigma_1(k)\}} \text{ if } k \in \{1, \ldots, p\}\\ \sigma_2\left(f^{\{k\}}\right) = f^{\{\sigma_2(k)\}} \text{ if } k \in \{p+1, \ldots, n\}.\end{cases} \tag{1}$$

Under this assumption, the polynomial system is equivariant with respect to the direct product $\mathcal{S}_{\{1,\ldots,p\}} \times \mathcal{S}_{\{p+1,\ldots,n\}}$. In what follows, we set

$$f^{\{\sigma(i)\}} := \sigma\left(f^{\{i\}}\right), \; \sigma \in \mathcal{S}_{\{1,\ldots,p\}} \times \mathcal{S}_{\{p+1,\ldots,n\}}, i = 1, \ldots, n. \tag{2}$$

In this work, we will study the resultant of such systems. As an application, we obtain a decomposition formula for the discriminant of an invariant multivariate homogeneous polynomial under the action of a direct product of two symmetric groups.

2 Resultant of a $\mathcal{S}_{\{1,\ldots,p\}} \times \mathcal{S}_{\{p+1,\ldots,n\}}$-Equivariant Polynomial System

Let R be a commutative ring, and denote by $R[x_1, \ldots, x_n]$ the ring of polynomials in $n \geqslant 2$ variables, which is graded with the usual weights: $\deg(x_i) = 1$ for all $i \in \{1, \ldots, n\}$. In this section, we consider a polynomial system of n homogeneous polynomials $f^{\{1\}}, \dots, f^{\{p\}}, f^{\{p+1\}}, \dots, f^{\{n\}}$ in $R[x_1, \dots, x_n]$ of same degree d which is equivariant to the direct product $\mathcal{S}_{\{1,\dots,p\}} \times \mathcal{S}_{\{p+1,\dots,n\}}$ of two subgroups of $\mathcal{S}_n$ with $1 \leqslant p < n$.

2.1 Partitions

Let $\lambda := (\lambda_1, \lambda_2, \dots, \lambda_{r_1})$ be a sequence such that $\lambda_1 \geqslant \cdots \geqslant \lambda_{r_1} > 0$. When $\sum_{i=1}^{r_1} \lambda_i = p$, we will say such a λ is a partition of p, and write $\lambda \vdash p$.

Given a partition $\lambda \vdash p$, its associated multinomial coefficient is defined as the integer

$$m_\lambda := \frac{1}{\prod_{j=1}^{n} s_j!}\binom{p}{\lambda_1, \lambda_2, \dots, \lambda_{r_1}} = \frac{p!}{(\prod_{j=1}^{p} s_j!)\lambda_1!\lambda_2!\cdots\lambda_{r_1}!}. \tag{3}$$

where s_j denotes the number of boxes having exactly j objects, $j \in [p]$ for the partition $\lambda \vdash p$.

Let $\Lambda = (\lambda, \lambda')$ be a couple of partition swhere $\lambda = (\lambda_1, \lambda_2, \dots, \lambda_{r_1}) \vdash p$ and $\lambda' = (\lambda'_1, \lambda'_2, \dots, \lambda'_{r_2}) \vdash q$; then, we will write $(\lambda, \lambda') \vdash (p, q)$ or $\Lambda \vdash (p, q)$. Given a couple of partitions $\Lambda = (\lambda, \lambda') \vdash (p, q)$ such that $p + q = n$, we consider the following homomorphism of algebras:

$$\begin{aligned} \rho_\Lambda : R[x_1, \dots, x_n] &\to R[y_1, \dots, y_{r_1}, y'_1, \dots, y'_{r_2}] \\ f(x_1, \dots, x_n) &\mapsto f(\underbrace{y_1, \dots, y_1}_{\lambda_1}, \dots, \underbrace{y_{r_1}, \dots, y_{r_1}}_{\lambda_{r_1}}, \underbrace{y'_1, \dots, y'_1}_{\lambda'_1}, \dots, \underbrace{y'_{r_2}, \dots, y'_{r_2}}_{\lambda'_{r_2}}). \end{aligned} \tag{4}$$

where $y_1, \dots, y_{r_1}, y'_1, \dots, y'_{r_2}$ are new indeterminates.
For two integers $i, j \in \{1, \dots, n\}$ and $\pi \in \mathcal{S}_{1,\dots,p} \times \mathcal{S}_{p+1,\dots,n}$ such that $\pi(l) = l$ if $l \notin \{i, j\}$ and $\pi(i) = j$, then

$$f^{\{i\}} - f^{\{j\}} = f^{\{i\}} - \pi(f^{\{i\}}) \in (x_i - x_j). \tag{5}$$

Therefore, the polynomial systems $f^{\{1\}}, \dots f^{\{p\}}$ and $f^{\{p+1\}} \dots f^{\{n\}}$ admit divided differences. From [2, Lemma 2.1] and (2) for any subsets $\{i_1, \dots, i_k\} \subset \{1, \dots, p\}$, $\{j_1, \dots, j_l\} \subset \{p+1, \dots, n\}$ and $\phi = (\pi, \sigma) \in \mathcal{S}_{1,\dots,p} \times \mathcal{S}_{p+1,\dots,n}$, we have

$$\phi(f^{\{i_1,\dots,i_k\}}) = f^{\{\pi(i_1),\dots,\pi(i_k)\}}, \text{ and } \phi(f^{\{j_1,\dots,j_l\}}) = F^{\{\sigma(j_1),\dots,\sigma(j_l)\}}. \tag{6}$$

Whenever $\rho_\Lambda(x_i) = \rho_\Lambda(x_j)$, by (5), we have

$$\rho_\Lambda(f^{\{i\}}) = \rho_\Lambda(f^{\{j\}}).$$

$$\begin{cases} f^{\{1\}} = ax_1^2 + bx_1^2 + bx_1x_2 + bx_1x_3 + cx_1^2 + cx_2^2 + cx_3^2 + x_4x_5 \\ f^{\{2\}} = ax_2^2 + bx_1x_2 + bx_2^2 + bx_2x_3 + cx_1^2 + cx_2^2 + cx_3^2 + x_4x_5 \\ f^{\{3\}} = ax_3^2 + bx_1x_3 + bx_2x_3 + bx_3^2 + cx_1^2 + cx_2^2 + cx_3^2 + x_4x_5 \\ f^{\{4\}} = px_4^2 + qx_5^2 \\ f^{\{5\}} = px_5^2 + qx_4^2. \end{cases}$$

This system $\{f^{\{1\}}, f^{\{2\}}, f^{\{3\}}, f^{\{4\}}, f^{\{5\}}\}$ is not equivariant with respect to the symmetric group $\mathcal{S}_5$; then, the formula of [2, Theorem 3.3] cannot help to split the resultant of that polynomial system. But this system is equivariant to the direct product $\mathcal{S}_{\{1,2,3\}} \times \mathcal{S}_{\{4,5\}}$. Then $f^{\{1\}}, f^{\{2\}}, f^{\{3\}}$ equivariant with respect to $\mathcal{S}_{\{1,2,3\}}$ and $f^{\{4\}}, f^{\{5\}}$ equivariant with respect to $\mathcal{S}_{\{4,5\}}$.

$$\operatorname{Res}\left(f^{\{1\}}, f^{\{2\}}, f^{\{3\}}, f^{\{4\}}, f^{\{5\}}\right) = \left(f^{\{1,2,3\}}\right)^{\mu} \operatorname{Res}\left(f^{\{1\}}_{(3),(2)}, f^{\{4\}}_{(3),(2)}\right)^{m_{(3)}m_{(2)}}$$

$$\times \operatorname{Res}\left(f^{\{1\}}_{(2,1),(2)}, f^{\{1,2\}}_{(2,1),(2)} f^{\{4\}}_{(2,1),(2)}\right)^{m_{(2,1)}m_{(2)}} \times \operatorname{Res}\left(f^{\{1\}}_{(3),(1,1)}, f^{\{4\}}_{(3),(1,1)} f^{\{4,5\}}_{(3),(1,1)}\right)^{m_{(3)}m_{(1,1)}}$$

$$\times \operatorname{Res}\left(f^{\{1\}}_{(2,1),(1,1)}, f^{\{1,2\}}_{(2,1),(1,1)} f^{\{4\}}_{(2,1),(1,1)} f^{\{4,5\}}_{(2,1),(1,1)}\right)^{m_{(2,1)}m_{(1,1)}}.$$

$f^{\{1\}}_{(3),(2)} = (a + 3b + 3c)x_1^2 + x_4^2,\ f^{\{4\}}_{(3),(2)} = (p + q)x_4^2,$
$f^{\{1\}}_{(2,1),(2)} = (a + 2b + 2c)x_1^2 + cx_2^2 + bx_1x_2 + x_4^2,$
$f^{\{1,2\}}_{(2,1),(2)} = (a + 2b)x_1 + (a + b)x_2,\ f^{\{4\}}_{(2,1),(2)} = (p + q)x_4^2,$
$f^{\{1\}}_{(3),(1,1)} = (a + 3b + 3c)x_1^2 + x_4x_5,\ f^{\{4\}}_{(3),(1,1)} = px_4^2 + qx_5^2,$
$f^{\{4,5\}}_{(3),(1,1)} = (p-q)x_4 + (p-q)x_5,\ f^{\{1\}}_{(2,1),(1,1)} = (a+2b+2c)x_1^2 + cx_2^2 + bx_1x_2 + x_4x_5,$
$f^{\{1,2\}}_{(2,1),(1,1)} = (a + 2b)x_1 + (a + b)x_2,\ f^{\{4\}}_{(2,1),(1,1)} = px_4^2 + qx_5^2,$
$f^{\{4,5\}}_{(2,1),(1,1)} = (p - q)x_4 + (p - q)x_5,\ f^{\{1,2,3\}} = a,\ \mu = 8.$

we have $\operatorname{Res}\left(f^{\{1\}}_{(3),(2)}, f^{\{4\}}_{(3),(2)}\right) = (3b + 3c + a)^2(p + q)^2$

$$\operatorname{Res}\left(f^{\{1\}}_{(2,1),(2)}, f^{\{1,2\}}_{(2,1),(2)} f^{\{4\}}_{(2,1),(2)}\right) = (p+q)^2(a^3 + 3a^2b + 3a^2c + 2ab^2 + 8abc + 6b^2c)^2$$

$$\operatorname{Res}\left(f^{\{1\}}_{(3),(1,1)}, f^{\{4\}}_{(3),(1,1)} f^{\{4,5\}}_{(3),(1,1)}\right) = (p + q)^2(p - q)^4(3b + 3c + a)^2$$

$$\operatorname{Res}\left(f^{\{1\}}_{(2,1),(1,1)}, f^{\{1,2\}}_{(2,1),(1,1)}, f^{\{4\}}_{(2,1),(1,1)}, f^{\{4,5\}}_{(2,1),(1,1)}\right)$$
$$= (p + q)^2(p - q)^4(a^3 + 3a^2b + 3a^2c + 2ab^2 + 8abc + 6b^2c)^{12}.$$

$$\begin{aligned}
&\operatorname{Res}\left(f^{\{1\}}, f^{\{2\}}, f^{\{3\}}, f^{\{4\}}, f^{\{5\}}\right) \\
&= a^8(a^3 + 3a^2b + 3a^2c + 2ab^2 + 8abc + 6b^2c)^{12}(p-q)^{16}(p+q)^{16} \\
&\quad (3b + 3c + a)^4.
\end{aligned}$$

3 Discriminant of a Homogeneous Polynomial Invariant Under Direct Product of Symmetric Groups

In this section, we will use Theorem 21 to develop a decomposition formula for the discriminant of an invariant homogeneous polynomial under the action of $\mathcal{S}_{\{1,\ldots,p\}} \times \mathcal{S}_{\{p+1,\ldots,n\}}$. Let $f \in R[x_1, \ldots, x_p, x_{p+1}, \ldots, x_n]$ of degree d be a homogeneous polynomial that is invariant under direct product $\mathcal{S}_{\{1,\ldots,p\}} \times \mathcal{S}_{\{p+1,\ldots,n\}}$ of symmetric groups with $1 \leqslant p < n$.

For all $\sigma_1 \in \mathcal{S}_{\{1,\ldots,p\}}$ and for all $\sigma_2 \in \mathcal{S}_{\{p+1,\ldots,n\}}$, we have:

$$(\sigma_1, \sigma_2)\Big(f\Big)(x_1, \ldots, x_p, x_{p+1}, \ldots, x_n) = f(x_{\sigma_1(1)}, \ldots x_{\sigma_1(p)}, x_{\sigma_2(p+1)}, \ldots, x_{\sigma_2(n)}). \quad (7)$$

$$= f(x_1, \ldots, x_p, x_{p+1}, \ldots, x_n) \quad (8)$$

We will denote the partial derivatives of F by

$$f^{\{i\}}(x_1, \ldots, x_p, x_{p+1}, \ldots, x_n) := \frac{\partial f}{\partial x_i}(x_1, \ldots, x_p, x_{p+1}, \ldots, x_n),\ i = 1, \ldots n.$$

The discriminant of F is defined by the equality

$$d^{a(n,d)}\operatorname{Disc}(f) = \operatorname{Res}\left(f^{\{1\}}, f^{\{2\}}, \ldots, f^{\{n\}}\right) \in \mathbb{U} \quad (9)$$

where

$$a(n,d) := \frac{(d-1)^n - (-1)^n}{d} \in \mathbb{Z}.$$

and that it is homogeneous of degree $n(d-1)^{n-1}$ see [1].

Lemma 31 *The set* $\{f^{\{1\}}, f^{\{2\}}, \ldots, f^{\{p\}}, f^{\{p+1\}}, \ldots, f^{\{n\}}\}$ *of partial derivatives of a* $\mathcal{S}_{\{1,\ldots,p\}} \times \mathcal{S}_{\{p+1,\ldots,n\}}$*-invariant homogeneous polynomial* f *is an equivariant polynomial system with respect to* $\mathcal{S}_{\{1,\ldots,p\}} \times \mathcal{S}_{\{p+1,\ldots,n\}}$.

Proof We will use the canonical inclusions $\mathcal{S}_{\{1,\ldots,p\}} \to \mathcal{S}_{\{1,\ldots,p\}} \times \mathcal{S}_{\{p+1,\ldots,n\}}$, $\sigma_1 \mapsto (\sigma_1, e_2)$ and $\mathcal{S}_{\{p+1,\ldots,n\}} \to \mathcal{S}_{\{1,\ldots,p\}} \times \mathcal{S}_{\{p+1,\ldots,n\}}$, $\sigma_2 \mapsto (e_1, \sigma_2)$, where e_1, e_2 are unit elements of $\mathcal{S}_{\{1,\ldots,p\}}$ and $\mathcal{S}_{\{p+1,\ldots,n\}}$, respectively. For all $i \in \{1, \ldots, p\}$ and

$\sigma_1 \in \mathcal{S}_{\{1,\ldots,p\}}$, we have $\sigma_1\left(f^{\{i\}}\right) = \sigma_1\left(\frac{\partial f}{\partial x_i}\right) = \frac{\partial(\sigma_1 f)}{\partial x_{\sigma_1(i)}} = \frac{\partial f}{\partial x_{\sigma_1(i)}} = f^{\{\sigma_1(i)\}}$. For all $j \in \{p+1,\ldots,n\}$ and $\sigma_2 \in \mathcal{S}_{\{p+1,\ldots,n\}}$,, $\sigma_2\left(f^{\{j\}}\right) = \sigma_2\left(\frac{\partial f}{\partial x_j}\right) = \frac{\partial(\sigma_2 f)}{\partial x_{\sigma_2(j)}} = \frac{\partial f}{\partial x_{\sigma_2(j)}} = f^{\{\sigma_2(j)\}}$. For all $k \in \{1,\ldots,p\}\sigma = (\sigma_1,\sigma_2) \in \mathcal{S}_{\{1,\ldots,p\}} \times \mathcal{S}_{\{p+1,\ldots,n\}}$, we have $\sigma\left(f^{\{k\}}\right) = \begin{cases} \sigma_1\left(f^{\{k\}}\right) = f^{\{\sigma_1(k)\}} \text{ if } k \in \{1,\ldots,p\} \\ \sigma_2\left(f^{\{k\}}\right) = f^{\{\sigma_2(k)\}} \text{ if } k \in \{p+1,\ldots,n\}. \end{cases}$ Hence $\sigma\left(f^{\{k\}}\right) \in \{f^{\{1\}},\ldots,f^{\{n\}}\}$, for all $k \in \{1,\ldots,p\}\sigma = (\sigma_1,\sigma_2) \in \mathcal{S}_{\{1,\ldots,p\}} \times \mathcal{S}_{\{p+1,\ldots,n\}}$. and the set of partial derivative of f form an equivariant polynomial system with respect to $\mathcal{S}_{\{1,\ldots,p\}} \times \mathcal{S}_{\{p+1,\ldots,n\}}$.

As a consequence of this lemma, Theorem 21 can be applied in order to decompose the resultant of the polynomials $f^{\{1\}}, f^{\{2\}},\ldots,f^{\{p\}}, f^{\{p+1\}},\ldots f^{\{n\}}$ and hence, by (9), to decompose the discriminant of the $\mathcal{S}_{\{1,\ldots,p\}} \times \mathcal{S}_{\{p+1,\ldots,n\}}$-invariant polynomial f.

Theorem 32 *Assume that $n \geqslant 2$ and $d \geqslant 2$. With the above notation, the following equalities hold:*

- *If $p < d$ and $q < d$ then:*

$$d^{a(n,d)}\mathrm{Disc}(f) = \prod_{\Lambda \vdash (p,q)} \mathrm{Res}\left(f_\Lambda^{\{1\}}, f_\Lambda^{\{1,2\}},\ldots,f_\Lambda^{\{1,2,\ldots,r_1\}}, f_\Lambda^{\{p+1\}}, f_\Lambda^{\{p+1,p+2\}},\ldots, f_\Lambda^{\{p+1,p+2,\ldots,p+r_2\}}\right)^{m_\lambda m_{\lambda'}}.$$

- *If $p \geqslant d$ and $q < d$ then:*

$$d^{a(n,d)}\mathrm{Disc}(f) = \left(f^{\{1,\ldots,d\}}\right)^\mu \times \prod_{\Lambda \vdash (p,q) r_1 < d} \mathrm{Res}\left(f_\Lambda^{\{1\}}, f_\Lambda^{\{1,2\}},\ldots,f_\Lambda^{\{1,2,\ldots,r_1\}}, f_\Lambda^{\{p+1\}}, f_\Lambda^{\{p+1,p+2\}},\ldots, f_\Lambda^{\{p+1,p+2,\ldots,p+r_2\}}\right)^{m_\lambda m_{\lambda'}}.$$

- *If $p < d$ and $q \geqslant d$ then:*

$$d^{a(n,d)}\mathrm{Disc}(f) = \left(f^{\{1,\ldots,p+d\}}\right)^\mu \times \prod_{\Lambda \vdash (p,q)\ r_2 < d} \mathrm{Res}\left(f_\Lambda^{\{1\}}, f_\Lambda^{\{1,2\}},\ldots,f_\Lambda^{\{1,2,\ldots,r_1\}}, f_\Lambda^{\{p+1\}}, F_\Lambda^{\{p+1,p+2\}},\ldots, f_\Lambda^{\{p+1,p+2,\ldots,p+r_2\}}\right)^{m_\lambda m_{\lambda'}}.$$

- *If $p \geqslant d$ and $q \geqslant d$ then:*

$$d^{a(n,d)}\mathrm{Disc}\,(f) = \left(f^{\{1,\ldots,d\}}\right)^{\mu} \times \left(f^{\{p+1,\ldots,p+d\}}\right)^{\mu}$$

$$\times \prod_{\Lambda \vdash (p,q) r_1<d,\ r_2<d} \mathrm{Res}\left(f_\Lambda^{\{1\}}, f_\Lambda^{\{1,2\}}, \ldots, f_\Lambda^{\{1,2,\ldots,r_1\}}, f_\Lambda^{\{p+1\}}, f_\Lambda^{\{p+1,p+2\}}, \ldots,\right.$$

$$\left. f_\Lambda^{\{p+1,p+2,\ldots,p+r_2\}}\right)^{m_\lambda m_{\lambda'}}.$$

where

$$\mu := n(d-1)^{n-1} - \sum_{\substack{(\lambda,\lambda')\vdash(p,q)\\ r_1<d,\ r_2<d}} m_\lambda m_{\lambda'} \left(\sum_{j=1}^{r_1} d_j + \sum_{j=1}^{r_2} d_j\right).$$

with

$$d_j = \frac{(d-1)\cdots(d-r_1)(d-1)\cdots(d-r_2)}{(d-j)}$$

Proof These formulas are obtained by specialization of the formulas given in Theorem 21 with the difference that the polynomials $f^{\{i\}}, i = 1, \ldots, n$ are of degree $d-1$ in our setting (and not of degree d as in Theorem 21).

Example 33 Consider a homogeneous polynomial of degree 4.

$$f := ax_1^4 + bx_1^2x_2^2 + ax_2^4 + cx_3^4 + x_3x_4^3 + x_3^3x_4 + cx_4^4$$

F is not symmetric polynomial but an invariant polynomial under the action of the direct product $\mathcal{S}_{\{1,2\}} \times \mathcal{S}_{\{3,4\}}$

Its partial derivatives are:
$$\begin{cases} f^{\{1\}} = 4ax_1^3 + 2bx_1x_2^2 \\ f^{\{2\}} = 4ax_2^3 + 2bx_1^2x_2 \\ f^{\{3\}} = 4cx_3^3 + x_4^3 + 3x_3^2x_4 \\ f^{\{4\}} = 4cx_4^3 + x_3^3 + 3x_3x_4^2 \end{cases}$$

$f^{\{1\}}$, $f^{\{2\}}$ equivariant with respect to $\mathcal{S}_{\{1,2\}}$ and $f^{\{3\}}$, $f^{\{4\}}$ equivariant with respect to $\mathcal{S}_{\{3,4\}}$

The formula given in Theorem 32 shows that

$$4^{\frac{3^4-(-1)^4}{4}}\mathrm{Disc}(f) = \mathrm{Res}\left(f_{(2),(2)}^{\{1\}}, f_{(2),(2)}^{\{3\}}\right) \times \mathrm{Res}\left(f_{(1,1),(2)}^{\{1\}}, f_{(1,1),(2)}^{\{1,2\}} f_{(1,1),(2)}^{\{3\}}\right)$$

$$\times \mathrm{Res}\left(f_{(2),(1,1)}^{\{1\}}, f_{(2),(1,1)}^{\{3\}} f_{(2),(1,1)}^{\{3,4\}}\right) \times \mathrm{Res}\left(f_{(1,1),(1,1)}^{\{1\}}, f_{(1,1),(1,1)}^{\{1,2\}} f_{(1,1),(1,1)}^{\{3\}} f_{(1,1),(1,1)}^{\{3,4\}}\right).$$

we have

- $\operatorname{Res}\left(f^{\{1\}}_{(2),(2)}, f^{\{3\}}_{(2),(2)}\right) = 512(2a+b)^3(c+1)^3$
- $\operatorname{Res}\left(f^{\{1\}}_{(1,1),(2)}, f^{\{1,2\}}_{(1,1),(2)} f^{\{3\}}_{(1,1),(2)}\right) = 8589934592a^6(2a+b)^3(2a-b)^6(c+1)^6$
- $\operatorname{Res}\left(f^{\{1\}}_{(2),(1,1)}, f^{\{3\}}_{(2),(1,1)} f^{\{3,4\}}_{(2),(1,1)}\right) = 262144(2a+b)^6(c-1)^3(8c^2+1)^6$

$$\operatorname{Res}\left(f^{\{1\}}_{(1,1),(1,1)}, f^{\{1,2\}}_{(1,1),(1,1)} f^{\{3\}}_{(1,1),(1,1)} f^{\{3,4\}}_{(1,1),(1,1)}\right)$$
$$= 73786976294838206464a^{12}(2a+b)^6(2a-b)^{12}(c-1)^6(8c^2+1)^{12}$$

$$\operatorname{Disc}(f) = 773712524553362671811952640a^{18}(c-1)^9(c+1)^9(8c^2+1)^{18}$$
$$(2a-b)^{18}(2a+b)^{18}.$$

Acknowledgments Gratitude is expressed to the projects of the African Centre of Excellence in Mathematical Sciences, Informatics and Applications (ACE-MSIA).

References

1. Busé, L., Jouanolou, J.-P.: On the discriminant scheme of homogeneous polynomials. Math. Comput. Sci. **8**(2), 175–234 (2014)
2. Busé, L., Karasoulou, A.: Resultant of an equivariant polynomial system with respect to the symmetric group. arXiv:1407.2799 v1[math.AC] 10 Juillet 2014
3. Dieudonné, J.A., Carrell, J.B.: Invariant Theory, Old and New. Academic Press, New York (1971)
4. Jouanolou, J.-P.: Le formalisme du résultant. Adv. Math. **90**(2), 117–263 (1991)

Color Image Filtering Using Convolution Fuzzy Neural Network

S. Anita Shanthi and G. Sathiyapriya

Keywords Convolution operation · Image filters · Hamming distance · Euclidean distance · Similarity index

1 Introduction

Anil K. Jain [1] discussed the implementation of various problems in image processing including restoration and enhancement of images, data compression, and filter design with examples. Image processing based on different structuring elements was developed by Anita Shanthi et al. [2]. Rishap Anand [3] dealt with several concepts of digital image processing. Egmont et al. [4] developed several applications of neural networks in image processing. Van De Ville [5] presented fuzzy filter to reduce heavy noise in images. Mishra et al. [6] developed several methods for color image contrast intensification operator. Wang et al. [7] proposed an effective method for image de-noising. Azad et al. [8] motivated the use of color in digital image processing. Albawi et al. [9] explained critical issues related to CNN and its applications in image classification. Chen et al.[10] proposed a CNN method to learn the perceptive features for identifying classic image processing operations. Nader et al. [11] introduced the effect of Gaussian noises and performed experimental analysis to reduce the effect of noise. Coady et al. [12] gave an overview of image filtering operations using edge detection, smooth filters, and its advantages. Mirmozzaffari [13] considered four filters for de-blurring and smoothing of images and performed a comparison analysis. Tomasi et al. [14] dealt with bilateral filtering Sultana et al. [15] discussed the advancements in image classification using CNN. Based on these concepts, color image filtering using CFNN is developed.

S. Anita Shanthi (✉) · G. Sathiyapriya
Department of Mathematics, Annamalai University, Annamalainagar, Tamil Nadu, India

H.-H. Leung et al. (eds.), *Recent Developments in Algebra and Analysis*,
Trends in Mathematics, https://doi.org/10.1007/978-3-031-37538-5_10

2 Convolutional Neural Network

One of the most popular deep neural networks is the convolution neural network. It is designated from mathematical linear operations between matrices called convolution. CNN has multiple layers including convolution layer, nonlinearity layer, pooling layer, and fully connected layer. The convolution and fully connected layers have parameter, whereas pooling and nonlinearity layers do not have parameters. The input layer consists of the given input image, which is in pixel values. Convolution is the mathematical operation performed with the filters to extract the features in the images as shown in Fig. 1.

Convolution Operations Convolution layer generates feature maps, from images. It contains filters that convert images, and these filters are called convolution filters. To understand the working of convolution process, 4 × 4 pixel image values are taken, which are operated upon by a 2 × 2 convolution filter as shown in Fig. 2. At the final stage, the 4 × 4 pixel image has been converted into a 3 × 3 pixel image. The feature map extracted depends on the convolution filter.

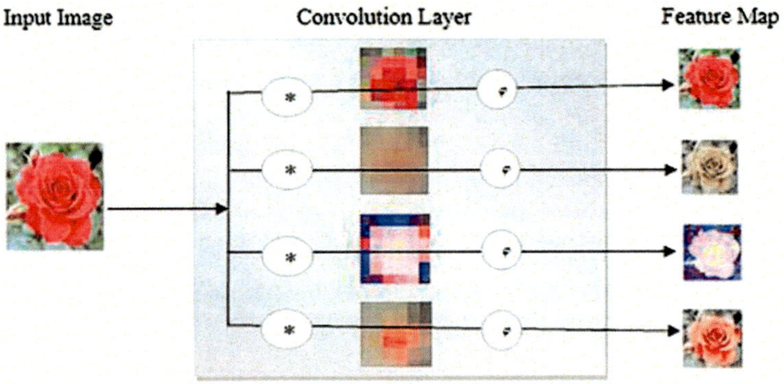

Fig. 1 CNN image features

1	2	5	1
6	4	2	0
8	0	21	5
4	2	4	1

$* \begin{bmatrix} 1 & 0 \\ 0 & 1 \end{bmatrix} =$

5		

1	2	5	1
6	4	2	0
8	0	21	5
4	2	4	1

$* \begin{bmatrix} 1 & 0 \\ 0 & 1 \end{bmatrix} =$

5	4	5
6	25	7
10	4	22

Fig. 2 Convolution operation

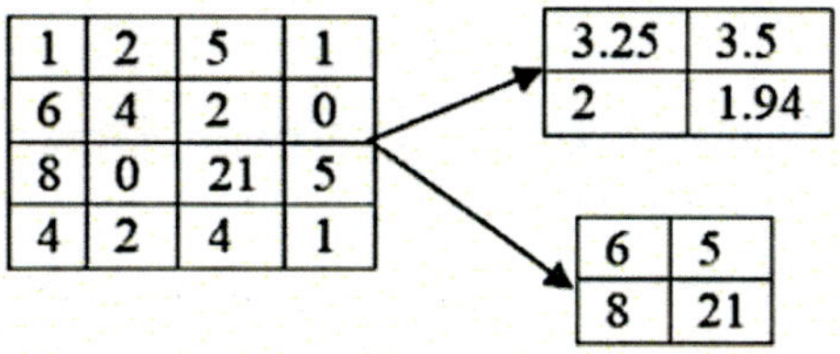

Fig. 3 Pooling operation

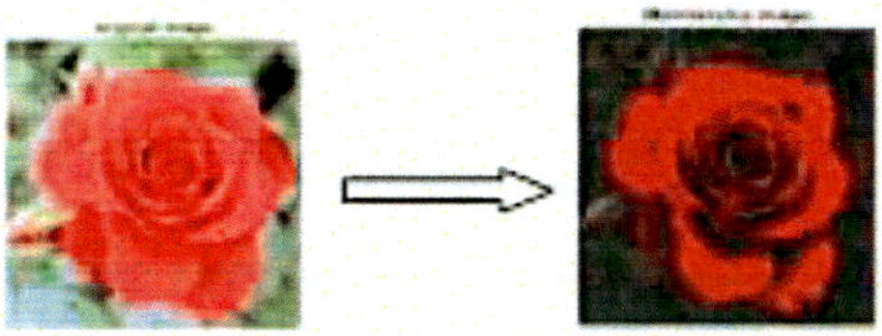

Fig. 4 Original image and its membership image

Pooling Operations It is used to reduce the size of the image. There are two types of pooling operations, which are max pooling and mean pooling. Mean pooling: finding the arithmetic mean of the convolution areas is mean pooling. Max pooling: this is the maximum value of the convolution area as shown in Fig. 3.

2.1 *Image Fuzzification*

Definition 1 Let I be the image that is represented as $m \times n$ matrix of pixel values x_{ij}. The image fuzzification membership function is defined by $\mu_{ij} = [\frac{1+(x_{ij})_{max}-(x_{ij})_{min}}{F_d}]^{F_e}$, where i and j represent ith row and jth column of the pixel values. F_e and F_d values differ for different images. The pixel value of an image is converted to fuzzy membership value matrix, and the corresponding membership images are found using MATLAB. Original image and its corresponding membership image are shown in Fig. 4.

2.2 *Types of Filters*

Mean filter:[3] A box/Mean is a low-pass filter that smoothens the image. The center pixel value is replaced by the average of all the values of its neighborhood N. $G(x, y) = \frac{1}{N} \sum_{(x,y)\in N}^{n} f(x, y)$, when n is the number of neighborhoods.

Median filter: [6] Median filter is a nonlinear filtering useful in reducing impulsive or salt and pepper noise. It preserves sharp edges.

Bilateral filter: [14] A bilateral filter is used for smoothing images and reducing noise while preserving edges.

Fig. 5 Gaussian membership images

Lab filter: Lab is a nonlinear transformation of RGB where the Euclidean distance between two colors is equal to their preceptual distances.

Noise filter: [12] Noise is always present in digital images during image acquisition, coding, transmission, and processing steps. Noise removal algorithm is the process of removing or reducing the noise from the image.

Unsharp filter: [3] Unsharp mask tool increases contrast so that the image is sharpened.

Standard filter: Standard filter specifies the neighborhood used to compute the standard deviation.

Gaussian filter: [6] Gaussian filter is a linear type of filter, which is based on Gaussian function and is much useful at separating frequencies. The (3×3) filter value is $\frac{1}{16}\begin{pmatrix} 1 & 2 & 1 \\ 2 & 4 & 2 \\ 1 & 2 & 1 \end{pmatrix}$.

Gaussian Membership Value Images

The Gaussian filter along with the membership values from 0 to 0.9 is applied to the original image and the resultant images obtained. These images are given in Fig. 5.

2.3 Example

Step 1 Pixel membership values of the image (Rose) is taken, and the Gaussian filter is applied for three colors RGB of the original image.

Step 2 Apply the Gaussian filter for Red image pixel membership values, and perform the convolution operation for red image.

$$\begin{pmatrix} 0.1295 & 0.1543 & 0.1513 & 0.1391 & 0.1369 & 0.1429 \\ 0.1219 & 0.14 & 0.1496 & 0.3397 & 0.2274 & 0.1325 \\ 0.1198 & 0.1328 & 0.1796 & 0.4804 & 1 & 0.1448 \\ 0.1210 & 0.1207 & 0.1537 & 0.4082 & 0.57 & 0.1369 \\ 0.1219 & 0.1213 & 0.1348 & 0.2847 & 0.1913 & 9.1474 \\ 0.1344 & 0.1459 & 0.1357 & 0.1387 & 0.1277 & 0.1474 \end{pmatrix} * \begin{pmatrix} 0.0625 & 0.125 & 0.0625 \\ 0.125 & 0.25 & 0.125 \\ 0.0625 & 0.125 & 0.0625 \end{pmatrix}$$

$$\begin{pmatrix} 0.1183 & 0.1712 & 0.2007 & 0.2905 \\ 0.1063 & 0.2076 & 0.4083 & 0.4673 \\ 0.1097 & 0.1931 & 0.3655 & 0.3916 \\ 0.0985 & 0.1485 & 0.2138 & 0.2085 \end{pmatrix}, \text{ which is filtered image matrix for Red.}$$

Similarly, apply the Gaussian filter for blue and green image pixel membership values, and perform the convolution operation for green image.

Step 3 Add the three convolution filtered matrices.

$$\begin{pmatrix} 0.1487 & 0.2616 & 0.4021 & 0.3204 \\ 0.1402 & 0.2334 & 0.3483 & 0.3185 \\ 0.1531 & 0.1892 & 0.5524 & 0.2401 \\ 0.1425 & 0.1769 & 0.2139 & 0.2045 \end{pmatrix}.$$

Step 4 Max pooling yields

$$\begin{pmatrix} 0.8252 & 1.0323 \\ 0.6079 & 1.1394 \end{pmatrix}.$$

Step 5 Weights are taken as trapezoidal fuzzy number indicating the features of the image.

$$\begin{pmatrix} 0.1 & 0.2 & 0.35 & 0.5 \\ 0 & 0.19 & 0.25 & 0.42 \\ 0.3 & 0.32 & 0.39 & 0.52 \\ 0.11 & 0.15 & 0.43 & 0.61 \end{pmatrix}.$$

Step 6 Peer-to-peer multiplication of the convolution filtered matrix, and the weight matrix gives the output image pixel values of 6×6 membership matrix considered.

$$\begin{pmatrix} 0.0573 & 0.165 & 0.3212 & 0.4394 \\ 0 & 0.1359 & 0.2503 & 0.4336 \\ 0.1415 & 0.1945 & 0.4444 & 0.4639 \\ 0.0532 & 0.08 & 0.2764 & 0.3972 \end{pmatrix}.$$

HSV Filter [2] HSV filter converts a color image into three channels, color (hue), brightness (value), and saturation (shades). It is useful for object detection. Different colors can be assigned to the background of an image in the HSV color space. In MATLAB, HSV is a three-dimensional matrix, which represents three components of HSV as shown in Fig. 6.

Fig. 6 Hue and saturation modified image

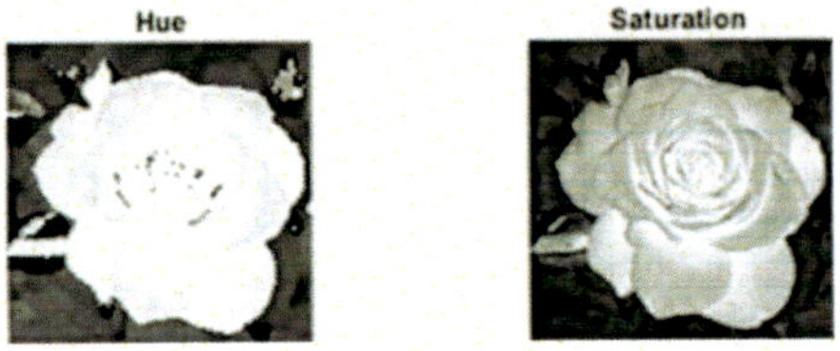

Fig. 7 Hue- and saturation-modified membership images

Fig. 8 Filtered and its membership images

Hue-modified image with different membership values and saturation membership value images in the range 0.1 to 0.9 are shown in Fig. 7.

2.4 *Filtered Images*

The different types of filtered images and its membership images are shown in Fig. 8.

3 Hamming and Euclidean Distances

Definition 2 Hamming distance on filtered images Hamming distance on filtered images shown in Fig. 9 is denoted by H_D and is defined as $H_D = \sum_{i,j} \left\| x_{ij} - x'_{ij} \right\|$, i, j= 1 to 256, where x_{ij} and x'_{ij} denotes pixel values of two filtered image matrices.

Definition 3 Similarity index on filtered images is denoted by $SI(H_D)$ and is defined as $SI(H_D) = \frac{1}{1+H_D}$.

Definition 4 Hamming distance on membership filtered images shown in Fig. 10 is denoted by μ_{H_D} and is defined as $\mu_{H_D} = \sum_{i,j} \left\| \mu(x_{ij}) - \mu(x'_{ij}) \right\|$, i, j= 1 to 256, where $\mu(x_{ij})$ and $\mu(x'_{ij})$ denote pixel values of two filtered image matrices.

Definition 5 Similarity index on membership filtered images is denoted by $\mu_{SI(H_D)}$ and is defined as $\mu_{SI(H_D)} = \frac{1}{1+\mu_{H_D}}$.

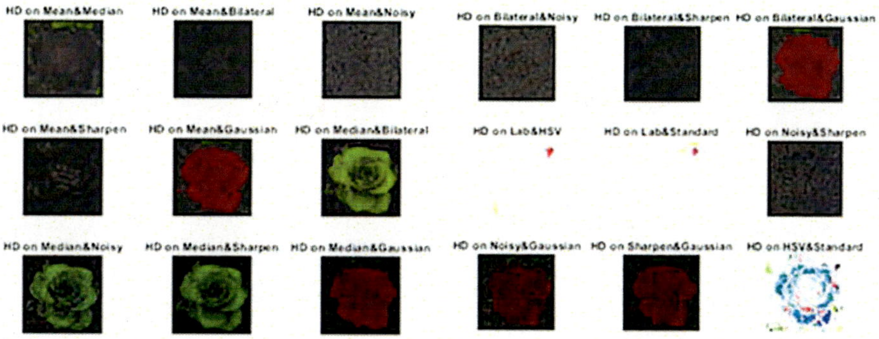

Fig. 9 Hamming distance of filtered images

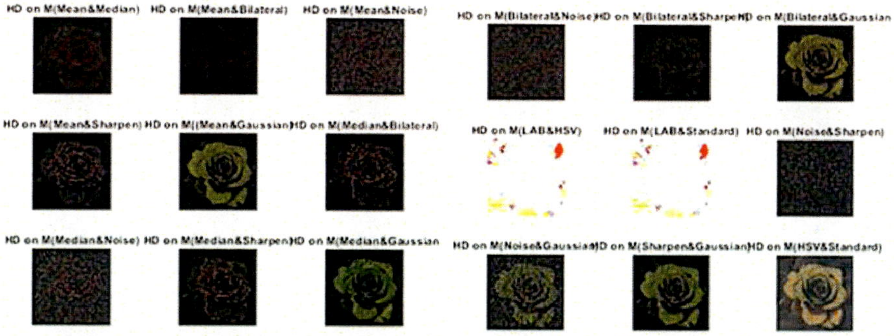

Fig. 10 Hamming distance on membership filtered images

Table 1 Similarity index based on hamming distance

x_{ij}	x'_{ij}	$SI(E_D)$	$\mu_{SI(E_D)}$
Mean	Median	0.0031	0.8379
Mean	Bilateral	0.0149	1.7934
Mean	Noise	0.0014	0.1619
Mean	Sharpen	0.0062	0.4968
Mean	Gaussian	0.0019	0.3777
Median	Bilateral	0.0077	0.8365
Median	Noise	0.0011	0.1537
Median	Sharpen	0.0051	0.4384
Median	Gaussian	0.0035	0.3928
Bilateral	Noise	0.0014	0.1697
Bilateral	Sharpen	0.0094	0.6149
Bilateral	Gaussian	0.0018	0.3703
Lab	HSV	0.0004	0.0006
Lab	Standard	0.0005	0.0006
Noise	Sharpen	0.0015	0.1262
Noise	Gaussian	0.0011	0.1274
Sharpen	Gaussian	0.0017	0.2943
Hsv	Standard	0.0020	0.0279

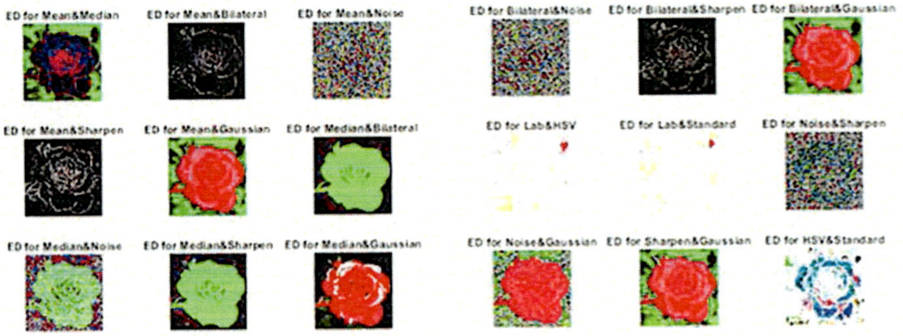

Fig. 11 Euclidean distance of filtered images

Similarity index on filtered images and its corresponding membership images are given in Table 1.

Definition 6 Euclidean distance on filtered images shown in Fig. 11 is denoted by E_D and is defined as $E_D = \sqrt{\sum_{i,j}(x_{ij} - x'_{ij})^2}, i, j = 1$ to 256.

Definition 7 Similarity index on filtered images is denoted by $SI(E_D)$ and is defined as $SI(E_D) = \frac{1}{1+E_D}$.

Definition 8 Euclidean distance on membership filtered images shown in Fig. 12 is denoted by μ_{E_D} and is defined as $\mu_{E_D} = \sqrt{\sum_{i,j} \mu((x_{ij}) - \mu(x'_{ij}))^2}$

Definition 9 Similarity index on membership filtered images is denoted by $\mu_{SI(H_D)}$ and is defined as $\mu_{SI(E_D)} = \frac{1}{1+\mu_{E_D}}$.

Similarity index on filtered images and its corresponding membership images are given in Table 2.

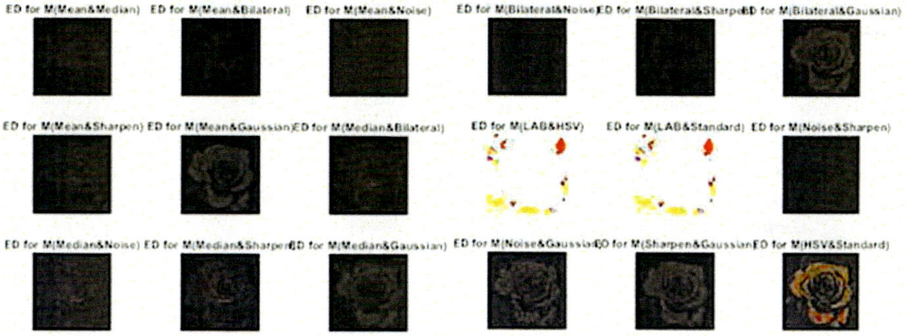

Fig. 12 Euclidean distance on membership filtered images

Table 2 Similarity index based on Euclidean distance

x_{ij}	x'_{ij}	$SI(E_D)$	$\mu_{SI(E_D)}$
Mean	Median	0.0001825	0.1433
Mean	Bilateral	0.0008829	0.9671
Mean	Noise	0.00009317	0.1149
Mean	Sharpen	0.0003891	0.2317
Mean	Gaussian	0.000114	0.0304
Median	Bilateral	0.000437	0.01438
Median	Noise	0.00007333	0.0654
Median	Sharpen	0.0003166	0.1105
Median	Gaussian	0.0002023	0.032
Bilateral	Noise	0.0000939	0.1235
Bilateral	Sharpen	0.0006161	0.2599
Bilateral	Gaussian	0.0001064	0.0303
Lab	HSV	0.00000374	0.0403
Lab	Standard	0.0000664	0.0403
Noise	Sharpen	0.000102	0.1262
Noise	Gaussian	0.0006721	0.0322
Sharpen	Gaussian	0.0000988	0.0319
Hsv	Standard	0.0000988	0.3

4 Conclusion

Different image filters are applied to original and membership images, and the corresponding images are found using MATLAB. Hamming and Euclidean distances between different combinations of original and membership filtered images are calculated and the similarity index values tabulated. It is found that using both Hamming and Euclidean distances, the similarity index between mean and bilateral filter images is maximum. This concept is useful in identifying diseases in the leaves of plants.

References

1. Jain, A.K.: Advances in mathematical models for image processing. Proc. IEEE **69**(5), 502–528 (1981). http://doi.org/doi:10.1109/PROC.1981.12021
2. Anita Shanthi, S., Valarmathi, R.: Image processing on fuzzy near sets. In: 2021 IEEE Madras Section Conference (MASCON), pp. 1–5 (2021)
3. Anand, R.: Digital Image Processing. Medtec Publication, New Gloucester (2015)
4. Egmont-Petersen, M., De Riddem, D., Handels, H.: Image processing with neural network – a review. Patt. Recog. **35**, 2279–2301 (2002). https://doi.org/10.1016/S0031-3203(01)00178-9
5. Van De Ville, D., Nachtegael, M., Van der Weken, D., Kerre, E.E., Philips, W., Lemahieu, I.: Noise reduction by fuzzy image filtering. IEEE Trans. Fuzzy Syst. **11**(4), 429–436 (2003). http://doi.org/doi:10.1109/TFUZZ.2003.814830
6. Mishra, P., Sinha, K.: A highly efficient color image contrast enhancement using fuzzy based contrast intensification operator. Adv. Res. Elect. Electron. Eng. **1**(4), 26–31 (2014)
7. Wang, M., Zheng, S., Li, X., Qin, X.: A new image denoising method based on Gaussian filter. In: Internation Conference on Information Science, Electronics and Electrical Engineering, Info IEEE, pp. 163–167 (2014). http://doi.org/doi:10.1109/InfoSEEE.2014.6948089
8. Azad, M.M., Mahedihasan, M., Naseer, M.: Color image processing in digital image. Int. J. New Technol. Res. **3**(3), 56–62 (2017)
9. Albawi, S., Mohammad, T.A., AL-Zawi, S.: Understanding of a Convolutional Neural Network. ICET2017, Antalya (2017). http://doi.org/doi:10.1109/ICEngTechnol.2017.8308186
10. Chen, B., Li, H., Luo, W.: Image Processing Operations Identification via Convolutional Neural Network, pp. 1–8. IEEE, Piscataway (2017). arXiv:1709.02908v1 [cs.MM] 9 Sep 2017. https://doi.org/10.1007/s11432-018-9492-6
11. Nader, J., Alqadi, Z.A., Zahran, B.: Analysis of color image filtering methods. Int. J. Comput. Appl. **174**(8), 12–17 (2017)
12. Coady, J., Newe, T., Riordan, A.O., Toal, D.: An overview of popular digital image processing filtering operations. In: 2019 Thirteenth International Conference on Sensing Technology. IEEE, Piscataway, pp. 1–5 (2019). http://doi.org/doi:10.1109/ICST46873.2019.9047683
13. Mirmozaffari, M.: Filtering in image processing. ENG Trans. **1**, 1–5 (2020)
14. Tomasi, C., Manduchi, R.: Bilateral Filtering for Gray and Color Images, pp. 839–846. http://doi.org/doi:10.1109/ICCV.1998.710815
15. Sultana, F., Sufian, A., Dutta, P.: Advancements in image classification using convolutional neural network, 2018. In: Fourth International Conference on Research in Computational Intelligence and Communication Networks (ICRICN), pp. 122–129 (2018). arXiv:1905.03288v1 [cs.CV] 8 May 2019. http://doi.org/doi:10.1109/ICRCICN.2018.8718718

Some Combinatorial Results for Partial and Full Symmetric Semigroups

F. Al-Kharousi, W. Alnadabi, and A. Umar

Keywords Partial symmetric semigroup · Full symmetric semigroup · Idempotents

1 Introduction and Preliminaries

Let $X_n = \{1, 2, \ldots, n\}$. A transformation $\alpha : \text{Dom}\,\alpha \subseteq X_n \to X_n$ is said to be total or full if $\text{Dom}\,\alpha = X_n$; otherwise, it is called strictly partial. Let $\mathcal{T}_n$ and $\mathcal{P}_n$ be the full and partial transformation semigroups on X_n, respectively. Howie found some notable combinatorial results in $\mathcal{T}_n$ [5], while Garba was interested in $\mathcal{P}_n$ [4]. Recently Laradji and Umar obtained some interesting results on these semigroups and some of their subsemigroups [6]. Umar in [10] computed and gathered together the combinatorial results in $\mathcal{P}_n$ and $\mathcal{T}_n$ and some of their subsemigroups and highlighted some open problems. Motivated by that paper, we compute some of the unknown results in $\mathcal{P}_n$ and $\mathcal{T}_n$. In this section we give necessary definitions. In Sect. 2 we compute the cardinalities of some equivalences defined by equalities of some parameters in $\mathcal{P}_n$ and $\mathcal{T}_n$. For basic definitions and standard concepts in (transformation) semigroup theory, we refer the reader to [3] and [10].

For any transformation $\alpha \in \mathcal{P}_n$, the *fix* and the *collapse* of α are denoted and defined by

The results in this chapter are from Wafa Alnadabi's MSc. thesis (2015) [1]

F. Al-Kharousi · W. Alnadabi
Department of Mathematics, Sultan Qaboos University, Muscat, Oman
e-mail: fatma9@squ.edu.om

A. Umar (✉)
Khalifa University, Abu Dhabi, UAE
e-mail: abdullahi.umar@ku.ac.ae

H.-H. Leung et al. (eds.), *Recent Developments in Algebra and Analysis*,
Trends in Mathematics, https://doi.org/10.1007/978-3-031-37538-5_11

$$F(\alpha) = \{x \in X_n : x\alpha = x\}$$

$$C(\alpha) = \{x \in X_n : x\alpha = y \text{ and } |y\alpha^{-1}| \geq 2\}$$

respectively. Let $b(\alpha) = |\text{Dom}\,\alpha|$, $h(\alpha) = |\text{Im}\,\alpha|$, $w^+(\alpha) = \max(\text{Im}\,\alpha)$, $f(\alpha) = |F(\alpha)|$ and $c(\alpha) = |C(\alpha)|$. Consider the natural equivalences on $\mathcal{P}_n$ defined by equalities of breadths, heights, waist, fix, and collapse. The intersection of these equivalences can be counted by the following combinatorial function: let S be a set of partial transformations of X_n; define the combinatorial function

$$F(n; k, m, p, q, r) = |\{\alpha \in S : w^+(\alpha) = k,\, f(\alpha) = m, h(\alpha) = p, c(\alpha) = q, b(\alpha) = r\}|$$

Here, we introduce a new notation to the combinatorial functions by putting the parameters $k, m, p, q,$ and r (ordered alphabetically) as subscripts. Then the six-parameter function $F(n; k, m, p, q, r)$ can be simply written as F_{kmpqr} and similarly any two-, three-, four-, or five-parameter function.

Stirling numbers of the second kind denoted by $S(n, k)$ are defined to be the number of partitions of $\{1, \ldots, n\}$ into k nonempty subsets and can be calculated by the explicit formula.

$$S(n, k) = \frac{1}{k!}\sum_{j=0}^{k}(-1)^{k-j}\binom{k}{j}j^n. \tag{1}$$

Lemma 1 ([2]) *For all natural numbers n and m, we have*

$$\sum_{i=0}^{n}\binom{n}{i}S(m, i)i! = n^m. \tag{2}$$

Definition 1 ([2]) An **r-associated Stirling number of the second kind** is the number of ways to partition a set of n objects into k subsets, with each subset containing at least r elements and is denoted by $S_r(n, k)$. The 2-associated Stirling numbers of the second kind array can be found in ([9], A008299).

Lemma 2 (Vandemonde's Convolution Identity [8]) *For all natural numbers m, n, and k,*

$$\sum_{i=0}^{n}\binom{n}{k-i}\binom{m}{i} = \binom{n+m}{k}.$$

2 Some Combinatorial Results in $\mathcal{T}_n$ and $\mathcal{P}_n$

Proposition 1 *[1, Proposition 3.1] Let $S = \mathcal{P}_n$. Then*

$$F_{km} = \binom{k}{m}(k+1)^{n-k}k^{k-m} - \binom{k-1}{m}(k-1)^{k-m-1}k^{n-k+1}.$$

Proof Note that $F_{km} = G_{km} - J_{km}$, where

$$G_{km} = |\{\alpha \in \mathcal{P}_n : f(\alpha) = m \text{ and } \operatorname{Im}\alpha \cap X_{n-k} = \emptyset\}|;$$

$$J_{km} = |\{\alpha \in \mathcal{P}_n : f(\alpha) = m \text{ and } \operatorname{Im}\alpha \cap X_{n-k+1} = \emptyset\}|.$$

For G_{km}, there are $\binom{k}{m}$ ways to select the m fixed points from the k elements. The $k-m$ elements in X_k are either in the domain or not. If they are in the domain, then they can map to any of the k elements except themselves. So, they have $(k-1)+1$ degrees of freedom. The remaining $n-k$ elements in $X_n \backslash X_k$ are either in the domain or not. If they are in the domain, then they can map to any of the k elements. So, they have $k+1$ degrees of freedom. Thus, we get,

$$G_{km} = \binom{k}{m}(k+1)^{n-k}k^{k-m}$$

By a similar argument, we find that

$$J_{km} = \binom{k-1}{m}(k-1)^{k-m-1}k^{n-k+1},$$

and the result follows directly.

Corollary 1 *[1, Corollary 3.15] Let $S = T_n$. Then*

$$F_{km} = \binom{k}{m}(k)^{n-k}(k-m)^{k-1} - \binom{k-1}{m}(k-2)^{k-m-1}(k-1)^{n-k+1}.$$

Following similar arguments as the ones used to obtain $F(n; p, m)$ in the semigroup of full transformations $\mathcal{T}_n$ [7], we can compute the following:

Theorem 1 *[1, Theorem 3.2] Let $S = \mathcal{P}_n$. Then,*

$$F_{kmpr} = \binom{k-1}{p-1}\binom{p}{m}\sum_{j=m}^{p}(-1)^{p+j}\binom{p-m}{p-j}$$

$$\sum_{i=m}^{j}\binom{j-m}{i-m}\binom{n-j}{r-i}j^{r-i}(j-1)^{i-m}.$$

Proof There are $\binom{k-1}{p-1}$ ways to select the images. Let

$$G(n; m, p, r) = |\{\alpha \in \mathcal{P}_n : \operatorname{Im}\alpha \cap X_{n-p} = \emptyset, b(\alpha) = r, h(\alpha) = p, \text{ and } f(\alpha) = m\}|.$$

Then $F(n; k, m, p, r) = \binom{k-1}{p-1} G(n; m, p, r)$. Using the *principle of inclusion-exclusion*, we find that

$$\begin{aligned} G(n; m, p, r) &= |A_{n-p}| - \binom{p}{1}|A_{n-p+1}| + \binom{p}{2}|A_{n-p+2}| - \dots \\ &\quad + (-1)^{p-m}\binom{p}{p-m}|A_{n-m}| \\ &= \sum_{j=0}^{p-m}(-1)^j\binom{p}{j}|A_{n-p+j}|, \end{aligned}$$

where $A_{n-p+j} = \{\alpha \in \mathcal{P}_n : \operatorname{Im}\alpha \cap X_{n-p+j} = \emptyset,\ b(\alpha) = r \text{ and } f(\alpha) = m\}$. Now,

$$|A_{n-p+j}| = \binom{p-j}{m}\sum_{i=0}^{p-j-m}\binom{p-j-m}{i}\binom{n-p+j}{r-m-i}(p-j)^{r-m-i}(p-j-1)^i.$$

Note that we have $\binom{p-j}{m}$ ways to choose the m fixed points from the $p-j$ images. Since these m fixed points are among the domain elements, then we are left with $r-m$ points to be chosen. We either choose them from the $p-j-m$ images or from the $n-p+j$ points, which are not in $\operatorname{Im}\alpha$. So, we have $\sum_{i=o}^{p-j-m}\binom{p-j-m}{i}\binom{n-p+j}{r-m-i}$ ways to choose the rest of the domain elements. The i points chosen from the $p-j-m$ elements have $p-j-1$ possible images, while the remaining $r-m-i$ points have $p-j$ possible images. Summing up we get

$$\begin{aligned} F_{kmpr} &= \binom{k-1}{p-1}\sum_{j=0}^{p-m}(-1)^j\binom{p}{j}\binom{p-j}{m} \\ &\quad \sum_{i=0}^{p-j-m}\binom{p-j-m}{i}\binom{n-p+j}{r-m-i}(p-j)^{r-m-i}(p-j-1)^i \\ &= \binom{k-1}{p-1}\sum_{j=m}^{p}(-1)^{p+j}\binom{p}{j}\binom{j}{m}\sum_{i=0}^{j-m}\binom{j-m}{i}\binom{n-j}{r-m-i}j^{r-m-i}(j-1)^i \\ &= \binom{k-1}{p-1}\sum_{j=m}^{p}(-1)^{p+j}\binom{p}{j}\binom{j}{m}\sum_{i=m}^{j}\binom{j-m}{i-m}\binom{n-j}{r-i}j^{r-i}(j-1)^{i-m}, \end{aligned}$$

and the result follows.

Corollary 2 *[1, Corollary 3.3(iii)] Let $S = \mathcal{P}_n$. Then,*

$$F_{kmp} = \binom{k-1}{p-1}\binom{p}{m}\sum_{j=m}^{p}(-1)^{p+j}\binom{p-m}{p-j}(j+1)^{n-j}j^{j-m}.$$

Proof By Theorem 1 we see that

$$\begin{aligned}
F_{kmp} &= \sum_{r=0}^{n} F_{kmpr} \\
&= \binom{k-1}{p-1}\binom{p}{m}\sum_{j=m}^{p}(-1)^{p+j}\binom{p-m}{p-j}\sum_{j=m}^{p}(-1)^{p+j}\binom{p-m}{p-j} \\
&\quad \sum_{i=m}^{j}\binom{j-m}{i-m}(j-1)^{i-m}\sum_{r=1}^{n}\binom{n-j}{r-i}j^{r-i} \\
&= \binom{k-1}{p-1}\binom{p}{m}\sum_{j=m}^{p}(-1)^{p+j}\binom{p-m}{p-j}j^{j-m}(j+1)^{n-j}.
\end{aligned}$$

Corollary 3 *[1, Corollary 3.3(i)] Let $S = \mathcal{P}_n$. Then,*

$$\begin{aligned}
F_{kmr} = \sum_{p=m}^{n}&\binom{k-1}{p-1}\binom{p}{m}\sum_{j=m}^{p}(-1)^{p+j}\binom{p-m}{p-j} \\
&\sum_{i=m}^{j}\binom{j-m}{i-m}\binom{n-j}{r-i}j^{r-i}(j-1)^{i-m}.
\end{aligned}$$

Corollary 4 *[1, Corollary 3.3(iv)] Let $S = \mathcal{P}_n$. Then,*

$$F_{kpr} = \binom{k-1}{p-1}\binom{n}{r}S(r,p)p!$$

Proof From Theorem 1, we get

$$\begin{aligned}
F_{kpr} &= \sum_{m=0}^{p} F_{kmpr} \\
&= \sum_{m=0}^{p}\binom{k-1}{p-1}\binom{p}{m}\sum_{j=m}^{p}(-1)^{p+j}\binom{p-m}{p-j}
\end{aligned}$$

$$\sum_{i=m}^{j}\binom{j-m}{i-m}\binom{n-j}{r-i}j^{r-i}(j-1)^{i-m}$$

$$=\binom{k-1}{p-1}\sum_{j=0}^{p}(-1)^{p+j}\binom{p}{j}\sum_{i=0}^{j}\binom{n-j}{r-i}j^{r-i}\sum_{m=0}^{i}\binom{j}{m}\binom{j-m}{i-m}(j-1)^{i-m}$$

$$=\binom{k-1}{p-1}\sum_{j=0}^{p}(-1)^{p+j}\binom{p}{j}\sum_{i=0}^{j}\binom{n-j}{r-i}\binom{j}{i}j^{r-i}\sum_{m=0}^{i}\binom{i}{i-m}(j-1)^{i-m}$$

$$=\binom{k-1}{p-1}\sum_{j=0}^{p}(-1)^{p+j}\binom{p}{j}\sum_{i=0}^{j}\binom{n-j}{r-i}\binom{j}{i}j^{r-i}j^{i}$$

$$=\binom{k-1}{p-1}\sum_{j=0}^{p}(-1)^{p+j}\binom{p}{j}j^{r}\sum_{i=0}^{j}\binom{n-j}{r-i}\binom{j}{i}$$

$$=\binom{k-1}{p-1}\binom{n}{r}S(r,p)p! \qquad \text{(by Eq. 1)}$$

Similarly from Theorem 1, we get the next corollary.

Corollary 5 *[1, Corollary 3.3(ii)] Let $S=\mathcal{P}_n$. Then,*

$$F_{mpr}=\binom{n}{p}\binom{p}{m}\sum_{j=m}^{p}(-1)^{p+j}\binom{p-m}{p-j}\sum_{i=m}^{j}\binom{j-m}{i-m}\binom{n-j}{r-i}j^{r-i}(j-1)^{i-m}.$$

By similar arguments as in Corollary 4, we deduce the next two corollaries.

Corollary 6 *[1, Corollary 3.5(i)] Let $S=\mathcal{P}_n$. Then,*

$$F_{kp}=\binom{k-1}{p-1}S(n+1,p+1)p!$$

Corollary 7 *[1, Corollary 3.5(ii)] and [10, Proposition 2.8] Let $S=\mathcal{P}_n$. Then,*

$$F_{kr}=\binom{n}{r}[k^r-(k-1)^r].$$

Corollary 8 *[1, Corollary 3.5(iii)] and [10, Proposition 2.2] Let $S=\mathcal{P}_n$. Then,*

$$F_{pr}=\binom{n}{p}\binom{n}{r}S(r,p)p!$$

Corollary 9 *[1, Corollary 3.4] Let $S = \mathcal{P}_n$. Then,*

$$F_{mp} = \binom{n}{p}\binom{p}{m}\sum_{j=m}^{p}(-1)^{p+j}\binom{p-m}{p-j}(j+1)^{n-j}j^{j-m}.$$

Corollary 10 *[1, Corollary 3.16(i)] Let $S = T_n$. Then*

$$F_{kmp} = \binom{k-1}{p-1}\binom{p}{m}\sum_{j=m}^{p}(-1)^{p+j}\binom{p-m}{p-j}(j)^{n-j}(j-1)^{j-m}.$$

Corollary 11 *[1, Corollary 3.17(i)] [7, Proposition 2.6]. Let $S = T_n$. Then,*

$$F_{mp} = \binom{n}{p}\binom{p}{m}\sum_{j=m}^{p}(-1)^{p+j}\binom{p-m}{p-j}(j)^{n-j}(j-1)^{j-m}.$$

Note that a partial transformation α is idempotent if and only if $\operatorname{Im}\alpha = F(\alpha)$. Using this fact, we can recover the formula for F_{kmr} and F_m in $E(\mathcal{P}_n)$ as follows:

Corollary 12 *[1, Corollary 4.3(ii)] Let $S = E(\mathcal{P}_n)$. Then,*

$$F_{kmr} = \binom{k-1}{m-1}\binom{n-m}{r-m}m^{r-m}.$$

Corollary 13 *[1, Corollary 4.6(ii)] Let $S = E(\mathcal{P}_n)$. Then,*

$$F_m = \binom{n}{m}(m+1)^{n-m}.$$

Corollary 14 *[1, Corollary 3.11] For $n \geq m \geq 0$,*

$$n^{n-m} = \sum_{p=m}^{n}\sum_{j=m}^{p}(-1)^{p+j}\binom{n-m}{p-m}\binom{p-m}{p-j}(j+1)^{n-j}j^{j-m}.$$

Proof By [10, Corollary 2.6], F_m in $\mathcal{P}_n$ is given by $\binom{n}{m}n^{n-m}$. So, we get

$$\begin{aligned} n^{n-m} &= \frac{F_m}{\binom{n}{m}} \\ &= \frac{\sum_{p=m}^{n} F_{mp}}{\binom{n}{m}} \end{aligned}$$

$$= \frac{\sum_{p=m}^{n} \binom{n}{p}\binom{p}{m} \sum_{j=m}^{p} (-1)^{p+j} \binom{p-m}{p-j} (j+1)^{n-j} j^{j-m}}{\binom{n}{m}}$$

$$= \frac{\sum_{p=m}^{n} \binom{n}{m}\binom{n-m}{p-m} \sum_{j=m}^{p} (-1)^{p+j} \binom{p-m}{p-j} (j+1)^{n-j} j^{j-m}}{\binom{n}{m}}$$

$$= \sum_{p=m}^{n} \sum_{j=m}^{p} (-1)^{p+j} \binom{n-m}{p-m} \binom{p-m}{p-j} (j+1)^{n-j} j^{j-m}.$$

Similarly, from [4, Corollary 2], we deduce the identity given in the next corollary.

Corollary 15 *[1, Corollary 3.12] For $n \geq p \geq 0$,*

$$S(n+1, p+1)p! = \sum_{m=0}^{p} \sum_{j=m}^{p} (-1)^{p+j} \binom{p}{m} \binom{p-m}{p-j} (j+1)^{n-j} j^{j-m}.$$

Lemma 3 *[1, Lemma 3.6] For all natural numbers n and k,*

$$S(n, k) = \sum_{j=0}^{n} \binom{n}{j} S_2(n-j, k-j).$$

Proof Note that $S(n, k)$ is the number of ways to partition n objects into k nonempty subsets. Let j be the number of one-element subsets. These can be selected in $\binom{n}{j}$ ways. The remaining $n-j$ elements will be partitioned into $k-j$ nonempty subsets, each with at least two elements in $S_2(n-j, k-j)$ ways. Thus,

$$S(n, k) = \sum_{j=0}^{n} \binom{n}{j} S_2(n-j, k-j).$$

Proposition 2 *[1, Proposition 3.7] Let $S = \mathcal{P}_n$. Then,*

$$F_{kpqr} = \binom{k-1}{p-1} \binom{n}{q} \binom{n-q}{r-q} S_2(q, p+q-r)p!$$

Proof There are $\binom{k-1}{p-1}$ ways to select the images and $\binom{n}{q}$ ways to select the collapse points. The remaining $r-q$ points of the domain can be selected in $\binom{n-q}{r-q}$ ways. Let j be the number of images that absorb the collapse. Then, since the $r-q$ points of the domain are not among the collapse, they must be adjoined one-to-one to the remaining $p-j$ images. So, $p-j=r-q$. Hence, $j=p+q-r$, and we can select those in $\binom{p}{p+q-r}$ ways. Now, we partition the q collapse points into $S_2(q, p+q-r)$ nonempty subsets, where each has at least two elements, and then permute them in $(p+q-r)!$ ways. The remaining $r-q$ pre-images and $r-q$ images can be matched in $(r-q)!$ ways. Thus,

$$\begin{aligned} F_{kpqr} &= \binom{k-1}{p-1}\binom{n}{q}\binom{n-q}{r-q}\binom{p}{p+q-r}S_2(q, p+q-r)(p+q-r)!(r-q)! \\ &= \binom{k-1}{p-1}\binom{n}{q}\binom{n-q}{r-q}S_2(q, p+q-r)p! \end{aligned}$$

Corollary 16 *[1, Corollary 3.8(i)] Let $S=\mathcal{P}_n$. Then,*

$$F_{kpq} = \binom{k-1}{p-1}\binom{n}{q}\sum_{r=q}^{n}\binom{n-q}{r-q}S_2(q, p+q-r)p!$$

Corollary 17 *[1, Corollary 3.8(iv)] Let $S=\mathcal{P}_n$. Then,*

$$F_{kpr} = \binom{k-1}{p-1}\binom{n}{r}S(r, p)p!$$

Proof By Proposition 2, we see that

$$\begin{aligned} F_{kpr} &= \sum_{q=0}^{r} F_{kpqr} \\ &= \sum_{q=0}^{r}\binom{k-1}{p-1}\binom{n}{q}\binom{n-q}{r-q}S_2(q, p+q-r)p! \\ &= \binom{k-1}{p-1}\binom{n}{r}p!\sum_{q=0}^{r}\binom{r}{r-q}S_2(q, p-(r-q)) \\ &= \binom{k-1}{p-1}\binom{n}{r}S(r, p)p! \text{ (by Lemma 3).} \end{aligned}$$

Corollary 18 *[1, Corollary 3.8(iii)] Let $S=\mathcal{P}_n$. Then,*

$$F_{kqr} = \binom{n}{r}\binom{r}{q}\sum_{p=0}^{r}\binom{k-1}{p-1}S_2(q, p+q-r)p!$$

Corollary 19 *[1, Corollary 3.8(ii)] Let $S = \mathcal{P}_n$. Then,*

$$F_{pqr} = \binom{n}{p}\binom{n}{q}\binom{n-q}{r-q} S_2(q, p+q-r)p!$$

Corollary 20 *[1, Corollary 3.9(i)] Let $S = \mathcal{P}_n$. Then,*

$$F_{kp} = \binom{k-1}{p-1} S(n+1, p+1)p!$$

Corollary 21 *[1, Corollary 3.10(i)] Let $S = \mathcal{P}_n$. Then,*

$$F_{kq} = \binom{n}{q}\sum_{p=0}^{r}\sum_{r=q}^{n}\binom{n-q}{r-q}\binom{k-1}{p-1} S_2(q, p+q-r)p!$$

Corollary 22 *[1, Corollary 3.9(ii)] Let $S = \mathcal{P}_n$. Then,*

$$F_{kr} = \binom{n}{r}[k^r - (k-1)^r].$$

Proof By Corollary 18, we get

$$\begin{aligned}
F_{kr} &= \sum_{q=0}^{r} F_{kqr} \\
&= \sum_{q=0}^{r}\binom{n}{r}\binom{r}{q}\sum_{p=0}^{r}\binom{k-1}{p-1} S_2(q, p+q-r)p! \\
&= \binom{n}{r}\sum_{p=0}^{r}\binom{k-1}{p-1}\sum_{q=0}^{r}\binom{r}{q} S_2(q, p+q-r)p! \\
&= \binom{n}{r}\sum_{p=0}^{r}\binom{k-1}{p-1} S(r, p)p! \text{ (by Lemmas 1 and 3)} \\
&= \binom{n}{r}\sum_{p=0}^{r}\left[\binom{k}{p} - \binom{k-1}{p}\right] S(r, p)p! \\
&= \binom{n}{r}[k^r - (k-1)^r].
\end{aligned}$$

Corollary 23 *[1, Corollary 3.10(iii)] Let $S = \mathcal{P}_n$. Then,*

$$F_{pq} = p!\binom{n}{p}\binom{n}{q}\sum_{j=0}^{n-q}\binom{n-q}{j}S_2(q, p-j).$$

By Corollary 19 and Lemma 3, we deduce the next corollary.

Corollary 24 *[1, Corollary 3.9(iii)] and [10, Proposition 2].*

$$F_{pr} = \binom{n}{p}\binom{n}{r}S(r, p)p!$$

Corollary 25 *[1, Corollary 3.10(ii)] Let $S = \mathcal{P}_n$. Then,*

$$F_{qr} = \binom{n}{r}\binom{r}{q}\sum_{p=0}^{r}\binom{n}{p}S_2(q, p+q-r)p!$$

Corollary 26 *[1, Corollary 3.10(iv)] Let $S = \mathcal{P}_n$. Then,*

$$F_q = \binom{n}{q}\sum_{p=0}^{n}p!\binom{n}{p}\sum_{j=0}^{n-q}\binom{n-q}{j}S_2(q, p-j).$$

By [10, Corollary 2.3], we deduce the identity given in the next corollary.

Corollary 27 *[1, Corollary 3.13] For $n \geq r \geq 0$*

$$n^r = \sum_{q=0}^{r}\sum_{p=0}^{r}\binom{r}{q}\binom{n}{p}S_2(q, p+q-r)p!$$

By [4, Corollary 2], we deduce the identity given in the next corollary.

Corollary 28 *[1, Corollary 3.14] For $n \geq p \geq 0$,*

$$S(n+1, p+1) = \sum_{q=0}^{n}\sum_{j=0}^{n-q}\binom{n}{q}\binom{n-q}{j}S_2(q, p-j).$$

Corollary 29 *[1, Corollary 3.16(ii)] Let $S = \mathcal{T}_n$. Then,*

$$F_{kpq} = \binom{k-1}{p-1}\binom{n}{q}S_2(q, p+q-n)p!$$

From Corollary 29, we deduce the following successive corollaries.

Corollary 30 *[1, Corollary 3.17(ii)] and [10, Proposition 2.11] Let $S = \mathcal{T}_n$. Then,*

$$F_{kp} = \binom{k-1}{p-1} S(n,p)p!$$

Corollary 31 *[1, Corollary 3.16(iii)] Let $S = \mathcal{T}_n$. Then,*

$$F_{pq} = \binom{n}{p}\binom{n}{q} S_2(p, p+q-n)p!$$

Corollary 32 *[1, Corollary 3.16(iv)] Let $S = \mathcal{T}_n$. Then,*

$$F_{kq} = \binom{n}{q} \sum_{p=0}^{n} \binom{k-1}{p-1} S_2(q, p+q-n)p!$$

References

1. Alnadabi, W.: Some combinatorial results in the full and partial symmetric semigroups and their sets of idempotents, MSc. thesis. Sultan Qaboos University, Oman (2015)
2. Comtet, L.: Advanced Combinatorics: The Art of Finite and Infinite Expansions. D. Reidel Publishing Company, Dordrecht (1974)
3. Ganyushkin, O., Mazorchuk, V.: Classical Finite Transformation Semigroups: An Introduction. Springer, London (2009)
4. Garba, G.U.: Idempotents in partial transformation semigroups. Proc. Roy. Soc. Edinburgh **116A**, 359–366 (1990)
5. Howie, J.M., McFadden, R.B.: Idempotent rank in finite full transformation semigroups. Proc. Roy. Soc. Edinburgh **114A**, 161–167 (1990)
6. Laradji, A., Umar, A.: Combinatorial results for semigroups of order-preserving full transformations. Semigroup Forum **72**, 51–62 (2006)
7. Laradji, A., Umar, A.: Some combinatorial properties of the symmetric monoid. Int. J. Algebra Comput. **21**(No.6), 857–865 (2011)
8. Riordan J.: Combinatorial Identities. John Wiley and Sons, New York (1968)
9. Sloane, N.J.A. (ed.) The online Encyclopedia of Integer Sequences (2011). Available at http://oeis.org/
10. Umar, A.: Some combinatorial problems in the theory of partial transformation semigroups. Algebra Discrete Math. J. **17**, 110–134 (2014)

Integrated Tomato Cultivation Using Backpropagation Neural Network on Bipolar Fuzzy Sets

S. Anita Shanthi and R. Preethi

Keywords Bipolar fuzzy set · Activation function · Backpropagation neural network

1 Introduction

Zadeh [1] introduced fuzzy sets. Zhang [2] introduced bipolar fuzzy set. Akram et al. [3] described the application of BFS in graph structures. Dongare et al. [4] proposed artificial neural network (ANN) as a tool for analysis of different parameters of a system. Wu et al. [5] introduced four characteristics of ANN and its application. Seenivasan et al. [6] dealt with deep learning. Svozil et al. [7] described the multilayer feed-forwarded neural network and also discussed the advantages and disadvantages of this network. Ishibuchi et al. [8] proposed multilayer feed-forward neural networks and also the learning algorithm of fuzzy neural network. Jin et al. [9] derived BP algorithm for fuzzy neural network. Li et al. [10] analyzed the characteristics and mathematical theory of BP neural network. Nawi et al. [11] proposed an algorithm by introducing the adaptive gain of the activation function and improved the learning speed of the conventional BP algorithm. Shihab [12] discussed an efficient and scalable technique for computer network security. Hegazy et al. [13] effectively improved the process of developing practical neural network. Chen et al. [14] proposed the privacy-preserving BPNN learning. Zheng et al. [15] developed a rockburst prediction model to select the evaluation factors based on the entropy weight gray relational BP neural network. Won et al. [16] proposed a method of recognition and prediction of nutrient deficiency in tomato plants based on deep neural network. Walgenbach et al. [17] determined the persistence of insecticides on tomato foliage and plant growth rate. Schmitz-Eiberger et al. [18] investigated on the influence of deficient calcium supply on tomato leaves. Based on these concepts, BP on BFNN is developed.

S. Anita Shanthi (✉) · R. Preethi
Department of Mathematics, Annamalai University, Chidambaram, Tamil Nadu, India

H.-H. Leung et al. (eds.), *Recent Developments in Algebra and Analysis*,
Trends in Mathematics, https://doi.org/10.1007/978-3-031-37538-5_12

2 Backpropagation on Bipolar Fuzzy Neural Network (BPBFNN)

BP is one of the major concepts of a NN. The input values x_m, weights (b^-_{mn}, b^+_{mn}) of the input layer(IL), and bias value b_a are feed-forwarded to find hidden layer (HL). Output $g(\phi)$ is found using sigmoid function. If output $\neq$ target, then the error is calculated. Weights and bias are revised and BP to achieve the target. Figure 1 represents BPBFNN.

Definition 1 Degree of BF set is defined as $d_{mn} = 1 - |b^-_{mn} - b^+_{mn}|$.

Definition 2 The entropy measure, $E_n = \frac{1}{p}\sum_{m=1}^{p} d_{mn}; n = 1, 2, \ldots, q$.

Definition 3 Using the entropy values, the HL weights are calculated and are defined as $W_n = \frac{1-E_n}{\sum_{n=1}^{q} 1-E_n}$.

The weights $W_n = (W_1, W_2, \ldots, W_q)$ satisfy $\sum_{n=1}^{q} W_n = 1$.

Definition 4 HL weight correction, $\nabla W_n = W_n - \eta\omega g(G_n)$ where η is the assumed learning rate, $g(G_n)$ is the output of the HL, and $\omega = (T - g(\phi)) * g(\phi) * (1 - g(\phi))$, where T is the targeted value and $g(\phi)$ is the calculated output.
IL weight correction,
Negative: $\nabla b^-_{mn} = b^-_{mn} - \eta\omega\zeta^-_n x_m$
where $\zeta^-_n = g(G^-_n) * (1 - g(G^-_n)) * \frac{W_n}{2}$ and x_m's are the input values.
Positive: $\nabla b^+_{mn} = b^+_{mn} - \eta\omega\zeta^+_n x_m$
where $\zeta^+_n = g(G^+_n) * (1 - g(G^+_n)) * \frac{W_n}{2}$.

Definition 5 The bias correction is $\nabla b_a = b_a - \eta\omega, a = 1, 2$.

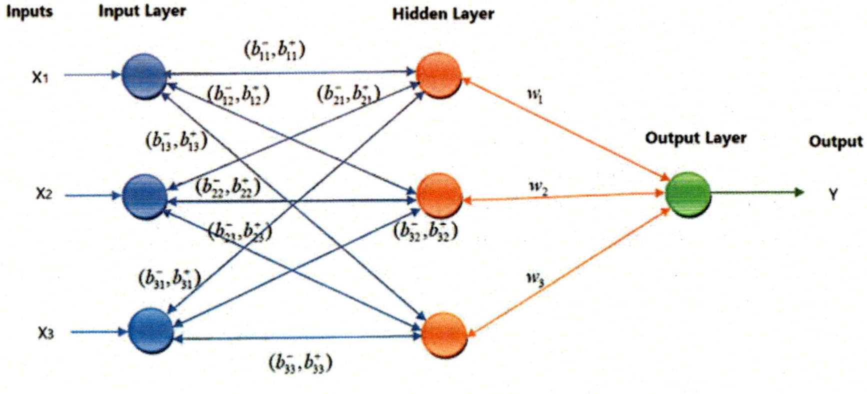

Fig. 1 BPNN using BF sets

3 Algorithm

Step 1: Set the targeted output $'T'$.

Step 2: Let the input values be x_1, x_2, x_3 and bias values b_1 and b_2.

Step 3: Determine IL weights as bipolar fuzzy set (b_{mn}^-, b_{mn}^+).

Step 4: Calculate weighted sum (WS) of positive and negative membership function of BF set,

$G_n^- = \sum_{m=1}^{p} |x_m b_{mn}^- - b_1|$ and $G_n^+ = \sum_{m=1}^{p} |x_m b_{mn}^+ + b_1|$.

Step 5: Applying sigmoid function, $g(G_n) = \frac{g(G_n^-)+g(G_n^+)}{2}$ is calculated, where $g(G_n^-) = \frac{1}{1+e^{-G_n^-}}$ and $g(G_n^+) = \frac{1}{1+e^{-G_n^+}}$

Step 6: Calculate the weights of HL,

$W_n = \frac{1-E_n}{\sum_{n=1}^{q}(1-E_n)}$ which satisfies $\sum_{n=1}^{q} W_n = 1$.

Step 7: For the output unit, the weighted sum of HL is,

$\phi = \sum W_n g(G_n) + b_2$.

Step 8: Use activation function $g(\phi) = \frac{1}{1+e^{-\phi}}$ to get the output.

If calculated output $\neq$ targeted output, then proceed further.

Step 9: To find error, $Er = \frac{1}{2}(T - g(\phi))^2$.

Step 10: For output unit, find weight and bias corrections using Definitions 4 and 5.

Repeat Steps 4 to 8. When calculated output = targeted output, end the process. Otherwise, repeat the process.

4 Application

Tomato is a fruit rich in vitamins and minerals. Nutrients are essential for plant reproduction, growth, and metabolism. Normal life cycle of a plant is incomplete without minerals. Nutrients such as nitrogen, potassium, and phosphorus are described below.

1. **Nitrogen:** Plants require a lot of nitrogen, in order to produce desired crop growth and obtain maximum benefits.
2. **Potassium:** It is an essential element for plant growth. It contributes to stem strength, disease resistance, and growth.
3. **Phosphorus:** It benefits the formation of new roots and is used in flower, fruit, and seed production.

Nitrogen-, potassium-, and phosphorus-deficient plants are shown in Fig. 2.

Insecticides control pests that affect plants.

1. **Carbosulfan:** It is used to control soil dwelling and foliar insect pests.

Fig. 2 Nutrient-deficient tomato plants

Fig. 3 Tomato plant affected by insects

2. **Abamectin:** It is a natural fermentation product for the control of mites, leaf miners, and fire ants.
3. **Acetamiprid:** It is a broad-spectrum insecticide used to regulate sucking-type insects.

Tomato plants affected by insects are shown in Fig. 3.

5 Example

Let $\alpha_1, \alpha_2, \alpha_3$ denote the amount of tomato seeds sown in three farms. Let (b_{mn}^-, b_{mn}^+) for $m, n = 1, 2, 3$ denote the nutrients nitrogen, potassium, and phosphorus applied to the farms. The returns in three farms are registered in the HL. W_1, W_2, W_3 denote the insecticides such as carbosulfan, abamectin, and acetamiprid sprayed to each farm. The total returns of three farms are found. If the returns obtained are not the targeted value, BP is carried out, and the error is determined. Then the optimal quantity of nutrients and insecticides to be supplied is estimated till the favorable returns are obtained.

Step 1: Assume the targeted output $T = 0.80$.
Step 2: The input values $\alpha_1 = 0.15; \alpha_2 = 0.03; \alpha_3 = 0.28$ and the bias values are $b_1 = 0.5$ and $b_2 = 0.65$.
Step 3: The bipolar fuzzy sets taken as the IL weights are represented in a matrix form.

$$\mathbb{BFM} = \begin{array}{c} \\ nitrogen \\ potassium \\ phosphorus \end{array} \begin{array}{c} \begin{array}{ccc} Field1 & Field2 & Field3 \end{array} \\ \begin{pmatrix} (-0.2, 0.58) & (-0.65, 0.1) & (-0.11, 0.63) \\ (-0.09, 0.79) & (-0.12, 0.69) & (-0.56, 0.3) \\ (-0.28, 0.7) & (-0.27, 0.6) & (-0.51, 0.08) \end{pmatrix} \end{array}$$

Step 4: Determine the negative and positive weighted sum.
$G_1^- = 0.3889, G_2^- = 0.3233, G_3^- = 0.3239.$
$G_1^+ = 0.8067, G_2^+ = 0.7037, G_3^+ = 0.6259.$
Step 5: Sigmoid activation function
$g(G_1^-) = 0.5960, g(G_2^-) = 0.5801, g(G_3^-) = 0.5802.$
$g(G_1^+) = 0.6914, g(G_2^+) = 0.6690, g(G_3^+) = 0.6515.$
$g(G_1) = \frac{g(G_1^-)+g(G_1^+)}{2} = 0.6437.$
Similarly, $g(G_2) = 0.6245, g(G_2) = 0.6159.$
Step 6: W_n is calculated using d_{mn} and E_n.
$d_{11} = 0.22, d_{21} = 0.12, d_{31} = 0.02.$
$d_{12} = 0.25, d_{22} = 0.19, d_{32} = 0.13.$
$d_{13} = 0.26, d_{23} = 0.14, d_{33} = 0.41.$
$E_1 = 0.12, E_2 = 0.57, E_3 = 0.81.$
Finally, $W_1 = \frac{0.88}{2.42} = 0.3636, W_2 = 0.3347, W_3 = 0.3016.$
Step 7: Calculate the weighted output.
$\alpha = g(G_1)W_1 + g(G_2)W_2 + g(G_3)W_3 + b_2$
$= 0.2340 + 0.2090 + 0.1857 + 0.65 = 1.2789.$
Step 8: Determine the output value, $g(\alpha) = 0.7822$.
Nutrients added to tomato farms at trial 1 are plotted graphically in Fig. 4.
If the output $\neq$ target, proceed further.
Step 9: $Er = 0.00015$.
Step 10: For output unit, Error
$\omega = -0.00302, \omega g(G_1) = -0.00194.$
Choose the learning rate $\eta = 0.9$.
$\eta\omega g(G_1) = -0.00175.$
Weight correction: $\nabla W_1 = 0.3653, \nabla W_2 = 0.3364, \nabla W_3 = 0.3033.$

Bias Correction: $\nabla b_2 = 0.6527.$
Step 11: For input, Error
$\zeta_1^- = 0.0437, \zeta_2^- = 0.0407, \zeta_3^- = 0.0367.$
$\zeta_1^+ = 0.0387, \zeta_2^+ = 0.0370, \zeta_3^+ = 0.0342.$
Weight Correction:
Negative:

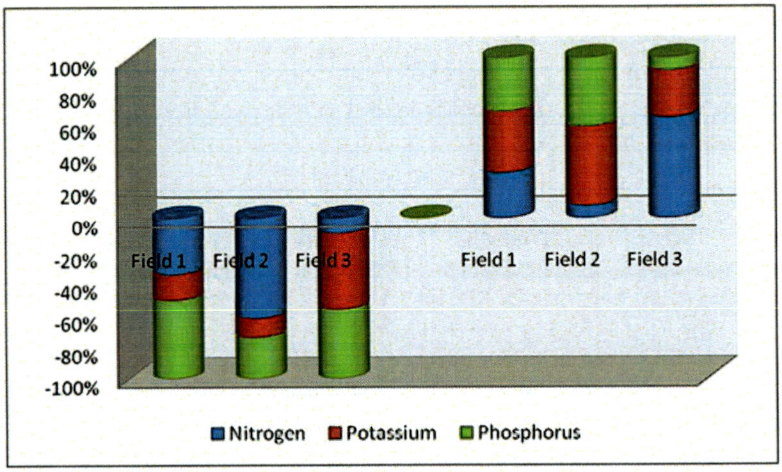

Fig. 4 Nutrients added to plants (trial 1)

$\nabla b_{11}^{-} = -0.19998$, $\nabla b_{21}^{-} = -0.09$, $\nabla b_{31}^{-} = -0.2799$.
$\nabla b_{12}^{-} = -0.64998$, $\nabla b_{22}^{-} = -0.12$, $\nabla b_{32}^{-} = -0.2699$.
$\nabla b_{13}^{-} = -0.1099$, $\nabla b_{23}^{-} = -0.56$, $\nabla b_{33}^{-} = -0.5099$.
Positive:
$\nabla b_{11}^{+} = 0.5800$, $\nabla b_{21}^{+} = 0.7900$, $\nabla b_{31}^{+} = 0.7000$.
$\nabla b_{12}^{+} = 0.1000$, $\nabla b_{22}^{+} = 0.6900$, $\nabla b_{32}^{+} = 0.6000$.
$\nabla b_{13}^{+} = 0.6300$, $\nabla b_{23}^{+} = 0.3000$, $\nabla b_{33}^{+} = 0.0800$.
Bias Correction: $\nabla b_1 = 0.5027$.

Step 12: Repeat Steps 4 to 8 after updating the weights.
Determine the WS. $G_1^{-} = |x_1 \nabla b_{11}^{-} + x_2 \nabla b_{21}^{-} + x_3 \nabla b_{31}^{-} + \nabla b_1| = 0.3916$, $G_2^{-} = 0.3260$, $G_3^{-} = 0.3266$
$G_1^{+} = 0.8094$, $G_2^{+} = 0.7064$, $G_3^{+} = 0.6286$.
Activation function, $f(G_1) = 0.6443$, $f(G_2) = 0.6252$, $f(G_3) = 0.6165$.
Weighted output, $\alpha = g(G_1)\nabla W_1 + g(G_2)\nabla W_2 + g(G_3)\nabla W_3 + b_2 = 1.2854$.
Determine the output, $g(\alpha) = 0.7833$.
Using the UW (updated weight), the output is 0.7833 which is $\neq$ target.
Nutrients added to tomato farms at trial 2 are plotted graphically in Fig. 5.
Repeating the process n=51 times UW for IL, For input layer,
$\nabla b_{11}^{-} = -0.1997$, $\nabla b_{21}^{-} = -0.09$, $\nabla b_{31}^{-} = -0.2795$,
$\nabla b_{12}^{-} = -0.6497$, $\nabla b_{22}^{-} = -0.12$, $\nabla b_{32}^{-} = -0.2695$,
$\nabla b_{13}^{-} = -0.1098$, $\nabla b_{23}^{-} = -0.56$, $\nabla b_{33}^{-} = -0.5095$.
$\nabla b_{11}^{+} = 0.5802$, $\nabla b_{21}^{+} = 0.7900$, $\nabla b_{31}^{+} = 0.7004$,
$\nabla b_{12}^{+} = 0.1011$, $\nabla b_{22}^{+} = 0.6900$, $\nabla b_{32}^{+} = 0.6004$,
$\nabla b_{13}^{+} = 0.6302$, $\nabla b_{23}^{+} = 0.3001$, $\nabla b_{33}^{+} = 0.0814$.
For output layer, $\nabla W_1 = 0.39032$, $\nabla W_2 = 0.3606$, $\nabla W_3 = 0.3268$.

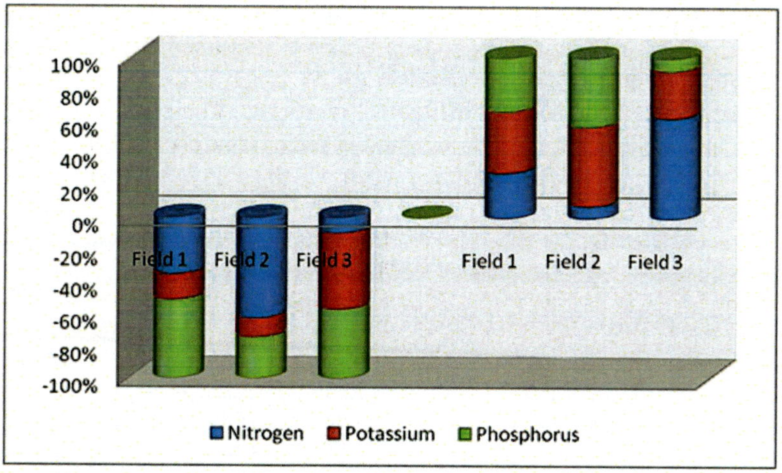

Fig. 5 Nutrients added to plants (trial 2)

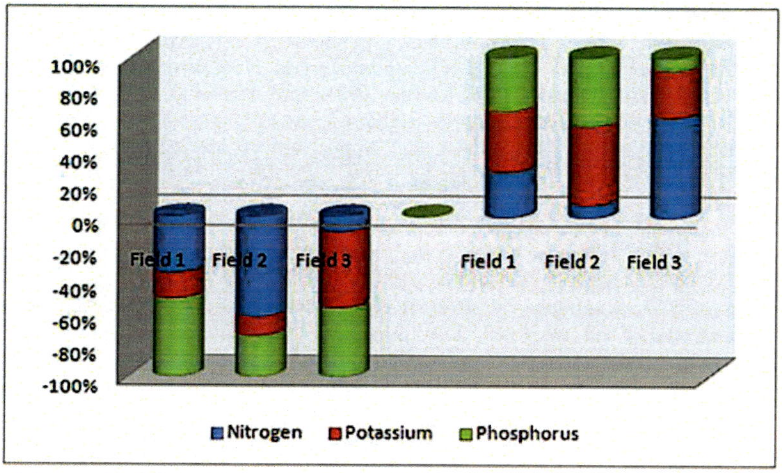

Fig. 6 Nutrients added to plants (trial n)

Bias values, $\nabla b_1 = 0.5423$, $\nabla b_2 = 0.6923$
Using these weights, the targeted output = 0.8018.
Nutrients added to tomato farms at trial n are plotted graphically in Fig. 6.

6 Conclusion

Taking the tomato seeds sown as input, nutrients as IL weight, and insecticides as HL weight, the total returns of three farms are registered. The total returns at the end of the first trial are 0.7822. If the returns obtained $\neq$ the targeted value, BP is carried out, and the error is determined. Then the optimal quantity of nutrients and insecticides to be supplied is registered. The returns after the UW are 0.7833. The process is repeated for 51 times, and finally, the targeted value, which is the favorable returns 0.8018, is reached. The correct amount of nutrients and insecticides that must be supplied in three farms is calculated using BPFNN so that the farmers attain favorable returns, which could be of great benefit to them.

Acknowledgments The authors are thankful to the reviewers for their valuable comments, which have greatly helped in improving the quality of the paper.

Conflict of Interest The authors declare that they have no conflict of interest.

References

1. Zadeh, L.A.: Fuzzy sets. Inf. Control **8**, 338–353 (1965). http://doi.org/10.2307/2272014
2. Zhang, W.R.: Bipolar fuzzy sets. Proc. FUZZ-IEEE **1**, 835–840 (1998). http://doi.org/10.1109/FUZZY.1998.687599
3. Akram, M., Akmal, R.: Application of bipolar fuzzy sets in graph structures. Appl. Comput. Intell. Soft Comput., 1–13 (2016). http://doi.org/10.1155/2016/5859080
4. Dongare, A.D., Kharde, R.R., Kachare, A.D.: Introduction to artificial neural network. Int. J. Eng. Innovat. Technol. **2**, 189–194 (2012)
5. Wu, Y., Feng, J.: Development and Application of Artificial Neural Network, pp. 1–12. Springer Science+Business Media, New York (2017). http://doi.org/10.1007/s11277-017-5224-x
6. Seenivasan, M., Chakravarthy, V.J.: Brain computing interface using deep learning for blind people. Int. J. Recent Technol. Eng. **8**(4), 8227–8230 (2019). http://doi.org/10.35940/ijrte.D8906.118419
7. Svozil, D., Kvasnicka, V., Pospichal, J.: Introduction to multi-layer feed-forward neural networks. Chemom. Intell. Lab. Syst. **39**, 43–62 (1997). http://doi.org/10.1016/S0169--7439(97)00061--0
8. Ishibuchi, H., Tanaka, H., Okada, H.: Fuzzy Neural Networks with Fuzzy Weights and Fuzzy Bias, pp. 1650–1665. IEEE (1993). http://doi.org/10.1109/ICNN.1993.298804
9. Jin, H., Quan, G., Linhui, C.: A Fuzzy Neural Network Based on Back Propagation, pp. 151–159. Springer, Berlin, Heidelberg (2012). http://doi.org/10.1007/978-3-540-72393-6_20
10. Li, J., Cheng, J., Shi, J., Huang, F.: Brief Introduction of Back Propagation (BP) Neural Network Algorithm and Its Improvement, pp. 553–558. Springer, Berlin, Heidelberg (2012). http://doi.org/10.1007/978-3-642-30223-7_87
11. Nawi, N.M., Ransing, R.S., Salleh, M.N.M., Ghazali, R., Hamid, N.A.: An Improved Back Propagation Neural Network Algorithm on Classification Problems, pp. 177–188. Springer, Berlin, Heidelberg (2010). http://doi.org/10.1007/978-3-642-17622-7_18
12. Shihab, K.: A backpropagation neural network for computer network security. J. Comput. Sci. **2**(9), 710–715 (2006)
13. Hegazy, T., Fazio, P., Moselhi, O.: Developing practical neural network applications using back-propagation. Microcomput. Civil Eng. **9**, 145–159 (1994). http://doi.org/10.1111/j.1467-8667.1994.tb00369.x
14. Chen. T., Zhong, S.: Privacy-Preserving back propagation neural network learning. IEEE Trans. Neural Netw. **20**(10), 1554–1564 (2009). http://doi.org/10.1109/TNN.2009.2026902
15. Zheng, Y., Zhong, H., Fang, Y., Zhang, W., Liu, K., Fang, J.: Rockburst prediction model based on entropy weight integrated with gray relational BP neural network. Adv. Civil Eng., 1–8 (2019). http://doi.org/10.1155/2019/3453614

16. Jae-Won, C., Trung, T.T., Thien, T.L.H.: A nutrient deficiency prediction method using deep learning on development of tomato fruits. In: International Conference on Fuzzy Theory and Its Applications iFUZZY, pp. 338–341. IEEE (2018). http://doi.org/10.1109/iFUZZY.2018.8751688
17. Walgenbach, J.F., Leidy, R.B., Sheets, T.J.: Persistence of insecticides on tomato foliage and implications for control of tomato fruit worm (Lepidoptera:Noctuidae). J. Econ. Entomol. **84**(3), 978–986 (1991). http://doi.org/10.1093/jee/84.3.978
18. Schmitz-Eiberger, M., Haefs, R., Noga, G.: Calcium deficiency - Influence on the anti oxidative defense system in tomato plant. J. Plant Physiol., 733–742 (2022). http://doi.org/10.1078/0176-1617-0621

Cryptographic Coding of Some Fibonacci Type Numbers to Determine Repeated Steps of Their Residues

T. Srinivas and K. Sridevi

Keywords Fibonacci · Lucas · Pell · Pell-Lucas · Jacobistal · Jacobistal-Lucas · Narayana numbers

1 Introduction

Number theory is the Queen of Pure Mathematics. We can apply the theory of numbers to our nature, in particular by observation in plant growth (patterns in leaves, seed distribution, development in flower petals, and branch mechanisms in the plants). In most of the cases, their mechanism is in the Fibonacci sequence of numbers, and some of them are represented below. From References [1–5].

We can introduce numbers of Fibonacci type of rth order of linear recurrence defined as follows:

$$a_n = c_1 a_{n-1} + +c_2 a_{n-2} + c_3 a_{n-3} + c_4 a_{n-4} + \ldots\ldots\ldots c_r a_{n-r}, \text{for n} \geq r, r \geq 2.$$

Some of them are represented according to their second order as follows:

Fibonacci numbers $\{1, 1, 2, 3, 5, 8, 13, 21 \ldots\ldots\ldots\}$ satisfy following recurrence relation

$$F_n = F_{n-1} + F_{n-2} \quad for\ n \geq 2, \quad with\ F_0 = 1, F_1 = 1.$$

Lucas numbers $\{2, 1, 3, 4, 7, 11, 18, 29, \ldots\ldots\ldots\}$ satisfy following recurrence relation

T. Srinivas (✉)
Department of H&S, Priyadarshini College of Engineering &Technology, Nellore, Andhra Pradesh, India

K. Sridevi
Department of Mathematics, Dr. B R Ambedkar Open University, Hyderabad, Telangana, India

H.-H. Leung et al. (eds.), *Recent Developments in Algebra and Analysis*,
Trends in Mathematics, https://doi.org/10.1007/978-3-031-37538-5_13

$$L_n = L_{n-1} + L_{n-2} \quad for\ n \geq 2, \quad with\ L_0 = 2, L_1 = 1.$$

Pell numbers {0,1,2,5,12,29 . . . ., } satisfy following recurrence relation

$$P_n = 2P_{n-1} + P_{n-2} \quad for\ n \geq 2, \quad with\ P_0 = 0, P_1 = 1.$$

Pell-Lucas numbers $\{1, 3, 7, 17, 41, 99, \ldots\}$ satisfy following recurrence relation

$$Q_n = 2Q_{n-1} + Q_{n-2} \quad for\ n \geq 2, \quad with\ Q_0 = 1, Q_1 = 3.$$

Jacobistal numbers $\{0, 1, 1, 3, 5, 11, \ldots\}$ satisfy following recurrence relation

$$J_n = J_{n-1} + 2J_{n-2} \quad for\ n \geq 2, \quad with\ J_0 = 0, J_1 = 1.$$

Jacobistal-Lucas numbers $\{2, 1, 5, 7, 17, \ldots\}$ satisfy the following recurrence relation

$$J_n = J_{n-1} + 2J_{n-2} \quad for\ n \geq 2, \quad with\ J_0 = 2, J_1 = 1.$$

Narayana numbers $\{0, 1, 1, 1, 2, 3, 4, \ldots\}$ satisfy following recurrence relation

$$N_n = N_{n-1} + N_{n-3} \quad for\ n \geq 3, \quad with\ N_0 = 1, N_1 = 1, N_2 = 1.$$

1.1 Fibonacci Numbers

Program

```
// fibonacci Numbers
    decimal[] fibonacci = new decimal[100];
    fibonacci[0] = 1; fibonacci[1] = 1;
    for (int i = 2; i < fibonacci.Length; i++)
    {
    fibonacci[i] = fibonacci[i - 1] + fibonacci[i-2];
    }
    Console.WriteLine("Fibonacci Numbers for n values
                                        2 to 100 \n");
    foreach (var item in fibonacci)
    {
            Console.Write($"{item},");
    }
```

```
Console.WriteLine("\n");
for (int n = 2; n <= 10; n++)
{
        Console.WriteLine("\nReminders of Fibonacci Numbers
        when divided by {0} \n",n);
        foreach (var item in fibonacci)
        {
           Console.Write("{0},", (item % n));
        }
        Console.WriteLine("\n");
 }
```

Reminders of Fibonacci numbers when divided by 2
1,1,0,1,

Reminders of Fibonacci numbers when divided by 3
1,1,2,0,2,2,1,0,1,1,2,0,2,2,1,0,1,1,2,0,2,2,1,0,1,1,2,0,2,2,1,0,1,1,2,0,2,2,1,0,1,1,2,0,2,2,1,0,1,1,2,0,2,2,1,0,1,1,2,0,2,2,1,0,1,1,2,0,2,2,1,0,1,1,2,0,2,2,1,0,1,1,2,0,2,2,1,0,1,1,2,0,2,2,1,0,1,1,2,0,

Reminders of Fibonacci numbers when divided by 4
1,1,2,3,1,0,1,1,2,3,1,0,1,1,2,3,1,0,1,1,2,3,1,0,1,1,2,3,1,0,1,1,2,3,1,0,1,1,2,3,1,0,1,1,2,3,1,0,1,1,2,3,1,0,1,1,2,3,1,0,1,1,2,3,1,0,1,1,2,3,1,0,1,1,2,3,1,0,1,1,2,3,1,0,1,1,2,3,1,0,1,1,2,3,1,0,1,1,2,3,

Reminders of Fibonacci numbers when divided by 5
1,1,2,3,0,3,3,1,4,0,4,4,3,2,0,2,2,4,1,0,1,1,2,3,0,3,3,1,4,0,4,4,3,2,0,2,2,4,1,0,1,1,2,3,0,3,3,1,4,0,4,4,3,2,0,2,2,4,1,0,1,1,2,3,0,3,3,1,4,0,4,4,3,2,0,2,2,4,1,0,1,1,2,3,0,3,3,1,4,0,4,4,3,2,0,2,2,4,1,0,

Reminders of Fibonacci numbers when divided by 6
1,1,2,3,5,2,1,3,4,1,5,0,5,5,4,3,1,4,5,3,2,5,1,0,1,1,2,3,5,2,1,3,4,1,5,0,5,5,4,3,1,4,5,3,2,5,1,0,1,1,2,3,5,2,1,3,4,1,5,0,5,5,4,3,1,4,5,3,2,5,1,0,1,1,2,3,5,2,1,3,4,1,5,0,5,5,4,3,1,4,5,3,2,5,1,0,1,1,2,3,

Reminders of Fibonacci numbers when divided by 7
1,1,2,3,5,1,6,0,6,6,5,4,2,6,1,0,1,1,2,3,5,1,6,0,6,6,5,4,2,6,1,0,1,1,2,3,5,1,6,0,6,6,5,4,2,6,1,0,1,1,2,3,5,1,6,0,6,6,5,4,2,6,1,0,1,1,2,3,5,1,6,0,6,6,5,4,2,6,1,0,1,1,2,3,5,1,6,0,6,6,5,4,2,6,1,0,1,1,2,3,1,1,2,3,5,1,6,0,6,6,5,4,2,6,1,0,1,1,2,3,5,1,6,0,6,6,5,4,2,6,1,0,1,1,2,3,5,1,6,0,6,6,5,4,2,6,1,0,1,1,2,3,5,1,6,0,6,6,5,4,2,6,1,0,1,1,2,3,5,1,6,0,6,6,5,4,2,6,1,0,1,1,2,3,5,1,6,0,6,6,5,4,2,6,1,0,1,1,2,3,

Reminders of Fibonacci numbers when divided by 8
1,1,2,3,5,0,5,5,2,7,1,0,1,1,2,3,5,0,5,5,2,7,1,0,1,1,2,3,5,0,5,5,2,7,1,0,1,1,2,3,5,0,5,5, 2,7,1,0,1,1,2,3,5,0,5,5,2,7,1,0,1,1,2,3,5,0,5,5,2,7,1,0,1,1,2,3,5,0,5,5,2,7,1,0,1,1,2,3, 5,0,5,5,2,7,1,0,1,1,2,3,

Reminders of Fibonacci numbers when divided by 9
1,1,2,3,5,8,4,3,7,1,8,0,8,8,7,6,4,1,5,6,2,8,1,0,1,1,2,3,5,8,4,3,7,1,8,0,8,8,7,6,4,1,5,6, 2,8,1,0,1,1,2,3,5,8,4,3,7,1,8,0,8,8,7,6,4,1,5,6,2,8,1,0,1,1,2,3,5,8,4,3,7,1,8,0,8,8,7,6, 4,1,5,6,2,8,1,0,1,1,2,3,

Reminders of Fibonacci numbers when divided by 10
1,1,2,3,5,8,3,1,4,5,9,4,3,7,0,7,7,4,1,5,6,1,7,8,5,3,8,1,9,0,9,9,8,7,5,2,7,9,6,5,1,6,7,3, 0,3,3,6,9,5,4,9,3,2,5,7,2,9,1,0,1,1,2,3,5,8,3,1,4,5,9,4,3,7,0,7,7,4,1,5,6,1,7,8,5,3,8,1, 9,0,9,9,8,7,5,2,7,9,6,5,

Integer modulo m (from 2 to 11) for Fibonacci numbers

m-modulo	Sequence of residues	Residues repeats	Nonresidues
2-modulo	(1,1,0,1,1,0,1,1,0, ...)	3 steps	–
3-modulo	**(1,1,2,0,2,2,1,0,1,1,2,0,2,2,1,0, 1,1,2,0,2,2,1,0,1,1** . . . . . .)	8 steps	–
4-modulo	(1,1,2,3,1,0,1,1,2,3,1,0,1,1,2,3 . . . .)	6 steps	–
5-modulo	**(1,1,2,3,0,3,3,1,4,0,4,4,3,2,0,2,2,4,1,0,1,1,2,** . . . . . .) **(3,0,3,3,1,4,0,4,4,3,2,0,2,2,4,1,0** . . . . . .)	20 steps	–
6-modulo	$\begin{pmatrix} 1,1,2,3,5,2,1,3,4,1,5,0, \\ 5,5,4,3,1 \\ ,4,5,3,2,5,1,0 \\ ,1,1,2,3,5,2,,3,2,5,1,0, \end{pmatrix}$	24 steps	–
7-modulo	**(1,1,2,3,5,1,6,0,6,6,5,4,2,6,1,0,1,1,2,3,** . . . .) **5,1,6,0,6,6,5,4,2,6,1,0,** . . . .)	16 steps	–
8-modulo	**(1,1,2,3,5,0,5,5,2,7,1,0,1,1,2,3,5,0,5,5,2, 7,1,0,1,1,2,3,** . . . .)	12 steps	4,6
9-modulo	**(1,1,2,3,5,8,4,3,7,1,8,0,8,8,7,6,4,1,5,6,2,8,1,0, 1,1,2,3,5,8,** . . . ..)	24 steps	–
10-modulo	**(1,1,2,3,5,8,3,1,4,5,9,4,3,7,0,7,7,4,1,5,6,1,7,8, 5,3,8,1,9,0,9,9,8,7,5,2,7,9,6,5,1,6,7,3,0,3,3,6,9, 5,4,9,3,2,5,7,2,9,1,0,1,1,2,3,5,8,3,,** . . . .)	60 steps	–

1.2 Lucas Numbers

Program:

```
// Lucas Numbers
```

```
decimal[] lucas = new decimal[100];
lucas[0] = 1; lucas[1] = 1;
for (int i = 2; i < lucas.Length; i++)
{
    lucas[i] = lucas[i - 1] + lucas[i-2];
}
Console.WriteLine("Lucas Numbers for n values 2 to 100 \n");
foreach (var item in lucas)
{
     Console.Write($"{item},");
}
Console.WriteLine("\n");
for (int n = 2; n <= 10; n++)
{
    Console.WriteLine("\nReminders of Lucas Numbers
    when divided by {0} \n", n);
    foreach (var item in lucas)
    {
         Console.Write("{0},",(item % n));
    }
    Console.WriteLine("\n");
                                        }
```

Reminders of Lucas numbers when divided by 2
1,1,0,1,1,0,1,1,0,1,1,0,1,1,0,1,1,0,1,1,0,1,1,0,1,1,0,1,1,0,1,1,0,1,1,0,1,1,0,1,1,0,1,1,0,1,1,
0,1,1,0,1,1,0,1,1,0,1,1,0,1,1,0,1,1,0,1,1,0,1,1,0,1,1,0,1,1,0,1,1,0,1,1,0,1,1,0,1,1,0,1,1,0,
1,1,0,1,1,0,1,1,0,1,1,0,1,

Reminders of Lucas numbers when divided by 3
1,1,2,0,2,2,1,0,1,1,2,0,2,2,1,0,1,1,2,0,2,2,1,0,1,1,2,0,2,2,1,0,1,1,2,0,2,2,1,0,1,1,2,
0,2,2,1,0,1,1,2,0,2,2,1,0,1,1,2,0,2,2,1,0,1,1,2,0,2,2,1,0,1,1,2,0,2,2,1,0,1,1,2,0,2,
2,1,0,1,1,2,0,2,2,1,0,1,1,2,0,

Reminders of Lucas numbers when divided by 4
1,1,2,3,1,0,1,1,2,3,1,0,1,1,2,3,1,0,1,1,2,3,1,0,1,1,2,3,1,0,1,1,2,3,1,0,1,1,
2,3,1,0,1,1,2,3,1,0,1,1,2,3,1,0,1,1,2,3,1,0,1,1,2,3,1,0,1,1,2,3,1,0,1,1,2,
3,1,0,1,1,2,3,1,0,1,1,2,3,1,0,1,1,2,3,1,0,1,1,2,3,

Reminders of Lucas numbers when divided by 5
1,1,2,3,0,3,3,1,4,0,4,4,3,2,0,2,2,4,1,0,1,1,2,3,0,3,3,1,4,0,4,4,3,2,0,2,2,4,1,
0,1,1,2,3,0,3,3,1,4,0,4,4,3,2,0,2,2,4,1,0,1,1,2,3,0,3,3,1,4,0,4,4,3,2,0,2,2,4,1,0,1,1,
2,3,0,3,3,1,4,0,4,4,3,2,0,2,2,4,1,0,

Reminders of Lucas numbers when divided by 6
1,1,2,3,5,2,1,3,4,1,5,0,5,5,4,3,1,4,5,3,2,5,1,0,1,1,2,3,5,2,1,3,4,1,5,0,5,5,
4,3,1,4,5,3,2,5,1,0,1,1,2,3,5,2,1,3,4,1,5,0,5,5,4,3,1,4,5,3,2,5,1,0,1,1,2,3,
5,2,1,3,4,1,5,0,5,5,4,3,1,4,5,3,2,5,1,0,1,1,2,3,

Reminders of Lucas numbers when divided by 7 1,1,2,3,5,1,6,0,6,6,5,4,2,6,1,0,1,1,2,3,5,1,6,0,6,6,5,4,2,6,1,0,1,1,2,3,5,1,6,0,6,6,5,4,2,6,1,0,1,1,2,3,5,1,6,0,6,6,5,4,2,6,1,0,1,1,2,3,5,1,6,0,6,6,5,4,2,6,1,0,1,1,2,3,5,1,6,0,6,6,5,4,2,6,1,0,1,1,2,3,

Reminders of Lucas numbers when divided by 8
1,1,2,3,5,0,5,5,2,7,1,0,1,1,2,3,5,0,5,5,2,7,1,0,1,1,2,3,5,0,5,5,2,7,1,0,1,1,2,3,5,0,5,5,2,7,1,0,1,1,2,3,5,0,5,5,2,7,1,0,1,1,2,3,5,0,5,5,2,7,1,0,1,1,2,3,5,0,5,5,2,7,1,0,1,1,2,3,5,0,5,5,2,7,1,0,1,1,2,3,

Reminders of Lucas numbers when divided by 9
1,1,2,3,5,8,4,3,7,1,8,0,8,8,7,6,4,1,5,6,2,8,1,0,1,1,2,3,5,8,4,3,7,1,8,0,8,8,7,6,4,1,5,6,2,8,1,0,1,1,2,3,5,8,4,3,7,1,8,0,8,8,7,6,4,1,5,6,2,8,1,0,1,1,2,3,5,8,4,3,7,1,8,0,8,8,7,6,4,1,5,6,2,8,1,0,1,1,2,3,

Reminders of Lucas numbers when divided by 10
1,1,2,3,5,8,3,1,4,5,9,4,3,7,0,7,7,4,1,5,6,1,7,8,5,3,8,1,9,0,9,9,8,7,5,2,7,9,6,5,1,6,7,3,0,3,3,6,9,5,4,9,3,2,5,7,2,9,1,0,1,1,2,3,5,8,3,1,4,5,9,4,3,7,0,7,7,4,1,5,6,1,7,8,5,3,8,1,9,0,9,9,8,7,5,2,7,9,6,5,

Integer modulo m (from 2 to 11) of Lucas numbers

m-modulo	Sequence of residues	Residues repeats	Nonresidues
2-modulo	{1, 1, 0, 1, 1, 0, 1, 1, 0, 1, 1, 0, 1, 1, 0, ..}	3 steps	–
3-modulo	{1, 1, 2, 0, 2, 2, 1, 0, 1, 1, 2, 0, 2, 2, 1, 0, }	8 steps	–
4-modulo	{1, 1, 2, 3, 1, 0, 1, 1, 2, 3, 1, 0, 1, 1, 2, 3, 1, 0, }	6 steps	–
5-modulo	{1, 1, 2, 3, 0, 3, 3, 1, 4, 0, 4, 4, 3, 2, 0, 2, 2, 4, 1, 0, 1, 1, 2, 3, 0, 3, 3}	20 steps	–
6-modulo	{1, 1, 2, 3, 5, 2, 1, 3, 4, 1, 5, 0, 5, 5, 4, 3, 1 , 4, 5, 3, 2, 5, 1, 0, 1, 1, 2, 3}	24 steps	–
7-modulo	{1, 1, 2, 3, 5, 1, 6, 0, 6, 6, 5, 4, 2, 6, 1, 0, 1, 1, 2, 3}	16 steps	–
8-modulo	{1, 1, 2, 3, 5, 0, 5, 5, 2, 7, 1, 0, 1, 1, 2, 3, 5, 0, 5, 5, 2, 7, 1, 0, 1, 1, 2, 3}	12 steps	4,6
9-modulo	{1, 1, 2, 3, 5, 8, 4, 3, 7, 1, 8, 0, 8, 8, 7, 6, 4, 1, 5, 6, 2, 8, 1, 0, 1, 1, 2, 3,}	24 steps	–
10-modulo	{1, 1, 2, 3, 5, 8, 3, 1, 4, 5, 9, 4, 3, 7, 0, 7, 7, 4, 1, 5, 6, 1, 7, 8, 5, 3, 8, 1, 9, 0, 9, 9, 8, 7 , 5, 2, 7, 9, 6, 5, 1, 6, 7, 3, 0, 3, 3, 6, 9, 5 , 4, 9, 3, 2, 5, 7, 2, 9, 1, 0, 1, 1, 2, 3}	60 steps	

1.3 Pell Numbers

Program

```
// Pell Numbers
   using System.Numerics;
   BigInteger[] pell = new BigInteger[100];
   pell[0] = 0; pell[1] = 1;
   for (int i = 2; i < pell.Length; i++)
   {
       pell[i] = (2 * pell[i - 1]) + pell[i-2];
   }
   Console.WriteLine("Pell Numbers for n values 2 to 100 \n");
   foreach (var item in pell)
   {
         Console.Write($"{item},");
   }
   Console.WriteLine("\n");
   for (int n = 2; n <= 10; n++)
   {
        Console.WriteLine("\nReminders of Pell Numbers when
        divided by {0} \n", n);
        foreach (var item in pell)
        {
        Console.Write("{0},",(item % n));
        }
        Console.WriteLine("\n");
    }
```

Integer modulo of Pell numbers

1.4 Jacobistal Numbers

Program

```
      // Jacobistal Numbers
        using System.Numerics;
        BigInteger[] jacobistal = new BigInteger[100];
        jacobistal[0] = 0; jacobistal[1] = 1;
        for (int i = 2; i < jacobistal.Length; i++)
{
            jacobistal[i] = jacobistal[i - 1] +
            (2 * jacobistal[i-2]);
}
Console.WriteLine("Jacobistal Numbers for n values 2 to 100
\n");
foreach (var item in jacobistal)
{
     Console.Write($"{item},");
}
```

n-modulo	Sequence of residues	Residues repeats	Nonresidues
2-modulo	$\begin{Bmatrix} 0, 1, 0, 1, 0, \\ 1, 0, 1 \ldots\ldots., \end{Bmatrix}$	2 steps	–
3-modulo	$\begin{Bmatrix} 0, 1, 2, 2, \\ 0, 2, 1, 1, 0 \\ , 1, 2, 2, \ldots., \end{Bmatrix}$	8 steps	–
4-modulo	$\begin{Bmatrix} 0, 1, 2, 1, 0, 1, \\ 2, 1, 0, 1, 2, 1, \\ 0, 1, 2, 1 \ldots., \end{Bmatrix}$	4 steps	3
5-modulo	$\begin{Bmatrix} 0, 1, 2, 0, 2, 4 \\ , 0, 4, 3, 0, 3, 1, \\ 0, 1, 2, 0 \ldots., \end{Bmatrix}$	12 steps	–
6-modulo	$\begin{Bmatrix} 0, 1, 2, 5, 0, 5, \\ 4, 1, 0, 1, 2, 5, \\ 0, 5, 4, \ldots., \end{Bmatrix}$	8 steps	3
7-modulo	$\begin{Bmatrix} 0, 1, 2, 5, 5, 1, \\ 0, 1, 2, 5, 5, 1 \end{Bmatrix}$	6 steps	3,4,6
8-modulo	$\begin{Bmatrix} 0, 1, 2, 5, 4, 5, 6, \\ 1, 0, 1, 2, \ldots. \end{Bmatrix}$	8 steps	3,7
9-modulo	{0, 1, 2, 5, 3, 2, 7, 7, 3, 4, 2, 8, 0 ,8, 7, 4, 6, 7, 2, 2, 6, 5, 7, 1, **0,1,2,5,3**}	24 steps	–
10-modulo	$\begin{Bmatrix} 0, 1, 2, 5, 2, 9, \\ 0, 9, 8, 5, , 8, 1, \\ 0, 1, 2, 5 \ldots \end{Bmatrix}$	12 steps	3,4,6,7

```
Console.WriteLine("\n");
for (int n = 2; n <= 10; n++)
{
      Console.WriteLine("\nReminders of Jacobistal
      Numbers when divided by {0} \n", n);
      foreach (var item in jacobistal)
      {
      Console.Write("{0},", (item % n));
      }
      Console.WriteLine("\n");
}
```

Integer modulo of Jacobistal numbers

n-modulo	Sequence of residues	Residues repeats	Nonresidues
2-modulo	{0, 1, 1, 1, 1, }	Residue "1" is repeated	-
3-modulo	{0, 1, 1, 0, 2, 2, 0, 1, 1, 0, 2, 2}	6 steps	-
4-modulo	{**0, 1, 1, 3, 1, 3, 1, 3, 1, 3, 1, 3**}	After two steps 1,3 are repeated	2
5-modulo	{0, 1, 1, 3, 0, 1, 1, 3, 0}	4 steps	2,4
6-modulo	{0, 1, 1, 3, 5, 5, 3, 1, 1, 3, 5, ...}	6 steps (except first residue)	2,4
7-modulo	{0, 1, 1, 3, 5, 4, 0, 1, 1, 3, ..}	6 steps	2,6
8-modulo	{0, 1, 1, 3, 5, 3, 5, 3, 5, ..}	After 5 steps, two residues 3,5 are repeated	2,4,6,7
9-modulo	{**0, 1, 1, 3, 5, 2, 3, 7, 4, 0, 8, 8, 6, 4, 7, 6, 2, 5, 0, 1, 1, 3,**}	18 steps	-
10-modulo	{**0, 1, 1, 3, 5, 1, 1, 3, 5, 1, 1, 3, 5, 1, 1, 3, 5**}	1,1,3,5 repeats	2,4,7,9

1.5 *Jacobistal-Lucas Numbers*

Program

```
// Jacobistal-Lucas Numbers
    using System.Numerics;
    BigInteger[] jacobistalLucas = new BigInteger[100];
    jacobistalLucas[0] = 2; jacobistalLucas[1] = 1;
    for (int i = 2; i < jacobistalLucas.Length; i++)
    {
        jacobistalLucas[i] = jacobistalLucas[i - 1]
        + (2 * jacobistalLucas[i-2]);
    }
    Console.WriteLine("Jacobistal-Lucas Numbers for n
    values 2 to 100 \n");
    foreach (var item in jacobistalLucas)
    {
          Console.Write($"{item},");
    }
    Console.WriteLine("\n");
```

Reminders of Narayana numbers when divided by 4
0,1,1,1,2,3,0,2,1,1,3,0,1,0,0,1,1,1,2,3,0,2,1,1,3,0,1,0,0,1,1,1,2,3,0,2,1,1,
3,0,1,0,0,1,1,1,2,3,0,2,1,1,3,0,1,0,0,1,1,1,2,3,0,2,1,1,3,0,1,0,0,1,1,1,2,
3,0,2,1,1,3,0,1,0,0,1,1,1,2,3,0,2,1,1,3,0,1,0,0,1,

Reminders of Narayana numbers when divided by 5
0,1,1,1,2,3,4,1,4,3,4,3,1,0,3,4,4,2,1,0,2,3,3,0,3,1,1,4,0,1,0,0,1,1,1,2,3,
4,1,4,3,4,3,1,0,3,4,4,2,1,0,2,3,3,0,3,1,1,4,0,1,0,0,1,1,1,2,3,4,1,4,3,4,
3,1,0,3,4,4,2,1,0,2,3,3,0,3,1,1,4,0,1,0,0,1,1,1,2,3,4,

Reminders of Narayana numbers when divided by 6
0,1,1,1,2,3,4,0,3,1,1,4,5,0,4,3,3,1,4,1,2,0,1,3,3,4,1,4,2,3,1,3,0,1,4,4,5,
3,1,0,3,4,4,1,5,3,4,3,0,4,1,1,5,0,1,0,0,1,1,1,2,3,4,0,3,1,1,4,5,0,4,3,3,1,
4,1,2,0,1,3,3,4,1,4,2,3,1,3,0,1,4,4,5,3,1,0,3,4,4,1,

Reminders of Narayana numbers when divided by 7
0,1,1,1,2,3,4,6,2,6,5,0,6,4,4,3,0,4,0,0,4,4,4,1,5,2,3,1,3,6,0,3,2,2,5,0,2,0,
0,2,2,2,4,6,1,5,4,5,3,0,5,1,1,6,0,1,0,0,1,1,1,2,3,4,6,2,6,5,0,6,4,4,3,0,4,0,
0,4,4,4,1,5,2,3,1,3,6,0,3,2,2,5,0,2,0,0,2,2,2,4,

Reminders of Narayana numbers when divided by 8
0,1,1,1,2,3,4,6,1,5,3,4,1,4,0,1,5,5,6,3,0,6,1,1,7,0,1,0,0,1,1,1,2,3,4,6,1,5,
3,4,1,4,0,1,5,5,6,3,0,6,1,1,7,0,1,0,0,1,1,1,2,3,4,6,1,5,3,4,1,4,0,1,5,5,6,3,
0,6,1,1,7,0,1,0,0,1,1,1,2,3,4,6,1,5,3,4,1,4,0,1,

Reminders of Narayana numbers when divided by 9
0,1,1,1,2,3,4,6,0,4,1,1,5,6,7,3,0,7,1,1,8,0,1,0,0,1,1,1,2,3,4,6,0,4,1,1,5,6,
7,3,0,7,1,1,8,0,1,0,0,1,1,1,2,3,4,6,0,4,1,1,5,6,7,3,0,7,1,1,8,0,1,0,0,1,1,
1,2,3,4,6,0,4,1,1,5,6,7,3,0,7,1,1,8,0,1,0,0,1,1,1,

Reminders of Narayana numbers when divided by 10
0,1,1,1,2,3,4,6,9,3,9,8,1,0,8,9,9,7,6,5,2,8,3,5,3,6,1,4,0,1,5,5,6,1,6,2,3,9,1,
4,3,4,8,1,5,3,4,9,2,6,5,7,3,8,5,8,6,1,9,5,6,5,0,6,1,1,7,8,9,6,4,3,9,3,6,5,8,4,
9,7,1,0,7,8,8,5,3,1,6,9,0,6,5,5,1,6,1,2,8,9,

2 Conclusion

This paper focused on to study cryptographic coding of linear recurrence relations of some Fibonacci type numbers to determine repeated steps of their residues for integer modulo from 2 to 10. It particularly focused to study repeated steps of residues of Fibonacci, Lucas, Pell, Pell-Lucas, Jacobistal, Jacobistal-Lucas, and Narayana numbers. In cryptographic coding, repeated steps of residues are useful for public sector, and nonresidues are useful for private sector. So we are focused to generate the Fibonacci type number repeated steps of residues under integer modulo from 2 to 10.

References

1. Bednarj, N.: ON (k,p)- Fibonacci numbers. J. MDPI Math. **9**(7), 727 (2021)
2. Hazra Matrix, A.K.: Algebra, Calculus and Generalised Inverse (2010)
3. Singh, B.: Advanced Abstract Algebra (2002)
4. David M.: Burton Elementary Number theory (2003)
5. Sridevi, K., Srinivas, T.: A new approach to define integer sequences of Fibonacci type numbers with using of third order linear recurrence relations. In: AIP Conference Proceedings 2385, 130005 (2022); https://doi.org/10.1063/5.0070691, 06 January 2022

On Permutation Distributive BI-Algebras

Nadia M. Ali Abbas, Shuker Khalil, and Enoch Suleiman

Keywords Cycles · Permutation · BI-algebras · Compatible relation · Quasi-associative · Distributive · Symmetric groups

1 Introduction

Imai and Iseki [1] investigated and studied the notion of BCK-algebra, and also they looked into several connections between d-algebras and BCK-algebras, where the concept of d-algebras is introduced by Neggers and Kim [2]. Many academics have intensively examined numerous generalizations of a B-algebra, and properties have been considered methodically. Next, the concept of B-algebras ([3]) is pioneered. The B-algebra is an algebra of type (2,0).

The notion of BI-algebra is shown by Saeid et al. [4]. They discuss the essential properties of BI-algebras as well as ideals and congruence relations. A BI-algebra is an extension of (dual) implication algebra. Alsalem [5] provides permutation sets. The permutations of symmetric and alternating groups are examined [6–13]. A permutation set is a nonclassical set such as fuzzy sets [14–21], soft sets [22–27], neutrosophic sets [28–33], and nano sets [34].

We looked at permutation quasi-associative BI-algebra, permutation BI-ideal, and permutation right (left) distributive BI-algebra. Moreover, we explored some new notions in permutation theory for the first time. We also examined permutation right compatible relation, permutation left compatible relation, and permutation compatible relation in permutation BI-algebra.

N. M. Ali Abbas
Ministry of Education, Directorate General of Education, Baghdad, Al-Kark, Iraq

S. Khalil (✉)
Department of Mathematics, College of Science, University of Basrah, Basrah, Iraq

E. Suleiman
Department of Mathematics, Federal University Gashua, Gashua, Yobe State, Nigeria

H.-H. Leung et al. (eds.), *Recent Developments in Algebra and Analysis*,
Trends in Mathematics, https://doi.org/10.1007/978-3-031-37538-5_14

2 Preliminaries

This section covers the fundamental principles as well as facts relevant to this topic.

Definition 2.1: [3] Let $X \neq \varnothing$ and 0 be a constant with a binary operation $*$. We say that $(X, *, \varnothing)$ is a *BI-algebra* if it satisfies the following conditions:

(a) $x * x = 0$.
(b) $x * (y * x) = x, \quad \forall x, y \in X$.

Definition 2.2: [5] For any permutation $\beta = \prod_{i=1}^{c(\beta)} \lambda_I$ in a symmetric group S_n, where $\{\lambda_i\}_{i=1}^{c(\beta)}$ is a composite of pairwise disjoint cycles $\{\lambda_i\}_{i=1}^{c(\beta)}$ where $\lambda_i = \left(t_1^I, t_2^I, \ldots, t_{\alpha_i}^i\right), 1 \leq i \leq c(\beta)$, for some $1 \leq \alpha_i$, $c(\beta) \leq n$. If $\lambda = (t_1, t_2, \ldots, t_k)$ is $k-$cycle in S_n, we define $\beta-$set as $\lambda^\beta = \{t_1, t_2, \ldots, t_k\}$ and is called $\beta-$set of cycle λ. So the $\beta-$sets of $\{\lambda_i\}_{i=1}^{c(\beta)}$ are defined by$\left\{\lambda_i^\beta = \{t_1^I, t_2^I, \ldots, t_{\alpha_i}^i\} \mid 1 \leq i \leq c(\beta)\right\}$.

3 Permutation BI-Algebras

In this section, we'll examine some of the core traits of permutation BI-algebras ($PBI - As$) and explore some fresh applications.

Definition 3.1: Let $X = \left\{\lambda_i^\beta\right\}_{i=1}^{c(\beta)}$ be a collection of β-sets, where β is a permutation in the symmetric group $G = S_n$. Then X is, namely, a permutation BI-algebra ($PBI - A$) if there exists a mapping $\# : X \times X \longrightarrow X$ such that (1) $\lambda_i^\beta \# \lambda_i^\beta = \{1\}$, (2) $\lambda_i^\beta \# \left(\lambda_j^\beta \# \lambda_i^\beta\right) = \lambda_i^\beta, \quad \forall \lambda_i^\beta, \lambda_j^\beta \in X$. We say that $\{1\}$ is the fixed element in X.

Example 3.2: Let (S_{12}, o) be a symmetric group and $\beta = \begin{pmatrix} 1 & 2 & 3 & 4 & 5 & 6 & 7 & 8 & 9 & 10 & 11 & 12 \\ 6 & 3 & 5 & 2 & 4 & 7 & 1 & 9 & 8 & 12 & 10 & 11 \end{pmatrix}$ be a permutation in S_{12}. Since $\beta = \begin{pmatrix} 1 & 2 & 3 & 4 & 5 & 6 & 7 & 8 & 9 & 10 & 11 & 12 \\ 6 & 3 & 5 & 2 & 4 & 7 & 1 & 9 & 8 & 12 & 10 & 11 \end{pmatrix} = (167)(2354)(89)$

(10 12 11). Therefore, we have $X = \left\{\lambda_i^\beta\right\}_{i=1}^{4} \cup \{1\} = \{\{1, 6, 7\}, \{2, 3, 5, 4\}, \{8, 9\}$, $\{10$, $\{10,12,11\}$, $\{1\}\}$. Define $\# : X \times X \longrightarrow X$ by $\#\left(\lambda_i^\beta, \lambda_j^\beta\right) = \lambda_i^\beta \# \lambda_j^\beta = \lambda_k^\beta$, where λ_k^β its cycle λ_k^β such that $\lambda_k = \begin{cases} \lambda_i o \lambda_j^{-1}, if\ i = j \\ \lambda_j, if\ i \neq j \end{cases}$, where λ_i and λ_j are cycles for λ_i^β and λ_j^β, respectively. Here, we have (i)$\lambda_i o \lambda_i^{-1} = (1) \rightarrow \lambda_i^\beta \# \lambda_i^\beta = \{1\}$, (ii)

when $i = j \to \lambda_i o\left(\lambda_i o\lambda_i^{-1}\right) = \lambda_i o(1)^{-1} = \lambda_i \to \lambda_i^\beta \# \left(\lambda_j^\beta \# \lambda_i^\beta\right) = \lambda_i^\beta$, also, if $i \neq j \Longrightarrow \lambda_i^\beta \# \left(\lambda_j^\beta \# \lambda_i^\beta\right) = \lambda_i^\beta \# \lambda_j^\beta = \lambda_i^\beta$. Then X is a $(PBI - A)$.

Definition 3.3: Let $(X, \#, \{1\})$ be a $(PBI - A)$. We introduce a relation "$\leq$" on X defined by $\lambda_i^\beta \leq \lambda_j^\beta$ if and only if $\lambda_i^\beta \# \lambda_j^\beta = \{1\}$. This relation "$\leq$" is not partially ordered. In other side, it is just reflexive.

Proposition 3.4: Let $(X, \#, \{1\})$ be a $(PBI - A)$. Then, (1) $\lambda_i^\beta \# \{1\} = \lambda_i^\beta$, (2) $\{1\} \# \lambda_i^\beta = \{1\}$, (3) $\lambda_i^\beta \# \lambda_j^\beta = \left(\lambda_i^\beta \# \lambda_j^\beta\right) \# \lambda_j^\beta$, (4) If $\lambda_j^\beta \# \lambda_i^\beta = \lambda_i^\beta, \forall \lambda_i^\beta, \lambda_j^\beta \in X$ then $X = \{\{1\}\}$, (5) If $\lambda_i^\beta \# \left(\lambda_j^\beta \# \lambda_k^\beta\right) = \lambda_j^\beta \# \left(\lambda_i^\beta \# \lambda_k^\beta\right), \forall \lambda_i^\beta, \lambda_j^\beta, \lambda_k^\beta \in X$, then $X = \{\{1\}\}$, (6) If $\lambda_i^\beta \# \lambda_j^\beta = \lambda_k^\beta$, then $\lambda_k^\beta \# \lambda_j^\beta = \lambda_k^\beta$ and $\lambda_j^\beta \# \lambda_k^\beta = \lambda_j^\beta$. (7) If $\left(\lambda_i^\beta \# \lambda_j^\beta\right) \# \left(\lambda_k^\beta \# \lambda_l^\beta\right) = \left(\lambda_i^\beta \# \lambda_k^\beta\right) \# \left(\lambda_j^\beta \# \lambda_l^\beta\right)$, then $X = \{\{1\}\}$ for all $\lambda_i^\beta, \lambda_j^\beta, \lambda_k^\beta, \lambda_l^\beta \in X$.

***Proof*:**

(1) Substituting $\lambda_j^\beta = \lambda_i^\beta$ in (2) of Definition 3.1, we have that $\lambda_i^\beta = \lambda_i^\beta \# \left(\lambda_i^\beta \# \lambda_i^\beta\right) = \lambda_i^\beta \# \{1\}$.

(2) Substituting $\lambda_j^\beta = \lambda_i^\beta$ and $\lambda_i^\beta = \{1\}$ in (2) of Definition 3.1, we have that $\{1\} = \{1\} \# \left(\lambda_i^\beta \# \{1\}\right) = \{1\} = \lambda_i^\beta$ (From (1)).

(3) Given that $\lambda_i^\beta, \lambda_j^\beta \in X$, we have that $\lambda_i^\beta \# \lambda_j^\beta = \left(\lambda_i^\beta \# \lambda_j^\beta\right) \# \left(\lambda_j^\beta \# \left(\lambda_i^\beta \# \lambda_j^\beta\right)\right) = \left(\lambda_i^\beta \# \lambda_j^\beta\right) \# \lambda_j^\beta$.

(4) If $\lambda_j^\beta \# \lambda_i^\beta = \lambda_i^\beta, \forall \lambda_i^\beta, \lambda_j^\beta \in X$, then $\lambda_i^\beta = \lambda_i^\beta \# \left(\lambda_j^\beta \# \lambda_i^\beta\right) = \lambda_i^\beta \# \lambda_i^\beta = \{1\}$. Thus, $X = \{\{1\}\}$.

(5) $\forall \lambda_i^\beta \in X$, we have that $\{1\} = \{1\} \# \left(\lambda_i^\beta \# \{1\}\right) = \lambda_i^\beta \# (\{1\} \# \{1\}) = \lambda_i^\beta \# \{1\} = \lambda_i^\beta$. Hence $X = \{\{1\}\}$.

(6) If $\lambda_i^\beta \# \lambda_j^\beta = \lambda_k^\beta$, then from (3), we have that $\lambda_k^\beta \# \lambda_j^\beta = \left(\lambda_i^\beta \# \lambda_j^\beta\right) \# \lambda_j^\beta = \lambda_i^\beta \# \lambda_j^\beta = \lambda_k^\beta$. Also, $\lambda_j^\beta \# \lambda_k^\beta = \lambda_j^\beta \# \left(\lambda_i^\beta \# \lambda_j^\beta\right) = \lambda_j^\beta$.

(7) If $\lambda_i^\beta \in X$, then we have that $\lambda_i^\beta = \lambda_i^\beta \# \{1\} = \left(\lambda_i^\beta \# \{1\}\right) \# \left(\lambda_i^\beta \# \lambda_i^\beta\right) = \left(\lambda_i^\beta \# \lambda_i^\beta\right) \# \left(\{1\} \# \lambda_i^\beta\right) = \{1\} \# \left(\{1\} \# \lambda_i^\beta\right) = \{1\} \# \{1\} = \{1\}$. Hence $X = \{\{1\}\}$.

Definition 3.5: A permutation BI-algebra X is, namely, permutation right distributive BI-algebra $(PRDBI - A)$, if $\left(\lambda_i^\beta \# \lambda_j^\beta\right) \# \lambda_k^\beta = \left(\lambda_i^\beta \# \lambda_k^\beta\right) \# \left(\lambda_j^\beta \# \lambda_k^\beta\right)$. Also, it is, namely, permutation left distributive BI-algebra $(PLDBI - A)$, if $\lambda_k^\beta \# \left(\lambda_i^\beta \# \lambda_j^\beta\right) = \left(\lambda_k^\beta \# \lambda_i^\beta\right) \# \left(\lambda_k^\beta \# \lambda_j^\beta\right)$, for all $\lambda_i^\beta, \lambda_j^\beta, \lambda_k^\beta \in X$.

Proposition 3.6: If $(X, \#, \{1\})$ is $(PLDBI - A)$, then $X = \{\{1\}\}$.

***Proof*:** Let $\lambda_i^\beta \in X$. Thus, substituting $\lambda_j^\beta = \lambda_i^\beta$ in (2) of Definition 3.1, we have that $\lambda_i^\beta = \lambda_i^\beta \# \left(\lambda_i^\beta \# \lambda_i^\beta\right) = \left(\lambda_i^\beta \# \lambda_i^\beta\right) \# \left(\lambda_i^\beta \# \lambda_i^\beta\right) = \{1\} \# \{1\} = \{1\}$.

Proposition 3.7: Assume that$(X, \#)$ is a permutation groupoid with $\{1\} \in X$. If it is such that (1) $\lambda_i^\beta \# \lambda_i^\beta = \{1\}$, (2) $\lambda_i^\beta \# \lambda_j^\beta = \lambda_i^\beta$, $\forall \lambda_i^\beta \neq \lambda_j^\beta \in X$. Then $(X, \#, \{1\})$ is a $(PRDBI - A)$.

***Proof*:** Let $\lambda_i^\beta \neq \lambda_j^\beta \neq \lambda_k^\beta \in X$. Then $\left(\lambda_i^\beta \# \lambda_j^\beta\right) \# \lambda_k^\beta = \lambda_i^\beta \# \lambda_j^\beta = \left(\lambda_i^\beta \# \lambda_k^\beta\right) \# \left(\lambda_j^\beta \# \lambda_k^\beta\right)$.

Proposition 3.8: Let $(X, \#, \{1\})$ be a $(PRDBI - A)$. Then (1) $\lambda_j^\beta \# \lambda_i^\beta \leq \lambda_j^\beta$, (2) $\left(\lambda_j^\beta \# \lambda_i^\beta\right) \# \lambda_i^\beta \leq \lambda_j^\beta$, (3) $\left(\lambda_i^\beta \# \lambda_k^\beta\right) \# \left(\lambda_j^\beta \# \lambda_k^\beta\right) \leq \lambda_i^\beta \# \lambda_j^\beta$. (4) If $\lambda_i^\beta \leq \lambda_j^\beta$, then $\lambda_i^\beta \# \lambda_k^\beta \leq \lambda_j^\beta \# \lambda_k^\beta$, (5) $\left(\lambda_i^\beta \# \lambda_j^\beta\right) \# \lambda_k^\beta \leq \lambda_i^\beta \# \left(\lambda_j^\beta \# \lambda_k^\beta\right)$, and (6) if $\lambda_i^\beta \# \lambda_j^\beta = \lambda_k^\beta \# \lambda_j^\beta$, then $\left(\lambda_i^\beta \# \lambda_k^\beta\right) \# \lambda_j^\beta = \{1\}$, $\forall \lambda_i^\beta, \lambda_j^\beta, \lambda_k^\beta \in X$.

***Proof*:** For all $\lambda_i^\beta, \lambda_j^\beta \in X$, we have that

(1) $\left(\lambda_j^\beta \# \lambda_i^\beta\right) \# \lambda_j^\beta = \left(\lambda_j^\beta \# \lambda_j^\beta\right) \# \left(\lambda_i^\beta \# \lambda_j^\beta\right) = \{1\} \# \left(\lambda_i^\beta \# \lambda_j^\beta\right) = \{1\}$. Thus $\lambda_j^\beta \# \lambda_i^\beta \leq \lambda_j^\beta$.

(2) $\left(\left(\lambda_j^\beta \# \lambda_i^\beta\right) \# \lambda_i^\beta\right) \# \lambda_j^\beta = \left(\left(\lambda_j^\beta \# \lambda_i^\beta\right) \# \lambda_j^\beta\right) \# \left(\lambda_i^\beta \# \lambda_j^\beta\right) = \left(\left(\lambda_j^\beta \# \lambda_j^\beta\right) \# \left(\lambda_i^\beta \# \lambda_j^\beta\right)\right) \# \left(\lambda_i^\beta \# \lambda_j^\beta\right) = \left(\{1\} \# \left(\lambda_i^\beta \# \lambda_j^\beta\right)\right) \# \left(\lambda_i^\beta \# \lambda_j^\beta\right) = \{1\} \# \left(\lambda_i^\beta \# \lambda_j^\beta\right) = \{1\}$. Thus, $\left(\lambda_j^\beta \# \lambda_i^\beta\right) \# \lambda_i^\beta \leq \lambda_j^\beta$.

(3) $\left(\left(\lambda_i^\beta \# \lambda_k^\beta\right) \# \left(\lambda_j^\beta \# \lambda_k^\beta\right)\right) \# \left(\lambda_i^\beta \# \lambda_j^\beta\right) = \left(\left(\lambda_i^\beta \# \lambda_j^\beta\right) \# \lambda_k^\beta\right) \# \left(\lambda_i^\beta \# \lambda_j^\beta\right) = \left(\left(\lambda_i^\beta \# \lambda_j^\beta\right) \# \left(\lambda_i^\beta \# \lambda_j^\beta\right)\right) \# \left(\lambda_k^\beta \# \left(\lambda_i^\beta \# \lambda_j^\beta\right)\right) = \{1\} \# \left(\lambda_k^\beta \# \left(\lambda_i^\beta \# \lambda_j^\beta\right)\right) = \{1\}$.

Thus, $\left(\lambda_i^\beta \# \lambda_k^\beta\right) \# \left(\lambda_j^\beta \# \lambda_k^\beta\right) \leq \lambda_i^\beta \# \lambda_j^\beta$.

(4) If $\lambda_i^\beta \leq \lambda_j^\beta$, then $\lambda_i^\beta \# \lambda_j^\beta = \{1\}$ and hence$\left(\lambda_i^\beta \# \lambda_k^\beta\right) \# \left(\lambda_j^\beta \# \lambda_k^\beta\right) = \left(\lambda_i^\beta \# \lambda_j^\beta\right) \# \lambda_k^\beta = \{1\} \# \lambda_k^\beta = \{1\}$. Hence $\lambda_i^\beta \# \lambda_k^\beta \leq \lambda_j^\beta \# \lambda_k^\beta$.

(5) From (1), we have $\lambda_i^\beta \# \lambda_k^\beta \leq \lambda_i^\beta$. It follows from (4) that $\left(\lambda_i^\beta \# \lambda_k^\beta\right) \# \left(\lambda_j^\beta \# \lambda_k^\beta\right) \leq \lambda_i^\beta \# \left(\lambda_j^\beta \# \lambda_k^\beta\right)$. Since X is $(PRDBI - A)$, we have that $\left(\lambda_i^\beta \# \lambda_j^\beta\right) \# \lambda_k^\beta \leq \lambda_i^\beta \# \left(\lambda_j^\beta \# \lambda_k^\beta\right)$.

(6) Let $\lambda_i^\beta \# \lambda_j^\beta = \lambda_k^\beta \# \lambda_j^\beta$. Since X is $(PRDBI-A)$, we get that $\left(\lambda_i^\beta \# \lambda_k^\beta\right) \# \lambda_j^\beta = \left(\lambda_i^\beta \# \lambda_j^\beta\right) \# \left(\lambda_k^\beta \# \lambda_j^\beta\right) = \left(\lambda_i^\beta \# \lambda_j^\beta\right) \# \left(\lambda_i^\beta \# \lambda_j^\beta\right) = \{1\}$. In other side, $\lambda_i^\beta \leq \lambda_j^\beta$ does not imply that $\lambda_k^\beta \# \lambda_i^\beta \leq \lambda_k^\beta \# \lambda_j^\beta$.

Proposition 3.9: Let $(X, \#, \{1\})$ be a $(PBI-A)$ with the condition. $\left(\lambda_k^\beta \# \lambda_i^\beta\right) \# \left(\lambda_k^\beta \# \lambda_j^\beta\right) = \lambda_j^\beta \# \lambda_i^\beta \forall \lambda_i^\beta, \lambda_j^\beta, \lambda_k^\beta \in X$. If $\lambda_i^\beta \leq \lambda_j^\beta$, then $\lambda_k^\beta \# \lambda_j^\beta \leq \lambda_k^\beta \# \lambda_i^\beta$.

***Proof*:** If $\lambda_i^\beta \leq \lambda_j^\beta$, then $\lambda_i^\beta \# \lambda_j^\beta = \{1\}$. Now $\left(\lambda_k^\beta \# \lambda_j^\beta\right) \# \left(\lambda_k^\beta \# \lambda_i^\beta\right) = \lambda_i^\beta \# \lambda_j^\beta = \{1\}$. Thus, $\lambda_k^\beta \# \lambda_j^\beta \leq \lambda_k^\beta \# \lambda_i^\beta$.

Definition 3.10: Let $(X, \#, \{1\})$ be a $(PBI-A)$. We say it has an *inclusion condition* if $\left(\lambda_i^\beta \# \lambda_j^\beta\right) \# \lambda_i^\beta = \{1\} \forall \lambda_i^\beta, \lambda_j^\beta \in X$. Note that any $(PRDBI-A)$ has the inclusion condition (see Proposition 3.8 (1)).

Definition 3.11: If $(X, \#, \{1\})$ is a $(PRDBI-A)$, then $(X, \#, \{1\})$ is, namely, a permutation quasi-associative BI-algebra $(PQABI-A)$.

Proposition 3.12: Let $(X, \#, \{1\})$ be a $(PRDBI-A)$, and then the induced relation "$\leq$" is a transitive relation.

***Proof*:** If $\lambda_i^\beta \leq \lambda_j^\beta$ and $\lambda_j^\beta \leq \lambda_k^\beta$, then from Proposition 3.4 (1) $\lambda_i^\beta \# \lambda_k^\beta = \left(\lambda_i^\beta \# \lambda_k^\beta\right) \# \{1\} = \left(\lambda_i^\beta \# \lambda_k^\beta\right) \# \left(\lambda_j^\beta \# \lambda_k^\beta\right) = \left(\lambda_i^\beta \# \lambda_j^\beta\right) \# \lambda_k^\beta = \{1\} \# \lambda_k^\beta = \{1\}$. Thus, $\lambda_i^\beta \leq \lambda_k^\beta$.

Definition 3.13: Assume that $(X, \#, \{1\})$ is a $(PBI-A)$, and $\varnothing \neq I \subseteq X$. The set I is, namely, an permutation BI-ideal $(PBI-I)$ of X if (1) $\{1\} \in I$, and (2) $\lambda_j^\beta \in I$ and $\lambda_i^\beta \# \lambda_j^\beta \in I$ imply that $\lambda_i^\beta \in I$ for all $\lambda_i^\beta, \lambda_j^\beta \in X$. Note that each one of $\{\{1\}\}$ and X is $(PBI-I)$ of X. Also, they are, namely, the *fixed ideal* and the *trivial ideal*, respectively. An $(PBI-I)$ I is, namely, a *proper* $(PBI-I)$ if $I \neq X$.

Lemma 3.14: If $\{I_i\}_{i \in \Lambda}$ is a family of $(PBI-Is)$ of X, then $\bigcap_{i\in\Lambda} I_i$ is a $(PBI-I)$ of X.

Proposition 3.15: If $(X, \#, \{1\})$ is a $(PBI-A)$, then $(I(X), \subseteq)$ is a complete lattice.

***Proof*:** The result follows from the fact that the set $I(X)$ is closed under arbitrary intersections.

Proposition 3.16: Let I be $(PBI-I)$ of $(PBI-A)$ $(X, \#, \{1\})$. If $\lambda_j^\beta \in I$ and $\lambda_i^\beta \leq \lambda_j^\beta$ then $\lambda_i^\beta \in I$.

***Proof*:** If $\lambda_j^\beta \in I$ and $\lambda_i^\beta \leq \lambda_j^\beta$, then $\lambda_i^\beta \# \lambda_j^\beta = \{1\} \in I$. Since $\lambda_j^\beta \in I$ and I is $(PBI-I)$, we have that $\lambda_i^\beta \in I$.

Definition 3.17: For all $\lambda_i^\beta, \lambda_j^\beta \in X$, we define $A\left(\lambda_i^\beta, \lambda_j^\beta\right) = \left\{\lambda_k^\beta \in X \mid \left(\lambda_k^\beta \# \lambda_i^\beta\right) \# \lambda_j^\beta = \{1\}\right\}$. Note that **(1)-** $\{1\}, \lambda_i^\beta \in A\left(\lambda_i^\beta, \lambda_j^\beta\right)$, **(2)-** $A\left(\lambda_i^\beta, \lambda_j^\beta\right) \neq A\left(\lambda_j^\beta, \lambda_i^\beta\right)$, and **(3)-** $A\left(\lambda_i^\beta, \{1\}\right) = \left\{\lambda_k^\beta \in X \mid \left(\lambda_k^\beta \# \lambda_i^\beta\right) \# \{1\} = \{1\}\right\} = \left\{\lambda_k^\beta \in X \mid \lambda_k^\beta \# \lambda_i^\beta = \{1\}\right\} = \left\{\lambda_k^\beta \in X \mid \left(\lambda_k^\beta \# \{1\}\right) \# \lambda_i^\beta = \{1\}\right\} = A\left(\{1\}, \lambda_i^\beta\right)$.

Proposition 3.18: If $(X, \#, \{1\})$ is a $(PRDBI - A)$, then $A\left(\lambda_i^\beta, \lambda_j^\beta\right)$ is $(PBI - I)$ of $(X, \#, \{1\})$ where $\lambda_i^\beta, \lambda_j^\beta \in X$.

***Proof*:** Let $\lambda_i^\beta \# \lambda_j^\beta \in A\left(\lambda_m^\beta, \lambda_n^\beta\right), \lambda_j^\beta \in A\left(\lambda_m^\beta, \lambda_n^\beta\right)$. Then $\left(\left(\lambda_i^\beta \# \lambda_j^\beta\right) \# \lambda_m^\beta\right) \# \lambda_n^\beta = \{1\}$ and $\left(\lambda_j^\beta \# \lambda_m^\beta\right) \# \lambda_n^\beta = \{1\}$. Since X is $(PRDBI - A)$, so $\{1\} = \left(\left(\lambda_i^\beta \# \lambda_j^\beta\right) \# \lambda_m^\beta\right) \# \lambda_n^\beta = \left(\left(\lambda_i^\beta \# \lambda_m^\beta\right) \# \left(\lambda_j^\beta \# \lambda_m^\beta\right)\right) \# \lambda_n^\beta = \left(\left(\lambda_i^\beta \# \lambda_m^\beta\right) \# \lambda_n^\beta\right) \# \left(\left(\lambda_j^\beta \# \lambda_m^\beta\right) \# \lambda_n^\beta\right) = \left(\left(\lambda_i^\beta \# \lambda_m^\beta\right) \# \lambda_n^\beta\right) \# \{1\} = \left(\lambda_i^\beta \# \lambda_m^\beta\right) \# \lambda_n^\beta$. Thus, $\lambda_i^\beta \in A\left(\lambda_m^\beta, \lambda_n^\beta\right)$. Therefore, $A\left(\lambda_m^\beta, \lambda_n^\beta\right)$ is a$(PBI - I)$ of $(X, \#, \{1\})$.

Proposition 3.19: Let $(X, \#, \{1\})$ be a $(PBI - A)$. Then, **(1)-**$A\left(\{1\}, \lambda_i^\beta\right) \subseteq A\left(\lambda_i^\beta, \lambda_j^\beta\right), \forall \lambda_i^\beta, \lambda_j^\beta \in X$, **(2)-**If $A\left(\{1\}, \lambda_j^\beta\right)$ is a $(PBI - I)$ and $\lambda_i^\beta \in A\left(\{1\}, \lambda_j^\beta\right)$, and then $A\left(\lambda_i^\beta, \lambda_j^\beta\right) \subseteq A\left(\{1\}, \lambda_j^\beta\right)$.

***Proof*:** **(1)** Let $\lambda_k^\beta \in A\left(\{1\}, \lambda_i^\beta\right)$. Then $\lambda_k^\beta \# \lambda_i^\beta = \left(\lambda_k^\beta \# \{1\}\right) \# \lambda_i^\beta = \{1\}$. Hence, $\left(\lambda_k^\beta \# \lambda_i^\beta\right) \# \lambda_j^\beta = \{1\} \# \lambda_j^\beta = \{1\}$. Thus, $\lambda_k^\beta \in A\left(\lambda_i^\beta, \lambda_j^\beta\right)$ and so $A\left(\{1\}, \lambda_i^\beta\right) \subseteq A\left(\lambda_i^\beta, \lambda_j^\beta\right)$. **(2)** Let $A\left(\{1\}, \lambda_j^\beta\right)$ be a $(PBI - I)$ and $\lambda_i^\beta \in A\left(\{1\}, \lambda_j^\beta\right)$. If $\lambda_k^\beta \in A\left(\lambda_i^\beta, \lambda_j^\beta\right)$, then $\left(\lambda_k^\beta \# \lambda_i^\beta\right) \# \lambda_j^\beta = \{1\}$. Hence, $\left(\left(\lambda_k^\beta \# \lambda_i^\beta\right) \# \{1\}\right) \# \lambda_j^\beta = \{1\}$. Therefore, $\lambda_k^\beta \# \lambda_i^\beta \in A\left(\{1\}, \lambda_j^\beta\right)$. Now, since $A\left(\{1\}, \lambda_j^\beta\right)$ and $\lambda_i^\beta \in A\left(\{1\}, \lambda_j^\beta\right)$, $\lambda_k^\beta \in A\left(\{1\}, \lambda_j^\beta\right)$. Thus, $A\left(\lambda_i^\beta, \lambda_j^\beta\right) \subseteq A\left(\{1\}, \lambda_j^\beta\right)$.

Proposition 3.20: If $(X, \#, \{1\})$ is a $(PBI - A)$, then $A\left(\{1\}, \lambda_i^\beta\right) = \cap_{\lambda_j^\beta \in X} A\left(\lambda_i^\beta, \lambda_j^\beta\right), \forall \lambda_i^\beta, \lambda_j^\beta \in X$.

***Proof*:** From Proposition 3.19 (1), we have that $A\left(\{1\},\lambda_i^\beta\right) \subseteq \cap_{\lambda_j^\beta \in X} A\left(\lambda_i^\beta,\lambda_j^\beta\right)$. If $\lambda_k^\beta \in \bigcap_{\lambda_j^\beta \in X} A\left(\lambda_i^\beta,\lambda_j^\beta\right)$, then $\lambda_k^\beta \in A\left(\lambda_i^\beta,\lambda_j^\beta\right) \forall \lambda_i^\beta,\lambda_j^\beta \in X$. Thus, $\lambda_k^\beta \in A\left(\{1\},\lambda_i^\beta\right)$. $\cap_{\lambda_j^\beta \in X} A\left(\lambda_i^\beta,\lambda_j^\beta\right) \subseteq A\left(\{1\},\lambda_i^\beta\right)$.

Proposition 3.21: If $\varnothing \neq I \subseteq X$, where $(X, \#, \{1\})$ is a $(PBI-A)$, then I is a $(PBI-I)$ of X if and only if $A\left(\lambda_i^\beta,\lambda_j^\beta\right) \subseteq I, \forall \lambda_i^\beta,\lambda_j^\beta \in I$.

***Proof*:** Assume that I is a $(PBI-I)$ of X and $\lambda_i^\beta,\lambda_j^\beta \in I$. If $\lambda_k^\beta \in A\left(\lambda_i^\beta,\lambda_j^\beta\right)$, then $\left(\lambda_k^\beta \# \lambda_i^\beta\right) \# \lambda_j^\beta = \{1\} \in I$. Since I is a$(PBI-I)$ and $\lambda_i^\beta,\lambda_j^\beta \in I$ we have that $\lambda_k^\beta \in I$. Thus, $A\left(\lambda_i^\beta,\lambda_j^\beta\right) \subseteq I$. Conversely, assume that $A\left(\lambda_i^\beta,\lambda_j^\beta\right) \subseteq I \forall \lambda_i^\beta,\lambda_j^\beta \in I$. Since $\left(\{1\} \# \lambda_i^\beta\right) \# \lambda_j^\beta = \{1\}$, $\{1\} \in A\left(\lambda_i^\beta,\lambda_j^\beta\right) \subseteq I$. Let $\lambda_m^\beta \# \lambda_n^\beta$ and $\lambda_n^\beta \in I$. Since $\left(\lambda_m^\beta \# \lambda_n^\beta\right) \# \left(\lambda_m^\beta \# \lambda_n^\beta\right) = \{1\}$, we consider that $\lambda_m^\beta \in A\left(\lambda_n^\beta, \lambda_m^\beta \# \lambda_n^\beta\right) \subseteq I$, i.e., $\lambda_m^\beta \in I$. Hence, I is a $(PBI-I)$ of X.

Proposition 3.22: If I is a $(PBI-I)$ of a $(PBI-A)(X, \#, \{1\})$, then $I = \cup_{\lambda_i^\beta,\lambda_j^\beta \in I} A\left(\lambda_i^\beta,\lambda_j^\beta\right)$.

***Proof*:** Let I be a $(PBI-I)$of X and $\lambda_k^\beta \in I$. Since $\left(\lambda_k^\beta \# \{1\}\right) \# \lambda_k^\beta = \lambda_k^\beta \# \lambda_k^\beta = \{1\}$. We have that $\lambda_k^\beta \in A\left(\{1\},\lambda_k^\beta\right)$. Then, $I \subseteq \cup_{\lambda_k^\beta \in I} A\left(\{1\},\lambda_k^\beta\right) \subseteq \cup_{\lambda_i^\beta,\lambda_j^\beta \in I} A\left(\lambda_i^\beta,\lambda_j^\beta\right)$. If $\lambda_k^\beta \in \bigcup_{\lambda_i^\beta,\lambda_j^\beta \in I} A\left(\lambda_i^\beta,\lambda_j^\beta\right)$, then there exists $\lambda_m^\beta, \lambda_n^\beta \in I$ such that $\lambda_k^\beta \in A\left(\lambda_m^\beta, \lambda_n^\beta\right)$. It follows from Proposition 3.21 that $\lambda_k^\beta \in I$, that is, $\cup_{\lambda_i^\beta,\lambda_j^\beta \in I} A\left(\lambda_i^\beta,\lambda_j^\beta\right) \subseteq I$.

Proposition 3.23: If I is a$(PBI-I)$ of a $(PBI-A)$ $(X, \#, \{1\})$,then $I = \cup_{\lambda_k^\beta \in I} A\left(\{1\},\lambda_k^\beta\right)$.

***Proof*:** Let I be a $(PBI-I)$of X and $\lambda_k^\beta \in I$. Since $\left(\lambda_k^\beta \# \{1\}\right) \# \lambda_k^\beta = \lambda_k^\beta \# \lambda_k^\beta = \{1\}$. We have that $\lambda_k^\beta \in A\left(\{1\},\lambda_k^\beta\right)$. Hence, $I \subseteq \cup_{\lambda_k^\beta \in I} A\left(\{1\},\lambda_k^\beta\right)$. If $\lambda_k^\beta \in \bigcup_{\lambda_k^\beta \in I} A\left(\{1\},\lambda_k^\beta\right)$, then there exists $\lambda_m^\beta \in I$ satisfies $\lambda_k^\beta \in A\left(\{1\},\lambda_m^\beta\right)$, which means that $\lambda_k^\beta \# \lambda_m^\beta = \left(\lambda_k^\beta \# \{1\}\right) \# \lambda_m^\beta = \{1\} \in I$. Since I is a $(PBI-I)$of X and $\lambda_m^\beta \in I$, we have that $\lambda_k^\beta \in I$. Thus $\cup_{\lambda_k^\beta \in I} A\left(\{1\},\lambda_k^\beta\right) \subseteq I$.

Definition 3.24: Let $(X, \#, \{1\})$ be a $(PRDBI-A)$ and let I be a $(PBI-I)$ of X and $\lambda_m^\beta \in X$. Define $I^l_{\lambda_m^\beta} = \left\{\lambda_i^\beta \in X \mid \lambda_i^\beta \# \lambda_m^\beta \in I\right\}$.

Proposition 3.25: If $(X, \#, \{1\})$ is a $(PRDBI - A)$, then $I^l_{\lambda_k^\beta}$ is the least $(PBI - I)$ of X containing I and λ_m^β.

***Proof*:** By (1) of Definition 3.1, we have that $\lambda_m^\beta \# \lambda_m^\beta = \{1\}$ for all $\lambda_m^\beta \in X$, i.e., $\lambda_m^\beta \in I^l_{\lambda_m^\beta}$ and so $I_{\lambda_m^\beta} \neq \varnothing$. Assume that $\lambda_i^\beta \# \lambda_j^\beta \in I^l_{\lambda_m^\beta}$ and $\lambda_j^\beta \in I^l_{\lambda_m^\beta}$. Then $\left(\lambda_i^\beta \# \lambda_j^\beta\right) \# \lambda_m^\beta \in I$ and $\lambda_j^\beta \# \lambda_m^\beta \in I$. Since X is $(PRDBI - A)$, we get $\left(\lambda_i^\beta \# \lambda_m^\beta\right) \# \left(\lambda_j^\beta \# \lambda_n^\beta\right) \in I$. Since $\lambda_j^\beta \# \lambda_m^\beta \in I$, we have that $\lambda_i^\beta \# \lambda_m^\beta \in I$ and so $\lambda_i^\beta \in I^l_{\lambda_m^\beta}$. Then, $I^l_{\lambda_m^\beta}$ is a $(PBI - I)$ of X. Let $\lambda_i^\beta \in I$. Since $\left(\lambda_i^\beta \# \lambda_m^\beta\right) \# \lambda_i^\beta = \left(\lambda_i^\beta \# \lambda_i^\beta\right) \# \left(\lambda_m^\beta \# \lambda_i^\beta\right) = \{1\} \# \left(\lambda_m^\beta \# \lambda_i^\beta\right) = \{1\} \in I$, and I is a $(PBI - I)$ of X, we have that $\lambda_i^\beta \# \lambda_m^\beta \in I$. Hence, $\lambda_i^\beta \in I_{\lambda_m^\beta}$. Thus, $I \subseteq I^l_{\lambda_m^\beta}$. Now, let J be a $(PBI - I)$ of X containing I and λ_m^β. Let $\lambda_i^\beta \in I^l_{\lambda_m^\beta}$. Then $\lambda_i^\beta \# \lambda_m^\beta \in I \subseteq J$. Since $\lambda_m^\beta \in J$ and J is a $(PBI-)$ of X, we have that $\lambda_i^\beta \in J$. Therefore, $I^l_a \subseteq J$.

Definition 3.26: Let I be a $(PBI - I)$ of a $(PBI - A)$ $(X, \#, \{1\})$ and $\lambda_m^\beta \in X$. We denote $I^r_{\lambda_m^\beta} = \left\{\lambda_i^\beta \in X \mid \lambda_m^\beta \# \lambda_i^\beta \in I\right\}$. Note that $I^r_{\lambda_m^\beta}$ is not always a $(PBI - I)$ of X.

Definition 3.27: Let $\varnothing \neq I \subseteq X$ and $(X, \#, \{1\})$ be $(PBI - A)$. We define a *binary relation* "$\sim_I$" by $\lambda_i^\beta \sim_I \lambda_j^\beta$ if and only if $\lambda_i^\beta \# \lambda_j^\beta \in I$ and $\lambda_j^\beta \# \lambda_i^\beta \in I$. The set $\left\{\lambda_j^\beta \mid \lambda_i^\beta \sim_I \lambda_j^\beta\right\}$ will be denoted by $\left[\lambda_i^\beta\right]_I$.

Proposition 3.28: The relation "$\sim_I$" on X is an equivalence, where I be a $(PBI - I)$ of a $(PRDBI - A)(X, \#, \{1\})$.

***Proof*:** We consider that $\lambda_i^\beta \# \lambda_i^\beta = \{1\} \in I$ (since I is a$(PBI - I)$ of X). Thus, $\lambda_i^\beta \sim_I \lambda_i^\beta$. Thus, $\sim_I$ is reflexive. If $\lambda_i^\beta \# \lambda_j^\beta = \{1\} \in I$, then $\lambda_j^\beta \# \lambda_i^\beta = \{1\}$ by Proposition 3.8 (3). Thus, $\lambda_i^\beta \sim_I \lambda_j^\beta \Longrightarrow \lambda_j^\beta \sim_I \lambda_i^\beta$ and so I is symmetric. Now, let $\lambda_i^\beta \sim_I \lambda_j^\beta$ and $\lambda_j^\beta \sim_I \lambda_i^\beta$. Then, $\lambda_i^\beta \# \lambda_j^\beta, \lambda_j^\beta \# \lambda_i^\beta \in I$, and $\lambda_j^\beta \# \lambda_k^\beta, \lambda_k^\beta \# \lambda_j^\beta \in I$. By Proposition 3.8 (3), we have that $\left(\lambda_i^\beta \# \lambda_k^\beta\right) \# \left(\lambda_j^\beta \# \lambda_k^\beta\right) \leq \lambda_i^\beta \# \lambda_j^\beta$. Since I is a $(PBI - I)$ and $\lambda_i^\beta \# \lambda_j^\beta \in I$, we have that $\left(\lambda_i^\beta \# \lambda_k^\beta\right) \# \left(\lambda_j^\beta \# \lambda_k^\beta\right) \in X$ and so $\lambda_i^\beta \# \lambda_k^\beta \in I$. Similarly, we have that $\lambda_k^\beta \# \lambda_i^\beta \in I$. Thus, $\lambda_i^\beta \sim_I \lambda_k^\beta$ and so $\sim_I$ is a transitive. Therefore, the relation $\sim_I$ on X is an equivalence.

Definition 3.29: A binary relation "θ" on $(PBI - A)$ $(X, \#, \{1\})$ is, namely,

(1) A permutation right compatible relation if $\lambda_i^\beta \theta \lambda_j^\beta$ and $\lambda_p^\beta \in X$, then $\left(\lambda_i^\beta \# \lambda_p^\beta\right) \theta \left(\lambda_j^\beta \# \lambda_p^\beta\right)$.

(2) A permutation left compatible relation if $\lambda_i^{\beta}\theta\lambda_j^{\beta}$ and $\lambda_q^{\beta} \in X$, then $\left(\lambda_q^{\beta} \# \lambda_i^{\beta}\right)\theta\left(\lambda_q^{\beta} \# \lambda_j^{\beta}\right)$.

(3) A permutation compatible relation if $\lambda_i^{\beta}\theta\lambda_j^{\beta}$ and $\lambda_p^{\beta}\theta\lambda_q^{\beta}$, then $\left(\lambda_i^{\beta} \# \lambda_p^{\beta}\right)$ $\theta\left(\lambda_j^{\beta} \# \lambda_q^{\beta}\right)$.

A permutation right (left) compatible equivalence relation on $(PBI - A)$ $(X, \#, \{1\})$ is, namely, a *permutation* right (left) *congruence relation* on X [They are abbreviated by *PRCR* (*PLCR*)].

Proposition 3.30: The equivalence relation "$\sim_I$" as stated in Proposition 3.28 is a (*PRCR*) on $(PBI - A)I$.

***Proof*:** If $\lambda_i^{\beta} \sim_I \lambda_j^{\beta}$ and $\lambda_p^{\beta} \in X$, then $\lambda_i^{\beta} \# \lambda_j^{\beta} \in I$ and $\lambda_j^{\beta} \# \lambda_i^{\beta} \in I$. From Proposition 3.8 (3), we have that $\left(\left(\lambda_i^{\beta} \# \lambda_p^{\beta}\right) \# \left(\lambda_j^{\beta} \# \lambda_p^{\beta}\right)\right) \# \left(\lambda_i^{\beta} \# \lambda_j^{\beta}\right) = \{1\} \in I$. Since I is a $(PBI - I)$ and $\lambda_i^{\beta} \# \lambda_j^{\beta} \in I$, we have that $\left(\lambda_j^{\beta} \# \lambda_p^{\beta}\right) \# \left(\lambda_i^{\beta} \# \lambda_p^{\beta}\right) \in I$. Therefore, $\left(\lambda_i^{\beta} \# \lambda_p^{\beta}\right) \sim_I \left(\lambda_j^{\beta} \# \lambda_p^{\beta}\right)$.

Proposition 3.31: Let I be a subset of a $(PBI - A)$ $(X, \#, \{1\})$ with $\{1\} \in I$. If I has the condition: If $\lambda_i^{\beta} \# \lambda_j^{\beta} \in I$, then $\left(\lambda_k^{\beta} \# \lambda_i^{\beta}\right) \# \left(\lambda_k^{\beta} \# \lambda_j^{\beta}\right) \in I$. Then $X = I$.

***Proof*:** Let $\lambda_i^{\beta} = \{1\}$ and $\lambda_j^{\beta} = \lambda_k^{\beta}$. Then $\{1\} \# \lambda_k^{\beta} = \{1\} \in I$ imply that $\left(\lambda_k^{\beta} \# \{1\}\right) \# \left(\lambda_k^{\beta} \# \lambda_k^{\beta}\right) = \lambda_k^{\beta} \# \{1\} = \lambda_k^{\beta} \in I$. Therefore, $X \subseteq I$ and so $I = X$.

Proposition 3.32: If $\sim_I$ is a (*PLCR*) on $(PRDBI - A)$ $(X, \#, \{1\})$, then $[\{1\}]_I$ is a $(PBI - I)$ of X.

***Proof*:** Obviously, $\{1\} \in [\{1\}]_I$. If λ_j^{β} and $\lambda_i^{\beta} \# \lambda_j^{\beta}$ are in $[\{1\}]_I$, then $\lambda_i^{\beta} \# \lambda_j^{\beta} \sim_I \{1\}$ and $\lambda_j^{\beta} \sim_I \{1\}$. It follows that $\lambda_i^{\beta} = \lambda_i^{\beta} \# \{1\} \sim_I \lambda_i^{\beta} \# \lambda_j^{\beta} \sim_I \{1\}$. Therefore, $\lambda_i^{\beta} \in [\{1\}]_I$.

4 Conclusion and Future Work

Certain new extensions of BI-algebras are introduced in this paper, and their properties are investigated using permutation sets and have been used to study numerous mathematical issues in recent work. Therefore, in future research, rather than using permutation sets, we shall also employ neutrosophic sets to broaden our concepts.

References

1. Imai, Y., Iseki, K.: On axiom system of propositional calculi, XIV. Proc. Japan Acad. **42**(1), 19–22 (1966)
2. Neggers, J., Kim, H.S.: On d-algebras. Math. Slovaca. **49**, 19–26 (1999)
3. Neggers, J., Kim, H.S.: On B-algebras. Mate. Vesnik. **54**, 21–29 (2002)
4. Saeid, A.B., Kim, H.S., Rezaei, A.: On BI-Agebras. Versita. **25**(1), 177–194 (2017)
5. Alsalem, S.: The Permutation Topological Spaces and their Bases. Basrah Journal of Science. **32**(1), 28–42 (2014)
6. Khalil, S.M., Rajah, A.: Solving the Class Equation $x^d = \beta$ in an Alternating Group for each $\beta \in H \cap C^\alpha$ and $n \notin \theta$. Arab J. Basic Appl. Sci. **10**, 42–50 (2011)
7. Khalil, S.M., Rajah, A.: Solving Class Equation $x^d = \beta$ in an Alternating Group for all $n \in \theta$ & $\beta \in H_n \bigcap C^\alpha$. Arab J. Basic Appl. Sci. **16**, 38–45 (2014)
8. Khalil, S.M., Suleiman, E., Torki, M.M.: Generated new classes of permutation I/B-algebras. J. Discrete Math. Sci. Cryptogr. **25**(1), 31–40 (2022)
9. Torki, M.M., Khalil, S.M.: New types of finite groups and generated algorithm to determine the integer factorization by excel. AIP Conf. Proc. **2290**, 040020 (2020)
10. Khalil, S.M., Hameed, F.: An algorithm for generating permutations in symmetric groups using soft spaces with general study and basic properties of permutations spaces. J. Theor. Appl. Inf. Technol. **96**, 2445–2457 (2018)
11. Khalil, S.M., Abbas, N.M.: Applications on new category of the symmetric groups. AIP Conf. Proc. **2290**, 040004 (2020)
12. Khalil, S.M., Abbas, N.M.A.: Characteristics of the number of conjugacy classes and P-regular classes in finite symmetric groups. IOP Conf. Ser. Mater. Sci. Eng. **571**, 012007 (2019). https://doi.org/10.1088/1757-899X/571/1/012007
13. Khalil, S.M., Suleiman, E., Ali Abbas, N.M.: New technical to generate permutation measurable spaces. In: *2021 1st Babylon International Conference on Information Technology and Science (BICITS)*, pp. 160–163 (2021). https://doi.org/10.1109/BICITS51482.2021.9509892
14. Khalil, S.M., Hameed, F.: Applications of fuzzy ρ-Ideals in ρ-algbras. Soft Comput. **24**, 13997–14004 (2020). https://doi.org/10.1007/s00500-020-04773-3
15. Khalil, S.M., Hasab, M.H.: Decision making using new distances of intuitionistic fuzzy sets and study their application in the universities, INFUS. Adv. Intell. Syst. Comput. **1197**, 390–396 (2020)
16. Khalil, S., A. Hassan A, H. Alaskar, W. Khan, Hussain A.: Fuzzy logical algebra and study of the effectiveness of medications for COVID-19. Mathematics. **9**(22), 28–38 (2021)
17. Khalil, S.M., Hassan, A.N.: New class of algebraic fuzzy systems using cubic soft sets with their applications. IOP Conf. Seri. Mater. Sci. Eng. **928**, 042019 (2020). https://doi.org/10.1088/1757-899X/928/4/042019
18. Khalil, S.M.: Dissimilarity fuzzy soft points and their applications. Fuzzy Inf. Eng. **8**, 281–294 (2016)
19. Khalil, S.M., Ulrazaq, M., Abdul-Ghani, S., Al-Musawi, A.F.: $\sigma-$Algebra and $\sigma-$Baire in Fuzzy Soft Setting. Adv. Fuzzy Syst., 10 (2018)
20. Khalil, S.M., Hassan, A.: Applications of fuzzy soft ρ−ideals in ρ−algebras. Fuzzy Inf. Eng. **10**, 467–475 (2018)
21. Abdul-Ghani, S.A., Khalil, S.M., Abd Ulrazaq, M., Al-Musawi, A.F.: New branch of intuitionistic fuzzification in algebras with their applications. Int. J. Math. Math. Sci. 5712676, 6 pages (2018)
22. Khalil, S.M., Abdul-Ghani, S.A.: Soft M-ideals and soft S-ideals in soft S-algebras. IOP Conf. Ser. J. Phys. **1234**, 012100 (2019)
23. Hasan, M.A., Khalil, S.M., Abbas, N.M.A.: Characteristics of the soft-(1, 2)-gprw–closed sets in soft bi-topological spaces. Conf. IT-ELA. **2020**., 9253110, 103–108 (2020)
24. Abbas, N.M.A., Khalil, S.M., Hamza, A.A.: On α^*-continuous and contra α^*-continuous mappings in topological spaces with soft setting. Int. J. Nonlinear Anal. Appl. **12**, 1107–1113 (2021)

25. Khalil, S.M., Hameed, F.: Applications on cyclic soft symmetric. IOP Conf. Ser. J. Phys. **1530**, 012046 (2020)
26. Hasan, M.A., Abbas, N.M.A., Khalil, S.M.: On soft α^*-open sets and soft contra α^*-continuous mappings in soft topological spaces. J. Interdiscip. Math. **24**, 729–734 (2021)
27. Al-Musawi, A.M., Khalil, S.M., Ulrazaq, M.A.: Soft (1,2)-strongly open maps in bi-topological spaces. IOP Conf. Ser. Mater. Sci. Eng. **571**, 012002 (2019). https://doi.org/10.1088/1757-899X/571/1/012002
28. Damodharan, K., Vigneshwaran, M., Khalil, S.M.: $N_{\delta^*g\alpha}$—continuous and irresolute functions in neutrosophic topological spaces. Neutrosophic Sets Syst. **38**(1), 439–452 (2020)
29. Ali Abbas, N.M., Khalil, S.M.: On new classes of neutrosophic continuous and contra mappings in neutrosophic topological spaces. IJNAA. **12**(1), 718–725 (2021)
30. Nivetha, A.R., Vigneshwaran, M., Abbas, N.M.A., Khalil, S.M.: On $N_{*g\alpha}$- continuous in topological spaces of neutrosophy. J. Interdiscip. Math. **24**(3), 677–685 (2021)
31. Ali Abbas, N.M., Khalil, S.M., Vigneshwaran, M.: The neutrosophic strongly open maps in neutrosophic bi-topological spaces. J. Interdiscip. Math. **24**(3), 667–675 (2021)
32. Khalil, S.M.: On neurosophic delta generated per-continuous functions in neutrosophic topological spaces. Neutrosophic Sets Syst. **48**, 122–141 (2022)
33. Imran, Q.H., Smarandache, F., et al.: On neutrosophic semi alpha open sets. Neutrosophic Sets Syst. **18**, 37–42 (2018)
34. Khalil, S.M., Abbas, N.M.A.: On nano with their applications in medical field. AIP Conf. Proc. **2290**, 040002 (2020)

A Note on Multiplicative Ternary Hyperring

Md. Salim Masud Molla

Keywords Multiplicative ternary hyperrings · Hyperideals · Regular equivalence relations · Quotient multiplicative ternary hyperrings · Isomorphism theorems

1 Introduction

A binary hyperoperation is a mapping $\circ : A \times A \to \wp^*(A)$, where A is a nonempty set and $\wp*(A)$ is a nonempty subsets of A. A hyperstructure is a multivalued algebra [5], expressed as $(A, \circ)$, and a nonempty set A assigned with the hyperoperation "$\circ$." In the beginning of the nineteenth century, hyperstructure theory moved forward by pathfinder F. Marty's [14] paper on hypergroups. Over the decades, there are several famous mathematicians and researchers like Corsini [3, 4, 6] who studied hypergroups, hypergraph, and hypermodules. De Salvo [9] studied hyperrings and hyperfields. Dehkordi and Davvaz [10] found out Γ-semihyperrings. S. Abdullah et al. [1] introduced Γ-hyperideals of Γ-semihyperring and developed the theory of hyperstructure and applied in different fields. In [7], P. Corsini et al. and, in [8] , B. Davvaz and V. L. Fotea mention their appositeness in different fields like cryptography, graph theory, computer science, etc. In [11], Krasner initiated Krasner hyperring $(K, +, \cdot)$ in which addition is a binary hyperoperation and "$\cdot$" is a binary operation, and both the distributive laws hold. In 1990, R. Rota [16] launched multiplicative hyperring $(H, +, \circ)$, where "+" is the usual binary operation and "$\circ$" is a binary hyperoperation, and $(h_1 + h_2) \circ h_3 \subseteq h_1 \circ h_3 + h_2 \circ h_3$, $h_1 \circ (h_2 + h_3) \subseteq h_1 \circ h_2 + h_1 \circ h_3$ holds, for all $h_1, h_2, h_3 \in H$.

Schar is a ternary algebraic structure initiated in 1924 by H. Pr*ü*fer [15]. In [12], D. H. Lehmer presented the commutative ternary groups, which are a special type of ternary algebraic structure known to be triplex. After that in 1971, W. G. Lister [13] initiated the concept of ternary ring T, a commutative group in which product is defined on three elements and right, center, and left distributive laws hold.

M. S. M. Molla (✉)
Department of Mathematics, Darjeeling Government College, Darjeeling, India

H.-H. Leung et al. (eds.), *Recent Developments in Algebra and Analysis*,
Trends in Mathematics, https://doi.org/10.1007/978-3-031-37538-5_15

In 2015, Md. Salim et al. [17] presented the concept of multiplicative ternary hyperring (MTH), which is a generalized concept of ternary ring and multiplicative hyperring. In [17], we studied the regular equivalence relation and strongly regular equivalence relation and analyzed them. Also we defined quotient MTH over a regular equivalence relation "τ" and analyzed it and gained three isomorphism theorems on MTH by using the regular equivalence relation "τ." In this note, we learn on hyperideals of MTH, and we demonstrate a one-to-one correspondence between the family of all hyperideals and the family of all regular equivalence relations of a MTH. Also, we obtain quotient MTH over a hyperideal and show that above two quotient MTHs are coincide. Lastly using the notion of hyperideal, we obtain three isomorphism theorems on a MTH.

Some earlier works of the author on the MTH may be found in [18, 19].

2 Preliminaries

A ternary hyperoperation is a mapping $\circ : A \times A \times A \longrightarrow \wp*(A)$, where $\wp*(A)$ is the class of all nonempty subset of the nonempty set A. The image of $(a_1, a_2, a_3) \in A \times A \times A$ will be denoted by $a_1 \circ a_2 \circ a_3$ (which is known to be a ternary hyperproduct of $a_1, a_2, a_3 \in A$).

Definition 1 ([18]) Let $(R, +, \circ)$ be a MTH, that is, an abelian group $(R, +)$ together with a ternary hyperproduct "$\circ$" satisfying the following conditions:

(i) $(\alpha \circ \beta \circ \gamma) \circ \delta \circ \eta = \alpha \circ (\beta \circ \gamma \circ \delta) \circ \eta = \alpha \circ \beta \circ (\gamma \circ \delta \circ \eta)$;

(ii) $(\alpha + \beta) \circ \gamma \circ \delta \subseteq \alpha \circ \gamma \circ \delta + \beta \circ \gamma \circ \delta$; $\alpha \circ (\beta + \gamma) \circ \delta \subseteq \alpha \circ \beta \circ \delta + \alpha \circ \gamma \circ \delta$; $\alpha \circ \beta \circ (\gamma + \delta) \subseteq \alpha \circ \beta \circ \gamma + \alpha \circ \beta \circ \delta$;

(iii) $(-\alpha) \circ \beta \circ \gamma = \alpha \circ (-\beta) \circ \gamma = \alpha \circ \beta \circ (-\gamma) = -(\alpha \circ \beta \circ \gamma)$ for all $\alpha, \beta, \gamma \in R$;

(iv) $0_R \circ \alpha \circ \beta = \alpha \circ 0_R \circ \beta = \alpha \circ \beta \circ 0_R = \{0_R\} \forall \alpha, \beta \in R$(absorbing property of 0_R), for all $\alpha, \beta, \gamma, \delta, \eta \in R$,

Definition 2 ([19]) Let $(I, +)$ be a subgroup of a MTH $(R, +, \circ)$ is said to be a right (resp. a lateral or a left) hyperideal of R if for $\alpha, \beta \in R$ and for all $i \in I$ such that $i \circ \alpha \circ \beta \subseteq I$(resp. $\alpha \circ i \circ \beta \subseteq I$ or $\alpha \circ \beta \circ i \subseteq I$).

The additive subgroup I of MTH $(R, +, \circ)$ is both a left hyperideal and a right hyperidel of R; then I is called both sided hyperideal of R.

If I is a right hyperideal, a left hyperideal, and a lateral hyperideal of MTH R, then I is said to be a hyperideal.

In [17], we sketch the regular equivalence and strongly equivalence relation on MTH.

Let τ be a equivalence relation over a MTH R and $\wp*(R) = \wp(R) \backslash \{0\}$, where $\wp(R)$ is the subsetes of R. Now $\overline{\tau}$ and $\overline{\overline{\tau}}$ are two equivalence relation marked by (i) $X\overline{\tau}Y$ holds if and only if for every $x \in X$, $\exists y \in Y$ such that $x\tau y$ retains and also for every $y' \in Y$ and $\exists x' \in X$ such that $x'\tau y'$ retains. (ii) For all $x \in X$ and $y \in Y$, $X\overline{\overline{\tau}}Y$ if an only if $x\tau y$ retains, for any $X, Y \in \wp*(R)$.

Definition 3 ([17]) Let $(R, +, \circ)$ be a MTH and τ be a equivalence relation on it. Then τ is regular equivalence relation if $x\tau y \Rightarrow (x+z)\tau(y+z)$, and $x\tau y, z\tau w, u\tau v$ implies $(x \circ z \circ u)\overline{\tau}(y \circ w \circ v)$ for all $x, y, z, w, u, v \in R$.

The relation τ is strongly regular equivalence relation if $x\tau y \Rightarrow (x+z)\tau(y+z)$, and $x\tau y, z\tau w, u\tau v$ implies $(x \circ z \circ u)\overline{\overline{\tau}}(y \circ w \circ v) \forall x, y, z, w, u, v \in R$.

Remark 1 ([17]) From the above definition, we have the following conditions:

$x\tau y$ implies $(x \circ z \circ w)\overline{\tau}(y \circ z \circ w)$, $(z \circ x \circ w)\overline{\tau}(z \circ y \circ w)$ and $(z \circ w \circ x)\overline{\tau}(z \circ w \circ y), \forall x, y, z, w \in R$.

$x\tau y$ implies $(x \circ z \circ w)\overline{\overline{\tau}}(y \circ z \circ w)$, $(z \circ x \circ w)\overline{\overline{\tau}}(z \circ y \circ w)$ and $(z \circ w \circ x)\overline{\overline{\tau}}(z \circ w \circ y), \forall x, y, z, w \in R$.

3 Hyperideals and Regular Equivalence Relations

Proposition 1 *Let I be a hyperideal of a MTH. Then there exists a regular equivalence relation τ over MTH such that $x\tau y$ if and only if $x - y \in I$.*

Proof Consider I be a hyperideal of a MTH $(R, +, \circ)$. Let τ be a relation marked by $x\tau y$ if and only if $x - y \in I$ for all $x, y \in R$. Obviously, τ is an equivalence relation on R.

Let $x\tau y$ hold for $x, y \in R$. Then $x - y \in I$ implis $(x + z) - (y + z) \in I$. This implies that $(x + z)\tau(y + z)$ holds for $x, y, z \in R$. Therefore, τ is a congruence on $(R, +)$.

Let $x, y \in R$ and $x\tau y$. Then $x - y \in I$. Then $x = y + i$, for some $i \in I$. Then $x \circ z \circ w = (y + i) \circ z \circ w \subseteq y \circ z \circ w + i \circ z \circ w \subseteq y \circ z \circ w + I$. So, for every $a \in x \circ z \circ w$, there exists an element $b \in y \circ z \circ w$ such that $a = b + i$ for some $i \in I$. Hence, $a - b \in I$. So $a\tau b$. Therefore, for each $a \in x \circ z \circ w$, there exists $b \in y \circ z \circ w$ such that $a\tau b$ holds. Also we can show that for any $b \in y \circ z \circ w$, there exists an element $a \in x \circ z \circ w$ such that $a\tau b$. Hence, $(x \circ z \circ w)\overline{\tau}(y \circ z \circ w)$. By a similar fashion, we get $(z \circ x \circ w)\overline{\tau}(z \circ y \circ w)$, $(z \circ w \circ x)\overline{\tau}(z \circ w \circ y)$. Thus, τ is regular equivalence relation over $(R, +, \circ)$.

Proposition 2 *Let τ be regular equivalence relation on a MTH $(R, +, \circ)$. Then there exists a hyperideal I of $(R, +, \circ)$ such that $x\tau y$ if and only if $x - y \in I$.*

Proof Suppose $(R, +, \circ)$ is a MTH and τ is a regular equivalence on the set R. Therefore τ determines a partition on R into disjoint equivalence classes. Let 0_τ be the equivalence class containing 0. Let $I = 0_\tau$. Now we verify I is a hyperideal of R. Let $x, y \in I = 0_\tau$. Then $x\tau 0$ and $y\tau 0$ hold. Since τ is congruence on R, $(x - y)\tau 0$ holds. So, $x - y \in I$.

Let $x \in 0_\tau$. Then $x\tau 0$ holds. Since τ is a regular equivalence relation on $(R, +, \circ)$, $(r_1 \circ r_2 \circ a)\overline{\tau}(r_1 \circ r_2 \circ 0)$ for all $r_1, r_2 \in R$, by Remark 1, i.e., $(r_1 \circ r_2 \circ x)\overline{\tau}\{0\}$. So for any $x \in r_1 \circ r_2 \circ x$, $x\tau 0$ holds. This implies that $x \in 0_\tau = I$. Therefore $r_1 \circ r_2 \circ x \subseteq I$. Likewise $r_1 \circ x \circ r_2 \subseteq I$ and $x \circ r_1 \circ r_2 \subseteq I$. So I is hyperideal of $(R, +, \circ)$. Lastly $a\tau y \Leftrightarrow (x - y)\tau 0 \Leftrightarrow x - y \in 0_\tau = I$.

Theorem 1 *Suppose $(R, +, \circ)$ is a MTH. Then there exists an inclusion preserving bijection between the collection of all hyperideals of $(R, +, \circ)$ and the collection of all regular equivalence relations on $(R, +, \circ)$.*

Proof Consider $\mathcal{I}$ to be a collection of all hyperideals of the MTH $(R, +, \circ)$ and $\mathcal{E}$ a collection of all regular equivalence relations on R. We define a mapping $\phi : \mathcal{E} \to \mathcal{I}$ by $\phi(\tau) = 0_\tau$. Let τ_1 and τ_2 be regular equivalence relations on $(R, +, \circ)$; then, $\phi(\tau_1) = \phi(\tau_2)$. This implies that $0_{\tau_1} = 0_{\tau_2}$. Now, for $x, y \in R$, $x\tau_1 y \Leftrightarrow (x = y)\tau_1 0 \Leftrightarrow (x - y) \in 0_{\tau_1} = 0_{\tau_2} \Leftrightarrow (x - y)\tau_2 0 \Leftrightarrow x\tau_2 y$. So, $\tau_1 = \tau_2$, i.e., ϕ is injective. Let $H \in \mathcal{I}$. Then there exists an equivalence relation τ which is regular on $(R, +, \circ)$, $x\tau y$ if and only if $x - Y \in H$. Let $x \in H$. Then $x \in H \Leftrightarrow x - 0 \in H \Leftrightarrow x\tau 0 \Leftrightarrow x \in 0_\tau$. Thus $\phi(\tau) = 0_\tau = H$. So ϕ is surjective. Let $\tau_1, \tau_2 \in \mathcal{E}$ be such that $\tau_1 \subseteq \tau_2$. Let $x \in 0_{\tau_1} \Rightarrow x\tau_1 0 \Rightarrow x\tau_2 0 \Rightarrow x \in 0_{\tau_2}$. So, $0_{\tau_1} \subseteq 0_{\tau_2}$. Thus ϕ is inclusion preserving.

Definition 4 ([17]) Let τ be a regular equivalence relation on a MTH $(R, +, \circ)$; then the MTH $(R/\tau, +, \circ)$ where $R/\tau = \{a_\tau : a \in R\}$, is known to be the quotient MTH of $(R, +, \circ)$ by τ, where $a_\tau + b_\tau = (a+b)_\tau$ and $a_\tau o b_\tau o c_\tau = \{x_\tau : x \in aoboc\}$ for $a, b, c \in R$.

Definition 5 ([17]) Let μ and ν be two regular equivalence relations on a MTH $(R, +, \circ)$ with $\mu \subseteq \nu$. We set a relation ν/μ on R/μ by : $a_\mu(\nu/\mu)b_\mu$ if and only if $a\nu b$ for $a, b \in R$.

Lemma 1 ([17]) *ν/μ is a regular equivalence relation on the quotient MTH $(R/\mu, +, \circ)$.*

Theorem 2 (Correspondence Theorem) *Suppose $(R, +, \circ)$ is a MTH and τ a regular equivalence relation over $(R, +, \circ)$. Then there exists an inclusion preserving bijection between the family of regular equivalence relation over $(R, +, \circ)$ containing τ and the family of regular equivalence relations on $(R/\tau, +, \circ)$.*

Proof Suppose τ is a regular equivalence relation on $(R, +, \circ)$. Let $\mathcal{M}$ be family of all regular equivalence relations on $(R, +, \circ)$ containing τ and $\mathcal{N}$ be the family of all regular equivalence relation on (R/τ). We consider a mapping $\psi : \mathcal{M} \longrightarrow \mathcal{N}$ by $\psi(\phi) = \phi/\tau$ where $\tau \subseteq \phi$. Let ϕ_1 and $\phi_2 \in \mathcal{M}$ be such that $\psi(\phi_1) = \psi(\phi_2)$. Then $\phi_1/\tau = \phi_2/\tau$. Then $a\phi_1 b \Leftrightarrow a_\tau(\phi_1/\tau)b_\tau \Leftrightarrow a_\tau(\phi_2/\tau)b_\tau \Leftrightarrow a\phi_2 b$. This implies that $\phi_1 = \phi_2$. So, ψ is a one-to-one mapping. Clearly ψ is surjective. Therefore ψ is a bijective mapping.

Lastly, let $\phi_1, \phi_2 \in \mathcal{M}$ be such that $\phi_1 \subseteq \phi_2$. Now $a_\tau(\phi_1/\tau)b_\tau \Rightarrow a\phi_1 b \subseteq a\phi_2 b \Rightarrow a_\tau(\phi_2/\tau)b_\tau$. This implies that $\phi_1/\tau \subseteq \phi_2/\tau$. Thus ψ is inclusion preserving.

Proposition 3 *Suppose $(R, +, \circ)$ is a MTH and H is a hyperideal of R. Let $R/H = \{a + H : a \in R\}$. Then $(R/H, +)$ is an abelian group. We define $(a + H) \circ (b + H) \circ (c + H) = \{p + H : p \in a \circ b \circ c\}$. Then with respect to the above ternary hyperoperation, R/H forms a MTH.*

Proof Let H be a hyperideal of $(R, +, \circ)$. Then we have the quotient group $(R/H, +)$. Obviously $(R/H, +)$ is an additive commutative group. Now we shall show that above defined multiplicative ternary hyperoperation is established. Let $x+H = x'+H, b+H = b'+H$ and $z+H = z'+H$. Let $p+H \in (x+H)\circ(y+H)\circ(z+H)$. Then $p \in x\circ y\circ z$. Again $x+H = x'+H \Rightarrow x-x' \in H \Rightarrow x = x'+h_1$ for some $h_1 \in H$. Similarly $y = y'+h_2$ and $z = z'+h_3$ for some $h_2, h_3 \in H$. Now $x\circ y\circ z = (x'+h_1)\circ(y'+h_2)\circ(z'+h_3) \subseteq x'\circ y'\circ z'+H$. This implies that $p = q+h$ where $q \in x'\circ y'\circ z'$. Therefore $p+H = q+H \in (x'+H)\circ(y'+H)\circ(z'+H)$. So, $(x+H)\circ(y+H)\circ(z+H) \subseteq (x'+H)\circ(y'+h)\circ(z'+H)$. Similarly $(x'+H)\circ(y'+H)\circ(z'+H) \subseteq (x+H)\circ(y+h)\circ(z+H)$. Consequently $(x+H)\circ(y+H)\circ(z+H) = (x'+H)\circ c(y'+H)\circ(z'+H)$ and ternary hyperoperation '$\circ$' is well defined. Let $a+H \in ((x+H)\circ(y+h)\circ(z+H))\circ(w+H)\circ(t+H)$ where $x, y, z, w, t \in R$. Then $a \in p\circ w\circ t$ where $p+H \in (x+H)\circ(y+H)\circ(z+H) \Rightarrow p \in x\circ y\circ z$. Then $a \in (x\circ y\circ z)\circ w\circ t = x\circ y\circ(z\circ w\circ t)$, so $a+H \in (x+H)\circ(y+H)\circ(b+H)$, where $b \in z\circ w\circ t$. So, $b+H \in (z+H)\circ(w+H)\circ(t+H)$. Thus $a+H \in (x+H)\circ(y+H)\circ((z+H)\circ(w+H)\circ(t+H))$. Thus $((x+H)\circ(y+H)\circ(z+H))\circ(w+H)\circ(t+H) \subseteq (x+H)\circ(y+H)\circ((z+H)\circ(w+H)\circ(t+H))$. Converse inclusion is similar. Hence $(x+H)\circ(y+H)\circ((z+H)\circ(w+H)\circ(t+H)) = ((x+H)\circ(y+H)\circ(z+H))\circ(w+H)\circ(t+H)$. Similarly $(x+H)\circ((y+H)\circ(z+H)\circ(w+H))\circ(t+H) = (x+H)\circ(y+H)\circ((z+H)\circ(w+H)\circ(t+H))$.

Similarly we can prove the distributive laws.

Now, $(x+H)\circ(y+H)\circ(0+H) = \{a+h : a \in x\circ y\circ 0 = \{0_R\}\} = 0+H$. Similarly $(x+H)\circ(0+H)\circ(y+H) = (0+H)\circ(x+H)\circ(y+H) = 0+H$.

Lastly, So, $a = -b$ for some $b \in x\circ y\circ z$, i.e., $a+H = -b+H = -(b+H) \in -(x+H)\circ(y+H)\circ(z+H)$. Thus $(x+H)\circ(y+H)\circ(-(z+H)) \subseteq -(x+H)\circ(y+H)\circ(z+H)$. Likewise we prove $-(x+H)\circ(y+H)\circ(z+H) \subseteq (x+H)\circ(y+H)\circ(-(z+H))$. Hence $-(x_+H)\circ(y+H)\circ(z+H) = (x+H)\circ(y+H)0(-(z+H))$. Comparably we can show $(x+H)\circ(-(y+H))\circ(z+H) = (-(x+H))\circ(y+H)\circ(z+H) = -((x+H)\circ(y+H)\circ(z+H))$. Hence, $(R/H, +, \circ)$ is MTH.

Theorem 3 *Let $(R, +, \circ)$ be a MTH and I be a hyperideal of $(R, +, \circ)$. Then there exists an inclusion preserving bijection from the family of all hyperideals on $(R, +, \circ)$ containing H and the family all of hyperideals on $(R/H, +, \circ)$.*

Proof Theorems 1 and 2 comply the proof.

Lemma 2 *Let $(R, +, \circ)$ be a MTH. Let τ be the regular equivalence relation on R and I be a hyperideal corresponding to τ, i.e $\alpha - \beta \in I$ if an only if $\alpha\tau\beta$. Then $\alpha_\tau = \alpha + I$ for all $\alpha \in R$.*

Proof Since I is the hyperideal corresponding to τ then $\alpha\tau\beta$ if and only if $\alpha - \beta \in I$. Now $a \in \alpha_\tau \Leftrightarrow a\tau\alpha \Leftrightarrow a - \alpha \in I$ implies and implied by $a \in \alpha + I$. So, $\alpha_\tau = \alpha + I$.

Proposition 4 *Let τ be regular equivalence relation on a MTH $(R, +, \circ)$. Then all the regular equivalence classes are equipotent.*

Proof Here we show that the regular equivalence classes a_τ and b_τ are equipotent for $a, b \in R$. By Lemma 2, $a_\tau = a + H$ and $b_\tau = b + H$, where H is the hyperideal corresponding to τ. Obviously the mapping f from $a+H \longmapsto H$ marked as follows $f(a+i) = i$ for all $i \in H$, is bijective. So $a + H$ and H are equipotent. Similarly $b + H$ and H are equipotent. Hence $a_\tau = a + H$ and $b_\tau = b + H$ are equipotent.

The above proposition enables us to get an example of an equivalence relation, which is not a regular equivalence relation.

Example 1 Consider the MTH $(\mathbb{Z}_A, +, \circ)$ induced by A, where $\mathbb{Z}$ is the set of all integers and A is any subset of $\mathbb{Z}$. Now $\{\mathbb{Z}_A \backslash \{1\}, \{1\}\}$ is a partition in $\mathbb{Z}_A$. So this gives an equivalence relation on $\mathbb{Z}_A$. Since the equivalence classes are not equipotent, the above equivalence relation is not a regular equivalence relation.

Remark 2 The above condition stated in Proposition 4 is a necessary condition but not sufficient.

Theorem 4 *Let $(R, +, \circ)$ be a MTH. Let τ be the regular equivalence relation on $(R, +, \circ)$ and H be the hyperideal on $(R, +, \circ)$ corresponding to τ. Then the quotient MTH's $(R/\tau, +, \circ)$ and $(R/H, +, \circ)$ coincide.*

Proof As τ is regular equivalence relation over $(R, +, \circ)$, $R/\tau = \{a_\tau : a \in R\} = \{a + H : a \in R\} = R/H$. Let $a, b \in R$. Now $a_\tau + b_\tau = (a \circ b)_\tau = (a + b) + H = (a + H) + (b + H)$. Again $a_\tau \circ b_\tau \circ c_\tau = \{x_\tau : x \in a \circ b \circ c\} = \{x + H : x \in a \circ b \circ c\} = (a + H) \circ (b + H) \circ (c + H)$ for every $a, b, c \in R$. Hence the quotient MTHs coincide.

In [17] we define the homomorphism and good homomorphism between two MTH. An epimorphism (resp. monomorphism) is a surjective (resp. injective) MTH homomorphism.

Proposition 5 *Consider $\pi : R \to T$ to be a MTH homomorphism from a MTH $(R, +, \circ)$ to another MTH $(T, +, \circ)$. Then the kernel of π, designed by $ker(\pi)$, and marked as $\{x \in R : \pi(x) = 0_T\}$, is a hyperideal of multiplicative ternary hyperring(MTH) R.*

Proof The mapping $\pi : R \to T$ is a group homomorphism, since π is MTH homomorphism. Thus $\pi(0_R) = 0_T \Rightarrow 0_R \in ker(\pi)$. Let $x, y \in ker(\pi)$; then $x - y \in ker(\pi)$. Now $x \in ker(\pi) \Rightarrow \pi(x) = 0_T \Rightarrow \pi(r_1 \circ r_2 \circ x) \subseteq \pi(r_1) \circ \pi(r_2) \circ \pi(x) = \{0_T\} \Rightarrow r_1 \circ r_2 \circ x \subseteq ker(\pi)$. Likewise, $r_1 \circ x \circ r_2 \subseteq ker(\pi)$ and $x \circ r_1 \circ r_2 \subseteq ker(\pi)$. Consequently, $ker(\pi)$ is a hyperideal of $(R, +, \circ)$.

Theorem 5 *Let $\pi : (R, +, \circ) \to (T, +, \circ)$ be an epimorphism from a MTH $(R, +, \circ)$ to another MTH $(T, +, \circ)$. Then $R/ker(\pi) \cong T$.*

Proof Since $\pi : R \to T$ is a MTH epimorphism. Then $\pi : (R, +) \to (T, +)$ is a group-epimorphism. Then $\Psi : (R/ker(\pi), +) \to (T, +)$, expressed by $\Psi(r + ker(\pi)) = \pi(r)$ for all $r \in R$ is group-isomorphism. For any $x, y \in R$, $\Psi((x + ker(\pi)) \circ (y + ker(\pi)) \circ (z + ker(\pi))) = \Psi(p + ker(\pi))$ (where $p \in x \circ y \circ z) =$

$\pi(p) \in \pi(x \circ y \circ z) \subseteq \pi(x) \circ \pi(y) \circ \pi(z) = \Psi(x + ker(\pi)) \circ \Psi(y + ker(\pi)) \circ \Psi(z + ker(\pi))$. Thus, Ψ is MTH isomorphism.

Theorem 6 *Let M and N be two hyperideals of a MTH $(R, +, \circ)$. Then $M/(M \cap n) \cong (M + N)/N$.*

Proof We construct a mapping $\Phi : M/(M \cap N) \to (M + N)/N$ by $\Phi(\alpha + (M \cap N)) = \alpha + N$ for all $\alpha \in M$. Let $\alpha, \beta \in M$. Now $(\alpha + M) \cap N = (\beta + M) \cap N \Leftrightarrow (\alpha - \beta) \in M \cap N \Leftrightarrow (\alpha - \beta) \in N \Leftrightarrow \alpha + N = \beta + N \Leftrightarrow \Phi(\alpha + (M \cap N)) = \Phi((\beta + M) \cap N)$. So, Φ is well defined and injective.

Now let $\alpha, \beta, \gamma \in M$. Then $\Phi((\alpha + (M \cap N)) + (\beta + (M \cap N))) = \Phi((\alpha + \beta) + (M \cap N)) = (\alpha + \beta) + N = (\alpha + N) + (\beta + N) = \Phi(\alpha + (M \cap N)) + \Phi(\beta + (M \cap N))$. Also $\Phi((\alpha + (M \cap N)) \circ (\beta + (M \cap N)) \circ (\gamma + (M \cap N))) = \Phi(p + (M \cap N))$ (where $p \in \alpha \circ \beta \circ \gamma$) $= p + N \subseteq (\alpha \circ \beta \circ \gamma) + N = (\alpha + N) \circ (\beta + N) \circ (\gamma + N) = \Phi(\alpha + (M \cap N)) \circ \Phi(\beta + (M \cap N)) \circ \Phi(\gamma + (M \cap N))$. Thus Φ is a homomorphism. It is obvious that Φ is surjective, so Φ is an epimorphism. Therefore by Theorem 5, we gain $M/(M \cap N) \cong (M + N)/N$.

Theorem 7 *Suppose M and N two hyperideals of MTH $(R, +, \circ)$ with $M \subseteq N$. Then $(R/M)/(N/M) \cong R/N$.*

Proof Consider $\Psi : R/M \to R/N$ by $\Psi(\alpha + M) = \alpha + N, \forall \alpha \in R$. Let $\alpha, \beta, \gamma \in R$. Then $\alpha + M = \beta + M$ implies $\alpha - \beta \in M \subseteq N$. This implies that $\alpha + N = \beta + N \Rightarrow \Psi(\alpha + M) = \Psi(\beta + N)$. So, definition of Ψ is established.

Let $\alpha, \beta, \gamma \in R$. Then $\Psi((\alpha+M)+(\beta+M)) = \Psi((\alpha+\beta)+M) = (\alpha+\beta)+N = (\alpha+N)+(\beta+N) = \Psi(\alpha+M)+\Psi(\beta+N)$. Also $\Psi(((\alpha+M)\circ(\beta+M)\circ(\gamma+M))) = \Psi(p + M)$(where $p \in \alpha \circ \beta \circ \gamma$) $= p + N \subseteq \alpha \circ \beta \circ \gamma + N = (\alpha + N) \circ (\beta + N) \circ (\gamma + N) = \Psi(\alpha + M) \circ \Psi(\beta + M) \circ \Psi(\gamma + M)$. Thus Ψ is a homomorphism. Obviously ψ is an epimorphism. Now $ker(\Psi) = \{\alpha + M \in R/M : \Psi(\alpha + M) = 0 + N\} = \{\alpha + M \in R/M : \alpha + N = 0 + N\} = \{\alpha + M \in R/M : \alpha \in N\} = N/M$. Therefore by Theorem 5, $(R/M)/(N/M) \cong R/N$.

References

1. Abdullah, S., Hila, K., Aslam, M.: On bi-Γ-hyperideals of Γ-semihypergroups. U.P.B. Sci. Bull. Ser. A. **74**(4), 79–90 (2012)
2. Castillo, J. R., Paradero-Vilela, J. S.: Quotient and Homomorphism in ternary hyperrings. Int. J. Math. Anal. **58**(8), 2845–2859 (2014)
3. Corsini, P.: Prolegomena of Hypergroup Theory, 2nd edn. Aviani editor, Italy (1993)
4. Corsini, P.: Hypergraphs and hypergroups. Algebra Universalis **35**, 548–555 (1996)
5. Corsini, P.: Hypergroups réguliers et hypermodules. Ann. Univ. Ferrara-Sez. VII-Sc. Mat. **20**, 121–235 (1975)
6. Corsini, P.: Sur les semihypergroupes. Atti Soc. Pelor. Sc. Mat. Fis. Nat. **XXVI**, 363–372 (1980)
7. Corsini, P., Leoreanu, V.: Applications of Hyperstructure Theory. Advances in Mathematics. Kluwer Academic, Dodrecht (2003)

8. Davvaz, B., Leoreanu-Fotea, V.: Hyperring Theory and Applications. International Academic Press, Palm Harbor, FL, USA (2007)
9. De Salvo, M.: Hyperrings and Hyperfields: Annales Scientifiques de Universite de Clermont-Ferrand II. No. 22 (1984)
10. Dehkordi, O., Davvaz, B.: A strong regular relation on Γ-semihyperrings. J. Sci. Islam. Repub. Iran **22**(3), 257–266 (2011)
11. Krasner, M.: A class of hyperrings and hyperfields. Int. J. Math, Sci. **6**, 307–311 (1983)
12. Lehmer, D.H.: A ternary analogue of Abelian groups. Am. J. Math. **54**(2), 329–338 (1932)
13. Lister, W.G.: Ternary rings. Trans. Am. Math. Soc. **154**, 37–55 (1971)
14. Marty, F.: Sur une generalization de la notion de group. 8th congress des Math. Scandenaves Stockholm, 45–49 (1934)
15. Prüfer, H.: Theorie der Abelschen Gruppen, I. Grundeigenschaften. Math. Z. **20**, 165–187 (1924)
16. Rota, R.: Strongly distribute multiplicative hyperrings. J. Geom. **39**(1), 130–138 (1990)
17. Salim, Md., Chanda, T., Dutta, T.K.: Regular equivalence and strongly regular equivalence on multiplicative ternary hyperring. J. Hyperstruct. **4**(1), 20–36 (2015)
18. Salim, Md., Dutta, T.K.: Regular multiplicative ternary hyperring. Ital. J. Pure Appl. Math. **37**, 77–88 (2017)
19. Salim, Md., Dutta, T.K.: Prime hyperideal of Multiplicative ternary hyperrings. Int. J. Algebra **10**(5), 207–219 (2016). https://doi.org/10.12988/ija.2016.6320

ℵeutrosophic ℵ-Ideals and ℵ-Filters of BF-Algebra

B. Satyanarayana and P. Rajani

Keywords BF-algebra · Subalgebra · Ideal · Neutrosophic N-BF-ideal · Neutrosophic N-BF-filter

1 Introduction

Zadeh [1] introduced the idea of fuzzy set theory. F. Smarandache launched the concept of ℵeutrosophic logic and set to deal with uncertainty as a generalization of the intuitionistic fuzzy set, paraconsistent set, and intuitionistic set [2]. Atanassov [3] unveiled the degree of nonmembership/falsehood (f) and elucidated the intuitionistic fuzzy set. Smarandache devised the term "ℵeutrosophic," which means knowledge of neutral thought, and this third/neutral represents the main difference between "fuzzy"/"intuitionistic fuzzy" logic/set and "ℵeutrosophic" logic/set. He introduced the degree of indeterminacy/neutrality as an independent component and developed the ℵeutrosophic set on three components (t, i, f = truth, indeterminacy, falsehood). Jun et al. [4] have introduced a new mapping which is called negative-valued mapping and built ℵ-structures to deal with negative information. Khan et al. [5] displayed the concept of ℵeutrosophic ℵ-structure (ℵℵS) and applied it to a semigroup. Walendziak[6] worked out BF-algebra which is a generic form of B-algebra and explored some characteristics of ideals and normal ideals in BF-algebra. Seok-ZunSong et al. [7] presented the concept of ℵeutrosophic ℵ-ideal in BCK-algebras and investigated numerous attributes. In this paper, we introduce the concept of ℵeutrosophic ℵ-BF-subalgebra (ℵℵSA), ℵeutrosophic ℵ-BF-ideal (ℵℵi), ℵeutrosophic N-positive implicative Bf-ideal (ℵℵPIi), ℵeutrosophic ℵ-

B. Satyanarayana (✉)
Department of Mathematics, Acharya Nagarjuna University, Guntur, Andhra Pradesh, India

P. Rajani
Seshadri Rao Gudlavalleru Engineering College, Seshadri Rao Knowledge Village, Gudlavalleru, Andhra Pradesh, India
e-mail: rajani@gecgudlavallerumic.in

H.-H. Leung et al. (eds.), *Recent Developments in Algebra and Analysis*,
Trends in Mathematics, https://doi.org/10.1007/978-3-031-37538-5_16

near BF-filter (ƝƝƝf), and Ɲeutrosophic Ɲ-BF-filter (ƝƝf) and examine numerous attributes. We consider an association among ƝƝSA, types of ƝƝi, and ƝƝf.

2 Preliminaries

Definition 2.1: A BF-algebra is a structure S:=(S≠Ø, ⋇,0)∈K(τ) satisfying

$$p \divideontimes p = 0 \tag{1}$$

$$p \divideontimes 0 = p \tag{2}$$

$$0 \divideontimes (p \divideontimes q) = q \divideontimes p, \forall p, q \in \mathrm{S} \tag{3}$$

Example 2.2: The following composition table is a BF-algebra (Table 1).

Definition 2.3: A relation "≤" on BF-algebra S:=(S≠Ø, ⋇,0) is defined as

$$(\forall p, q \in \mathrm{S})\ (p \leq q \text{ iff } p \divideontimes q = 0) \tag{4}$$

Definition 2.4: Consider a BF-algebra S:=(S≠Ø,⋇, 0). M(≠Ø)⊆S is a subalgebra if

$$p \divideontimes q \in \mathrm{M}, \forall p, q \in \mathrm{M} \tag{5}$$

Definition 2.5: Consider a BF-algebra S:=(S≠Ø, ⋇, 0). M(≠Ø)⊆S is an ideal if

$$0 \in \mathrm{M} \tag{6}$$

and

$$(\forall p, q \in \mathrm{S})\ (p \divideontimes q \in \mathrm{M}, q \in \mathrm{M} \Rightarrow p \in \mathrm{M}) \tag{7}$$

Definition 2.6: Consider a BF-algebra S:=(S≠Ø, ⋇, 0). M(≠Ø)⊆S is a positive implicative BF-ideal if (6) holds and satisfies

Table 1 (S={0, 1, 2, 3},⋇,0)

⋇	0	1	2	3
0	0	1	2	3
1	1	0	3	2
2	2	3	0	1
3	3	2	1	0

$$(\forall p, q, r \in S), ((p ⵅ q) ⵅ r) \in M, q ⵅ r \in M \Rightarrow p ⵅ r \in M \quad (8)$$

Example 2.7: Let R be the set of real numbers. Then, S=(R, ⵅ, 0) where ⵅ is given by

$$p ⵅ q = \begin{cases} p \text{ if } q = 0 \\ q \text{ if } p = 0 \\ 0 \text{ otherwise} \end{cases}$$

is a BF-algebra.

M=R^+ U{0} is a positive implicative BF-ideal of S.

Definition 2.8: Consider a BF-algebra S:=(S≠Ø, ⵅ, 0). M(≠Ø)⊆S is a near BF-filter if

$$(\forall p, q \in S)\ (q \in M \Rightarrow p ⵅ q \in M) \quad (9)$$

Example 2.9: M=R^+ U{0}is a near BF-filter of S for the example defined in 2.5.

Definition 2.10: Consider a BF-algebra S:=(S≠Ø, ⵅ, 0). M(≠Ø)⊆S is a BF-filter if (6) holds and

$$(\forall p, q \in S)\ (p ⵅ q \in M, p \in M \Rightarrow q \in M) \quad (10)$$

Example 2.11: For the BF-algebra in Table 1, M={0,1} is a BF-filter of S.

3 Neutrosophic N-Concept on BF-Algebra

Let γ(S,[−1,0]) be the family of negative-valued mappings from a set S to [−1,0] (called N-mapping on S). An N-structure is denoted by (S, g) of S and g is a N-mapping on S. A NNS over a universe S ≠Ø is

$$S_N = \frac{S}{(Y_N, I_N, \mathbb{N}_N)} = \left\{ \frac{p}{(Y_N(p), I_N(p), \mathbb{N}_N(p))} / p \in S \right\}$$

where Y_N, I_N, and $\mathbb{N}_N$ are N-mappings on S, which are called the negative truth membership mapping, the negative indeterminacy membership mapping, and the negative falsity membership mapping, respectively, on S.

A NNS S_N over S holds

$$(\forall p \in S)\left(-3 \le Y_N(p) + I_N(p) + \mathbb{N}_N(p) \le 0\right)$$

Let us represent ∀p,q∈[−1,0], pvq denotes max{p,q}, and pΛq denotes min{p,q}.

Definition 3.1: A ℕℕS $S_ℕ$ of a BF-algebra S:=(S≠Ø, ⋇, 0) is a ℕℕSA of S if

$$Y_{\mathbb{N}}(p \ast q) \le \vee \left\{Y_{\mathbb{N}}(p), Y_{\mathbb{N}}(q)\right\} \quad (11)$$

$$I_{\mathbb{N}}(p \ast q) \ge \wedge \left\{I_{\mathbb{N}}(p), I_{\mathbb{N}}(q)\right\} \quad (12)$$

$$\mathbb{N}_{\mathbb{N}}(p \ast q) \le \vee \left\{\mathbb{N}_{\mathbb{N}}(p), \mathbb{N}_{\mathbb{N}}(q)\right\} \quad (13)$$

for all p,q∈S

Example 3.2: The following table $S_ℕ$ is a ℕℕSA of Table 1 (Table 2).

Definition 3.3: A ℕℕS $S_ℕ$ of a BF-algebra S:=(S≠Ø, ⋇, 0) is a ℕℕi of S if

$$Y_{\mathbb{N}}(0) \le Y_{\mathbb{N}}(p) \le \vee \left\{Y_{\mathbb{N}}(p \ast q), Y_{\mathbb{N}}(q)\right\} \quad (14)$$

$$I_{\mathbb{N}}(0) \ge I_{\mathbb{N}}(p) \ge \wedge \left\{I_{\mathbb{N}}(p \ast q), I_{\mathbb{N}}(q)\right\} \quad (15)$$

$$\mathbb{N}_{\mathbb{N}}(0) \le \mathbb{N}_{\mathbb{N}}(p) \le \vee \left\{\mathbb{N}_{\mathbb{N}}(p \ast q), \mathbb{N}_{\mathbb{N}}(q)\right\} \quad (16)$$

for all p,q∈S

Example 3.4: The following table $S_ℕ$ is a ℕℕi of Table 1 (Table 3).

Definition 3.5: A ℕℕS $S_ℕ$ of a BF-algebra S:=(S≠Ø, ⋇, 0) is a ℕℕPIi of S if

$$Y_{\mathbb{N}}(0) \le Y_{\mathbb{N}}(p), I_{\mathbb{N}}(0) \ge I_{\mathbb{N}}(p), \mathbb{N}_{\mathbb{N}}(0) \le \mathbb{N}_{\mathbb{N}}(p) \quad (17)$$

Table 2 ℕℕSA of Table 1 (S={0, 1, 2, 3}, ⋇, 0)

	0	1	2	3
$Y_ℕ$	-0.8	-0.8	-0.8	-0.8
$I_ℕ$	-0.1	-0.8	-0.9	-0.9
$ℕ_ℕ$	-0.8	-0.4	-0.4	-0.6

Table 3 ℕℕi of Table 1 (S={0, 1, 2, 3}, ⋇, 0)

	0	1	2	3
$Y_ℕ$	-0.7	-0.2	-0.6	-0.2
$I_ℕ$	-0.1	-0.8	-0.9	-0.9
$ℕ_ℕ$	-0.8	-0.4	-0.4	-0.6

$$Y_N\ (p \mathbin{x} r) \le \vee \left\{ Y_N\ ((p \mathbin{x} q) \mathbin{x} r), Y_N\ (q \mathbin{x} r) \right\} \tag{18}$$

$$I_N\ (p \mathbin{x} r) \ge \wedge \left\{ I_N\ ((p \mathbin{x} q) \mathbin{x} r), I_N\ (q \mathbin{x} r) \right\} \tag{19}$$

$$\mathbb{N}_N\ (p \mathbin{x} r) \le \vee \left\{ \mathbb{N}_N\ ((p \mathbin{x} q) \mathbin{x} r), \mathbb{N}_N\ (q \mathbin{x} r) \right\} \tag{20}$$

for all $p,q,r \in S$

Example 3.6: The following composition table is a BF-algebra (Table 4).
The NNS S_N of S is NNPIi as shown below (Table 5):

Definition 3.7: A NNS S_N of a BF-algebra S:=(S≠∅, x, 0) is a NNNf of S if

$$(\forall p, q \in S)\ \left(Y_N\ (p \mathbin{x} q) \le Y_N\ (q) \right) \tag{21}$$

$$(\forall p, q \in S)\ \left(I_N\ (p \mathbin{x} q) \ge I_N\ (q) \right) \tag{22}$$

$$(\forall p, q \in S)\ \left(\mathbb{N}_N\ (p \mathbin{x} q) \le \mathbb{N}_N\ (q) \right) \tag{23}$$

Example 3.8: The following table S_N is a NNNf of Table 4 (Table 6).

Definition 3.9: A NNS S_N of a BF-algebra S:=(S≠∅, x, 0) is a NNf of S if

$$(\forall p \in S)\ \left(Y_N\ (0) \le Y_N\ (p) \right), \left(I_N\ (0) \ge I_N\ (p) \right), \left(\mathbb{N}_N\ (0) \le \mathbb{N}_N\ (p) \right) \tag{24}$$

Table 4 (S={0, 1, 2}, x, 0)

x	0	1	2
0	0	1	2
1	1	0	0
2	2	0	0

Table 5 NNPIi of Table 4 (S={0, 1, 2}, x, 0)

	0	1	2
Y_N	-1	-1	-1
I_N	-0.2	-0.2	-0.2
$\mathbb{N}_N$	-0.5	-0.5	-0.5

Table 6 NNNf of Table 4 (S={0, 1, 2}, x, 0)

	0	1	2
Y_N	-0.9	-0.9	-0.9
I_N	-0.5	-0.5	-0.5
$\mathbb{N}_N$	-0.8	-0.8	-0.8

$$Y_{\mathbb{N}}(q) \leq \vee \left\{ Y_{\mathbb{N}}(p \ast q), Y_{\mathbb{N}}(p) \right\} (\forall p, q \in S) \tag{25}$$

$$I_{\mathbb{N}}(q) \geq \wedge \left\{ I_{\mathbb{N}}(p \ast q), I_{\mathbb{N}}(p) \right\} (\forall p, q \in S) \tag{26}$$

$$\mathbb{N}_{\mathbb{N}}(q) \leq \vee \left\{ \mathbb{N}_{\mathbb{N}}(p \ast q), \mathbb{N}_{\mathbb{N}}(p) \right\} (\forall p, q \in S) \tag{27}$$

Example 3.10: The following table $S_{\mathbb{N}}$ is a $\mathbb{N}$Nf of Table 4 (Table 7).

Definition 3.11: Consider a $\mathbb{N}$NS $S_{\mathbb{N}}$ of a BF-algebra S:=(S $\neq \emptyset$, $\ast$, 0) and λ_1, λ_2, $\lambda_3 \in [-1,0]$ with $-3 \leq \lambda_1 + \lambda_2 + \lambda_3 \leq 0$ with the following:

$$Y_{\mathbb{N}}^{\lambda_1} = \left\{ p \in S / Y_{\mathbb{N}}(p) \leq \lambda_1 \right\},$$

$$I_{\mathbb{N}}^{\lambda_2} = \left\{ p \in S / I_{\mathbb{N}}(p) \geq \lambda_2 \right\},$$

$$\mathbb{N}_{\mathbb{N}}^{\lambda_3} = \left\{ p \in S / \mathbb{N}_{\mathbb{N}}(p) \leq \lambda_3 \right\}.$$

Then

$$S_{\mathbb{N}}(\lambda_1, \lambda_2, \lambda_3) = \left\{ p \in S / Y_{\mathbb{N}}(p) \leq \lambda_1, I_{\mathbb{N}}(p) \geq \lambda_2, \mathbb{N}_{\mathbb{N}}(p) \leq \lambda_3 \right\}$$

is the $(\lambda_1, \lambda_2, \lambda_3)$ – level set of $S_{\mathbb{N}}$.

Note: From the Definition 3.11, it is obvious that $S_{\mathbb{N}}(\lambda_1, \lambda_2, \lambda_3) = Y_{\mathbb{N}}^{\lambda_1} \cap I_{\mathbb{N}}^{\lambda_2} \cap \mathbb{N}_{\mathbb{N}}^{\lambda_3}$.

Definition 3.12: For any fixed numbers, λ_Y, $\lambda_{\mathbb{N}} \in [-1,0)$, $\lambda_I \in (-1,0]$, and a non-empty subset G of a BF-algebra S:=(S $\neq \emptyset$, $\ast$, 0), a $\mathbb{N}$NS $S_{\mathbb{N}}{}^{G}$ over S are defined to be

$$S_{\mathbb{N}}{}^{G} := \left\{ \frac{p}{Y_{\mathbb{N}}{}^{G}(p), I_{\mathbb{N}}{}^{G}(p), \mathbb{N}_{\mathbb{N}}{}^{G}(p)} / p \in S \right\},$$

Table 7 $\mathbb{N}$Nf of Table 4 (S={0, 1, 2}, $\ast$, 0)

	0	1	2
$Y_{\mathbb{N}}$	-0.7	-0.5	-0.5
$I_{\mathbb{N}}$	-0.5	-0.5	-0.5
$\mathbb{N}_{\mathbb{N}}$	-0.7	-0.4	-0.5

where $Y_N{}^G$, $I_N{}^G$, $\mathbb{N}_N{}^G$ are N mappings on S which are shown below:

$$Y_N{}^G : S \to [-1, 0], \text{ as } Y_N{}^G(p) = \begin{cases} \lambda_Y, & \text{if } p \in G \\ 0, & \text{otherwise} \end{cases}$$

$$I_N{}^G : S \to [-1, 0], \text{ as } I_N{}^G(p) = \begin{cases} \lambda_I, & \text{if } p \in G \\ -1, & \text{otherwise} \end{cases}$$

and

$$\mathbb{N}_N{}^G : S \to [-1, 0], \text{ as } \mathbb{N}_N{}^G(p) = \begin{cases} \lambda_N, & \text{if } p \in G \\ 0, & \text{otherwise} \end{cases}$$

Theorem 3.13: Every NNPIi of a BF-algebra is a NNi.

Proof: Let NNS S_N be a NNPIi of a BF-algebra S:=(S ≠ Ø, ⁎, 0). Then

$$\begin{pmatrix} Y_N(0) \leq Y_N(p) \\ I_N(0) \geq I_N(p) \\ \mathbb{N}_N(0) \leq \mathbb{N}_N(p) \end{pmatrix}, \forall p \in S \qquad \text{by (17)}$$

and

i. $Y_N(p) = Y_N(p⁎0) \leq \vee\{Y_N((p⁎q)⁎0), Y_N(q⁎0)\}$ by (2) & (18)

Therefore, $Y_N(p) \leq \vee\{Y_N(p⁎q), Y_N(q)\}$ ∀*p,q*∈S.
Similar proof follows for I_N and $\mathbb{N}_N$ also.
Hence, S_N is NNi of S.
Note: Converse of the theorem need not be true, i.e., every NNi need not be NNPIi.

Example 3.14: The following table S_N is a NNi but not NNPIi of Table 4 (Table 8).
(By (18),$Y_N(0⁎1) = Y_N(1) = -0.5 > \vee\{Y_N(0⁎1)⁎1), Y_N(1⁎1)\} = Y_N(0) = -0.7$)

Theorem 3.15: Let a NNS S_N be a NNf of a BF-algebra S:=(S≠Ø,⁎, 0) and then the sets.

Table 8 NNi but not NNPIi

	0	1	2
Y_N	-0.7	-0.5	-0.5
I_N	-0.5	-0.5	-0.5
$\mathbb{N}_N$	-0.7	-0.4	-0.5

(i) $S_{\mathcal{N}_{Y_{\mathcal{N}}}} = \{p \in S / Y_{\mathcal{N}}(p) = Y_{\mathcal{N}}(0)\}$.

(ii) $S_{\mathcal{N}_{I_{\mathcal{N}}}} = \{p \in S / I_{\mathcal{N}}(p) = I_{\mathcal{N}}(0)\}$.

(iii) $S_{\mathcal{N}_{\mathbb{N}_{\mathcal{N}}}} = \{p \in S / \mathbb{N}_{\mathcal{N}}(p) = \mathbb{N}_{\mathcal{N}}(0)\}$ are BF-filters of S.

Proof: Suppose a $\mathcal{N}$NS $S_{\mathcal{N}}$ is a $\mathcal{N}$Nf of a BF-algebra $S:=(S\neq\emptyset, \varkappa, 0)$. It is obvious that $0 \in S_{\mathcal{N}_{Y_{\mathcal{N}}}} \cap S_{\mathcal{N}_{I_{\mathcal{N}}}} \cap S_{\mathcal{N}_{\mathbb{N}_{\mathcal{N}}}}$.

Let p, q ∈ S be such that $p, p\varkappa q \in S_{\mathcal{N}_{Y_{\mathcal{N}}}} \cap S_{\mathcal{N}_{I_{\mathcal{N}}}} \cap S_{\mathcal{N}_{\mathbb{N}_{\mathcal{N}}}}$.

This implies $Y_{\mathcal{N}}(p)=Y_{\mathcal{N}}(p\varkappa q)=Y_{\mathcal{N}}(0)$, $I_{\mathcal{N}}(p)=I_{\mathcal{N}}(p\varkappa q)=I_{\mathcal{N}}(0)$, $\mathbb{N}_{\mathcal{N}}(p)=\mathbb{N}_{\mathcal{N}}(p\varkappa q)=\mathbb{N}_{\mathcal{N}}(0)$ and

(i) $Y_{\mathcal{N}}(0)=\vee\{Y_{\mathcal{N}}(p), Y_{\mathcal{N}}(p\varkappa q)\}\geq Y_{\mathcal{N}}(q)\geq Y_{\mathcal{N}}(0)$ by (24) and (25)

- $Y_{\mathcal{N}}(q) = Y_{\mathcal{N}}(0)$
- $q \in S_{\mathcal{N}_{Y_{\mathcal{N}}}}$

Similar proof follows for (ii) and (iii) also.
Hence, $S_{\mathcal{N}_{Y_{\mathcal{N}}}}, S_{\mathcal{N}_{I_{\mathcal{N}}}}, S_{\mathcal{N}_{\mathbb{N}_{\mathcal{N}}}}$ are BF-filters of S.

Theorem 3.16: Every $\mathcal{N}$NSA of a BF-algebra satisfies (17).

Proof: Proof is straightforward by (1) and Definition (3.1).

Theorem 3.17: Let a $\mathcal{N}$NSA $S_{\mathcal{N}}$ of a BF-algebra $S:=(S\neq\emptyset, \varkappa, 0)$ satisfying $\forall p,q \in S$

$$(p \varkappa q \neq 0) \Rightarrow \begin{pmatrix} Y_{\mathcal{N}}(p) \leq Y_{\mathcal{N}}(q) \\ I_{\mathcal{N}}(p) \geq I_{\mathcal{N}}(q) \\ \mathbb{N}_{\mathcal{N}}(p) \leq \mathbb{N}_{\mathcal{N}}(q) \end{pmatrix} \tag{28}$$

and then $S_{\mathcal{N}}$ is a $\mathcal{N}\mathcal{N}$Nf of S.

Proof: Suppose that a $\mathcal{N}$NSA $S_{\mathcal{N}}$ of a BF-algebra $S:=(S\neq\emptyset, \varkappa, 0)$ holds (28). Then

$$\begin{pmatrix} Y_{\mathcal{N}}(0) \leq Y_{\mathcal{N}}(p) \\ I_{\mathcal{N}}(0) \geq I_{\mathcal{N}}(p) \\ \mathbb{N}_{\mathcal{N}}(0) \leq \mathbb{N}_{\mathcal{N}}(p) \end{pmatrix} (\forall p \in S) \qquad \text{by (Theorem 3.16)}$$

Case (i): If $p \varkappa q = 0$, then it is obvious that $S_{\mathcal{N}}$ is a $\mathcal{N}\mathcal{N}$Nf of S.
Case (ii): If $p \varkappa q \neq 0$, then

(i) $Y_N(p \ast q) \le \vee\{Y_N(p), Y_N(q)\}(\forall p,q \in S)$ by (11)

- $Y_N(p \ast q) \le Y_N(q)(\forall p,q \in S)$ by (28)

Similar proof follows for I_N and $\mathbb{N}_N$ also.
Hence, S_N is a NNNf of S.

Theorem 3.18: If a NNS S_N of a BF-algebra S:=(S≠Ø, ∗, 0) satisfying ∀p,q,r∈ S

$$(r \le p \ast q) \Rightarrow \begin{pmatrix} Y_N(q) \le \vee\left\{Y_N(r), Y_N(p)\right\} \\ I_N(q) \ge \wedge\left\{I_N(r), I_N(p)\right\} \\ \mathbb{N}_N(q) \le \vee\left\{\mathbb{N}_N(r), \mathbb{N}_N(p)\right\} \end{pmatrix}, \tag{29}$$

then S_N is a NNf of S.

Proof: Suppose a NNS S_N of a BF-algebra S:=(S≠Ø, ∗, 0) holds (29). Then, we have $p \ast (p \ast 0) = 0$, and by (1) and (2)

$$\begin{pmatrix} Y_N(0) \le Y_N(p) \\ I_N(0) \ge I_N(p) \\ \mathbb{N}_N(0) \le \mathbb{N}_N(p) \end{pmatrix} (\forall p \in S) \qquad \text{by (4) and (29)}$$

and also $(p \ast q) \ast (p \ast q) = 0$. by (1)

$$\begin{pmatrix} Y_N(q) \le \vee\{Y_N(p \ast q), Y_N(p) \\ I_N(q) \ge \wedge\left\{I_N(p \ast q), I_N(p)\right\} \\ \mathbb{N}_N(q) \le \vee\{\mathbb{N}_N(p \ast q), \mathbb{N}_N(p) \end{pmatrix} (\forall p, q \in S) \qquad \text{by (4) (29)}$$

Hence, S_N is a NNf of S.

Theorem 3.19: Let a NNS S_N of a BF-algebra S:=(S ≠ Ø, ∗, 0) be a NNPIi. Then, $Y_N^{\lambda_1}, I_N^{\lambda_2}, \mathbb{N}_N^{\lambda_3}$ are positive implicative BF-ideals of S, $\forall \lambda_1, \lambda_2, \lambda_3 \in [-1,0]$ with $-3 \le \lambda_1 + \lambda_2 + \lambda_3 \le 0$ whenever they are non-empty.

Proof: Suppose a NNS S_N of a BF-algebra S:=(S ≠ Ø, ∗, 0) is a NNPIi. Let λ_1, λ_2, $\lambda_3 \in [-1,0]$ with $-3 \le \lambda_1 + \lambda_2 + \lambda_3 \le 0$.

Let $p \in Y_N^{\lambda_1}$, $q \in I_N^{\lambda_2}$, and $r \in \mathbb{N}_N^{\lambda_3}$ for some p, q, r ∈ S. Then

$$\begin{pmatrix} Y_{\mathbb{N}}(0) \leq Y_{\mathbb{N}}(p) \leq \lambda_1 \\ I_{\mathbb{N}}(0) \geq I_{\mathbb{N}}(p) \geq \lambda_2 \\ \mathbb{N}_{\mathbb{N}}(0) \leq \mathbb{N}_{\mathbb{N}}(p) \leq \lambda_3 \end{pmatrix} (\forall p \in S)\text{ .Thisimplies} \qquad \text{by (17)}$$

$$0 \in Y_{\mathbb{N}}^{\lambda_1} \cap I_{\mathbb{N}}^{\lambda_2} \cap \mathbb{N}_{\mathbb{N}}^{\lambda_3}$$

Let $((p \ast q) \ast r) \in Y_{\mathbb{N}}^{\lambda_1} \cap I_{\mathbb{N}}^{\lambda_2} \cap \mathbb{N}_{\mathbb{N}}^{\lambda_3}$, $q \ast r \in Y_{\mathbb{N}}^{\lambda_1} \cap I_{\mathbb{N}}^{\lambda_2} \cap \mathbb{N}_{\mathbb{N}}^{\lambda_3}$. Then

$$\begin{pmatrix} Y_{\mathbb{N}}((p \ast q) \ast r) \leq \lambda_1, Y_{\mathbb{N}}(q \ast r) \leq \lambda_1 \\ I_{\mathbb{N}}((p \ast q) \ast r) \geq \lambda_2, I_{\mathbb{N}}(q \ast r) \geq \lambda_2 \\ \mathbb{N}_{\mathbb{N}}((p \ast q) \ast r) \leq \lambda_3, \mathbb{N}_{\mathbb{N}}(q \ast r) \leq \lambda_3 \end{pmatrix}$$

and

(i) $Y_{\mathbb{N}}(p \ast r) \leq \vee\{Y_{\mathbb{N}}(p \ast q) \ast r), Y_{\mathbb{N}}(q \ast r)\} \leq \lambda_1$ by (18)

- $p \ast r \in Y_{\mathbb{N}}^{\lambda_1}$

Similar proof follows for $I_{\mathbb{N}}$ and $\mathbb{N}_{\mathbb{N}}$ also.
Hence, $Y_{\mathbb{N}}^{\lambda_1}$, $I_{\mathbb{N}}^{\lambda_2}$, $\mathbb{N}_{\mathbb{N}}^{\lambda_3}$ are positive implicative BF-ideals of S.

Theorem 3.20: Let a NNS $S_{\mathbb{N}}$ of a BF-algebra $S:=(S \neq \emptyset, \ast, 0)$, and let λ_1, λ_2, λ_3 $\in[-1,0]$ be such that $-3 \leq \lambda_1 + \lambda_2 + \lambda_3 \leq 0$. If $S_{\mathbb{N}}$ is a NNPIi of S, then the non-empty set $(\lambda_1, \lambda_2, \lambda_3)$ is a level set of $S_{\mathbb{N}}$ and is a positive implicative BF-ideal of S.

Proof: Suppose a NNS $S_{\mathbb{N}}$ of a BF-algebra $S:=(S \neq \emptyset, \ast, 0)$ is a NNPIi of S.
Let $\lambda_1, \lambda_2, \lambda_3 \in[-1,0]$ with $-3 \leq \lambda_1 + \lambda_2 + \lambda_3 \leq 0$.

$$(\lambda_1, \lambda_2, \lambda_3) \text{ --level set of } S_{\mathbb{N}} \text{ is } S_{\mathbb{N}}\ (\lambda_1, \lambda_2, \lambda_3)$$
$$= \left\{p \in S / Y_{\mathbb{N}}(p) \leq \lambda_1, I_{\mathbb{N}}\ (p) \geq \lambda_2, \mathbb{N}_{\mathbb{N}}\ (p) \leq \lambda_3\right\}.$$

Since

$$\begin{pmatrix} Y_{\mathbb{N}}(0) \leq Y_{\mathbb{N}}(p) \leq \lambda_1 \\ I_{\mathbb{N}}(0) \geq I_{\mathbb{N}}(p) \geq \lambda_2 \\ \mathbb{N}_{\mathbb{N}}(0) \leq \mathbb{N}_{\mathbb{N}}(p) \leq \lambda_3 \end{pmatrix} (\forall p \in S)\text{ , wehave} \qquad \text{by (17)}$$

$$0 \in S_N(\lambda_1, \lambda_2, \lambda_3)$$

Let p, q, r ∈ S be such that $(p{\ast}q) \ast r \in S_N(\lambda_1, \lambda_2, \lambda_3)$ and $(q{\ast}r) \in S_N(\lambda_1, \lambda_2, \lambda_3)$. Then

$$\begin{pmatrix} Y_N((p{\ast}q) \ast r) \le \lambda_1, Y_N(q{\ast}r) \le \lambda_1 \\ I_N((p{\ast}q) \ast r) \ge \lambda_2, I_N(q{\ast}r) \ge \lambda_2 \\ \mathbb{N}_N((p{\ast}q) \ast r) \le \lambda_3, \mathbb{N}_N(q{\ast}r) \le \lambda_3 \end{pmatrix}$$

and

$$\begin{pmatrix} Y_N(p \ast r) \le \vee\{Y_N((p \ast q) \ast r), Y_N(q \ast r) = \lambda_1 \\ I_N(p \ast r) \ge \wedge\left\{I_N((p \ast q) \ast r), I_N(q \ast r)\right\} = \lambda_2 \\ \mathbb{N}_N(p \ast r) \le \vee\{\mathbb{N}_N((p \ast q) \ast r), \mathbb{N}_N(q \ast r) = \lambda_3 \end{pmatrix} (\forall p, q, r \in S)$$

by (18), (19), and (20).

This implies $p{\ast}r \in S_N(\lambda_1, \lambda_2, \lambda_3)$ ∀p, q, r ∈ S.
Hence, S_N is a positive implicative BF-ideal of S.

Corollary 3.21: Let a S_N be a NNS of a BF-algebra S:=(S ≠ Ø, ∗, 0) and let λ_1, λ_2, λ_3 ∈[−1,0] with $-3 \le \lambda_1 + \lambda_2 + \lambda_3 \le 0$. If S_N is a NNSA of S, then the non-empty $(\lambda_1, \lambda_2, \lambda_3)$ – level set of S_N is a subalgebra of S.

Proof: Proof is straightforward by Definitions (3.11) and (3.1).

Theorem 3.22: Let S_N be a NNS of a BF-algebra S:=(S ≠ Ø, ∗, 0) and assume that $\mathbf{Y_N}^{\lambda_1}, \mathbf{I_N}^{\lambda_2}, \mathbb{N}_N^{\lambda_3}$ are subalgebra of S, ∀ λ_1, λ_2, λ_3 ∈[−1,0] with $-3 \le \lambda_1 + \lambda_2 + \lambda_3 \le 0$, then S_N is a NNSA of S.

Proof: Suppose S_N be a NNS of a BF-algebra S:=(S ≠ Ø, ∗, 0) and assume that $Y_N^{\lambda_1}$, $I_N^{\lambda_2}$, $\mathbb{N}_N^{\lambda_3}$ are subalgebras of S, ∀ λ_1, λ_2, λ_3 ∈[−1,0] with $-3 \le \lambda_1 + \lambda_2 + \lambda_3 \le 0$.

Let p q∈Sbe ∋ $Y_N(p{\ast}q) > \lambda_1 > \vee\{Y_N(p), Y_N(q)\}$ for some λ_1∈[−1,0].
This implies p, q $\in Y_N^{\lambda_1}$ but $p{\ast}q \notin Y_N^{\lambda_1}$, which is a contradiction.
Therefore, $Y_N(p{\ast}q) \le \vee\{Y_N(p), Y_N(q)\}$ ∀ p, q∈ S.

Let p q∈S be∋ $I_N(p{\ast}q) < \lambda_2 < \wedge\{I_N(p), I_N(q)\}$
where $\lambda_2 = ½\,\{I_N(p{\ast}q) + \wedge\{I_N(p), I_N(q)\}$.

This implies p, q $\in I_N^{\lambda_2}$ but $p{\ast}q \notin I_N^{\lambda_2}$, which is a contradiction.
Therefore, $I_N(p{\ast}q) \ge \wedge\{I_N(p), I_N(q)\}$ ∀ p, q∈ S.
Similarly, suppose $\mathbb{N}_N(q) > \lambda_3 \ge \vee\{\mathbb{N}_N(p{\ast}q), \mathbb{N}_N(p)\}$ for some p, q∈ S and λ_3 ∈[−1,0].

Then p, q $\in \mathbb{N}_{\mathcal{N}}^{\lambda_3}$ but pxq $\notin \mathbb{N}_{\mathcal{N}}^{\lambda_3}$, which is a contradiction.
Therefore, $\mathbb{N}_{\mathcal{N}}(p \ast q) \leq \vee\{\mathbb{N}_{\mathcal{N}}(p), \mathbb{N}_{\mathcal{N}}(q)\}\ \forall$ p, q$\in$ S.
Hence, $S_{\mathcal{N}}$ is $\mathcal{NN}$SA of S.

Theorem 3.23: Given a non-empty subset G of a BF-algebra of S:=(S $\neq \emptyset$, $\ast$, 0), a $\mathcal{NN}$S $S_{\mathcal{N}}{}^{G}$ over S is a $\mathcal{NN}$PIi of S iff G is a positive implicative BF-ideal of S.

Proof: Suppose that G is a positive implicative BF-ideal of a BF-algebra S:=(S $\neq$ $\emptyset$, $\ast$, 0). Since, 0$\in$G, we have

$$Y_{\mathcal{N}}{}^{G}(0) = {}_{\lambda}y \leq Y_{\mathcal{N}}{}^{G}(p)$$

$$I_{\mathcal{N}}{}^{G}(0) = {}_{\lambda}I \geq I_{\mathcal{N}}{}^{G}(p)$$

$$\mathbb{N}_{\mathcal{N}}{}^{G}(0) = \lambda_{\mathcal{N}} \leq \mathbb{N}_{\mathcal{N}}{}^{G}(p), \forall p \in S. \qquad \text{by Definition (3.12)}$$

$\forall$ p, q, r $\in$ S, we consider the following four cases:

Case (i): If(p $\ast$ q) $\ast$ r $\in$ G and q $\ast$ r $\in$ G, then p $\ast$ r $\in$ G.

Hence, $Y_{\mathcal{N}}{}^{G}(p \ast r) \leq \vee\{ Y_{\mathcal{N}}{}^{G}((p \ast q) \ast r), Y_{\mathcal{N}}{}^{G}(q \ast r)\}$.
Similarly, $I_{\mathcal{N}}{}^{G}$ and$\mathbb{N}_{\mathcal{N}}{}^{G}$ also by Definitions (3.12) and (3.5).

Case (ii): If (p $\ast$ q) $\ast$ r $\in$ G and q $\ast$ r $\notin$ G are valid, then $Y_{\mathcal{N}}{}^{G}(q \ast r) = 0$,

$$I_{\mathcal{N}}{}^{G}(q \ast r) = -1 \text{ and } \mathbb{N}_{\mathcal{N}}{}^{G}(q \ast r) = 0.$$

Thus, $Y_{\mathcal{N}}{}^{G}(p \ast r) \leq 0 = \vee\{ Y_{\mathcal{N}}{}^{G}((p \ast q) \ast r), Y_{\mathcal{N}}{}^{G}(q \ast r)\}$.
Similarly, $I_{\mathcal{N}}{}^{G}$ and $\mathbb{N}_{\mathcal{N}}{}^{G}$.

Case (iii): If (p $\ast$ q) $\ast$ r $\notin$ G and q $\ast$ r $\in$ G which is similar to case (ii), and for case (iv), if (p $\ast$ q) $\ast$ r $\notin$ G and q $\ast$ r $\notin$ G, then

$$Y_{\mathcal{N}}{}^{G}(p \ast r) \leq \vee\left\{ Y_{\mathcal{N}}{}^{G}((p \ast q) \ast r), Y_{\mathcal{N}}{}^{G}(q \ast r)\right\}.$$

Similarly, $I_{\mathcal{N}}{}^{G}$ and $\mathbb{N}_{\mathcal{N}}{}^{G}$ also.
Hence, $S_{\mathcal{N}}{}^{G}$is a $\mathcal{NN}$PIi of S.
Conversely, suppose that $S_{\mathcal{N}}{}^{G}$ is a $\mathcal{NN}$PIi of S then by (3.19)
$\left(Y_{\mathcal{N}}{}^{G}\right)^{\frac{\lambda_Y}{2}} = G$, $\left(I_{\mathcal{N}}{}^{G}\right)^{\frac{\lambda_I}{2}} = G$ and $\left(\mathbb{N}_{\mathcal{N}}{}^{G}\right)^{\frac{\lambda_{\mathbb{N}}}{2}} = G$ are positive implicative BF-ideals of S.

4 Conclusion

In the study, NNS, NNSA, NNi, NNPIi, NNNf, and NNf of a BF-algebra are introduced and proved that every NNPIi is a NNi and so on. Finally, level set of a NNS is also presented. In the future, the scope of this study could be expanded to include Neutrosophic sets within specific algebraic contexts.

References

1. Zadeh, L.A.: Fuzzy sets. Information. Control. **8**, 338–353 (1965)
2. Smarandache, F.: A Unifying Field in Logics: Neutrosophic Logic. Neutrosophy, Neutrosophic Set, Neutrosophic Probability. American Research Press, Rehoboth (1999)
3. Atanassov, K.: Intuitionistic fuzzy sets. Fuzzy Sets Syst. **20**, 87–96 (1986)
4. Jun, Y.B., Lee, K.J., Song, S.Z.: N-ideals of BCK/BCI-Algebras. J. Chungcheong Math. Soc. **22**, 417–437 (2009)
5. Khan, M., Anis, S., Smarandache, F., Jun, Y.B.: Neutrosophic N -structures and their applications in semigroups. Ann. Fuzzy Math. Inform. **14**, 583–598 (2017)
6. Walendziak, A.: On BF-Algebras. Math. Slovaca. **57**(2), 119–128 (2007)
7. Song, S.-Z.: FlorentinSmarandache and Young Bae Jun: Neutrosophic Commutative N-Ideals in BCK-Algebras. Information. **8**, 130 (2017)

Study of MBJ-Neutrosophic Level Sets in β-Ideal

P. Muralikrishna and M. Surya

Keywords β-Algebra · Ideal · MBJ-neutrosophic set · MBJ-neutrosophic normed level subsets

1 Introduction

A β-algebra is a development of BCK/BCI algebras by Neggers and Kim [9]. Further this beta-algebra was studied by many researches to explore their ideas. Zadeh [15] tackled the uncertainty and modeling of real and scientific problems, by a structure known as the fuzzy set (FS), by defining the membership grade in closed 0,1. Intuitionistic fuzzy set (IFS) by including nonmembership grade in the fuzzy set the idea came out from Atanassov [3]. Also, the concept called interval valued IFS (IVIFS) was initiated from FS which defines the membership and nonmembership grades into terms of intervals. Later, an adding component called indeterminate membership valued is included in between membership and nonmembership function and was named as neutrosophic set (NS). Here the components are read as truth-mem, indeterminacy-mem, and falsity-mem by Smarandache [11]. Again, this concept is being extended by giving the interval in indeterminate membership function and resulted in MBJ-neutrosophic set (MBJ-NS), and those components are truth-mem, interval-valued indeterminacy-mem, and a non-mem function.

Menger [6] started to analyze the content called t-norm that is triangular norm for the membership grand, and for the nonmembership, it is treated as t-conorm or s-norm. Later to discuss in the MBJ-neutrosophic set, the middle component takes interval valued triangular norm, and it is represented as $\bar{t}$-norm.

Rosenfeld [10] initiated the correlation of fuzzy with different algebraic structures. This correlation concept gathered research with a great interest to move

P. Muralikrishna · M. Surya (✉)
PG & Research Department of Maths, Muthurangam Government Arts College (A), Vellore, Tamil Nadu, India
e-mail: pmkrishna@rocketmail.com; suryamano95@gmail.com

H.-H. Leung et al. (eds.), *Recent Developments in Algebra and Analysis*, Trends in Mathematics, https://doi.org/10.1007/978-3-031-37538-5_17

forward with different kinds of fuzzy and algebra, and thus this concept also motivates me to define the concept by considering a MBJ-neutrosophic set, and a beta-algebra arises with MBJ-neutrosophic beta-ideal. Further, norm is introduced here and is named as normed MBJ-neutrosophic β-ideals of β-algebras, and its results are studied.

2 Preliminaires

This segregation presents the necessary definition in the sequel.

Definition 2.1 [5]: Υ is a β-algebra with a scalar value zero, operations $+ and -$, holds

(i) $\sigma - 0 = \sigma$.
(ii) $(0 - \sigma) + \sigma = 0$.
(iii) $(\sigma - \imath) - \iota = \sigma - (\imath + q)$ $\forall \sigma, \imath, \iota \in \Upsilon$.

Example 2.2: Υ is a collection of 0, 1, 2, and 3 with $+ and -$ tabled using Cayley's.

+	0	1	2	3
0	0	1	2	3
1	1	0	3	2
2	2	3	1	0
3	3	2	0	1

-	0	1	2	3
0	0	1	3	2
1	1	0	2	3
2	2	3	0	1
3	3	2	1	0

Definition 2.3 [1, 2, 8]: Let $\mathfrak{k}$ of Υ is a subset along with a constant and two operations $+ and -$, respectively, are the β – ideal of Υ, if

(i) $0 \in \mathfrak{k}$.
(ii) $\sigma + \imath \in \mathfrak{k}$.
(iii) $\sigma - \imath$ & $\imath \in \mathfrak{k}$ then $\sigma \in \mathfrak{k}$ $\forall$ σ, $\imath \in \Upsilon$.

Example 2.4: Υ is a collection of e, ρ, ϱ, ς with $+ and -$ tabled using Cayley's.
The subset $\mathfrak{k}_1 = \{ e, \ \varsigma \}$ of Υ is an β-ideal of Υ.

+	e	ρ	ϱ	ς
e	e	ρ	ϱ	ς
ρ	ρ	ς	e	ϱ
ϱ	ϱ	e	ς	ρ
ς	ς	ϱ	ρ	e

-	e	ρ	ϱ	ς
e	0	ϱ	ρ	ς
ρ	ρ	e	ς	ϱ
ϱ	ϱ	ς	e	ρ
ς	ς	ρ	ϱ	e

Definition 2.5: $\mathfrak{A} = \{\sigma, \tau_A(\sigma) : \sigma \in \mathrm{Y}\}$of Υ, $\tau : \Upsilon \to [0,1]$, $\tau(\sigma)$ refers the membership grade of $\sigma \in \Upsilon$.

Definition 2.6 [12]: $\mathfrak{A}$ on Υ consists of three components, namely, truth represented as τ_T, an indeterminacy as ς_I, and a falsity ϱ_F where τ_T,ς_I, ϱ_F is a mapping from Υ to [0,1] and $\mathfrak{A}$ is named as NS and structured in the form of $\mathfrak{A} = \{< \sigma, \ \tau_T(\sigma), \varsigma_I(\sigma), \ \zeta_F(\sigma) > / \sigma \in \mathrm{Y}\}$.

Definition 2.7 [13, 14]: An MBJ-NS $\mathfrak{A}$ in Υ is of the form $\mathfrak{A} = \{< \sigma, \ \tau_T(\sigma), \varsigma_I(\sigma), \ \zeta_F(\sigma) > / \sigma \in \mathrm{Y}\}$ where τ_T, $\zeta_F : \Upsilon \to [0,1]$ and $\overline{\varsigma}_I : \mathrm{Y} \to D[0,1]$ with τ_T as truth, $\overline{\varsigma}_I$ as an intermediate interval valued, and ζ_F denotes the falsity.

Definition 2.8 [4, 7]: $\mathfrak{I} : [0,1] \times [0,1] \to [0,1]$ defines the triangular norm (– norm), if

(i) $\mathfrak{I}(\sigma, 1) = \sigma$.
(ii) $\mathfrak{I}(\sigma, \imath) = \mathfrak{I}(\imath, \sigma)$.
(iii) $\mathfrak{I}(\mathfrak{I}(\sigma, \imath), \iota) = \mathfrak{I}(\sigma, \mathfrak{I}(\imath, \iota))$.
(iv) $\mathfrak{I}(\sigma, \imath) \leq \mathfrak{I}(\sigma, \iota)$ if $\imath \leq q$ $\forall \sigma, \imath, \iota \in [0,1]$

min($\sigma, \imath$) is denoted as $\mathfrak{I}_M(\sigma, \imath)$, $\sigma, \imath$ as $\mathfrak{I}_P(\sigma, \imath)$, and the Lukasiewicz $\mathfrak{I}$ – norm $\mathfrak{I}_L(\sigma, \imath) = \max(\sigma + \imath - 1, 0)$ $\forall o, \imath \in [0,1]$.

Definition 2.9: $\overline{\mathfrak{I}} : [0,1] \times [0,1] \to D[0,1]$ states the definition of i – v – triangular norm ($\overline{\mathfrak{I}}$– norm), if

(i) $\overline{\mathfrak{I}}(\overline{\sigma}, \overline{1}) = \overline{\sigma}$.
(ii) $\overline{\mathfrak{I}}(\overline{\sigma}, \overline{\imath}) = \overline{\mathfrak{I}}(\overline{\imath}, \overline{\sigma})$.
(iii) $\overline{\mathfrak{I}}\left(\overline{\mathfrak{I}}(\overline{\sigma}, \overline{\imath}), \overline{\iota}\right) = \overline{\mathfrak{I}}\left(\overline{\sigma}, \overline{\mathfrak{I}}(\overline{\imath}, \overline{\iota})\right)$.
(iv) $\overline{\mathfrak{I}}(\overline{\sigma}, \overline{\imath}) \leq \overline{\mathfrak{I}}(\overline{\sigma}, \overline{\iota})$ if $\overline{\imath} \leq \overline{\iota}$ $\forall \overline{\sigma}, \overline{\imath}, \overline{\iota} \in D[0,1]$.

$\overline{\mathfrak{I}}_M(\overline{\sigma}, \overline{\imath}) = \mathrm{rmin}(\overline{\sigma}, \overline{\imath})$, the product $\overline{\mathfrak{I}}_P(\overline{\sigma}, \overline{\imath}) = \overline{\sigma}.\overline{\imath}$ and the Lukasiewicz $\overline{\mathfrak{I}}$ – norm $\overline{\mathfrak{I}}_L(\overline{\sigma}, \overline{\imath}) = \mathrm{rmax}\left(\overline{\sigma} + \overline{\imath} - \overline{1}, 0\right) \forall \overline{\sigma}, \overline{\imath} \in D[0,1]$.

Definition 2.10: $\mathfrak{S} : [0,1] \times [0,1] \to [0,1]$ defines the triangular conorm ($\mathfrak{I}$ – conorm), if

(i) $\mathfrak{S}(\sigma, 1) = \sigma$.
(ii) $\mathfrak{S}(\sigma, \imath) = \mathfrak{S}(\imath, \sigma)$.
(iii) $\mathfrak{S}(\mathfrak{S}(\sigma, \imath), \iota) = \mathfrak{S}(\sigma, \mathfrak{S}(\imath, \iota))$.

MBJ-neutrosophic beta-ideal of a beta-algebra, a norm is introduced here and defines a definition of normed MBJ-neutrosophic beta-ideal, and this definition gives way to discuss the relevant results. One can move this to any other algebraic structures.

References

1. Ansari, M.A.A., Chandramouleeswaran, M.: Fuzzy β – subalgebra of β – algebras. Int. J. Math. Sci. Eng. Appl. **5**(7), 239–249 (2013)
2. Ansari, M.A.A., Chandramouleeswaran, M.: Fuzzy β – ideals of β – algebras. Int. J. Math. Sci. Eng. Appl. **5**(1), 1–10 (2014)
3. Atanassov, K.T.: Intuitionisitic fuzzy sets, fuzzy sets and systems. J. Math, Appl. **30**(1), 87–96 (1986)
4. Borumand Saeid, A., Muralikrishna, P., Hemavathi, P.: Bi – normed intuitionistic fuzzy β – ideals of β – algebras. J. Uncertain Syst. **13**(1), 42–55 (2019)
5. Jun, Y.B., Kim: β – subalgebras and related topics. Commun. Korean Math. Soc. **27**(2), 243–255 (2012)
6. Menger, K.: Stastistival metrics. Proc. Natl. Acad. Sci. USA. **8**(535), –537
7. Muralikrishna, P., Sujatha, K., Chandramouleeswaran, M.: (S, T) – normed intuitionistic fuzzy β – subalgebras. Bull. Soc. Math. Banja Laka. ISSN 0354 – 5792(o), ISSN 1986 – 521X (p)
8. Muralikrishna, P., Surya, M.: MBJ – Neutrosophic β – Ideals of β – algebra. Neutrosophic Sets Syst. **35**, 99–118 (2020)
9. Negger, J., Kim, H.S.: On β – algebra. Math. Slovaca. **52**(5), 517–530 (2002)
10. Rosenfeld, A.: Fuzzy groups. J. Math. Anal. Appl. **3593**, 512–517 (1971)
11. Smarandache, F.: Neutrosophic set, a generalization of intuitionistic fuzzy sets. Int. J. Pure Appl. Math. **24**(5), 287–297 (2005)
12. Song, S.Z., Smarandache, F., Jim, Y.B.: Neutrosophic commutative N – ideals in BCK – algebra. Information. **8**, 130 (2017)
13. Surya, M., Muralikrishna, P.: On MBJ – neutrosophic β – subalgebra. Neutrosophic Sets Syst. **28**, 216–227 (2019)
14. Talkalloo, M.M., Jun, Y.B.: MBJ – neutrosophic structures and its applications in BCK\BCI – algebra. Neutrosophic Sets Syst. **23**, 72–84 (2018)
15. Zadeh, L.A.: Fuzzy sets. Inf. Control. **8**(3), 338–353 (1965)

Maximal Solution of Tropical Linear Systems by Normalization Method

B. Amutha and R. Perumal

Keywords Semiring · Tropical semiring · Solution of linear system · Maximal solution

1 Introduction

Imre Simon [2], a Brazilian mathematician and computer scientist, was the first who brought tropical geometry into the literature. French mathematicians coined the term "tropical" to recognize Simon's efforts in applying min-plus algebra to optimization theory. In tropical geometry, tropical semirings play a significant role. Semirings with an underlying carrier set, that is, a subset of the set of real numbers and a binary operation of addition as maximum or minimum, product as addition, have been devised and reinvented numerous times in diverse fields of research since the late 1950s [3]. There are two tropical semirings, depending on the operation. One is the minimum tropical semiring, while the other one is maximum. In the minimum tropical semiring, an addition of two elements will be a minimum of that two elements and multiplication of two elements obtained by adding them. Min-plus semiring is another name for this algebraic structure. Similarly in maximum tropical semiring, addition of two elements will be the maximum of two elements, and the tropical product is a sum of the elements. It is also called as max-plus semiring [4]. Examples of max-plus semirings are $(\mathbb{R} \cup (-\infty), \bigoplus, \bigodot)$, $(\mathbb{Z}^+ \cup (-\infty), \bigoplus, \bigodot)$. The tropical semiring $(\mathbb{Z}^+ \cup (-\infty), \bigoplus, \bigodot)$ was introduced by Simons. Max-plus semiring is isomorphic to a min-plus semiring, and both are idempotent semirings [5]. Working with tropical semirings is appealing because of its simplicity and resemblance to algebraic geometry [9]. As a result, the ease of use and applicability might be inspiring. The tropical semiring structure is used in a variety of fields, including computer science, linear algebra, number theory, automata theory, etc.

B. Amutha · R. Perumal (✉)
Department of Mathematics, College of Engineering and Technology, Faculty of Engineering and Technology, SRM Institute of Science and Technology, Kattankulathur, Tamil Nadu, India
e-mail: ab5552@srmist.edu.in; perumalr@srmist.edu.in

H.-H. Leung et al. (eds.), *Recent Developments in Algebra and Analysis*, Trends in Mathematics, https://doi.org/10.1007/978-3-031-37538-5_18

[3, 8, 12]. Tropical semirings are also used in language theory, control theory, and operation research [3]. Tropical semirings are playing an important role in linear algebra, especially in solving the linear systems [10, 11, 13]. We intend to decide the behavior of some matrices over the tropical semiring. Tropical addition is denoted as $\bigoplus$ and the tropical product as $\bigodot$. In this chapter, we are concentrating on the maximum tropical semiring [8].

2 Preliminaries

A semiring S is a non-empty set with two binary operations, say addition and multiplication, that guarantees the conditions that, $(S, +)$ has the identity element 0 and it is commutative monoid; $(S, \cdot)$ is a monoid that has a single identity element which is 1; multiplication distributes over addition, that is, $a\ (b + c) = ab + ac$ and $(b + c)\ a = ba + ca$, $\forall\ a, b, c \in S$, $a.0 = 0.a = 0\ \forall\ a \in S$; and an element 1 is not equal to zero [1, 10]. A semiring S is said to be an idempotent semiring if $\forall\ a \in S, a + a = a$ [5]. A semiring is said to be zero-sum-free if $a + b = 0 \implies a = b = 0$. The maximum tropical semiring is the semiring $R = (S \cup (-\infty), \bigoplus, \bigodot)$, since the operations $\bigoplus$ and $\bigodot$ denoted the maximum tropical addition and maximum tropical multiplication, respectively, since S is a semiring and R should satisfy the following properties that commutative under the tropical addition, i.e., $a \bigoplus b = b \bigoplus a\ \forall\ a, b \in R$. It satisfies the associative property under the tropical addition and tropical multiplication i.e., $(a \bigoplus b) \bigoplus c = a \bigoplus (b \bigoplus c)$ and $(a \bigodot b) \bigodot c = a \bigodot (b \bigodot c)\ \forall\ a, b, c \in R$, It satisfies the property that multliplication distributes over addition *i.e.* $a \bigodot (b \bigoplus c) = (a \bigodot b) \bigoplus (a \bigodot c)$ $\forall\ a, b, c \in R$, property of existence of additive identity *i.e.* $\exists\ e \in R, \forall\ a \in R$ such that $e \bigoplus a = a \bigoplus e = a$ (since the additive identity is $-\infty$), and it never has an additive inverse [10, 20]. Similarly in minimum tropical semiring, instead of maximum we have to choose minimum [11, 19]. Maximum tropical semiring is a idempotent semiring, and all idempotent semirings are zero-sum-free [5]. Suppose that there is a semiring, say S; we denote the set of all $m \times n$ matrices over the semiring as $M_{m \times n}(S)$ and we denoting every ijth element of $P \in M_{m \times n}(S)$ matrix as p_{ij}; transpose of the matrix P is denoted as P^T. Let $P = (p_{ij}) \in M_{m \times n}(S)$, $Q = (q_{ij}) \in M_{m \times n}(S)$, $T = (t_{ij}) \in M_{n \times l}(S)$ and $\alpha \in S$. Addition of two matrices generally calculated by $P + Q = ((p_{ij}) + (q_{ij}))_{m \times n}$ and similarly product of two matrices PT can be calculated by,

$$\sum_{i=1}^{n} ((p_{ik})(t_{kj}))_{m \times l}$$

and $\alpha P = (\alpha(p_{ij}))_{m \times n}$.

Similarly in the max-plus semiring, addition of two tropical matrices, $P \bigoplus Q$, can be calculated by $(max((p_{ij}), (q_{ij})))_{m \times l}$, and the multiplication of two tropical matrices P⊙T is calculated by

$$max((p_{ik}) + (t_{kj}))_{m\times l}$$

and $\alpha \bigodot P = (\alpha + (p_{ij}))_{m\times n}$. A system $P \bigodot x = q$ is said to be a tropical system if all the entries of the system from the tropical semiring $R = (S \cup (\pm\infty), \bigoplus, \bigodot)$ [16, 17]. A matrix $P \in M_{m\times n}(S)$ is said to be a tropical matrix if all the elements of a matrix from the tropical semiring $R = (S \cup (\pm\infty), \bigoplus, \bigodot)$[5]. A matrix P is said to be maximum tropical matrix if all the elements of the matrix from the maximum tropical semiring $R = (S \cup (-\infty), \bigoplus, \bigodot)$. A matrix P is said to be minimum tropical matrix if all the elements of the matrix from the minimum tropical semiring $R = (S \cup (\infty), \bigoplus, \bigodot)$. Let $S = \mathbb{R}$ be the extended real number system under the max-plus algebra, and let P and Q be $m \times n$ matrices over the extended real numbers under the operation of maximum tropical semirings, where $P = (p_{ij})_{m\times n}$ and $Q = (q_{ij})_{m\times n}$ and (p_{ij}), (q_{ij}) are the ij^{th} entries of P and Q, respectively, $P \leq Q \leftrightarrow (p_{ij}) \leq (q_{ij}) \; \forall \; i, j$ [11, 18]. A matrix $P = (p_{ij})$ is said to be regular if $(p_{ij}) \neq \pm\infty$. A vector $b \in S^m$ is said to be a normal vector or regular vector if $b_j \neq -\infty \; \forall \; j \in m$ [10]. Since we have considered max-plus semiring, if we consider the min-plus algebra, then in the regular vector, each entries $b_j \neq \infty$ $\forall \; j \in m$ [11]. A solution x^* of the tropical system $P \bigodot x = q$ is called as the maximal solution if $x \leq x^*$ for any other solution x [10, 11]. A linear system $P \bigodot x = q$ is said to be a tropical linear system if the elements of the linear system are all from any one of the tropical semirings [14, 15].

3 Main Results

A linear system $P \bigodot x = q$ is said to be a maximum if the coefficients of the linear systems from the maximum tropical semirings [7]. We know that there are different methods to solving the linear equations [6]. In this chapter, we have used the method of normalization [10, 11]. Consider the system of equation $P \bigodot x = q$. $P = (p_{ij}) \in M_{m\times n} \; (S/(-\infty))$, $Q = (q_{ij}) \in M_{m\times n}(S/(-\infty))$ since $(S/(x))$ denote all the values of R except x, $q = (q_j)$ is a regular vector $1 \leq j \leq m$, and j^{th} column of P matrix denoted as P_j. We begin this section with some basic definitions, and then we discuss the general maximal solution of the particular matrices. Let us assume the tropical semirings $\mathbf{T} = (\mathbb{Z}^+ \cup (-\infty), \bigoplus, \bigodot)$ where $\mathbb{Z}^+$ denoting the set of all natural numbers, **V**=$(\mathbb{R} \cup (-\infty), \bigoplus, \bigodot)$ where $\mathbb{R}$ is a set of all real numbers, **W**=$(\mathbb{Z} \cup (-\infty), \bigoplus, \bigodot)$ where $\mathbb{Z}$ is a set of all integers.

Theorem 1 *The linear system $P \bigodot x = q$ has solution if and only if every row of associated normalized matrix U contains at least one element, which is column minimum.*

3.1 Analyzing the Maximal Solution of the Tropical Linear Systems with Natural Matrix

Definition 1 A matrix $P \in M_{m\times n}(\mathbf{T})$ is said to be a natural matrix if the entries of the P matrix are continuously written with the natural numbers in the way followed by the row or column. Types of natural matrix are:

- Row natural matrix
- Column natural matrix

Definition 2 A matrix $P \in M_{m\times n}(\mathbf{T})$ is said be a row natural matrix if it is in the form of

$$\begin{bmatrix} 1 & 2 & 3 & \ldots\ldots & n \\ n+1 & n+2 & n+3 & \ldots\ldots & 2n \\ 2n+1 & 2n+2 & 2n+3 & \ldots\ldots & 3n \\ \vdots & \vdots & \vdots & \vdots & \vdots \\ \vdots & \vdots & \vdots & \vdots & \vdots \\ .. & .. & .. & .. & m.n \end{bmatrix}$$

Definition 3 A matrix $P \in M_{m\times n}(\mathbf{T})$ is said be column natural matrix, if it is in the below form,

$$\begin{bmatrix} 1 & m+1 & 2m+1 & \ldots & .. \\ 2 & m+2 & 2m+2 & \ldots & .. \\ 3 & m+3 & 2m+3 & \ldots & .. \\ \vdots & \vdots & \vdots & \vdots & \ldots \\ \vdots & \vdots & \vdots & \vdots & \ldots \\ m & 2m & 3m & .. & n.m \end{bmatrix}$$

Theorem 2 *Let $P \in M_{m\times m}(\mathbf{T}/(-\infty))$ be a column natural matrix and $P \bigodot x = q$ the linear system over the tropical semiring $(\mathbf{T}/(-\infty))$. If the $m \times 1$ regular vector q is of the form $q_i = m^2 + i$, $1 \le i \le m$, then the linear system $P \bigodot x = q$ has a solution with the maximal solution*

$$x^* = \begin{bmatrix} m^2 \\ m^2 - m \\ m^2 - 2m \\ m^2 - 3m \\ \vdots \\ m \end{bmatrix}$$

Proof Given P is a column natural matrix over the tropical semiring $\mathbf{T}/(-\infty)$

$$\begin{bmatrix} 1 & m+1 & 2m+1 & 3m+1 & .. & .. & m(m-1)+1 \\ 2 & m+2 & 2m+2 & 3m+2 & .. & .. & m(m-1)+2 \\ 3 & m+3 & 2m+3 & 3m+3 & .. & .. & .. \\ \vdots & \vdots & \vdots & \vdots & .. & .. & .. \\ \vdots & \vdots & \vdots & \vdots & .. & .. & .. \\ m & 2m & 3m & 4m & .. & m(m-1) & m(m-1)+m \end{bmatrix} \begin{bmatrix} x_1 \\ x_2 \\ x_3 \\ x_4 \\ \vdots \\ \vdots \\ x_m \end{bmatrix} = \begin{bmatrix} m^2+1 \\ m^2+2 \\ m^2+3 \\ \vdots \\ \vdots \\ m^2+m \end{bmatrix}$$

Since $\hat{P_1} = \frac{m+1}{2}$, $\hat{P_2} = \frac{3m+1}{2}, \ldots\ldots, \hat{P_m} = \frac{2m^2-m+1}{2}$, $\hat{q} = \frac{2m^2+m+1}{2}$ finally U matrix is a zero matrix. Clearly every row of U matrix has at least one element, which is the minimum element in any one of the columns. By Theorem 1, the given system has a solution. The maximal solution of given system obtained by $x_j{}^* = y_j{}^* - \hat{P_j} + \hat{q}$

$$x^* = \begin{bmatrix} m^2 \\ m^2 - m \\ m^2 - 2m \\ m^2 - 3m \\ \vdots \\ m \end{bmatrix}$$

Theorem 3 *Let $P \in M_{m\times m}(\mathbf{T}/(-\infty))$ be a row natural matrix and $P \bigodot x = q$ linear system over the tropical semiring $(\mathbf{T}/(-\infty))$. Since $\mathbf{T} = (\mathbb{Z}^+ \cup (-\infty), \bigoplus, \bigodot)$. If the $m\times 1$ regular vector q is of the form $q_i = m^2+im$, $1 \le i \le m$ then the linear system $P \bigodot x = q$ has a solution with the maximal solution $x_i^* = m^2 + m - i$, for $1 \le i \le m$.*

Proof Given a row natural matrix

$$\begin{bmatrix} 1 & 2 & 3 & 4 & .... & m \\ m+1 & m+2 & m+3 & m+4 & .... & 2m \\ 2m+1 & 2m+2 & 2m+3 & 2m+4 & .... & 3m \\ \vdots & \vdots & \vdots & \vdots & .... & .. \\ \vdots & \vdots & \vdots & \vdots & .... & .. \\ (m-1)m+1 & (m-1)m+2 & (m-1)m+3 & .. & .... & m(m-1)+m \end{bmatrix} \begin{bmatrix} x_1 \\ x_2 \\ x_3 \\ x_4 \\ \vdots \\ \vdots \\ x_m \end{bmatrix}$$

$$= \begin{bmatrix} m^2+m \\ m^2+2m \\ m^2+3m \\ \vdots \\ \vdots \\ 2m^2 \end{bmatrix}$$

Since $\hat{P_1} = (\frac{m^2-m+2}{2})$, $\hat{P_2} = (\frac{m^2-m+4}{2})$, $\hat{P_3} = (\frac{m^2-m+6}{2}), \ldots\ldots, \hat{P_m} = (\frac{m^2-m+2m}{2}))$, $\hat{q} = \frac{3m^2+m}{2}$, now the matrix U has all of its entries zero $\Longrightarrow$ All the rows contain at least one column minimum element. By Theorem 1, given system has a solution. The general form of the maximal solution is

$$x^* = \begin{bmatrix} m^2 + m - 1 \\ m^2 + m - 2 \\ m^2 + m - 3 \\ m^2 + m - 4 \\ \vdots \\ m^2 + m - m \end{bmatrix}$$

3.2 Analysis of the Maximal Solution of the Tropical Linear Systems with J-Matrix

Definition 4 Let P be a $m \times n$ matrix, and it is named as **J**-matrix if all the entries of the P matrix are only j.

$$\begin{bmatrix} j & j & j & .. & j \\ j & j & j & .. & j \\ j & j & j & .. & j \\ \vdots & \vdots & \vdots & \vdots & j \\ j & j & j & .. & j \end{bmatrix}$$

Theorem 4 *Let $P \in M_{m\times m}$ be a **J**-matrix and $P \bigodot x = q$ a linear system over the tropical semiring $\mathbf{V}/(-\infty)$ where $\mathbf{V}$=($\mathbb{R} \cup (-\infty)$, $\bigoplus$, $\bigodot$) with the $m \times 1$ normal vector q of the form*

$$q = \begin{bmatrix} q_1 \\ q_2 \\ q_3 \\ \vdots \\ \vdots \\ q_m \end{bmatrix}$$

then

1. *$q_i = q_j$ for all $1 \le i, j \le m$ if and only if the system has solution.*
2. *$q_i \neq q_j$ for some $1 \le i, j \le m$ if and only if the system has no solution.*

Proof Given matrix is a **J**-matrix over the tropical semiring $\mathbf{V}/(-\infty)$

$$\begin{bmatrix} j & j & j & .. & j \\ j & j & j & .. & j \\ j & j & j & .. & j \\ : & : & : & : & j \\ j & j & j & .. & j \end{bmatrix} \begin{bmatrix} x_1 \\ x_2 \\ x_3 \\ : \\ : \\ x_m \end{bmatrix} = \begin{bmatrix} q_1 \\ q_2 \\ q_3 \\ : \\ : \\ q_m \end{bmatrix}$$

Since we have $\hat{P_1} = j$, $\hat{P_2} = j$, $\hat{P_3} = j$, ..., $\hat{P_m} = j$, $\hat{q} = (\frac{q_1+q_2+...+q_m}{m}) = k$. In U matrix, all the entries are equal in first row, all the entries are equal in second row, and similarly, this condition holds till for the last row.

1. Assume that $q_i = q_j = k$, for all $1 \leq i, j \leq m$; then

$$q_1 - (\frac{q_1 + q_2 + \ldots + q_m}{m}) = \ldots.. = q_m - (\frac{q_1 + q_2 + \ldots + q_m}{m}) = 0$$

now clearly verify that all the elements are column minimum elements. Since every row of U matrix has a column minimum element as 0. By Theorem 1, the system has a solution. To prove the converse part, assume that the system has a solution. By the method of contradiction, suppose that $q_i \neq q_j$ for some $1 \leq i, j \leq m$; then clearly we know that minimum element among ${q_k}^s$ where $1 \leq k \leq m$ can be either q_i or q_j for some $1 \leq i, j \leq m$; then that minimum element will be placed in the same row. All other rows have no column minimum element. By Theorem 1, the system has no solution, which is the contradiction to our assumption that system has a solution. So q_k^s should be equal for every $1 \leq k \leq m$. The general form of the maximal solution for this case will be $x_i^* = -j + k, \forall\, 1 \leq i \leq m$.
2. Assume $q_i \neq q_j$ for some $1 \leq i, j \leq m$; then minimum element can be one of the values of q_k^s where $1 \leq k \leq m$. The column minimum element will be placed in any one of the rows of U matrix. Other rows cannot have the column minimum element. By Theorem 1, that implies system has no solution. Conversely, let us assume that system has no solution. We can say that some row of the U matrix does not contain any column minimum element. Suppose $q_i = q_j$ for all i and j; then, by first part of Theorem 4, the system has a solution, which is a contradiction.

3.3 Analysis of the Maximal Solution of the Tropical Linear Systems with γ-Diagonal Matrix

Theorem 5 *Let $P \in M_{m\times m}$ be a γ-diagonal matrix and $P \bigodot x = q$ a linear systems over the tropical semiring $\mathbf{V}/(-\infty)$ where $\mathbf{V}$=($\mathbb{R} \cup (-\infty)$, $\bigoplus$, $\bigodot$) with the normal vector $q_i = \gamma, \forall 1 \leq i \leq m$ then $U = -\tilde{P}$ and the system has a solution.*

Proof Given is a γ-diagonal matrix,

$$\begin{bmatrix} \gamma & 0 & 0 & .. & 0 \\ 0 & \gamma & 0 & .. & 0 \\ 0 & 0 & \gamma & .. & 0 \\ \vdots & \vdots & \vdots & \vdots & \vdots \\ 0 & 0 & 0 & .. & \gamma \end{bmatrix} \begin{bmatrix} x_1 \\ x_2 \\ x_3 \\ \vdots \\ \vdots \\ x_m \end{bmatrix} = \begin{bmatrix} \gamma \\ \gamma \\ \gamma \\ \vdots \\ \vdots \\ \gamma \end{bmatrix}$$

when we compare the normalized matrix $\tilde{P}$ and associated normalized matrix $\Longrightarrow$ $U = -\tilde{P}$. Now we want to prove that the system has a solution. Associated normalized matrix has only two elements $(\frac{\gamma}{m})$ and $-(\gamma - \frac{\gamma}{m})$.

Case 1:

If $(\frac{\gamma}{m}) < -(\gamma - \frac{\gamma}{m})$, then $(\frac{\gamma}{m})$ is the column minimum element in every column. Also we know that every row and every column has an entry $(\frac{\gamma}{m})$, so every row has atleast one column minimum element. Hence the system always has a solution. Now the maximal solution of this system is

$$x^* = \begin{bmatrix} \gamma \\ \gamma \\ \gamma \\ \gamma \\ \vdots \\ \gamma \end{bmatrix}$$

Case 2:

If $-(\gamma - \frac{\gamma}{m}) < (\frac{\gamma}{m})$ then $-(\gamma - \frac{\gamma}{m})$ be the column minimum element in every column. Also we know that every row and every column has an entry $-(\gamma - \frac{\gamma}{m})$. So that every row has at least one column minimum element $\Longrightarrow$ System has a solution. In this case the maximal solution is

$$x^* = \begin{bmatrix} 0 \\ 0 \\ 0 \\ 0 \\ \vdots \\ 0 \end{bmatrix}$$

3.4 Analysis of the Maximal Solution of the Tropical Linear Systems with Circulant Matrix

Theorem 6 *Let $P \in M_{m\times m}(\mathbf{V}(-\infty))$ be a circulant matrix and $P \bigodot x = q$ a linear systems over the tropical semiring $\mathbf{V}/(-\infty)$ where $\mathbf{V}=(\mathbb{R} \cup (-\infty), \bigoplus, \bigodot)$ with the $m \times 1$ normal vector q of the form $q = C_j$, where C_j is a jth column of the circulant matrix; then the following conditions hold:*

1. *$\hat{P_i} = \hat{P_j} = \hat{q}, \forall\, 1 \le i, j \le m$*
2. *$\tilde{P}$ is also circulant matrix.*
3. *System has a solution.*
4. *$x^* = y^*$*

Proof

1. Given $P \in M_{m\times m}$ is a circulant matrix over the tropical semiring $\mathbf{V}/(-\infty)$. We know that $\hat{P_j} = (\frac{p_{1j}+p_{2j}+\ldots+p_{mj}}{m}), \forall j \in m$. Clearly every row of circulant matrix has every element from $c_i{}^s, \forall\, 0 \le i \le m-1$ exactly once and every column of the circulant matrix has every element from $c_i{}^s, \forall\, 0 \le i \le m-1$ exactly once. Sum of the entries in every columns is equal. Let the column sum of the circulant matrix be r. When calculating the $\hat{P_j}$, the $\hat{P_j} = \frac{r}{m} = k\ \forall\ j \in 1, 2, \ldots m$. So we conclude that $\hat{P_i} = \hat{P_j} = \hat{q} = k, \forall\, 1 \le i, j \le m$.
2. For the given system $P \bigodot x = q$, the normalized system is $\tilde{P} \bigodot y = \tilde{q}$
We know that by the first part of Theorem 6, we know $\hat{P_i} = \hat{P_j} = \hat{q}$. Assume that $\hat{P_i} = \hat{P_j} = \hat{q} = k$

$$\begin{bmatrix} c_0-k & c_{m-1}-k & c_{m-2}-k & .. & c_1-k \\ c_1-k & c_0-k & c_{m-1}-k & .. & c_2-k \\ c_2-k & c_1-k & c_0-k & .. & c_3-k \\ \vdots & \vdots & \vdots & \vdots & \vdots \\ \vdots & \vdots & \vdots & \vdots & c_{m-1}-k \\ c_{m-1}-k & c_{m-2}-k & c_2-k & .. & c_0-k \end{bmatrix} \begin{bmatrix} y_1 \\ y_2 \\ y_3 \\ \vdots \\ \vdots \\ y_m \end{bmatrix} = \begin{bmatrix} c_0-\hat{q} \\ c_1-\hat{q} \\ c_2-\hat{q} \\ \vdots \\ \vdots \\ c_{m-1}-\hat{q} \end{bmatrix}$$

This normalized matrix satisfies all the conditions of a circulant matrix. We can conclude that $\tilde{P}$ is also a circulant matrix.
3. After finding the normalized matrix when we are finding the associated normalized matrix, we are getting U matrix as,

$$U = \begin{bmatrix} \tilde{q_1}-(c_0-k) & \tilde{q_1}-(c_{m-1}-k) & \tilde{q_1}-(c_{m-2}-k) & .. & \tilde{q_1}-(c_1-k) \\ \tilde{q_2}-(c_1-k) & \tilde{q_2}-(c_0-k) & \tilde{q_2}-(c_{m-1}-k) & .. & \tilde{q_2}-(c_2-k) \\ \tilde{q_3}-(c_2-k) & \tilde{q_3}-(c_1-k) & \tilde{q_3}-(c_0-k) & .. & \tilde{q_3}-(c_3-k) \\ \vdots & \vdots & \vdots & \vdots & \vdots \\ \vdots & \vdots & \vdots & \vdots & \tilde{q_{m-1}}-(c_{m-1}-k) \\ \tilde{q_m}-(c_{m-1}-k) & \tilde{q_m}-(c_{m-2}-k) & \tilde{q_m}-(c_2-k) & .. & \tilde{q_m}-(c_0-k) \end{bmatrix}$$

if $q = C_j$, j^{th} column of U matrix is zero, and in the jth column, all elements are column minimum elements. We have at least one column minimum element in every row of the associated normalized matrix U. By Theorem 1, we can conclude that the system has a solution. Maximal solution of this system depending upon the y^*. For each value of y^*, we can find different maximal solution.

4. We know that

$$x^* = \begin{bmatrix} y_1{}^* - \hat{P}_1 + \hat{q} \\ y_2{}^* - \hat{P}_2 + \hat{q} \\ y_3{}^* - \hat{P}_3 + \hat{q} \\ \vdots \\ y_n{}^* - \hat{P}_n + \hat{q} \end{bmatrix}$$

By the first part of Theorem 6, we have $\hat{P}_i = \hat{P}_j = \hat{q}, \forall\, 1 \le i, j \le m$

$$x^* = \begin{bmatrix} y_1{}^* \\ y_2{}^* \\ y_3{}^* \\ \vdots \\ y_n{}^* \end{bmatrix} = y^*$$

hence $x^* = y^*$.

Notes and Comments To determine the solutions of tropical linear systems, we employed the normalization method in this article. We talked about the conditions in tropical systems and came up with a unique solution, many solutions, and no solution. We used normalized method to determine the maximal solution of the linear equations over the tropical semirings. We worked on some special matrices and studied the general form of the maximal solution of that special matrices. We have also given several theorems about the general maximal solutions of specific linear systems over the tropical semirings.

References

1. Golan, J.S.: Semirings and Their Applications. Springer Science and Business Media, New York (2013)
2. Pin, J.-E.: Tropical Semirings. Cambridge University Press, Cambridge (1998)
3. Aceto, L., Esik, Z., Ingólfsdóttir, A.: Equational theories of tropical semirings. Theor. Comput. Sci. (2003). https://doi.org/10.1016/S0304-3975(02)00864-2
4. Maclagan, D., Sturmfels, B.: Introduction to tropical geometry. Graduate Studies in Mathematics. American Mathematical Society, Providence, RI (2009)
5. Akian, M., Gaubert, S., Guterman, A.: Linear independence over tropical semirings and beyond. Contemp. Math. (2009). https://doi.org/10.48550/arXiv.0812.3496

6. Heidergott, B., Jan Olsder, G., van der Woude, J.: Max Plus at Work Modelling and Analysis of Synchronized System: A Course on Max-Plus Algebra and Its Application. Princeton University Press, Priceton and Oxford (2006)
7. Butkovič, P.: Max-Linear Systems: Theory and Algorithms. Springer Science and Business Media, New York (2010)
8. Omanović, A., Oblak, P., Curk, T.: Application of tropical semiring for matrix factorization. Uporabna Informatika (2020). https://doi.org/10.31449/upinf.99
9. Izhakian, Z.: Basics of linear algebra over the extended tropical semiring. Contemp. Math. American Mathematical Society, Providence, RI (2009)
10. Olia, F., Ghalandarzadeh, S., Amiraslani, A., Jamshidvand, S.: Solving linear systems over tropical semirings through normalization method and its applications. J. Algebra Appl. (2020). https://doi.org/10.1142/S0219498821501590
11. Jamshidvand, S., Ghalandarzadeh, S., Amiraslani, A., Olia, F.: On the maximal solution of a linear system over tropical semirings. Math. Sci. **14**, 147–157 (2020). https://doi.org/10.1007/s40096-020-00325-w
12. Grigoriev, D.: Complexity of solving tropical linear systems. Comput. Complex. (2013). http://doi.org/10.1007/s00037-0120053-5
13. Davydow, A.: New algorithms for solving tropical linear systems. St. Petersbg. Math. J. **28**(6), 727–740 (2017). https://doi.org/10.1090/spmj/1470
14. Grigoriev, D., Podolskii, V.V.: Complexity of tropical and min-plus linear prevarieties. Comput. Complex. **24**, 31–64 (2015). http://doi.org/10.1007/s00037-0130077-5
15. Lorenzo, E., De La Puente, M.J.: An algorithm to describe the solution set of any tropical linear system A⊙x= B⊙x. Linear Algebra Appl. **435**, 884–901 (2011). https://doi.org/10.1016/j.laa.2011.02.014
16. Butkovič, P., Zimmermann, K.: A strongly polynomial algorithm for solving two-sided linear systems in max-algebra. Discrete Appl. Math. **154**, 437–446 (2006). https://doi.org/10.1016/j.dam.2005.09.008
17. Baccelli, F., Hasenfuss, S., Schmidt, V.: Transient and stationary waiting times in (max,+)-linear systems with Poisson input. Queue. Syst. **26**, 301–342 (1997). https://doi.org/10.1023/A:1019141510202
18. Bemporad, A., Borrelli, F., Morari, M.: Min-max control of constrained uncertain discrete-time linear systems. IEEE Trans. Automat. Contr. **26**, 301–342 (2003). http://doi.org/10.1109/TAC.2003.816984
19. Butkovič, P., MacCaig, M.: On the integer max-linear programming problem. Discrete Appl. Math. **162**, 128–141 (2014). https://doi.org/10.1016/j.dam.2013.08.007
20. Akian, M., Gaubert, S., Kolokoltsov, V.: Solutions of max-plus linear equations and large deviations. In: Proceedings of the 44th IEEE Conference on Decision and Control. IEEE (2005). http://doi.org/10.1109/CDC.2005.1583420

A Theoretical Perception on Interval Valued Fuzzy β-Subalgebraic Topology

P. Muralikrishna, P. Hemavathi, K. Palanivel, and R. Vinodkumar

Keywords β-Algebra · Fuzzy β-algebra · Topology · Interval valued fuzzy · β-Subalgebra · Fuzzy β-subalgebraic topological space

1 Introduction

A F_S has been originated by Zadeh, and later the concept of linguistic variable was described [13, 14]. Following that, other researchers applied the F_S in numerous directions and scientific fields. Chang [5] introduced F_S as a concept for generalized topology. The idea of FT subsystem on a TM-algebra was established by Annalakshmi et al. [1]. The big idea of fuzzy β_subalgebraic_TS has been initiated by Chandramouleeswaran et al. [3]. Chanduraty et al. proposed the idea of FT on F_S [4]. Foster [6] proposed the idea of FT groups in which the results on homomorphic images and inverse images, product, and quotients of FT groups were investigated. Kandil et al. [8] dealt the separation axioms in i_v FTS. Moreover, the relation between induced FTS and separation axioms on a given i_v FTS has been studied.

P. Muralikrishna
PG & Research Department of Mathematics, Muthurangam Govt. Arts College (Autonomous), Vellore, Tamilnadu, India

P. Hemavathi (✉)
Department of Mathematics, Saveetha School of Engineering, SIMATS, Thandalam, Tamilnadu, India
e-mail: hemavathip.sse@saveetha.com

K. Palanivel
Department of Mathematics, School of Advanced Sciences, Vellore Institute of Technology (VIT), Vellore, Tamilnadu, India

R. Vinodkumar
Department of Mathematics, Prathyusha Engineering College, Aranvoyalkuppam, Thiruvallur, Tamilnadu, India

H.-H. Leung et al. (eds.), *Recent Developments in Algebra and Analysis*, Trends in Mathematics, https://doi.org/10.1007/978-3-031-37538-5_19

The notion of FTS involving Boolean algebraic structures was introduced by Sharma et al. [10], and the author described the properties of Boolean algebraic FTS. Veerappan Chandrasekar et al. [11] presented the concepts on fuzzy_ e_open sets fuzzy_ e_continuity and fuzzy _e_compactness in intuitionistic_ FTS. Neggers and Kim first suggested the idea of β_algebra in 2002 [9]. On β_subalgebras, Jun et al. [12] covered a number of related topics in 2001. The idea of fuzzy β_subalgebras of β_algebras was proposed in 2013 by the authors of [2] applied the F_Ss into β_algebra. In 2015 [7] Hemavathi et.al initiated the notion of i_v_f_β_subalgebras. This paper explores the notion of i_v_f_β_subalgebraic TS on β_algebras by bringing together the ideas of FTS and i_v_f_β_algebras and proving some of their features.

2 Preliminaries

Some fundamental definitions that are necessary for the sequel are recalled in this section.

Definition 2.1: For any non-empty set Ξ, a F_S in Ξ, we define $\psi : \Xi \to [0, 1]$. For every $\mathfrak{L} \in \Xi$, $\psi(\mathfrak{L})$ with $0 \leq \psi(\mathfrak{L}) \leq 1$ is known as the membership_ value of $\mathfrak{L}$ in Ξ.

Definition 2.2: Any non-empty set Ξ with the binary operations -,+ and constant 0 is known as a β_aβlgebra if (i) $\mathfrak{L} - 0 = \mathfrak{L}$ (ii) $(0 - \mathfrak{L}) + \mathfrak{L} = 0$ (iii) $(\mathfrak{L} - \xi) - \varpi = \mathfrak{L} - (\varpi + \xi)$ $\quad \forall \mathfrak{L}, \xi, \varpi \in \Xi$.

Definition 2.3: Consider a non-empty subset Å of a β_algebra $(\Xi, +, -, 0)$ is said to be a β_subalgebra of Ξ, if (i) $\mathfrak{L} + \xi \in$ Å (ii) $\mathfrak{L} - \xi \in$ Å for all $\mathfrak{L}, \xi \in$ Å.

Definition 2.4: For any β_algebra Ξ, a F_S ψ is referred to as fuzzy_ β_subalgebra of Ξ, if $\forall \mathfrak{L}, \xi \in \Xi$.

(i) $\psi(\mathfrak{L} + \xi) \geq \min\{\psi(\mathfrak{L}), \psi(\xi)\}$and (ii) $\psi(\mathfrak{L} - \xi) \geq \min\{\psi(\mathfrak{L}), \psi(\xi.)\}$

Example 2.5 Suppose a β_algebra $(\Xi, +, -, 0)$ with the Cayley's table.

+	0	ϵ_1	ϵ_2	ϵ_3
0	0	ϵ_1	ϵ_2	ϵ_3
ϵ_1	ϵ_1	0	ϵ_3	ϵ_2
ϵ_2	ϵ_2	ϵ_3	ϵ_1	0
ϵ_3	ϵ_3	ϵ_2	0	ϵ_1

-	0	ϵ_1	ϵ_2	ϵ_3
0	0	ϵ_1	ϵ_3	ϵ_2
ϵ_1	ϵ_1	0	ϵ_2	ϵ_3
ϵ_2	ϵ_2	ϵ_3	0	ϵ_1
ϵ_3	ϵ_3	ϵ_2	ϵ_1	0

Let us describe the F_S $\psi : \Xi \to [0, 1]$ such that

$$\psi(\mathfrak{L}) = \begin{cases} 1 & \mathfrak{L} = 0 \\ 0.5 & \mathfrak{L} = \epsilon_1 \\ 0 & \mathfrak{L} = \epsilon_2, \epsilon_3 \end{cases}$$

Thus, ψ is a fuzzy algebra in Ξ.

Definition 2.6: If $\overline{\psi}_1$ and $\overline{\psi}_2$ are two i_v F_Ss of Ξ, then the intersection of $\overline{\psi}_1$ and $\overline{\psi}_2$, expressed by $\overline{\psi}_1 \cap \overline{\psi}_2$, is defined by $\left(\overline{\psi}_1 \cap \overline{\psi}_2\right)(\mathfrak{L}) = rmin\left\{\overline{\psi}_1(\mathfrak{L}), \overline{\psi}_2(\mathfrak{L})\right\}\ \forall \mathfrak{L} \in \Xi$.

Definition 2.7: If $\overline{\psi}_1$and $\overline{\psi}_2$ are two i_v F_Ss of Ξ, then the $\cup$ of $\overline{\psi}_1$ and $\overline{\psi}_2$, denoted by $\overline{\psi}_1 \cup \overline{\psi}_2$, is defined by $\left(\overline{\psi}_1 \cup \overline{\psi}_2\right)(\mathfrak{L}) = rmax\left\{\overline{\psi}_1(\mathfrak{L}), \overline{\psi}_2(\mathfrak{L})\right\}\ \forall \mathfrak{L} \in \Xi$.

Definition 2.8: If Ξ has two i_v F_Ss $\overline{\psi}_1$ and $\overline{\psi}_2$, then $\overline{\psi}_1$ is regarded as a subset of $\overline{\psi}_2$, represented by $\overline{\psi}_1 \subseteq \overline{\psi}_2$, if $\overline{\psi}_1(\mathfrak{L}) \leq \overline{\psi}_2)(\mathfrak{L})\ \forall \mathfrak{L} \in \Xi$.

Definition 2.9: Let $\overline{\psi}$ be an i_v F_S in Ξ and f be a function from Ξ to Y. The image of $\overline{\psi}$ is defined as

$$\overline{\psi}(\xi) = \begin{cases} \underset{\mathfrak{L} \in f^{-1}(\xi)}{\text{rsup}}\ \overline{\lambda}(\mathfrak{L}), if\ \ f^{-1}(\xi) \neq \varnothing \\ [0, 0] \qquad\qquad otherwise \end{cases}$$

Consider an i_v F_S $\overline{\lambda}$ in Y. Then, we define the inverse function f^{-1} as$\overline{\lambda}_{f^{-1}}(\mathfrak{L}) = \overline{\lambda}(\text{f}(\mathfrak{L}))$ for all $\mathfrak{L} \in \Xi$.

Definition 2.10: Let Ξ be a universe. The collection Γ is called an i_v FT set of F_Ss in Ξ if

(i) $\breve{0}$ and $\breve{1} \in \Gamma$ and $\breve{0} = \psi(x) = 0$ and $\breve{1} = \psi(x) = 1$.
(ii) $\psi_H, \psi_K \in \Gamma$ *then* $\psi_{(H \cap K)} \in \Gamma$.
(iii) $\psi_{H_i} \in \Gamma$ *for each* $i \in \Lambda$, then $\bigcup_\Lambda \psi_{H_i} \in \Gamma$ with an index set Λ.

Remark 2.11: For any universe Ξ with an i_v F_S $\overline{\Gamma}$ of Ξ, $\left(\Xi, \overline{\Gamma}\right)$ is called an i_v FTS and each member of $\overline{\Gamma}$ is named as $\overline{\Gamma}$open F_S in Ξ.

Definition 2.12: For an i_v FTS $\left(\Xi, \overline{\Gamma}\right)$ and $\overline{\zeta} \in \overline{\Gamma}$. A i_v FS $\overline{\upsilon} \in \overline{\Gamma}$ is said to have a neighborhood (NBD) of $\overline{\zeta}$ if $\exists$ a $\overline{\Gamma}$open _F_S $\overline{\Theta}$ with $\overline{\zeta} \subset \overline{\Theta} \subset \overline{\upsilon}$,that is, $\overline{\zeta}(\mathfrak{L}) \leq \overline{\Theta}(\mathfrak{L}) \leq \overline{\upsilon}(\mathfrak{L})\, \forall \mathfrak{L} \in \Xi$.

Definition 2.13: Consider a i_v FTS $\left(\Xi, \overline{\Gamma}\right)$, and $\overline{\text{H}}$, $\overline{\text{K}}$ are two i_v F_Ss in it and $\overline{\text{H}} \supset \overline{\text{K}}$. Then, $\overline{\text{K}}$ is said to have an interior of $\overline{\text{H}}$ if $\overline{\text{H}}$ is a NBD of $\overline{\text{K}}$. The union of all i_v F_Ss of $\overline{\text{H}}$ is also an interior of $\overline{\text{H}}$ and is represented by $\overline{\text{H}}^0$.

3 Interval Valued Fuzzy β_Subalgebraic Topology

The concept of i_v_f β_subalgebraic TS is presented in this section, along with several associated findings.

Definition 3.1: Let Ξ be a β_algebra. If there is a family $\overline{\Gamma}$ of i_v F β_subalgebras that meets the conditions listed below, then $\left(\Xi, \overline{\Gamma}\right)$ is said to be an i_v_f β-subalgebraic TS on Ξ.

(i) $\breve{0}$ and $\breve{1} \in \overline{\Gamma}$ and $\breve{0} = \overline{\chi}(\mathfrak{L}) = \overline{0}$ and $\breve{1} = \overline{\chi}(\mathfrak{L}) = \overline{1}$.
(ii) If $\overline{\chi}_H, \overline{\chi}_K \in \overline{\Gamma}$ *then* $\overline{\chi}_{(H\cap K)} \in \overline{\Gamma}$.
(iii) If $\chi_{H_i} \in \overline{\Gamma}$ *for each* $i \in \Lambda$, then $\bigcup_\Lambda \overline{\chi}_{H_i} \in \overline{\Gamma}$ where Λ indicates an IS.

Each element in $\overline{\Gamma}$ is referred to as a $\overline{\Gamma}$open - i_v F_S in β_algebra of Ξ.

Example 3.2: For a β_algebra given in the Example 2.5, consider the fuzzy_ β_subalgebras, $\overline{\chi}_i : \Xi \to \mathfrak{D}[0,1]$, i = 1, 2, 3, 4, and 5 be given

$$\overline{\chi}_1(\mathfrak{L}) = \begin{cases} [0.3, 0.7] : & \mathfrak{L} = 0 \\ [0.2, 0.5] : & \mathfrak{L} = \epsilon_1, \epsilon_2 \\ [0.1, 0.3] : & \mathfrak{L} = \epsilon_3 \end{cases} \quad \overline{\chi}_2(\mathfrak{L}) = \begin{cases} [0.4, 0.6] : & \mathfrak{L} = 0 \\ [0.3, 0.5] : & \mathfrak{L} = \epsilon_1, \epsilon_2 \\ [0.2, 0.4] : & \mathfrak{L} = \epsilon_3 \end{cases}$$

$$\overline{\chi}_3(\mathfrak{L}) = \begin{cases} [0.3, 0.5] : & \mathfrak{L} = 0 \\ [0.2, 0.3] : & \mathfrak{L} = \epsilon_1, \epsilon_2 \\ [0.1, 0.2] : & \mathfrak{L} = \epsilon_3 \end{cases} \quad \overline{\chi}_4(\mathfrak{L}) = \begin{cases} [0.4, 0.5] : & \mathfrak{L} = 0 \\ [0.3, 0.4] : & \mathfrak{L} = \epsilon_1, \epsilon_2 \\ [0.1, 0.2] : & \mathfrak{L} = \epsilon_3 \end{cases}$$

$$\overline{\chi}_5(\mathfrak{L}) = \begin{cases} [0.4, 0.6] : & \mathfrak{L} = 0 \\ [0.3, 0.4] : & \mathfrak{L} = \epsilon_1, \epsilon_2 \\ [0.2, 0.3] : & \mathfrak{L} = \epsilon_3 \end{cases}$$

Then the collection $\tau = \left\{\tilde{0}, \tilde{1}, \overline{\chi}_1, \overline{\chi}_2, \overline{\chi}_3, \overline{\chi}_4, \overline{\chi}_5\right\}$ is an i_v_f β_subalgebras on Ξ.

Thus, $(\Xi, \overline{\Gamma})$ is an i_v_f β_subalgebraic TS on Ξ.

Definition 3.3: Suppose that $(\Xi, \overline{\Gamma})$ be an i_v_f β_subalgebraic TS. Consider an i_v F_S $\overline{\chi}$ in $\overline{\Gamma}$. An i_v F_S $\overline{\varphi} \in \overline{\Gamma}$ is known as NBD of $\overline{\chi}$ if ∃a $\overline{\Gamma}$_open i_v F set $\overline{\Theta}$ with $\overline{\chi} \subset \overline{\Theta} \subset \overline{\varphi}$ ie $\overline{\chi}(\mathfrak{L}) \leq \overline{\Theta}(\mathfrak{L}) \leq \overline{\varphi}(\mathfrak{L})\ \forall \mathfrak{L} \in \Xi$.

Example 3.4: Let us assume the i_v_f β_subalgebraic TS as in the example 3.2. $\overline{\chi}_3$ is a i_v_f NBD of i_v F_S $\overline{\chi}_1$, for $\overline{\chi}_1(\mathfrak{L}) \leq \overline{\chi}_2(\mathfrak{L}) \leq \overline{\chi}_3(\mathfrak{L})$.

Definition 3.5: Consider i_v_f sets $\overline{H}$ and $\overline{K}$ in an i_v_f β_subalgebraic TS $(\Xi, \overline{\Gamma})$. If $\overline{H} \supset \overline{K}$ and $\overline{H}$ is a NBD of $\overline{K}$, then we can say $\overline{K}$ is having an interior of $\overline{H}$. The union of all interior i_v F_S of $\overline{H}$ is also an interior of $\overline{H}$ and is represented by $\overline{H}^0$.

Example 3.6: For any i_v_f β_subalgebraic TS on Ξ given in Example 3.2.
$\overline{\chi}_2$ is an i_v_f NBD of a i_v F sets $\overline{\chi}_1, \overline{\chi}_4, \overline{\chi}_5$.
That is, $\overline{\chi}_1, \overline{\chi}_4, \overline{\chi}_5$ are i_v F interiors of $\overline{\chi}_2$.

$$\overline{\chi}_2{}^0 = \cup\overline{\left\{\chi_1, \overline{\chi}_4, \overline{\chi}_5\right\}}$$

$$= \text{rmax}\ \left\{\overline{\chi}_1(\mathfrak{L}), \overline{\chi}_4(\mathfrak{L}), \overline{\chi}_5(\mathfrak{L})\right\} = \overline{\chi}_1(\mathfrak{L}).$$

Theorem 3.7: Let $(\Xi, \overline{\Gamma})$ be an i_v_f β_subalgebraic TS on Ξ. Then, an i_v F β_subalgebra of $\overline{\mathfrak{Y}}$ is $\overline{\tau_{\text{open}}}$ if and only if for each i_v F set $\overline{\mathfrak{E}}$ contained in $\overline{\mathfrak{Y}}$, $\overline{\mathfrak{Y}}$ is an i_v_f NBD of $\overline{\mathfrak{E}}$.

Proof

Assume an i_v_f β_subalgebra of $\overline{H}$ is $\overline{\Gamma}$open.
Consider an i_v_f β_subalgebra $\overline{K}$ contained in $\overline{H}$.
Here $\overline{H}$ is open, and $\overline{K} \subset \overline{H}$.
Hence, $\overline{H}$ is i_v_f NBD of $\overline{K}$.
Alternatively, for each i_v_f β_subalgebra $\overline{K}$ contained in $\overline{H}$,
$\overline{H}$ is an i_v_f NBD of $\overline{K}$.
For $\overline{H} \subset \overline{K}$, by our assumption,
$\overline{H}$ is an i_v_f NBD of $\overline{H}$.
Then, there will be an i_v_f _open set $\overline{\Theta}$ with $\overline{H} \subset \overline{\Theta} \subset \overline{H}$.
Thus, $\overline{H} = \overline{\Theta}$ and $\overline{H}$ are $\overline{\Gamma}$_open in $(\Xi, \overline{\Gamma})$.

Theorem 3.8: Consider an i_v_f β_subalgebraic TS $(\Xi, \overline{\Gamma})$ on Ξ. For an i_v_f β_subalgebra $\overline{H}$ on Ξ,

(1) $\overline{H}^0$ is the largest i_v_f _open set contained in $\overline{H}$.
(2) $\overline{H} = \overline{H}^0$ if and only if the i_v_f β_subalgebra $\overline{H}$ is open.

Proof

(1) Suppose that $(\Xi, \overline{\Gamma})$ be an i_v_f β_subalgebraic TS on X.
Consider an i_v_f β_subalgebra $\overline{H}$ be in Ξ.
From the definition of i_v_f interior,
$\overline{H}^0$ is again an i_v_f interior_set of $\overline{H}$.
Hence, there exist an $\overline{\Gamma}$_ i_v_f _open set $\overline{\Theta}$ with $\overline{H}^0 \subset \overline{\Theta} \subset \overline{H}$.
On the other hand, $\overline{\Theta}$ is an fuzzy interior_set of H, $\overline{\Theta} \subset \overline{\mathfrak{Y}}^0$ and so, $\overline{H}^0 = \overline{\Theta}$.
Hence, $\overline{H}^0$ is the largest i_v_f _open set contained in $\overline{H}$.
(2) Assume that the i_v_f set $\overline{H}$ is open.

If $\overline{H}$ is open, then $\overline{H} \subset \overline{H}^0$, for $\overline{H}$ is an i_v_f interior set of $\overline{H}$ and so $\overline{H} = \overline{H}^0$.

Alternatively, suppose $\overline{H} = \overline{H}^0$.

By definition of i_v_f interior, the union of every i_v_f interior sets of $\overline{H}$ is also i_v_f interior of $\overline{H}$ and is denoted by $\overline{H}^0$.

Hence, $\overline{H}$ is a NBD of $\overline{H}^0$.

An i_v_f set $\overline{H}$ is $\overline{\Gamma}$_open.

Definition 3.9: On the β_algebras, Ξ and Y, consider i_v_f β_subalgebraic TS $(\Xi, \overline{\Gamma})$ and $(Y, \overline{\chi})$, respectively. The function f: $(\Xi, \overline{\Gamma}) \rightarrow (Y, \overline{\chi})$ is said to have a $\mathfrak{T}$ continuous function if the inverse of any $\overline{\chi}$_open i_v F_S of Y is $\overline{\Gamma}$_open i_v_f set of Ξ .

Theorem 3.10: For the β_algebras, Ξ and Y, consider i_v_f β_subalgebraic TS $(\Xi, \overline{\Gamma})$, and $(Y, \overline{\chi})$, respectively, f is $\mathfrak{T}$continuous if and only if the inverse image of every closed i_v F_S set is closed.

Proof

Assume that f is $\mathfrak{T}$continuous.

That is, the inverse of every $\overline{\varrho}$_open i_v F_S is $\overline{\Gamma}$open.

Choose $\overline{\varrho}^0$ be the set of closed F_S in Y

Thus, $\overline{\chi}_{f^{-1}(\overline{\varrho}')}(\mathfrak{L}) = \overline{\chi}'_{\overline{\varrho}}(f(\mathfrak{L}))$

$$= \overline{\chi}_{\overline{\varrho}'}(f(\mathfrak{L}))$$

$$= 1 - \overline{\chi}_{\overline{\varrho}}(f(\mathfrak{L}))$$

$$= 1 - \overline{\chi}_{f^{-1}(\overline{\varrho})}(\mathfrak{L})$$

$$= \overline{\chi}'_{f^{-1}(\overline{\varrho}')}(\mathfrak{L})$$

$$\Rightarrow \quad f^{-1}(\overline{\varrho}') = \left\{f^{-1}(\overline{\varrho})\right\}' \quad \forall \mathfrak{r} \in \Xi.$$

Because f is $\mathfrak{T}$ _continuous, the inverse of each closed i_v F_S is closed.

Alternatively, for the set $\overline{\chi}$ of open i_v F_S in Y,

$$\chi_{f^{-1}(\overline{\varrho})}(\mathfrak{L}) = \overline{\chi}_{\overline{\varrho}}(f(\mathfrak{L}))\, \forall \mathfrak{L} \in \Xi.$$

Since the inverse of every closed i_v F_S is closed.

∴ The inverse of every open i_v F_S is open and so f is $\mathfrak{T}$ _continuous.

This completes the proof.

Theorem 3.11: Assume two i_v_f β_subalgebraic TS $(\Xi, \overline{\Gamma})$ and $(Y, \overline{\chi})$ on the β_algebras, Ξ and Y correspondingly. Then, for each i_v F_S $\overline{H}$ in Ξ, the inverse of every NBD of f($\overline{H}$) is a NBD of $\overline{H}$ if for each i_v F set $\overline{H}$ in Ξ and each NBD $\overline{\varsigma}$ of f($\overline{H}$), there is NBD $\overline{\omega}$ of $\overline{H}$ with f($\overline{\omega}$) $\subset \overline{\varsigma}$.

Proof
Let $\overline{H}$ be the i_v F_S of Ξ.

Let $\mathfrak{U}$, $\mathfrak{J}$ *be* the family of NBDs of i_v F_S and their image.

Let $A \in \mathcal{A}$.

Take $\overline{\varsigma} \in \mathfrak{J}$, $\overline{\omega} \in \mathfrak{U}$ is the NBD of $f\left(\overline{H}\right)$ and $\overline{H}$.

So, the inverse of every NBD of f($\overline{H}$) is a NBD of $\overline{H}$.

$$\therefore f\left(\overline{\omega}\right) = f\left(f^{-1}\left(\overline{\varsigma}\right)\right) \tag{1}$$

If $f^{-1}(\xi)$ is not empty,

$$\overline{\chi}_{f(f^{-1}(\overline{\varsigma}))}\left(\xi\right) = rsup_{z \in f^{-1}(\xi)} \overline{\chi}_{f^{-1}(\overline{\varsigma})}\left(\varpi\right)$$

$$= rsup_{z \in f^{-1}(\xi)} \left\{ \overline{\chi}_{\overline{\varsigma}}\left(f\left(\varpi\right)\right)\right\}$$

$$= \overline{\chi}_{\overline{\varsigma}}\left(\xi\right) \quad \forall \xi \in Y.$$

If $f^{-1}(\xi)$ is not empty,

$$\overline{\chi}_{f(f^{-1}(\overline{\varsigma}))}\left(\xi\right) = 0$$

$$\therefore \overline{\chi}_{f(f^{-1}(\overline{\varsigma}))}\left(\xi\right) \leq \overline{\chi}_{(\overline{\varsigma})}\left(\xi\right) \quad \forall \xi \in Y.$$

$$\therefore f\left(f^{-1}\left(\overline{\varsigma}\right)\right) \subset \overline{\varsigma} \tag{2}$$

From (1) and (2) $\Rightarrow$ $f\left(\overline{\omega}\right) \subset \overline{\varsigma}$.

Conversely, let V be a NBD of $f\left(\overline{H}\right)$.

Since there is a NBD $\overline{\omega}$ of $\overline{H}$ such that $f\left(\overline{\omega}\right) \subset \overline{\varsigma}$,

$$\text{Hence, } \left(f^{-1}\left(f\left(\overline{\omega}\right)\right) \subset \left(f^{-1}\left(\overline{\varsigma}\right)\right)\right. \tag{3}$$

$$\overline{\chi}_{(f^{-1}(f(\overline{\omega})))}\left(\mathfrak{x}\right) = \overline{\chi}_{f(\overline{\omega})}\left(\mathrm{f}\left(\mathfrak{L}\right)\right)$$

$$= rsup_{\ \varpi \in f^{-1}(f(\mathfrak{L}))} \left\{ \overline{\chi}_{\overline{\omega}}\left(\varpi\right)\right\}$$

$$\geq \overline{\chi}_{\overline{\omega}}\left(\mathfrak{L}\right) \quad \forall \mathfrak{L} \in \mathrm{X}.$$

$$\therefore \overline{\omega} \subset f^{-1}\left(f\left(\overline{\omega}\right)\right) \tag{4}$$

From (3) and (4) $\Rightarrow \overline{\omega} \subset f^{-1}\left(f\left(\overline{\omega}\right)\right) \subset \left(f^{-1}\left(\overline{\varsigma}\right)\right)$,
$\left(f^{-1}\left(\overline{\varsigma}\right)\right)$ is a NBD of $\overline{\omega}$.

Theorem 3.12: On the β_algebras, Ξ and Y, respectively, consider i_v_f β_subalgebraic TS (Ξ, $\overline{\Gamma}$) and (Y, $\overline{\chi}$). If f is $\mathfrak{T}$ continuous, then for each i_v F_S $\overline{H}$ in Ξ, the inverse of every NBD of $f\left(\overline{H}\right)$is a NBD of $\overline{H}$.

Proof

Assume that $\mathcal{A}$ be the i_v F_Sof Ξ.

Suppose that $\mathfrak{J}$ be the family of NBD of i_v F_S on $\mathcal{A}$.

Let $\overline{H} \in \mathcal{A}$ and $\overline{\varsigma} \in \mathfrak{J}$.

That is, $\overline{A}$ is an i_v F_S in Ξ and $\overline{\varsigma}$ NBD of $f\left(\overline{H}\right)$.

Hence, there is an open NBD $\overline{\varpi}$ of $f\left(\overline{H}\right)$ with, $f\left(\overline{H}\right) \subset \overline{\varpi} \subset \overline{\varsigma}$.

$$\Rightarrow f^{-1}\left(f\left(\overline{H}\right)\right) \subset \left(f^{-1}\left(\overline{\varsigma}\right)\right) \tag{5}$$

Because f is $\mathfrak{T}$continuous, $f^{-1}\left(\overline{\omega}\right)$ is open.

$$\overline{\chi}_{\left(f^{-1}\left(f\left(\overline{H}\right)\right)\right)}(\mathfrak{L}) = \overline{\chi}_{f\left(\overline{H}\right)}(\mathrm{f}(\mathfrak{L}))$$

$$= rsup_{\varpi \in f^{-1}(f(\mathfrak{L}))}\left\{\overline{\chi}_{\overline{H}}(\varpi)\right\}$$

$$\geq \overline{\chi}_{\overline{H}}(\mathfrak{L}) \quad \forall \mathfrak{L} \in \Xi.$$

$$\therefore \overline{A} \subset f^{-1}\left(f\left(\overline{H}\right)\right) \tag{6}$$

From (5) and (6) $\Rightarrow \overline{H} \subset f^{-1}\left(f\left(\overline{\omega}\right)\right) \subset \left(f^{-1}\left(\overline{\varsigma}\right)\right)$,
$\left(f^{-1}\left(\overline{\varsigma}\right)\right)$ is a NBD of $\overline{H}$.

4 Conclusion and Future Scope

This article is intended to exhibit the new approach on interval valued fuzzy β_subalgebraic topology in a different dimension. Some of the absorbing results of fuzzy β_subalgebraic topology incorporated with an interval valued fuzzy sets were explored. Further, this thought can be extended to various kinds of algebraic structures in future work.

References

1. Annalakshmi, M., Chandramouleeswaran, M.: Fuzzy topological subsystem on a TM- algebra. Int. J. Pure Appl. Math. **94**(3), 439–449 (2014)
2. Aub Ayub Anasri, M., Chandramouleeswaran, M.: Fuzzy β−subalgebras of β−algebras. Int. J. Math. Sci. Engg. Appls. **5**(7), 239–249 (2013)

3. Chandramouleeswaran, M., Muralikrishna, P., Sujatha, K.: Fuzzy $\beta-$subalgebraic topological spaces. Commun. Appl. Anal. **20**, 307–315 (2016)
4. Chanduraty, A.K., Das, P.: Some results on fuzzy Topology on fuzzy sets. Fuzzy Sets Syst. **58**, 331–336 (1993)
5. Chang, C.L.: Fuzzy Topological spaces. J. Math. Anal. Appl. **24**, 182–190 (1968)
6. Foster, D.H.: Fuzzy Topological groups. J. Math. Anal. Appl. **67**(2), 549–564 (1979)
7. Hemavathi, P., Muralikrishna, P., Palanivel, K.: A note on interval valued fuzzy β -subalgebras. Glob. J. Pure Appl. Math. **11**(4), 2553–2560 (2015)
8. Kandil, A., Tantawy, O., Yakout, M., Saleh, S.: Separation axioms in interval valued fuzzy topological spaces. Appl. Math. Inf. Sci. **5**(2), 1–19 (2011)
9. Neggers, J., Sik, K.H.: On β- algebras. Math. Solvaca. **52**(5), 517–530 (2002)
10. Sharma, P.K.: On fuzzy topological spaces involving Boolean algebraic structures. J. Math. Comput. Sci. **15**, 252–260 (2015)
11. Chandrasekar, V., Sobana, D., Vadivel, A.: Fuzzy e-open Sets, Fuzzy e-continuity and Fuzzy e-compactness in Intuitionistic Fuzzy Topological spaces. Sahand Commun. Math. Anal. **12**(1), 131–153 (2018)
12. Young Bae Jun, K.: β- subalgebras and related topics. Commun. Korean Math. Soc. **27**(2), 243–255 (2012)
13. Zadeh, L.A.: fuzzy sets. Inf. Control. **8**(3), 338–353 (1965)
14. Zadeh, L.A.: The concept of a linguistic variable and its application to approximate reasoning. I. Inf. Sci. **8**, 199–249 (1975)

A Specific Key Sharing Protocol Among Multiuser Using Noncommutative Group for Telecare Medicine Information System

Girija Murugan and Uma Kaliyappan

Keywords Telecare medicine information system · Authentication · Braid decomposition problem (BDP) · Twin conjugacy search problem (TCSP)

1 Introduction

Communication algorithms such as public key cryptosystems are based on hard problems. The cryptographers are always hunting for hard problems to increase the diversity of public key cryptography. The braid group is used as the implementation group of twin conjugacy theory, but all of Chen et al. [1] results can be applied to any non-commutative group, as long as two exchangeable subgroups are contained in the group. Also, for the first time, Chen et al. define several security assumptions related to the conjugacy search problem and analyze the security of the CSP-hEIG scheme under the security assumptions [1]. As a cryptographic stage, braid organizations are a characteristic longing among noncommutative organizations. Despite the fact that the conjugacy seeks for issue is thought to be troublesome, the word issue is known to make some polynomial memories algorithmic arrangement. Since they are individuals from a braid bunch, finding subgroups that excursion with individuals from different subgroups is straightforward. Artin [2] started the interlace organizations, which are noncommutative twist-free partnerships. They're of cryptographic interest since calculations and statis-spasms capacity can be performed decently accurately; however, they're sufficiently mind-boggling that it seems improbable that they have any astounding hidden structure from the outset. The name "mesh bunch" is completely proper, as the nth plait association Bn is depicted as a fixed of "n-twists."

The conjugacy issue in twist organizations' administration is the establishment for some advanced cryptosystems, and late measurements uncovered that the issue

G. Murugan · U. Kaliyappan (✉)
Department of Mathematics, School of Advanced Sciences, VIT, Vellore, Tamil Nadu, India
e-mail: kuma@vit.a.in

H.-H. Leung et al. (eds.), *Recent Developments in Algebra and Analysis*, Trends in Mathematics, https://doi.org/10.1007/978-3-031-37538-5_20

is more conceivable than many individuals anticipated [3–5]. The mission of public-key cryptography is to communicate straight over open channels with the goal that a vindictive intruder can't procure any secret records regardless of whether he has direct associations with the sent messages [6]. Numerous residents have researched noncommutative logarithmic frameworks in the desire for finding another option, and twist organizations seemed to hold a ton of promise [7–9]. Following a basic examination, it was found that twist organizations might have a lot of design and that a portion of similar systems for proficient calculations could be utilized to go after interlace based total conventions [6, 10–13].

Current encryption ordinarily utilizes complex limits dependent absolutely upon mistaken and much of the time testing to check probabilistic contentions as insurance verifications. Compositional plan and investigation are one of a handful of procedures that might in any way whatsoever license the test of complicated frameworks. Here, one determines, by means of notable structure ideas, the well-being of a bigger contraption from the security of individual parts. This exposition adds to the assortment of ability in this subject. Our canvases spend significant time in the composability of secure key substitute, quite possibly of the most fundamental cryptographic work. To report convention well-being, there are much of the time two procedures. The reproduction worldview is the dream for models very much like the overall organization design and others [14]. The elective methodology reproduces security through games. Test system essentially based security presents coordinated, stylishly interesting ways of characterizing insurance and habitually permits formed convention security to be resolved precisely.

Further, the resulting frameworks can include complexities which might be difficult to understand furthermore; because of the simulation's strict protection requirements, many critical jobs can't be effectively and securely found out. Furthermore, simulation systems [15] are generally surely insufficient for the exam of modern protocols of realistic relevance, in part due to the fact such techniques do not adhere to the very rigorous constraints demanded through simulation-primarily based safety. This is probably considered the number one driving pressure for the ongoing development. The closing ultimate preference is to apply formalists stimulated by using video games [16]. Even though they have much less strict regulations, the derived keys often have a terrific degree of safety and cannot be outstanding from random keys. Even though we're aware of the extent of security required for key-alternate methods when employed independently by way of commonplace recreation-based models, there may be sadly no solidly set up warranty nation of their blending with different sports. This gap is filled by our work.

That is (in all likelihood) now not the case, despite the fact. Inside the prolonged model, we exhibit that a session matching method exists if a primary swap etiquette is preparable with any harmonic primary rules and safety is mounted using a particular form of black-area discount. We accentuate that the outer meeting matching does not affect the imperative thing substitute conventions as it best impacts the convention for key trade and no extraordinary key makes use of. Ultimately, we draw attention to the truth that combining relaxed key change strategies with any symmetrical key techniques could probable seem tough.

That's what the plain contention is if the key innovation convention "messes up" by utilizing approach to wrongly rehashing a couple of levels of the key settlement convention, and the synthesis could quickly wind up unreliable.

Expect, for instance, that a nonce is scrambled utilizing the pristine meeting key as a feature of the critical substitute strategy and that the essential convention step is for a festival to legitimate away screen the meeting key after getting an encoded message in an uncommon condition. Replaying the past message from the significant thing trade fragment needs to then think twice about convention's typical security. In any case, this line of pondering is defective. A key trade convention's ability to refresh the genuine key with a chosen irregular key accurately decouples the two territories. This is implied through utilizing key lack of definition.

1.1 Literature Review

Girl et al. [17] proposed that the evolved scheme may be utilized to at ease special safety threats like replay and password cracking. In early 2015, Amin and Biswas [18] counseled that the lady scheme is risky to offline guessing passwords and assaults. Furthermore, they deployed a more advantageous authentication-primarily based scheme. Zhang and Zhu [19] advanced a scheme primarily based on a key agreement authentication protocol. Ostad-Sharif et al. [20] also elaborated on the telecare remedy-based records community for showcasing the efficiency of the authentication scheme. The usage of a key exchange protocol to create symmetric keys on the way to finally be used in a comfortable channel protocol is an instance of a normal application. We cope with protection definitions for each stand-alone and composite protocols inside the context of conventional sport-primarily based environments.

2 Protocol Key Exchange

A key trade procedure empowers two neighborhood meetings utilizing long haul character keys to agree on a fast-meeting key. Asymmetric long haul key personalities are thought about (symmetric extended keys can be treated likewise.) We "accomplice" two meetings utilizing a meeting recognizable proof worth. This not entirely settled by the key trade instrument.

The key trade convention ascertains this worth. One can in any case utilize a meeting distinguishing proof to put together cooperating with respect to thoughts like matching exchange, as shown by Bellare et al. [21]. One can likewise accomplish a comparable, though not indistinguishable, thought by utilizing the message record. The meeting key for collaborated meetings should be indistinguishable from irregular and should be processed by the two meetings. Also, just two meetings ought to at any point have a similar meeting ID while utilizing two-party

Step 2

U_2 gains the key k_1 and determine the elements $x_1^*, x_2^* \in LB_l$

Compute $X_1^* = x_1^* g\left(x_1^*\right)^{-1}$ *and* $X_2^* = x_2^* g\left(x_2^*\right)^{-1}$

Mutual key is $\left(X_1^*, X_2^*, g\right)$ and special key is $\left(x_1^*, x_2^*\right)$

Elect an aimless element $y*$ in RB_r

Figure out $Y* = y* g(y*)^{-1}$, $Z_1^* = y^* X_1^* (y^*)^{-1}$ *and* $Z_2^* = y^* X_2^* (y^*)^{-1}$

We get a key $k_2 = H\ (Y^*, Z^*) = y^* X_1^* g X_2^* (y^*)^{-1}$

Now concatenate the keys k_1, k_2 *compile* k_{12}

We get $k_{12} = y^* X_1^* k_1 X_2^* (y^*)^{-1}$

Exactly, customer deliver the keys k_1, k_2, k_{12} to the terminal user.

Step 3

Now the ultimate user hit the crucial role in our protocol that it can't be ended without the user U_3 which is clients. Scholarly we rerun the stride1.

That randomly selects elements $x_1^{**}, x_2^{**} \in LB_l$

Compute $X_1^{**} = x_1^{**} g\left(x_1^{**}\right)^{-1}$ *and* $X_2^{**} = x_2^{**} g\left(x_2^{**}\right)^{-1}$

Social key is $\left(X_1^{**}, X_2^{**}, g\right)$ and exclusive key is $\left(x_1^{**}, x_2^{**}\right)$

Accept an irregular element $y**$ in RB_r

Figure out $Y** = y** g(y**)^{-1}$, $Z_1^{**} = y^{**} X_1^{**} (y^{**})^{-1}$ *and* $Z_2^{**} = y^{**} X_2^{**} (y^{**})^{-1}$

We get a key $k_3 = H\ (Y^{**}, Z^{**}) = y^{**} X_1^{**} g X_2^{**} (y^{**})^{-1}$

Now compute the keys k_{13}*and* k_{23} and dispatch to the users U_1, U_2.

Step 4

The three users U_1, U_2, U_3 compute their mysterious keys $S(U_1), S(U_2), S(U_3)$.

That is, $k_{123} = S(U_1) = S(U_2) = S(U_3)$.

4.2 *Correctness*

$$S\left(U_1\right) = \left(yX_1gX_2y^{-1}\right)\left(y^* X_1^* g (y^*)^{-1} X_2^*\right)\left(y^{**} X_1^{**} g \left(y^{**}\right)^{-1} X_2^{**}\right)$$
$$= \left(X_1 X_1^* X_1^{**}\right) Y \left(X_2 X_2^* X_2^{**}\right).$$

Similarly, we get (U_2) *and* $S(U_3)$.

5 Conclusion

Cryptosystems have woven a great deal of exploration and fervor into the subject, and it immediately became evident that they were unreliable. At first, the issue of taking advantage of interlace bunches was believed to be sufficiently troublesome to develop a solid cryptosystem. It rolls out minor improvements to the encryption cycle to the first CSP conspire. As a matter of fact, the primary consequence of this article is the expansion of the finish of the dual Diffie-Hellman issue for universal periodic gatherings proposed by David Money to the twin conjugacy search issue for general noncommutative gatherings. And furthermore utilizing this twin conjugacy technique, we tackle an extraordinary key dividing convention between multiclients.

References

1. Chen, X., You, W., & Li, W.: The twin conjugacy search problem and applications. ArXiv, abs/1806.03078, (2018)
2. Artin, E.: Theory of braids. Ann. Math. **48**, 101–126 (1947). https://doi.org/10.2307/1969218
3. Birman, J.S., Ko, K.H., Lee, J.S.: A new approach to the word and conjugacy problems in the braid groups. Adv. Math. **139**, 322–353 (1997)
4. Birman, J.S., Ko, K.H., Lee, S.J.: The infimum, supremum, and geodesic length of a braid conjugacy class. Adv. Math. **164**, 41–56 (2001)
5. Gebhardt, V.: A new approach to the conjugacy problem in Garside groups. Math. eJ. (2003)
6. Koblitz, N., Menezes, A.J.: A survey of public-key cryptosystems. SIAM Rev. **46**(4), 599–634 (2004)
7. Anshel, I., Anshel, M., Goldfeld, D.: An algebraic method for public-key cryptography. Math. Res. Lett. **6**, 287–291 (1999)
8. Ko, K.H., Lee, S.J., Cheon, J.H., Han, J.W., Kang, J., Park, C.: New public-key cryptosystem using braid groups. In: Bellare, M. (ed.) Advances in Cryptology ? CRYPTO 2000. CRYPTO 2000. Lec- Ture Notes in Computer Science, vol. 1880. Springer, Berlin, Heidelberg (2000). https://doi.org/10.1007/3-540-44598-610
9. Sibert, H., Dehornoy, P., Girault, M.: Entity authentication schemes using braid word reduction. IACR Cryptol. ePrint Arch. **2002**, 187 (2006)
10. Franco, N., Gonzâlez-Meneses, J.: Conjugacy problem for braid groups and Garside groups. J. Algebra. **266**, 112–132 (2003). https://doi.org/10.1016/S0021-8693(03)00292-8
11. Hughes, J.: A linear algebraic attack on the AAFG1 braid group cryptosystem. In: Batten, L., Seberry, J. (eds.) Information Security and Privacy. ACISP 2002 Lecture Notes in Computer Science, vol. 2384. Springer, Berlin, Heidelberg (2002). https://doi.org/10.1007/3-540-45450-015
12. Hughes, J.P., & Tannenbaum, A.R.: Length-Based Attacks for Certain Group Based Encryption Rewriting Systems. ArXiv, cs.CR/0306032, (2003)
13. Lee, S.J., Lee, E.: Potential weaknesses of the commutator key agreement protocol based on braid groups. In: Knudsen, L.R. (ed.) Advances in Cryptology ? EUROCRYPT 2002. EUROCRYPT 2002. Lecture Notes in Computer Science, vol. 2332. Springer, Berlin, Heidelberg (2002). https://doi.org/10.1007/3-540-46035-72
14. Dagdelen, Ö., Fischlin, M.: Security analysis of the extended access control protocol for machine readable travel documents. In: International Conference on Information Security, pp. 54–68. Springer, Berlin, Heidelberg (2010)
15. Hanzlik, L., Loss, J., Wagner, B.: Token meets wallet: formalizing privacy and revocation for FIDO2. Crypto. ePrint Arch. (2022)
16. Rausch, D., Küsters, R., Chevalier, C.: Embedding the UC model into the IITM model. In: Annual International Conference on the Theory and Applications of Cryptographic Techniques, pp. 242–272. Springer, Cham (2022)
17. Giri, D., Maitra, T., Amin, R., Srivastava, P.D.: An efficient and robust RSA-based remote user authentication for telecare medical information systems. J. Med. Syst. **39**(1), 1–9 (2015)

18. Amin, R., Biswas, G.P.: An improved RSA based user authentication and session key agreement protocol usable in TMIS. J. Med. Syst. **39**(8), 1–14 (2015)
19. Zhang, L., Zhu, S.: Robust ECC-based authenticated key agreement scheme with privacy protection for telecare medicine information systems. J. Med. Syst. **39**(5), 1–11 (2015)
20. Ostad-Sharif, A., Abbasinezhad-Mood, D., Nikooghadam, M.: A robust and efficient ECC-based mutual authentication and session key generation scheme for healthcare applications. J. Med. Syst. **43**(1), 1–22 (2019)
21. Bellare, M., Rogaway, P.: Entity authentication and key distribution. In: Annual International Cryptology Conference, pp. 232–249. Springer, Berlin, Heidelberg (1993)

Some Analytic and Arithmetic Properties of Integral Models of Algebraic Tori

Armen G. Bagdasaryan

Keywords Linear algebraic group · Algebraic torus · Integral models · Arithmetic properties · Integer-valued polynomials · Factorization · Generators

1 Introduction

The last decades have witnessed the growing interest in the analysis of various properties of linear algebraic groups over local and global fields, the origins of which can be traced back to the works of Lagrange and Gauss. While this subject remains an area of active research, there is a particular interest in the arithmetic properties of linear algebraic groups over fields of an arithmetic nature that are not global, such as function fields of curves over various classes of fields, including p-adic fields and number fields. These recent developments rely on a combination of methods from the theory of algebraic groups and arithmetic geometry [17].

Let $\Bbbk$ be a field of characteristic 0, and suppose $\Bbbk$ is the field of fractions of an integral domain $\mathbb{O}$, and let G be a linear algebraic group over $\Bbbk$. An *integral model* of G is a group scheme $\mathcal{X}$ over $\mathbb{O}$ such that $\mathcal{X} \otimes_{\mathbb{O}} \Bbbk \cong G$. The integral forms of G always exist and can be easily constructed. An *algebraic torus* $\mathbb{T}$ is an algebraic $\Bbbk$-group that $\mathbb{T} \otimes_{\Bbbk} L \cong \mathbb{G}^d_{m,L}$ for a finite Galois extension $L/\Bbbk$, where $\mathbb{G}_m$ is the multiplicative group and $d = \dim \mathbb{T}$ is the dimension of $\mathbb{T}$ [3]. The smallest among such extensions L is called the *minimal splitting field* of $\mathbb{T}$, $\Bbbk \subseteq L \subseteq \overline{\Bbbk}$, $\overline{\Bbbk}$ is the algebraic closure of $\Bbbk$. The projections $\chi_i : \mathbb{T} \otimes_{\Bbbk} L \to \mathbb{G}_{m,L}$, $i = 1, 2, \ldots, d$ generate a free abelian group $\hat{T}$, the group of rational characters of $\mathbb{T}$, that may be regarded as a Γ-module of rank d over $\mathbb{Z}$, where $\Gamma = \mathrm{Gal}(L/\Bbbk)$.

The necessity of a study of integral models in $\Bbbk$-group G (specifically, $\Bbbk$-torus) is due to some principal problems in arithmetic and analysis on a group G (torus $\mathbb{T}$), such as reduction of algebraic groups modulo of a prime, computation of class

A. G. Bagdasaryan (✉)
American University of the Middle East, Egaila, Kuwait
e-mail: armen.bagdasaryan@aum.edu.kw

H.-H. Leung et al. (eds.), *Recent Developments in Algebra and Analysis*,
Trends in Mathematics, https://doi.org/10.1007/978-3-031-37538-5_21

numbers and Tamagawa measure, and application of Siegel-Tamagawa formulas [14, 17]. The algebras of integer-valued functions defined on a torus naturally appear when studying integral models of the algebraic torus. The study of properties of these algebras is important for classification of various integral forms of tori. Another problem related to analysis of integral models is to determine the generators of algebras defined on the torus, which is the main object of this work.

Notations As is customary, $\mathbb{Q}$, $\mathbb{Q}_p$, $\mathbb{Z}$, and $\mathbb{Z}_p$ denote the field of rational numbers, the field of p-adic numbers, the ring of rational integers, and the ring of p-adic integers. We let $\mathbb{N}$ and $\mathbb{N}_0$ denote the set of positive and nonnegative integers. We denote by R^* the group of invertible elements in a commutative ring R.

2 Main Results

2.1 Algebras as Integral Models of Algebraic Tori

Consider the algebra

$$\mathcal{A}_1 = \{f \in \mathbb{Q}_p[t, t^{-1}] \mid f\left(\mathbb{Z}_p^*\right) \subseteq \mathbb{Z}_p\}, \tag{1}$$

where $\mathcal{X}_1 = \mathrm{Spec}\mathcal{A}_1$ is an affine scheme [12, 13]. The algebra $\mathcal{A}_1$ is an integral model for the algebraic torus $\mathbb{T} = \mathbb{G}_m$, that is, $\mathbb{T} \cong \mathcal{X}_1 \otimes_{\mathbb{Z}_p} \mathbb{Q}_p$, and $\mathbb{G}_m$ is the multiplicative group over the ring $\mathbb{Z}_p$. If considered as a ring, $\mathcal{A}_1$ is an example of integer-valued ring, being extensively studied [7, 15, 16, 19]. We determine the generators of such rings and study their factorization properties [2, 9].

To find the generators of $\mathcal{A}_1$, we search for polynomials $Q_n(t)$ and numbers $t_n \in \mathbb{Z}_p^*$ with the following properties:

1. $\deg(Q_n(t)) = n$
2. $Q_n(t) \in \mathcal{A}_1$
3. $Q_n(t_n) \in \mathbb{Z}_p^*$ or $\frac{1}{p} Q_n(t) \notin \mathcal{A}_1$.

We first prove the existence of the polynomials $Q_n(t)$ by constructing $Q_n(t)$ when $n = \varphi\left(p^k\right)$, where φ is the Euler function.

Theorem 1 *The polynomials*

$$Q_{\varphi(p^k)}(t) = \frac{1}{p^{\alpha_k}} \prod_{\substack{i=1 \\ (i,p)=1}}^{p^{k-1}-1} (t - i),$$

where

$$\alpha_k = 1 + p + p^2 + \cdots + p^{k-1} = \frac{p^k - 1}{p - 1},$$

belong to the algebra $\mathcal{A}_1$ *and the polynomials* $\frac{1}{p} Q_{\varphi(p^k)}(t) \notin \mathcal{A}_1$.

Proof Since $t \in \mathbb{Z}_p^*$ and $i \in \mathbb{Z}_p^*$, we have

$$t = r_0 + r_1 p + \cdots + r_{k-1} p^{k-1} + r_k p^k + \cdots \quad (0 < r_0 \le p-1,\ 0 \le r_i \le p-1),$$

$$i = s_0 + s_1 p + \cdots + s_{k-1} p^{k-1} \quad (0 < s_0 \le p-1,\ 0 \le s_i \le p-1).$$

The number p^ℓ divides $(t-i)$ when $r_i = s_i$ for $i = 0, \ldots, \ell-1$, $r_\ell \neq s_\ell$. Fix t and choose i to run over reduced residue class modulo p^k; write $i \in \mathcal{R}$. Then there are in total $(p-1)p^{k-(\ell+1)}$ expressions $(t-i)$ that are divisible by p^ℓ for $\ell < k$, since for all $i \in \mathcal{R}$

$$r_0 = s_0,\ r_1 = s_1, \ldots, r_{\ell-1} = s_{\ell-1},\ r_\ell = s_\ell$$

and for $j \ge \ell + 1$, the numbers s_j are arbitrarily chosen. Note that when $r_j = s_j$ for $j = 0, 1, \ldots, k-1$, there exists unique $i \in \mathcal{R}$ such that $p^k \mid (t-i)$, and when $r_k = r_{k+1} = \cdots = r_{k+m} = 0$, $p^{k'} \mid (t-i)$, $k' > k$. Therefore, p^{α_k} divides $\prod_{i \in \mathcal{R}} (t-i)$, where

$$\begin{aligned}\alpha_k =& k + (p-1)(k-1) + p(p-1)(k-2) + \cdots + (p-1)p^{k-(\ell+1)}\ell + \\ &+ \cdots + (p-1)p^{k-2} \\ =& 1 + p + p^2 + \cdots + p^{k-1} = \frac{p^k - 1}{p - 1}.\end{aligned}$$

However, for $t' = p^k + 1$, $p^{\alpha_k + 1} \nmid \prod_{i \in \mathcal{R}} (t' - i)$, which completes the proof. □

To construct the numbers t_n, we choose them in ascending order from the reduced residue class modulo p^{k+1}. We set $n > \varphi(p^k)$; for instance, $t_n = p^k + 1$ when $n = \varphi(p^k) + 1$. Similar to the case of polynomials, for $n = \varphi(p^k)$, we first consider for arbitrary n the products

$$\mathcal{P}(t) = \prod_{\substack{i=1 \\ (i,p)=1}}^{p^k - 1} (t - i) \prod_{j=\varphi(p^k)+1}^{n} (t - t_j). \tag{2}$$

We determine a number ν such that $p^\nu \mid \mathcal{P}$. We have

$$n = n_k \varphi\left(p^k\right) + n_{k-1} \varphi\left(p^{k-1}\right) + \cdots + n_1 \varphi(p) + n_0,$$

where $n_k > 0$ and $0 \leq n_i < p$, $n_0 < p-1, i = 1, 2, \ldots, k-1$. From construction of n, we note the following: among the numbers i and t_j in (2), there are n_k reduced residue classes modulo p^k having the digits $0, 1, \ldots, n_k - 1$ at p^k. Among numbers with the digit n_k at p^k, there are n_{k-1} reduced residue classes modulo p^{k-1} having, respectively, the digits $0, 1, \ldots, n_{k-1} - 1$ at p^{k-1}, and so on. Hence, the numbers i run over n_k reduced residue classes modulo p^k, then over remaining n_{k-1} reduced residue classes modulo p^{k-1}, etc. Further, from the proof of Theorem 1, it follows that $p^{\nu_n} \mid \mathcal{P}$, where $\nu_n = n_k\alpha_k + n_{k-1}\alpha_{k-1} + \cdots + n_2\alpha_2 + n_1\alpha_1$. By construction of t_n, we have $t_{n+1} = \left(n_k p^k + n_{k-1} p^{k-1} + \cdots + n_1 p + n_0\right) + 1$. Then upon substitution $t = t_{n+1}$ into (2), using the proof scheme of Theorem 1, one finds that since $n_k \neq 0$, $p^{\alpha_k} \mid \prod(t_{n+1} - j)$ when j runs over any reduced residue class modulo p^k in (2). Similarly, if there is $n_u \neq 0$, then $p^{\alpha_u} \mid \prod(t_{n+1} - j)$ when j runs over any reduced residue class modulo p^u in (2). This shows that for $t = t_{n+1}$, it holds that $p^{\nu_n+1} \nmid \mathcal{P}$.

Thus, we have constructed the numbers t_n and polynomials

$$Q_n(t) = \frac{1}{p^{\nu_n}} \prod_{\substack{i=1 \\ (i,p)=1}}^{p^k-1} (t-i) \prod_{\varphi(p^k)<j\leq n} (t-t_j)$$

that possess the abovementioned properties, and moreover $Q_n(t_k) = 0$ for $k \leq n$.

Theorem 2 *Let* $\varphi\left(p^k\right) < n \leq \varphi\left(p^{k+1}\right)$ *and*

$$Q_n(t) = \frac{1}{p^{\nu_n}} \prod_{\substack{i=1 \\ (i,p)=1}}^{p^k-1} (t-i) \prod_{\varphi(p^k)<j\leq n} (t-t_j).$$

Moreover, if $n = n_k\varphi\left(p^k\right) + n_{k-1}\varphi\left(p^{k-1}\right) + \cdots + n_1\varphi(p) + n_0$, *then* $\nu_n = n_k\alpha_k + n_{k-1}\alpha_{k-1} + \cdots + n_2\alpha_2 + n_1\alpha_1$, *where* $0 \leq n_i < p$, $n_0 < p-1$, $n_k \neq 0$, *and* $\alpha_i = \dfrac{p^i - 1}{p-1}$ *for* $0 < i \leq k$. *Then*

$$\mathcal{A}_1 = \mathbb{Z}_p[t, t^{-1}, Q_1(t), Q_2(t), \ldots, Q_n(t), \ldots].$$

Proof Suppose $f(t) \in \mathbb{Q}[t, t^{-1}]$. Then it can be written $f(t) = \dfrac{1}{t^n} g(t)$. We have that $f(t) \in \mathcal{A}_1 \Leftrightarrow g(t) \in \mathcal{A}_1$. Let $\deg(g) = d$. The polynomials $Q_\ell(t)$ are defined for any degree ℓ, so we can write

$$g(t) = s_d Q_d(t) + s_{d-1} Q_{d-1}(t) + \cdots + s_1 Q_1(t) + s_0.$$

Since $g(t) \in \mathcal{A}_1$, we have that $g(t_n) \in \mathbb{Z}_p$ for $t_n \in \mathbb{Z}_p^*$. Then the proof is continued by induction over t. □

Theorem 3 *Let $Q'_k(t) = Q_{\varphi(p^k)}(t)$. Then it holds that*

$$\mathcal{A}_1 = \mathbb{Z}_p\left[t, t^{-1}, Q'_1(t), Q'_2(t), \ldots, Q'_k(t), \ldots\right].$$

Moreover, if

$$\mathcal{A}_1^{(k)} = \mathbb{Z}_p\left[t, t^{-1}, Q'_1(t), Q'_2(t), \ldots, Q'_k(t)\right]$$

then $pQ'_{k+1}(t) \in \mathcal{A}_1^{(k)}$, but $Q'_{k+1}(t) \notin \mathcal{A}_1^{(k)}$.

Proof The proof is by induction over degree d of polynomial $Q'_k(t)$. □

Consider an algebraic torus $\mathbb{T} = \mathbb{G}_m^n$ in $\mathbb{Q}_p$.

Theorem 4 *Suppose the algebra*

$$\mathcal{A}_1 = \{f(t) \in \mathbb{Q}_p[t_1, t_2, \ldots, t_n, t_1^{-1}, t_2^{-1}, \ldots, t_n^{-1}] \mid f\left(\mathbb{Z}_p^{*n}\right) \subseteq \mathbb{Z}_p\}$$

and let $Q_{\alpha_1,\alpha_2,\ldots,\alpha_n}(t_1, t_2, \ldots, t_n)$ be the polynomials such that

$$Q_{\alpha_1,\alpha_2,\ldots,\alpha_n}(t_1, t_2, \ldots, t_n) = Q_{\alpha_1}(t_1)Q_{\alpha_2}(t_2)\cdots Q_{\alpha_n}(t_n),$$

where $(\alpha_1, \alpha_2, \ldots, \alpha_n) \in \mathbb{N}_0^n$. Then it holds that

$$\mathcal{A}_1 = \mathbb{Z}_p[t_1, t_2, \ldots, t_n, t_1^{-1}, t_2^{-1}, \ldots, t_n^{-1}, \ldots, Q_{\alpha_1,\alpha_2,\ldots,\alpha_n}(t_1, t_2, \ldots, t_n), \ldots].$$

Proof The degrees of polynomials $Q(\cdot)$ run over all possible values α_i in the set of vectors $(\alpha_1, \alpha_2, \ldots, \alpha_n) \in \mathbb{N}_0^n$. Then any polynomial $f(t) \in \mathbb{Q}_p[t_1, t_2, \ldots, t_n]$ has the expansion

$$f(t) = \sum a_{\alpha_1,\alpha_2,\ldots,\alpha_n} Q_{\alpha_1,\alpha_2,\ldots,\alpha_n(t)},$$

where $a_{\alpha_1,\alpha_2,\ldots,\alpha_n} \in \mathbb{Q}_p$. The theorem is then proved, like in proof of Theorem 2, by showing that $f \in \mathcal{A}_1$ if and only if $a_i \in \mathbb{Z}_p$, $i = 1, 2, \ldots, n$. □

2.2 *The Ring of Integer-Valued Polynomials and Factorization Properties*

The problems of factorization and non-unique factorization in polynomial rings, as well as elasticity of factorization, have recently been studied by many authors [1, 4, 5, 9, 10, 18]. For the ring Int($\mathbb{Z}$) it was shown [11] that $\rho(\text{Int}(\mathbb{Z})) = \infty$ and proved [5] that the ring Int($\mathbb{Z}$) is fully elastic. We consider the ring of integer-valued p-adic polynomials $\mathrm{Int}(\mathbb{Z}_p) = \{f \in \mathbb{Q}_p[t] \mid f(\mathbb{Z}_p) \subseteq \mathbb{Z}_p\}$.

Theorem 5 *The following statements are true for the ring* Int($\mathbb{Z}_p$)*:*

1. *it is atomic;*
2. *it is bounded factorization domain;*
3. *it is non-factorial.*

Theorem 6 *The elasticity of* Int($\mathbb{Z}_p$) *is infinite, that is,* $\rho(\mathrm{Int}(\mathbb{Z}_p) = \infty$.

Proof It is known [8] that $\rho(\mathrm{Int}(\mathbb{Z}_{p^n}[t]) = \infty$. To show this, an irreducible polynomial $f(t)$ was constructed, and then irreducibility of the polynomial $f^q(t) + p^{n-1}$, where q is a large prime number, was proved. For the polynomial $g(t)$

$$g(t) = \left(f^q(t) + p^{n-1}\right)^2 \tag{3}$$

$$= f^q(t)\left(f^q(t) + 2p^{n-1}\right) \tag{4}$$

it follows that factorization (3) has two irreducibles, whereas factorization (4) has $q + 1$ irreducibles. In view of the Hensel's lemma and its modification [11], two factorizations of $g(t)$ can be lifted to a factorization in the ring Int($\mathbb{Z}_p$). The existence of the two factorizations of $g(t)$ in Int($\mathbb{Z}_p$) proves the theorem. □

2.3 Algebras of Integer-Valued Functions

Let $\mathbb{Q}_p$ be the field of p-adic numbers, $\mathcal{O} = \mathcal{O}_{\mathbb{Q}_p}$ be the ring of integers of the field $\mathbb{Q}_p$, and $\mathcal{O}^*$ the group of invertible elements in $\mathbb{Q}_p$, and $f = \mathcal{O}/\mathcal{O}^*$ is the norm field. Suppose $\mathbb{T}$ is an algebraic torus over $\mathbb{Q}_p$, and $\mathcal{X}$ is a group scheme over $\mathcal{O}$ such that $\mathbb{Q}_p$-groups $\mathcal{X} \otimes_{\mathcal{O}} \mathbb{Q}_p$ and $\mathbb{T}$ are isomorphic, $\mathcal{X} = \mathrm{Spec}A$, $A \subset \mathbb{Q}_p[\mathbb{T}]$.

Low-Dimensional Tori We consider the algebra $\mathcal{A} = \{f \in \mathbb{Q}_p[\mathbb{T}] \mid f(\mathcal{U}) \subset \mathcal{O}\}$, where $\mathcal{U} = \mathcal{U}_{\mathbb{Q}_p}$ is the maximal compact subgroup of the group $\mathbb{T}(\mathbb{Q}_p)$, and find its generators for one- and two-dimensional tori (as some of them are direct products of split and 1-dim tori). The 1-dim torus $\mathbb{T} = \mathbb{G}_m$, represented by $\mathbb{Z}[x, x^{-1}]$, a special case of split tori $\mathbb{T} = \mathbb{G}_m^d$, was studied in previous sections. There exist two 1-dimensional tori. So consider the 1-dim torus $\mathbb{T} = R^{(1)}_{L/\mathbb{Q}_p}(\mathbb{G}_m)$, the kernel of the norm mapping $N : R_{L/\mathbb{Q}_p}(\mathbb{G}_m) \to \mathbb{G}_{m,\mathbb{Q}_p}$, where $\mathbb{G}_{m,\mathbb{Q}_p} = \mathbb{G}_m \otimes_{\mathbb{Z}} \mathbb{Q}_p$, $L = \mathbb{Q}_p(\sqrt{d})$, R stands for the Weil functor of restriction of scalars, and $(d, p) = 1$. In this case, the generators of $\mathcal{A}$ are reduced to the generators of the algebra $\mathcal{A}_1 = \{f \in \mathbb{Q}_p[x, y] \mid f(U) \subseteq \mathcal{O}\}$, where $U = \{(x, y) \in \mathcal{O}^2 \mid x^2 - dy^2 = 1\}$. Let $U^{(*)} = \{(\pm x_1, \pm y_1), (\pm x_2, \pm y_2), \ldots, (\pm x_\nu, \pm y_\nu)\}$ be the set of solutions of the congruence $x^2 - dy^2 \equiv 1 \pmod{p}$. Suppose $(x_i, y_i) \in U^{(*)}$ is a solution of the given congruence. Then, for $y_i' = y_i + y_i^* p$, where y_i^* is in the complete residue system modulo p, there exists a unique solution x_i' of $(x_i')^2 - d(y_i')^2 \equiv 1 \pmod{p^2}$

that satisfies $(x_i') \equiv x_i \pmod{p}$, which follows from the Hensel lemma. Applying the lemma further, one constructs solutions modulo p^{k+1} based on the solutions modulo p^k, thus obtaining an element of U.

Let $U_k^{(*)}$ be the set of solutions of $x^2 - dy^2 \equiv 1 \pmod{p^k}$; then one has $|(U_{k+1}^{(*)})| = p\,|(U_k^{(*)})|$. Denote $\theta_1 = U_1^{(*)}$, $\theta_k = U_k^{(*)}/U_{k-1}^{(*)}$ for $k \leq 2$, $|\theta_1| = \nu$, $\theta_k = |U_k^{(*)}| - |U_{k-1}^{(*)}| = \tilde{\varphi}(k) = \varphi(p^{k-1})\nu$, φ is the Euler function.

Theorem 7 *Let* $|U_k^{(*)}| < \ell \leq |U_{k+1}^{(*)}|$ *and*

$$P_\ell(x) = \frac{1}{p^{\nu_\ell}} \prod_{(x_i, y_i) \in U_k^{(*)}} (x - x_i) \prod_{\substack{(x_j, y_j) \in \theta_{n+1} \\ |(U_k^{(*)})| < j \leq \ell}} (x - x_j),$$

$$Q_\ell(y) = \frac{1}{p^{\nu_\ell}} \prod_{(x_i, y_i) \in U_k^{(*)}} (y - y_i) \prod_{\substack{(x_j, y_j) \in \theta_{n+1} \\ |(U_k^{(*)})| < j \leq \ell}} (y - y_j).$$

Moreover, if $\ell = \ell_k \tilde{\varphi}(k) + \cdots + \ell_1 \tilde{\varphi}(1) + \ell_0$, *then* $\nu_\ell = \ell_k \alpha_k + \cdots + \ell_1 \alpha_1$, *where* $0 \leq \ell_i < p$, $\alpha_i = \frac{p^i - 1}{p-1}$, *for* $0 < i \leq \omega$; *also,* $\ell_\omega \neq 0$ *and* $0 \leq \ell_0 < \nu$. *Then*

$$\mathcal{A}_1 = \mathbb{O}[x, y, P_1(x), Q_1(x), \ldots, P_n(x), Q_n(x), \ldots].$$

Proof We prove the theorem for polynomials $Q_k(y)$. We verify that $Q_k(y)$ possess the properties: (1) $\deg(Q_\ell) = \ell$, (2) $Q_\ell \in \mathcal{A}_1$, (3) $\frac{1}{p} Q_\ell(y_{\ell+1}) \notin \mathcal{A}_1$. We first consider the polynomials when $\ell = |(U_k^{(*)})|$. For arbitrary $Q_k(y)$, the properties are verified in a similar way by ordering the elements of θ_k. Analogously, these properties are proved for $P_k(x)$. □

Quasi-Split Tori We consider the algebra $\mathcal{A}_0 = \{f \in \mathbb{Q}_p[\mathbb{T}] \mid f(u) \in \mathbb{O},\ \forall u \in \mathcal{X}(\mathbb{O})\}$, where $\mathcal{X}(\mathbb{O}) = \mathrm{Hom}(A, \mathbb{O})$, $\mathcal{X} = \mathrm{Spec} A$, for which we construct the generators in case of quasi-split tori defined as follows. Let $\Gamma = \mathrm{Gal}(L/\mathbb{Q}_p)$, $\hat{T}$ be a Γ-module of characters of $\mathbb{T}$, $\{\chi_i\}_{i=1}^d$ the basis of $\hat{T}$, and the group Γ which acts on $\hat{T}$ by the permutations. Suppose Γ_1 is a stabilizer of the element χ_1, then $\hat{T} \cong \mathbb{Z} \otimes_{\Gamma_1} \mathbb{Z}[\Gamma]$, and $\mathbb{T} \cong R_{F/\mathbb{Q}_p}(\mathbb{G}_m)$, where $F = L^{\Gamma_1}$ is the extension of $\mathbb{Q}_p$ with respect to an integral basis $\{\omega_i\}_{i=1}^d$ and the ring of integers $\mathbb{O}_F$ of F. We have decomposition $\chi_1 = x_1\omega_1 + \cdots + x_d\omega_d$. For the integral model of $\mathbb{T}$, $A = \mathbb{O}[x_1, x_2, \ldots, x_d, y^{-1}]$, and $y = \chi_1 \cdots \chi_d$ is a norm form with degree d in variables $x_1, \ldots, x_d$. Thus, our aim is to find the generators of

$$\mathcal{A}_0 = \{f \in \mathbb{Q}_p[\mathbb{T}] = \mathbb{Q}_p \otimes_{\mathbb{O}} A \mid f(u) \in \mathbb{O},\ \forall u \in \mathbb{O}_F^*\},$$

where $\mathbb{O}_F^* = \mathcal{X}(\mathbb{O}) = \mathrm{Hom}(A, \mathbb{O})$.

Lemma 1 *It holds* $\mathcal{A}_0 = \{f \in \mathbb{Q}_p[x_1, \dots, x_d, y^{-1}] \mid f(u) \in \mathcal{O},\ \forall u \in \mathcal{O}_F^*\}$.

Proof The proof immediately follows from the fact that y is a norm mapping that on the elements of $\mathcal{O}_F^*$ takes on the values in $\mathcal{O}^*$. □

One has $\mathcal{O}_F^* = \{(x_1, x_2, \dots, x_d) \mid \forall i, x_i \in \mathcal{O}$ and $\exists i_0, x_{i_0} \in \mathcal{O}^*\}$.

Theorem 8 *Let* $\mathcal{A}_0$ *be the algebra defined as above. The polynomials*

$$Q_{n_j}(t_j) = \frac{1}{p^{s_{n_j}}} \prod_{i=0}^{p^{k_j}-1} (t_j - i) \prod_{p^{k_j} \le i_n \le n_j} (t_j - t_{i_n}^*),$$

where t_j *is a variable,* $j = 1, 2, \dots d$, *and* $n_j = n_{j,k_j} p^{k_j} + n_{j,k_j-1} p^{k_j-1} + \dots + n_{j,1} p + n_{j,0}$, $0 \le n_{j,m} < p$, $1 \le m \le k_j - 1$, $n_{j,k_j} \ne 0$, $s_{n_j} = n_{j,k_j}\alpha_{k_j} + n_{j,k_j-1}\alpha_{k_j-1} + \dots + n_{j,1}\alpha_1$, $\alpha_i = 1 + p + p^2 + \dots + p^{i-1} = \frac{p^i-1}{p-1}$, *are in* $\mathcal{A}_0$. *Moreover, together with the polynomials*

$$Q'_{[k]}(t_1, t_2, \dots t_d) = \prod_{j=1}^{d} \left(\prod_{\substack{i=1 \\ (i,p)=1}}^{p^k-1} (t_j - i) \right) \frac{1}{p^{\alpha_k}}.$$

they form the generators of $\mathcal{A}_0$.

Proof We give the sketch of the proof. Using the inclusion $\mathcal{O}^* \subset \mathcal{O}$, we get $Q_{n_j}(t_j) \in \mathcal{A}_0$ and $Q_{n_j}(t_j)$ are integer-valued on $\mathcal{O}$. The necessity of adding $Q'_{[k]}$ is due to the representation of $u \in \mathcal{O}_F^*$ as $u = u_1\omega_1 + \dots + u_d\omega_d$, where $u_i \in \mathcal{O}$ and there exists i_0 such that $u_{i_0} \in \mathcal{O}^*$. Hence, when the values of i run over the reduced residue system modulo p, for the variable x_{i_0}, one can find an element that belongs to the same class with u_{i_0}. Therefore, p divides $\prod_{\substack{i=1 \\ (i,p)=1}}^{p-1} (x_{j_0} - i)$. Considering the reduced residue systems modulo p^k, one finds the values α_k. □

3 Future Research and Conclusion

The results of this work may lead to a study of algebras of integer-valued functions for arbitrary algebraic tori, since any algebraic torus is embedded into a quasi-split torus. There are nine different non-isomorphic two-dimensional tori. Some of those tori are represented as direct product of split and one-dimensional norm tori, which can be studied using the obtained results. Further study of algebras may be performed in several directions, for instance, the study of arithmetic and algebras of tori defined over more general fields and function fields of one or two variables. The construction of Artin L-functions for one- and two-dimensional tori over the field of rational numbers is another interesting topic for future research. Moreover, using the Artin L-functons of algebraic torus, one can compute its global ζ-function and the Tamagawa numbers as well. In addition, due to a certain property of any algebraic torus, two tori are isomorphic if and only if their minimal splitting fields coincide, and split groups are conjugate in $\mathrm{GL}(n, \mathbb{Z})$, which can be employed for classification of all n-dimensional tori. Also, finding the generators of algebras of integer-valued functions is important for classification of integral models of algebraic tori, including tori of small dimensions.

References

1. Anderson, D.D., Anderson D.F.: Elasticity of factorizations in integral domains. J. Pure Appl. Algebra **80**, 217–235 (1992)
2. Anderson D.F., Gotti, F.: Finite and bounded factorization domains. In: Badawi, A., Coykendall, J. (eds.) Rings, Monoids and Module Theory. Springer Proceedings in Mathematics and Statistics, vol. 382, pp. 7–57. Springer, Singapore (2021)
3. Bitan, R.: The discriminant of an algebraic torus. J. Number Theory **131**, 1657–1671 (2011)
4. Cahen, P.-J., Chabert, J.-L.: Elasticity for integral-valued polynomials. J. Pure Appl. Algebra **103**, 303–311 (1995)
5. Chapman, S., McClain, B.: Irreducible polynomials and full elasticity in rings of integer-valued polynomials. J. Algebra **293**, 595–610 (2005)
6. Eisenbud, D.: Commutative Algebra. Springer, New York (2004)
7. Elliott, J.: Integer-valued polynomials on commutative rings and modules. Commun. Algebra **46**, 1121–1137 (2018)
8. Frisch, S.: A construction of integer-valued polynomials with prescribed sets of lengths of factorizations. Monatsh. Math. **171**, 341–350 (2013)
9. Geroldinger, A., Halter-Koch, F.: Non-unique Factorizations: Algebraic, Combinatorial and Analytic Theory. Chapman & Hall/CRC, Boca-Raton (2006)
10. Halter-Koch, F.: Elasticity of factorizations in atomic monoids and integral domains. J. Theór. Nombres Bordeaux **7**, 367–385 (1995)
11. Hasse, H.: Number Theory. Springer, Berlin (2002)
12. Kriz, I., Kriz, S.: Introduction to Algebraic Geometry. Birkhäuser, Cham (2021)
13. Milne, J.S.: Algebraic Groups. Cambridge University Press, Cambridge (2017)
14. Ono, T.: Arithmetic of algebraic tori. Ann. Math. **74**, 101–139 (1961)
15. Peruginelli, G.: Integer-valued polynomials over matrices and divided differences. Monatsh. Math. **173**, 559–571 (2014)
16. Peruginelli, G.: Integral-valued polynomials over sets of algebraic integers of bounded degree. J. Number Theory **137**, 241–255 (2014)
17. Rapinchuk, A.S., Rapinchuk, I.A.: Linear algebraic groups with good reduction. Res. Math. Sci. **7**, 28 (2020)
18. Valenza, R.J.: Elasticity of factorizations in number fields. J. Number Theory **36**, 212–218 (1990)
19. Werner, N.: Integer-valued polynomials over matrix rings. Commun. Algebra **40**, 4717–4726 (2012)

On Algebraic Characteristics of μ-Anti-Q-Fuzzy Subgroups

M. Premkumar, M. Mohamed Althaf, A. Prasanna, Rachanna Kanabur, Rajneesh Sharma, and Atul Sarojwal

Keywords μ-Anti- Q—fuzzy subgroup (μ-anti- Q—FSG) · μ-Anti- Q—fuzzy normal subgroup (μ-anti- Q—FNSG) · μ-Anti- Q—fuzzy coset (μ-anti- Q—FCS) · μ-Anti- Q—fuzzy left coset (μ-anti- Q—FLCS) · μ-Anti- Q—fuzzy right coset (μ-anti- Q—FRCS)

Mathematics Subject Classification MSC2020-zbMATH-03B52

1 Introduction

Zadeh [24] launched the FSs in 1965. Later on, Rosenfeld [21] invented the theory of FGs in 1971. The idea of level FSG was innovated by Das [8] in 1981. Liu [13] described the FISG in 1982. Mukherje et al. [18] introduced the concept of FCSs in 1984. Bhattacharya [4] explored the considerable characterizations of FSG in 1987. Choudhury [6] propagated the belief of fuzzy homomorphisms between the two groups, and in 1988 he studied its implications for FSG. Akgul [3] meditated the notion of level subgroups of FNSGs and their homomorphisms

M. Premkumar (✉)
Department of Mathematics, Sathyabama Institute of Science and Technology (Deemed To Be University), Chennai, Tamilnadu, India

M. M. Althaf · A. Prasanna
Department of Mathematics, Jamal Mohamed College (Autonomous), (Affiliated to Bharathidasan University), Tiruchirappalli, Tamilnadu, India

R. Kanabur
Department of Mathematics, BLDEA'S Commerce BHS Arts and TGP Science College, Jamkhandi, Karanataka, India

R. Sharma
Department of Civil Engineering, Government Engineering College, Jhalawar, Rajasthan, India

A. Sarojwal
Department of Electrical Engineering, FET, MJP Rohilkhand University, Bareilly, Uttar Pradesh, India

H.-H. Leung et al. (eds.), *Recent Developments in Algebra and Analysis*, Trends in Mathematics, https://doi.org/10.1007/978-3-031-37538-5_22

in 1988. Mashou [14] discussed multiple consequential plats of FNSGs in 1990. Biswas [5] commenced the opinion of anti-FSG in 1990. Dixit et al. [9] studied the union of FSGs in 1990. Gupta [11] developed many classical t-operators in 1991. Kumar et al. [12] explored the FNSG, fuzzy direct product, and fuzzy quotients in 1992. Malik et al. [15] investigated the normality of FSGs in 1992. Filep [10] established the structure and construction of FSGs of group in 1992. Chakraborty and Khare [7] examined the behavior of the composition of fuzzy homomorphism and proved the fundamental theorem of homomorphism in 1993. Asaad and Zaid [2] proposed the study of FSGs of nilpotent groups in 1993. Ajmal [1] described the homomorphism of FSGs, fuzzy quotient groups, and correspondence theorem in 1994. Morsi et al. [17] examined fuzzy quotient groups and level structures in 1994. Mishref [16] described the normal, subnormal, and composition series of FNSGs in 1995. C.T. Nagaraj and M. Premkumar [19] introduced the concept on fundamental attributes on homomorphism of μ-anti- fuzzy subgroups in 2021. Ray [20] developed key features of the product of two FSb and FSGs in 1999. Sharma [22] expounded α-anti FSGs in 2012. Further, the basic algebraic properties of α-FSGs are determined by the coequal authors [23] in 2013.

In Sect. 3, we utilize the study of this phenomenon to define μ-anti- Q−FSG and prove that each anti- Q−FS is μ-anti- Q−FSG. Moreover, we investigate μ-anti- Q−fuzzy version of some fundamental outcomes of pure mathematics. Additionally, we innovate the concepts of μ-anti- Q−fuzzy cosets, μ-anti- Q−FNSG.

2 Preliminaries

Definition 2.1 [24] A FS $\breve{\underset{.}{A}}$ of a nonempty set P is a function

$$\breve{\underset{.}{A}}: P \to [0, 1].$$

Definition 2.2 [3] Let $\breve{\underset{.}{A}}$ be FSb of a group H. Then $\breve{\underset{.}{A}}$ is said to a FSG if $\breve{\underset{.}{A}}(u^{-1}v) \geq \min\{\breve{\underset{.}{A}}(u), \breve{\underset{.}{A}}(v)\}$, for all $u, v \in H$.

Definition 2.3 [5] Let $\breve{\underset{.}{A}}$ be FSb of a group H. Then $\breve{\underset{.}{A}}$ is said to an anti-FSG if $\breve{\underset{.}{A}}(u^{-1}v) \leq \max\{\breve{\underset{.}{A}}(u), \breve{\underset{.}{A}}(v)\}$, for all $u, v \in H$.

Definition 2.4 [19] Let d' be a nonempty set and also $\breve{\underset{.}{A}}$ be the FSb of d' and $\mu \in [0, 1]$. Then FS $\breve{\underset{.}{A}}$ is bellowed μ-anti-FSb afore d' and is defined as

$$\breve{\underset{.}{A}}_{\mu}(m) = S_{\check{d}}\left\{\breve{\underset{.}{A}}(m), 1 - \mu\right\}, \quad \text{for all } m \in \check{d}$$

3 Algebraic Properties on μ-Anti- $Q-$FSGs

In this chapter, we define aspects of μ-Anti- $Q-$FSG and μ-Anti- $Q-$FNSG. We show that every anti- $Q-$FSG (anti- $Q-$FNSG) is additionally μ-anti- $Q-$FS (μ-Anti- $Q-$FNS). The opinion of μ-anti- $Q-$fuzzy coset (μ-Anti- $Q-$FCS) is discussed deeply during this section. Further, we innovate the notion of factor group with regard to μ-anti- $Q-$FNS.

Definition 3.1 Let $\tilde{H}$ be a collection and $\breve{A}$ be a $Q-$FSb of $\tilde{H}$,$\mu \in [0,1]$ *and* $q \in Q$, and then $\breve{A}$ signifies μ-Anti- Q−FSG if

(i) $\breve{A}_\mu(mn,q) \le \max\{\breve{A}_\mu(m,q), \breve{A}_\mu(n,q)\}$, for all $m,n \in \tilde{H}$ *and* $q \in Q$
(ii) $\breve{A}_\mu(m^{-1},q) = \breve{A}_\mu(m,q)$.

Proposition 3.2 If $\breve{A} : \tilde{H} \longrightarrow [0,1]$ is a μ- Anti - $Q-$FSG of $\tilde{H}$, then

(i) $\breve{A}_\mu(m,q) \le \breve{A}_\mu(e,q)$, $\forall\, m \in \tilde{H}$,$q \in Q$ and where $e \in \tilde{H}$.
(ii) $\breve{A}_\mu(mn^{-1},q) = \breve{A}_\mu(e,q) \Rightarrow \breve{A}_\mu(m,q) = \breve{A}_\mu(n,q)$, $\forall m,n \in \tilde{H}$ *and* $q \in Q$.

Proof

(i) $\breve{A}_\mu(e,q) = \breve{A}_\mu(mm^{-1},q) \le \max\{\breve{A}_\mu(m,q), \breve{A}_\mu(m^{-1},q)\}$
$= \max\{\breve{A}_\mu(m,q), \breve{A}_\mu(m,q)\} = \breve{A}_\mu(m,q)$.

Hence, $\breve{A}_\mu(e,q) \le \breve{A}_\mu(m,q)$, for all $m \in \tilde{H}$.

(ii) $\breve{A}_\mu(m,q) = \breve{A}_\mu(mn^{-1}n,q) \le \max\{\breve{A}_\mu(mn^{-1},q), \breve{A}_\mu(n,q)\}$

$$= \max\left\{\breve{A}_\mu(e,q), \breve{A}_\mu(n,q)\right\} = \breve{A}_\mu(n,q).$$

Hence, $\breve{A}^\mu(m,q) \le \breve{A}_\mu(n,q)$.
Similarly, $\breve{A}_\mu(n,q) \le \breve{A}_\mu(m,q)$.

$$\breve{A}_\mu(n,q) = \breve{A}_\mu(m,q), \quad \text{for all } m,n \in \tilde{H} \text{ and } q \in Q.$$

The following findings explain that the every $Q-$FSG of the group is μ- $Q-$FSG of the group.

Theorem 3.3 Every μ-anti-FSG of $\tilde{H}$ is a μ-Anti- $Q-$FSG of $\tilde{H}$.

Proof Let $\breve{A}$ be an μ-anti- $Q-$FSG of $\tilde{H}$ and m & n in II elements in $\tilde{H}$.

$$\breve{A}_\mu(mn,q) = S_p\left\{\breve{A}(mn,q), 1-\mu\right\} \le S_p\left\{\max\left\{\breve{A}(m,q), \breve{A}(n,q)\right\}, 1-\mu\right\}$$

$$= \max\left\{S_p\left\{\breve{A}(m,q), 1-\mu\right\}, S_p\left\{\breve{A}(n,q), 1-\mu\right\}\right\} = \max\left\{\breve{A}_\mu(m,q), \breve{A}_\mu(n,q)\right\}$$

Hence, $\breve{A}_\mu(mn,q) \le \max\{\breve{A}^\mu(m,q), \breve{A}^\mu(n,q)\}$.

Definition 3.8 Let $\breve{A}$ be a μ-anti-Q−FSG of a group $^{\tau}H$ and $\mu \in [0, 1]$. $\forall m \in {}^{\tau}H$, the μ-anti- Q−FLCS of $\breve{A}$ in $^{\tau}H$ is represented by m$\breve{A}_{\mu}$.

$$m\breve{A}_{\mu}(g, q) = S_p\left\{\breve{A}\left(m^{-1}, q\right), \mu\right\}, \quad \forall m, g \in {}^{\tau}\text{H and } q \in Q.$$

μ-Anti- Q−FRCS of $\breve{A}$ in $^{\tau}H$ by $\breve{A}_{\mu}m$ and is defined as

$$(\breve{A}_{\mu}m,)(g, q) = S_p\left\{\breve{A}\left(gm^{-1}, q\right), \mu\right\}, \quad \text{for all } m, g \in {}^{\tau}\text{H and } q \in Q.$$

Definition 3.9 Let $\breve{A}$ be a μ-anti- Q−FSG of $^{\tau}H$ & $\mu \in [0, 1]$, and then $\breve{A}$ is named as μ-anti- Q−FNSG of $^{\tau}H$ iff $m\breve{A}_{\mu} = \breve{A}_{\mu}m$, for all $m \in {}^{\tau}H$.

Theorem 3.10 Every anti- Q−FNSG of $^{\tau}H$ is a μ-anti- Q−FNSG of $^{\tau}H$.

Proof Let $\breve{A}$ be an anti- Q−FNSG of $^{\tau}H$. $\forall\, m \in {}^{\tau}H$ *and* $q \in Q$.
$m\breve{A} = \breve{A}m$, $\Rightarrow (m\breve{A})(g, q) = (\breve{A}m)(g, q)$, for any $g \in {}^{\tau}H$.
Then, we have $\breve{A}(m^{-1}g, q) = \breve{A}(gm^{-1}, q) \Rightarrow S_p\{\breve{A}(m^{-1}g, q), \mu\}$
$= S_p\{\breve{A}(gm^{-1}, q), \mu\}$
$\Rightarrow (m\breve{A}_{\mu})(g, q) = (\breve{A}_{\mu}m)(g, q)$. Hence, $m\breve{A}_{\mu} = \breve{A}_{\mu}m$, for all $m \in {}^{\tau}H$ *and* $q \in Q$.
$\Rightarrow \breve{A}$ is μ-Anti- Q−FNSG of $^{\tau}H$.

Example 3.11 Let dihedral group of degree 3 with finite presentation
$^{\tau}\text{H} = D_3 = \; < m, n : \; m^3 = n^2 = e, nm = m^2n >$. Define the FSG of D_3 by

$$\breve{A}(y, q) = \begin{cases} 0.05, & \text{if } y \in < n > \\ 0.3, & \text{otherwise} \end{cases}$$

Take $\mu = 0$, and we have

$$(y\breve{A}_{\mu})(g, q) = S_p\left\{\breve{A}\left(y^{-1}g, q\right), 1 - \mu\right\} = S_p\left\{\breve{A}\left(y^{-1}g\right), 1\right\} = 1$$

$$= S_p\left\{\breve{A}\left(gy^{-1}, q\right), 1\right\} = S_p\left\{\breve{A}\left(gy^{-1}, q\right), 1 - \mu\right\} = \breve{A}_{\mu}y(g, q)$$

Hence, $y\breve{A}_{\mu} = \breve{A}_{\mu}y$, for all $y \in {}^{\tau}H$ *and* $q \in Q$. This shows that $\breve{A}$ is 0-anti-Q−FNSG of $^{\tau}H$.

Now, $\breve{A}(m^2(mn), q) = \breve{A}(m^3n, q) = \breve{A}(n, q) = 0.05$

$$\breve{A}\left((mn)m^2, q\right) = \breve{A}(m(nm)m, q) = \breve{A}\left(m\left(m^2n\right)m, q\right)$$

$$= \breve{A}\left(\left(m^3n\right)m, q\right) = \breve{A}(nm, q) = 0.3$$

$\Rightarrow \breve{A}$ is not μ -Anti- Q−FNSG of $^{\tau}H$.

Theorem 3.12 Let $\breve{A}$ be μ-anti- Q−FNSG of $\tilde{H}$, and then $\breve{A}_\mu(n^{-1}mn) = \breve{A}_\mu(m)$ or equivalently, $\breve{A}_\mu(mn, q) = \breve{A}_\mu(nm, q), \forall m, n \in \tilde{H}$.

Proof Since $\breve{A}$ be μ-anti- Q−FNSG of a $\tilde{H}$.

$\therefore\ m\breve{A}_\mu = \breve{A}_\mu m$, holds $\forall\ m \in \tilde{H}$ *and* $q \in Q$.

This implies that $(m\breve{A}_\mu)(n^{-1}, q) = \breve{A}_\mu m)(n^{-1}, q),\ n^{-1} \in \tilde{H}$

$$\Rightarrow S_p\left\{\breve{A}\left(m^{-1}n^{-1}, q\right), 1-\mu\right\} = S_p\left\{\breve{A}\left(n^{-1}m^{-1}, q\right), 1-\mu\right\}$$

which implies that $\breve{A}_\mu((nm)^{-1}, q) = \breve{A}_\mu((mn)^{-1}, q)$

as $\breve{A}$ is μ-anti- Q−FSG of $\tilde{H}$ so $\breve{A}_\mu(g^{-1}, q) = \breve{A}_\mu(g, q)$, for all $g \in \tilde{H}$ *and* $q \in Q$.

$\Rightarrow \breve{A}_\mu(nm, q) = \breve{A}_\mu(mn, q)$.

Definition 3.13 Let $\breve{A}$ be a μ-anti- Q−FNSG of a group $\tilde{H}$. Define $\tilde{H}_{\breve{A}_\mu} = \left\{m \in \tilde{H} \text{ and } q \in Q : \breve{A}_\mu(m, q) = \breve{A}_\mu(e, q)\right\}$, where $e \in \tilde{H}$. It follows that explaination that the set $\tilde{H}_{\breve{A}_\mu}$ is in fact a FNSG of $\tilde{H}$.

Theorem 3.14 Let $\breve{A}$ be a μ-anti- Q−FNSG of $\tilde{H}$, and then $\tilde{H}_{\breve{A}_\mu} \lhd \tilde{H}$.

Proof Obviously, $\tilde{H}_{\breve{A}_\mu} \neq \emptyset$ as $e \in \tilde{H}_{\breve{A}_\mu}$

Let $m, n \in \tilde{H}_{\breve{A}_\mu}$ *and* $q \in Q$ be any element.

We have

$$\breve{A}\left(mn^{-1}, q\right) \le \max\left\{\breve{A}_\mu(m, q), \breve{A}_\mu(n, q)\right\} = \max \breve{A}_\mu(e, q), \breve{A}_\mu(e, q)\} = \breve{A}_\mu(e, q)$$

$$\breve{A}_\mu\left(mn^{-1}, q\right) \le \breve{A}_\mu(e, q)\ \textit{but}\ \breve{A}_\mu\left(mn^{-1}, q\right) \ge \breve{A}_\mu(e, q).$$

Therefore, $\breve{A}_\mu(mn^{-1}, q) = \breve{A}_\mu(e, q), \Rightarrow \left(mn^{-1}, q\right) \in \tilde{H}_{\breve{A}_\mu}$.

Hence, $\tilde{H}_{\breve{A}_\mu}$ is a subgroup of $\tilde{H}$.

Let $m \in \tilde{H}_{\breve{A}_\mu}, n \in \tilde{H}$. We have $\breve{A}_\mu(n^{-1}mn, q) = \breve{A}_\mu(m, q) = \breve{A}_\mu(e, q)$

$\Rightarrow \left(n^{-1}mn, q\right) \in \tilde{H}_{\breve{A}_\mu}$. Consequently, $\tilde{H}_{\breve{A}_\mu} \lhd \tilde{H}$.

Theorem 3.15 Let $\breve{A}$ be the μ-anti- Q−FNSG of $\tilde{H}$, then

$$\left(m\breve{A}_\mu, q\right) = \left(n\breve{A}_\mu, q\right) \text{ iff } \left(m^{-1}n, q\right) \in \tilde{H}_{\breve{A}_\mu}$$

$$\left(\breve{A}_\mu m, q\right) = \left(\breve{A}_\mu n, q\right) \text{ iff } \left(mn^{-1}, q\right) \in \tilde{H}_{\breve{A}_\mu}$$

Proof

(i) $m\breve{A}_{\mu} = n\breve{A}_{\mu} \forall m, n \in \tilde{H}$ *and* $q \in Q$.

$$\breve{A}_{\mu}\left(m^{-1}n, q\right) = S_p\left\{\breve{A}\left(m^{-1}n, q\right), \mu\right\} \text{ for all } m, n \in \tilde{H} \text{ and } q \in Q$$

$$= \left(m\breve{A}_{\mu}\right)(n, q) = \left(n\breve{A}_{\mu}\right)(n, q) = S_p\left\{\breve{A}\left(n^{-1}n, q\right), \mu\right\} = S_p\left\{\breve{A}(e, q), \mu\right\}$$

$$= \breve{A}_{\mu}(e, q).$$

$\Rightarrow m^{-1}n \in \tilde{H}_{\breve{A}_{\mu}}$.

Converse, let $m^{-1}n \in \tilde{H}_{\breve{A}_{\mu}}$. $\Rightarrow \breve{A}_{\mu}(m^{-1}n, q) = \breve{A}_{\mu}(e, q)$.

For any element, $r \in \tilde{H}_{\breve{A}_{\mu}}$, $\left(m\breve{A}_{\mu}\right)(r, q) = S_p\left\{\breve{A}\left(m^{-1}r, q\right), \mu\right\} = \breve{A}_{\mu}\left(m^{-1}r, q\right)$

$$= \breve{A}_{\mu}\left(\left(m^{-1}n\right)\left(n^{-1}r\right), q\right) \leq \max\left\{\breve{A}_{\mu}\left(m^{-1}n, q\right), \breve{A}_{\mu}\left(n^{-1}r, q\right)\right\}$$

$$= \max\left\{\breve{A}_{\mu}(e, q), \breve{A}_{\mu}\left(n^{-1}r, q\right)\right\} = \breve{A}_{\mu}\left(n^{-1}r, q\right) = \left(n\breve{A}_{\mu}\right)(r, q).$$

Interchanging the role of m and n

$$\left(m\breve{A}_{\mu}\right)(r, q) = \left(n\breve{A}_{\mu}\right)(r, q), \quad \text{for all } r \in \tilde{H} \text{ and } q \in Q.$$

Consequently, $(m\breve{A}_{\mu}, q) = (n\breve{A}_{\mu}, q)$.

(ii) It can be proven as analogous to part (i).

Definition 3.16 Let $\breve{A}$ be a μ-anti- $Q-$FNSG of $\tilde{H}$. The set of all μ-anti- $Q-$FCSs of $\breve{A}$ $\tilde{H}/\breve{A}_{\mu}$ forms a H with respect to * defined by $\left(\left(m\breve{A}_{\mu}\right)_{*}\left(n\breve{A}_{\mu}\right), q\right) = (m_{*}n, q)\breve{A}_{\mu}$, where $m\breve{A}_{\mu}, n\breve{A}_{\mu} \in \tilde{H}/\breve{A}_{\mu}$, $m, n \in \tilde{H}$ and $q \in Q$ This group is called the factor group of $\tilde{H}$ with regard to μ-Anti- $Q-$FNSG $\breve{A}_{\mu}$.

Theorem 3.17 Let $\tilde{H}/\breve{A}_{\mu}$ form a group w.r.t the above stated binary arithmetic $_{*}$.

Proof Let $\breve{A}_{\mu}m_1 = \breve{A}_{\mu}m_2$ *and* $\breve{A}_{\mu}n_1 = \breve{A}_{\mu}n_2$, for some $m_1, m_2, n_1, n_2 \in \tilde{H}$ *and* $q \in Q$ Let $g \in \tilde{H}$ *and* $q \in Q$ be any element of $\tilde{H}$.

$$(\breve{A}_{\mu}(m_2 n_2, q))(g) = S_p\left\{\breve{A}\left(g(m_2 n_2)^{-1}, q\right), \mu\right\} = S_p\left\{\breve{A}\left(g n_2^{-1} m_2{}^{-1}, q\right), \mu\right\}$$

$$= S_p\left\{\breve{A}\left(\left(g n_2^{-1}\right) m_2{}^{-1}, q\right), \mu\right\} = \breve{A}_{\mu} m_2\left(g n_2^{-1}, q\right) = \breve{A}_{\mu} m_1\left(g n_2^{-1}, q\right)$$

$$= S_p\left\{\breve{A}\left(\left(g n_2^{-1}\right) m_1{}^{-1}, q\right), \mu\right\}$$

$$= S_p\left\{\breve{A}\left(\left(m_1^{-1}\right)\left(g n_2^{-1}\right), q\right), \mu\right\} = \breve{A}_{\mu} n_2\left(m_1^{-1} g, q\right) = \breve{A}_{\mu} n_1\left(m_1^{-1} g, q\right)$$

$$= S_p\left\{\breve{A}\left(\left(m_1^{-1} g\right) n_1^{-1}\right), \mu\right\}$$

$$= S_p\left\{\breve{A}\left(n_1^{-1}\left(m_1^{-1} g\right), q\right), \mu\right\} = S_p\left\{\breve{A}\left(\left(n_1^{-1} m_1^{-1}\right)(g), q\right), \mu\right\}$$

$$= S_p\left\{\breve{A}\left((m_1 n_1)^{-1}(g), q\right), \mu\right\}$$

$$= S_p\left\{\breve{A}\left((g)(m_1 n_1)^{-1}, q\right), \mu\right\} = \left(\breve{A}_{\mu}(m_1 n_1), q\right)(g).$$

Therefore, $*$ is well defined. Clearly, the set $\tilde{H}/\breve{A}_{\mu}$ satisfies $*$.

Moreover, $(\breve{A}_{\mu})(m\breve{A}_{\mu}, q) = ((e\breve{A}_{\mu}) * (m\breve{A}_{\mu}), q) = (e * m, q)\breve{A}_{\mu} = m\breve{A}_{\mu}$which implies that $\breve{A}_{\mu}$is identity of $\tilde{H}/\breve{A}_{\mu}$. The inverse of each element $m\breve{A}^{\mu} \in \tilde{H}/\breve{A}_{\mu}$ is $m^{-1}\breve{A}_{\mu} \in \tilde{H}/\breve{A}_{\mu}$ as $((m^{-1}\breve{A}_{\mu},) * (m\breve{A}_{\mu}), q) = (m^{-1} * m, q)\breve{A}_{\mu} = e\breve{A}_{\mu} = \breve{A}_{\mu}$.

Consequently, $(\tilde{H}/\breve{A}_{\mu}, *)$ is $\tilde{H}$.

Theorem 3.18 Let $\breve{A}$ be a μ-anti- Q−FNSG of a group $\tilde{H}$. Then ∃a natural onto homomorphism bwl $\tilde{H}$ and $\tilde{H}/\breve{A}_{\mu}$ which can be described as

$$f(m, q) = \breve{A}_{\mu} m, \quad m \in \tilde{H} \text{ and } q \in Q$$

Proof f is homomorphism as if for $m, n \in \tilde{H}$ *and* $q \in Q$ we have

$$f(mn, q) = \left(\breve{A}_{\mu} mn, q\right) = \left(\breve{A}_{\mu} m, q\right)\left(\breve{A}_{\mu} n, q\right) = f(m, q) f(n, q)$$

Obviously f is onto as well. Consequently, f is an onto homomorphism from $\tilde{H}$ to $\tilde{H}/\breve{A}_{\mu}$.

Further, $\ker f = \{m \in {}^{\tau}\text{Ḥ} \; and \; q \in Q : f(m, q) = \breve{A}_{\mu}(e, q)\} = \{m \in G and \; q \in Q : (\breve{A}_{\mu} m, q) = (\breve{A}_{\mu} e, q)\}$

$$ker f = \left\{ m \in {}^{\tau}\text{Ḥ} \text{and } q \in Q : \left(me^{-1}, q \right) \in {}^{\tau}\text{H}_{\breve{A}_{\mu}} \right\}$$

$$= \left\{ m \in {}^{\tau}\text{Ḥ} \text{and } q \in Q : \; (m, q) \in {}^{\tau}\text{H}_{\breve{A}_{\mu}} \right\} = {}^{\tau}\text{H}_{\breve{A}_{\mu}}.$$

Theorem 3.19 Let $\breve{A}$ be the μ-anti- Q−FNSG of ${}^{\tau}H$, and then ${}^{\tau}\text{Ḥ}/\breve{A}_{\mu} \cong {}^{\tau}\text{Ḥ}/{}^{\tau}\text{H}_{\breve{A}_{\mu}}$.

Proof We have ${}^{\tau}\text{Ḥ}/{}^{\tau}\text{H}_{\breve{A}_{\mu}}$ which is well defined.

Let $f : {}^{\tau}\text{Ḥ}/\breve{A}_{\mu} \to {}^{\tau}\text{Ḥ}/{}^{\tau}\text{H}_{\breve{A}_{\mu}}$ defined as $f\left(m\breve{A}_{\mu}, q\right) = m{}^{\tau}\text{H}_{\breve{A}_{\mu}}$, for any $m\breve{A}_{\mu} \in {}^{\tau}\text{Ḥ}/\breve{A}_{\mu}$ and $q \in Q$.

f is well defined because if $m\breve{A}_{\mu} = n\breve{A}_{\mu}$ this implies that $m{}^{\tau}\text{H}_{\breve{A}_{\mu}} = n{}^{\tau}\text{H}_{\breve{A}_{\mu}}$

Hence, $f(m\breve{A}_{\mu}, q) = f(n\breve{A}_{\mu}, q)$

This also shows that f is injective. Also, for each $m{}^{\tau}\text{H}_{\breve{A}_{\mu}} \in {}^{\tau}\text{Ḥ}/{}^{\tau}\text{H}_{\breve{A}_{\mu}}$ there exist$m\breve{A}_{\mu} \in {}^{\tau}\text{Ḥ}/\breve{A}_{\mu}$such that $f\left(m\breve{A}_{\mu}, q\right) = m{}^{\tau}\text{H}_{\breve{A}_{\mu}}$.Thus, f is surjective.

Further, for each $m\breve{A}_{\mu}, n\breve{A}_{\mu} \in {}^{\tau}\text{Ḥ}/\breve{A}^{\mu}$ and $q \in Q$, we have

$$f\left(m\breve{A}_{\mu}n\breve{A}_{\mu}, q\right) = f\left((mn)\,\breve{A}_{\mu}, q\right) = (mn, q)\,{}^{\tau}\text{H}_{\breve{A}_{\mu}} = \left(m{}^{\tau}\text{H}_{\breve{A}_{\mu}}, q\right)\left(n{}^{\tau}\text{H}_{\breve{A}_{\mu}}, q\right)$$

$$= f\left(m\breve{A}_{\mu}, q\right) f\left(n\breve{A}_{\mu}, q\right)$$

Hence, $f(m\breve{A}_{\mu}n\breve{A}_{\mu}, q) = f(m\breve{A}_{\mu}, q)f(n\breve{A}_{\mu}, q)$. Thus, f is a homomorphism.
Consequently, ${}^{\tau}\text{Ḥ}/\breve{A}_{\mu} \cong {}^{\tau}\text{Ḥ}/{}^{\tau}\text{H}_{\breve{A}_{\mu}}$.

4 Conclusion

In present work, the ideas of μ-anti- Q−FSG and μ-Anti- Q−FCS of a given group are delineated. Moreover, the opinion of μ-anti- Q−FNSG has been innovated, and we have established several fundamental characteristics of this notion.

Acknowledgments The authors would like to thank the Department of Biotechnology (DBT), New Delhi, on this paper. The third author (A. Prasanna) was supported by the DBT Star College Scheme Grant and the Management of Jamal Mohamed College, Tiruchirappalli.

References

1. Ajmal, N.: Homomorphism of groups, correspondence theorem and fuzzy quotient groups. Fuzzy Sets Syst. **61**, 329–339 (1994)
2. Asaad, M., Zaid, S.A.: Fuzzy subgroups of nilpotent groups. Fuzzy Sets Syst. **60**, 321–323 (1993)
3. Akgul, M.: Some properties of fuzzy groups. J. Math. Anal. Appl. **133**, 93–100 (1988)
4. Bhattacharya, P.: Fuzzy subgroups: some characterizations. J. Math. Anal. Appl. **128**, 241–252 (1987)
5. Biswas, R.: Fuzzy subgroups and anti-fuzzy subgroups. Fuzzy Sets Syst. **35**, 121–124 (1990)
6. Choudhury, F.P., Chakraborty, A.B., Khare, S.S.: A note on fuzzy subgroups and fuzzy homomorphism. J. Math. Anal. Appl. **131**, 537–553 (1988)
7. Chakraborty, A.B., Khare, S.S.: Fuzzy homomorphism and algebraic structures. Fuzzy Sets Syst. **51**, 211–221 (1993)
8. Das, P.S.: Fuzzy groups and level subgroups. J. Math. Anal. Appl. **84**, 264–269 (1981)
9. Dixit, V.N., Kumar, R., Ajmal, N.: Level subgroups and union of fuzzy subgroups. Fuzzy Sets Syst. **37**, 359–371 (1990)
10. Filep, L.: Structure and construction of fuzzy subgroups of a group. Fuzzy Set Syst. **51**, 105–109 (1992)
11. Gupta, M.M., Qi, J.: Theory of T-norms and fuzzy inference methods. Fuzzy Sets Syst. **40**, 431–450 (1991)
12. Kumar, I.J., Saxena, P.K., Yadav, P.: Fuzzy normal subgroups and fuzzy quotients. Fuzzy Sets Syst. **46**, 121–132 (1992)
13. Liu, W.J.: Fuzzy invariant subgroups and fuzzy ideals. Fuzzy Sets Syst. **8**, 133–139 (1982)
14. Mashour, A.S., Ghanim, H., Sidky, F.I.: Normal fuzzy subgroups. Ser. Math. **20**, 53–59 (1990)
15. Malik, D.S., Mordeson, J.N., Nair, P.S.: Fuzzy normal subgroups in fuzzy subgroups. J. Korean Math. Soc. **29**, 1–8 (1992)
16. Mishref, M.A.: Normal fuzzy subgroups and fuzzy normal series of finite groups. Fuzzy Sets Syst. **72**, 397–383 (1995)
17. Morsi, N.N., Yehia, S.E.B.: Fuzzy quotient groups. Inf. Sci. **81**, 177–191 (1994)
18. Mukherjee, N.P., Bhattacharya, P.: Fuzzy normal subgroups and fuzzy cosets. Inf. Sci. **34**, 225–239 (1984)
19. Nagaraj, C.T., Premkumar, M.: On fundamental attributes on homomorphism of μ-anti- fuzzy subgroups. J. Univ. Shanghai Sci. Technol. **23**(10), 793–798 (2021)
20. Ray, A.K.: On product of fuzzy subgroups. Fuzzy Sets Syst. **105**, 181–183 (1999)
21. Rosenfeld, A.: Fuzzy groups. J. Math. Anal. Appl. **35**, 512–517 (1971)
22. Sharma, P.K.: α-Anti Fuzzy subgroups. Int. Rev. Fuzzy Math. **7**, 47–58 (2012)
23. Sharma, P.K.: α-Fuzzy subgroups. Int. J. Fuzzy Math. Syst. **3**, 47–59 (2013)
24. Zadeh, L.A.: Fuzzy sets. Inf. Control. **8**, 338–353 (1965)

Multi-criteria Decision-Making with Bipolar Intuitionistic Fuzzy Soft Expert Sets

S. Anita Shanthi, N. Nishadevi, and S. Sampathu

Keywords Bipolar intuitionistic fuzzy soft expert set · Score function · Entropy weights · VIKOR method

1 Introduction

The concept of fuzzy set was introduced by Zadeh [1]. Atanassov [2] extended this concept to IFS. Bipolar fuzzy set was introduced by Zhang et al. [3]. Abdullah [4] developed the concept of bipolar fuzzy soft set. The concepts of soft expert and fuzzy soft expert sets were introduced by Alkhazaleh [5, 6]. Enginoglu et al. [7] constructed a MCDM method for modified soft expert sets. Geetharamani et al. [8] proposed fuzzy expert decision set model. Seenivasan et al. [9] designed a robust fuzzy ranking approach. Al-Qudah et al. [10] defined AND and OR operators an BFSES and applied it to MCDM problem. Chandran et al. [11] proposed IFSES theory and defined some basic operations using this concept.

Opricovic introduced VIKOR method to arrive at a compromise solution for MCDM method. Tara et al. [12] applied VIKOR method to select the best Magnesium alloy for automobile industry. Nurmuslimah et al. [13] analyzed the extent of damage of an amplifier using VIKOR method. VIKOR method was first introduced in fuzzy environment by Opricovic [14]. Yang et al. [15] developed a MCDM problem in IF environment. Hagar et al. [16] proposed neutrosophic VIKOR method. Fang et al. [17] proposed AHP-VIKOR-MRM method with PF information. Anita et al. [18] dealt with VIKOR method on BIFSS. Based on these concepts, the theory of BIFSES is developed.

S. Anitha Shanthi (✉) · N. Nishadevi
Department of Mathematics, Annamalai University, Chidambaram, Tamil Nadu, India

S. Sampathu
Sri Muthukumaran College of Education, Chennai, Tamil Nadu, India

H.-H. Leung et al. (eds.), *Recent Developments in Algebra and Analysis*,
Trends in Mathematics, https://doi.org/10.1007/978-3-031-37538-5_23

2 VIKOR Method Based on BIFSES

Definitions of BIFSES and BIFSE decision matrix are given in [19].

Definition 1 A BIFSES $BES = \{x, (\mu^n_{uv}(x), \mu^p_{uv}(x)), (\nu^n_{uv}(x), \nu^p_{uv}(x)) : x \in X\}$ where $\mu^n_{uv}{=}\mu^n_{uv}(x)$, $\mu^p_{uv}{=}\mu^p_{uv}(x)$, $\nu^n_{uv}{=}\nu^n_{uv}(x)$, $\nu^p_{uv}{=}\nu^p_{uv}(x)$.
The pair $(\mu^n_{uv}, \mu^p_{uv}), (\nu^n_{uv}, \nu^p_{uv})$ is often denoted by $\beta{=}(m, n), (o, p)$. The score function of BIFSES is $\beta{=}\frac{m^2+n^2+o^2+p^2}{4}$. Using this score function, score matrix is constructed.

Definition 2 The positive and negative ideal solutions in BIFSES is defined by
$PI = \max E_{uv}$, if $v \in k^1$, $\min E_{uv}$, if $v \in k^2$
$NI = \min E_{uv}$, if $v \in k^1$, $\max E_{uv}$, if $v \in k^2$
where k^1 denotes the collection of benefit criteria and k^2 denotes the collection of cost criteria.

Definition 3 Given a BIFSES,
$BES{=}\{x, [\mu^n_{uv}(x), \mu^p_{uv}(x)], [\nu^n_{uv}(x), \nu^p_{uv}(x)]; x \in X\}$,
$\lambda_{uv} = 1 - |\mu^n_{uv} - \nu^n_{uv} + \mu^p_{uv} - \nu^p_{uv}|$,
The entropy E_v is

$$E_v = \frac{1}{m}\sum_{u=1}^{m} \lambda_{uv}.$$

The entropy weight

$$w_v = \frac{1-E_v}{\sum_{v=1}^{l}(1-E_v)}$$

The weight vector $W = (w_1, w_2, \ldots, w_v)$ satisfies $w_1 + w_2 + \ldots + w_v = 1$.

Definition 4 Given BIFSES, the group utility value BM_u of an alternative is defined by $BM_u{=}\sum_{v=1}^{l}(\frac{PI-E_v}{PI-NI})$, where PI and NI are positive and negative ideal solutions.

Definition 5 Given BIFSES, the individual regret value BN_u of an alternative is defined by
$BN_u = \max_v(w_v \frac{PI-E_v}{PI-NI})$.

Definition 6 The value of BO_u in BIFSES is defined by
$BO_u{=}(Bt)\frac{BM_u-BM^*}{BM'-BM^*}$+(1-Bt)$\frac{BN_u-BN^*}{BN'-BN^*}$,
where $BM^* = \min_u\{BM_u\}$, $BM' = \max_u\{BM_u\}$, $BN^* = \min_u\{BN_u\}$,
$BN' = \min_u\{BN_u\}$.

$Bt \in [0, 1]$ represents weight of maximum group utility, and $(1 - Bt)$ represents the weight of individual regret value. Minimum value of BO_u is better alternative.

Definition 7 Compromise solution: The best alternative should satisfy the following conditions.

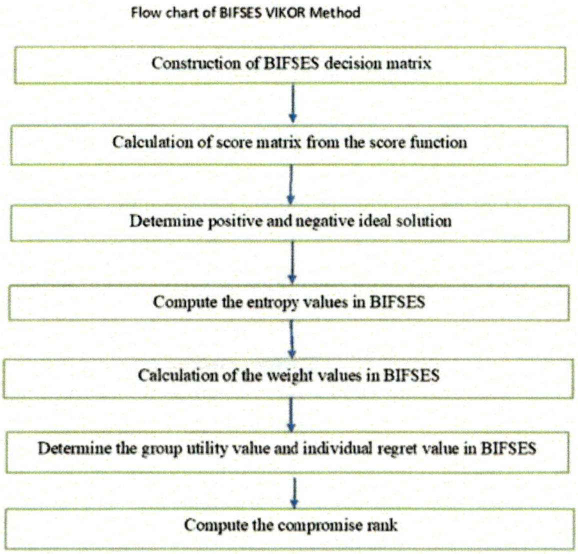

Fig. 1 Flowchart of BIFSES VIKOR method

Condition1: Acceptable Advantage
The first alternative EA on compromise list BO_u for $Bt = 0.5$ has sufficient advantage in to the next best ranked alternative EA,
if $BO_u(EA_2) - BO_u(EA_1) \geq \frac{1}{1-d}$, where d denotes the number of alternatives.
Condition2: Acceptable Stability
Alternative EA^1 is ranked by using BM_u and BN_u.
Case1: If the Condition1 is not satisfied , then the compromise solution of the alternatives are first and second rank alternatively.
Case2: If the Condition2 is not satisfied, then the compromise solution is $EA_1, EA_2, \ldots, EA_m$.

Flowchart of BIFSES VIKOR method is given in Fig. 1.

3 Procedure

The procedure of VIKOR method is as follows:

Step 1. Construction of BIFSES decision matrix.
Step 2. Calculate score matrix from Definition 1.
Step 3. Determine positive and negative ideal solutions from criteria by using Definition 2.
Step 4. Compute entropy and weight values of BIFSES by using Definition 3.
Step 5. Determine the BIFSE group utility, BIFSE individual regret, and BO_u values by Definitions 4, 5, and 6.
Step 6. Compute the compromising ranking list by considering two conditions given in Definition 7.

4 Illustrative Example

Let $\{EA_1, EA_2, EA_3, EA_4\}$ represent four different fans companies (Fig. 2). The quality of fans produced by these companies depends on the parameters $E = \{e_1, e_2\}$ where e_1 = speed and e_2 = energy efficiency.

Company EA_1: Fans produced by this company have high performance even at a low voltage. Also it is built with a high lift angle that provides uniform airflow across the room.

Company EA_2: Fans manufactured by this company are designed with aerodynamic technology and have wider blades. It is capable of enhancing energy conservation.

Company EA_3: Fans made by this company have powerful dominant motor, comprising of a double ball bearing, and it functions smoothly without much noise.

Company EA_4: Fans manufactured by this company are very speedy. It has powerful copper motor and delivers an efficient flow of cool air.

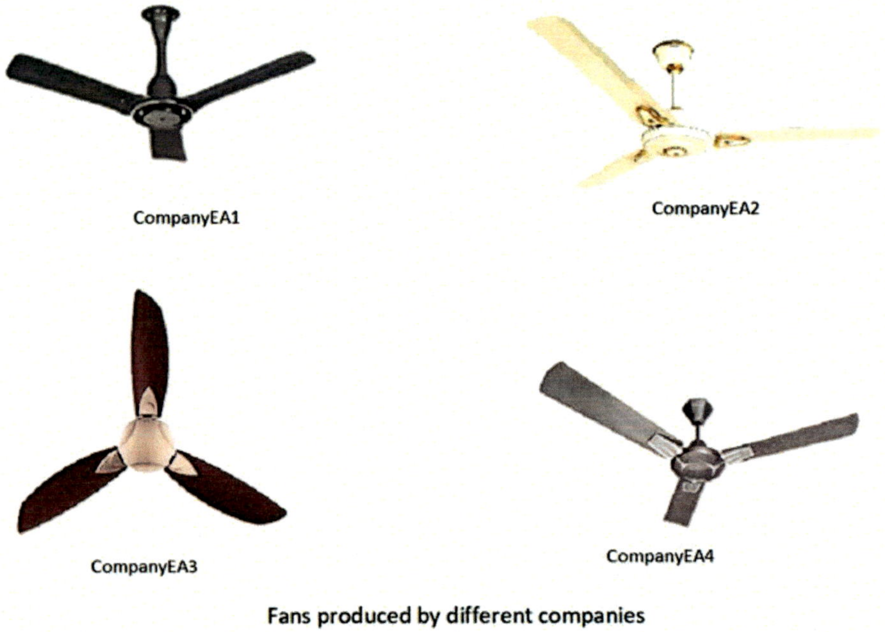

Fig. 2 Fans produced by different companies

Table 1 Decision matrix for agree

U	$(e_1, x_1, 1)$	$(e_1, x_2, 1)$
EA_1	$[-0.25, 0.49], [-0.45, 0.23]$	$[-0.32, 0.63], [-0.18, 0.19]$
EA_2	$[-0.18, 0.37], [-0.29, 0.14]$	$[-0.18, 0.38], [-0.26, 0.14]$
EA_3	$[-0.19, 0.61], [-0.25, 0.11]$	$[-0.48, 0.36], [-0.34, 0.27]$
EA_4	$[-0.52, 0.40], [-0.38, 0.19]$	$[-0.34, 0.15], [-0.12, 0.16]$
U	$(e_2, x_1, 1)$	$(e_2, x_2, 1)$
EA_1	$[-0.58, 0.28], [-0.26, 0.27]$	$[-0.31, 0.54], [-0.28, 0.26]$
EA_2	$[-0.47, 0.26], [-0.25, 0.16]$	$[-0.05, 0.21], [-0.02, 0.03]$
EA_3	$[-0.42, 0.57], [-0.51, 0.28]$	$[-0.20, 0.69], [-0.06, 0.17]$
EA_4	$[-0.27, 0.28], [-0.15, 0.26]$	$[-0.17, 0.33], [-0.08, 0.16]$

Table 2 Decision matrix for dis-agree

U	$(e_1, x_1, 0)$	$(e_1, x_2, 0)$
ED_1	$[-0.38, 0.25], [-0.01, 0.01]$	$[-0.25, 0.31], [-0.16, 0.17]$
ED_2	$[-0.36, 0.59], [-0.24, 0.19]$	$[-0.02, 0.21], [-0.01, 0.02]$
ED_3	$[-0.11, 0.39], [-0.33, 0.09]$	$[-0.59, 0.44], [-0.32, 0.25]$
ED_4	$[-0.14, 0.21], [-0.03, 0.07]$	$[-0.27, 0.22], [-0.10, 0.11]$
U	$(e_2, x_1, 0)$	$(e_2, x_2, 0)$
ED_1	$[-0.59, 0.52], [-0.24, 0.25]$	$[-0.27, 0.52], [-0.26, 0.24]$
ED_2	$[-0.45, 0.24], [-0.23, 0.14]$	$[-0.03, 0.19], [-0.02, 0.01]$
ED_3	$[-0.39, 0.55], [-0.49, 0.26]$	$[-0.17, 0.67], [-0.04, 0.15]$
ED_4	$[-0.25, 0.26], [-0.13, 0.21]$	$[-0.15, 0.31], [-0.06, 0.14]$

Step 1. Decision matrix for Agree is given in Table 1:
Decision matrix for Dis-Agree (Table 2):

Step 2. $BIFSE$ score matrix for $agree$ is as follows:

$$SM(EA) = \begin{pmatrix} 0.0118 & 0.1077 & 0.0685 & 0.0604 \\ 0.0164 & 0.0224 & 0.0501 & 0.0113 \\ 0.0834 & 0.0429 & 0.0407 & 0.1209 \\ 0.0625 & 0.0245 & 0.0153 & 0.0264 \end{pmatrix}$$

$BIFSE$ score matrix for $dis - agree$ is as follows:

$$SM(DA) = \begin{pmatrix} 0.0517 & 0.0260 & 0.1246 & 0.0545 \\ 0.0960 & 0.0110 & 0.0469 & 0.0091 \\ 0.0118 & 0.0942 & 0.0367 & 0.1134 \\ 0.0145 & 0.0248 & 0.0173 & 0.0238 \end{pmatrix}$$

Step 3. Best value and weakest value of $BIFSE$ for agree and $Dis - Agree$ given in Tables 3 and 4, respectively.

Table 3 Best value

U	$(e_1, x_1, 1)$	$(e_1, x_2, 1)$	$(e_2, x_1, 1)$	$(e_2, x_2, 1)$
$EA*$	0.0834	0.1077	0.0686	0.1209
EA'	0.0118	0.0224	0.0153	0.0113

Table 4 Weakest value

U	$(e_1, x_1, 1)$	$(e_1, x_2, 1)$	$(e_2, x_1, 1)$	$(e_2, x_2, 1)$
$ED*$	0.0960	0.0942	0.1246	0.1134
ED'	0.0118	0.0110	0.0173	0.0091

Table 5 BM_u, BN_u, BO_u values for agree

U	BM_u	BN_u	$BO_u(d = 0.5)$
EA_1	0.4622	0.3471	0.6631
EA_2	0.8302	0.3248	0.9384
EA_3	0.2841	0.1660	0
EA_4	0.7201	0.2257	0.5640

Table 6 BM_u, BN_u, BO_u values for dis-agree

U	BM_u	BN_u	$BO_u(d = 0.5)$
ED_1	0.4483	0.1742	0
ED_2	0.6081	0.3085	0.6115
ED_3	0.5225	0.3254	0.5778
ED_4	0.9253	0.3151	0.9658

Table 7 Average of BM_u, BN_u, BO_u

U	BM_u	BN_u	$BO_u(d = 0.5)$
EA_1	0.4552	0.2606	0.3315
EA_2	0.7191	0.3166	0.7749
EA_3	0.4033	0.2454	0.2889
EA_4	0.8227	0.2704	0.7649

Step 4. The $BIFSE$ entropy values of $agree$ are given below:
$EA_1 = 0.642$, $EA_2 = 0.775$, $EA_3 = 0.767$, $EA_4 = 0.785$.
The $BIFSE$ entropy values of $dis - agree$ are given below:
$ED_1 = 0.760$, $ED_2 = 0.907$, $ED_3 = 0.822$, $ED_4 = 0.772$.
Calculated weight values based on the criteria for $agree$ are
$w_1 = 0.3471$, $w_2 = 0.2184$, $w_3 = 0.2257$, $w_4 = 0.2087$.
Calculated weight values based on the criteria for $disagree$ are
$w_1 = 0.3254$, $w_2 = 0.1254$, $w_3 = 0.2407$, $w_4 = 0.3085$.

Step 5. Compute the values of BM_u, BN_u and BO_u for agree (Table 5)
Compute the values of BM_u, BN_u and BO_u for dis-agree (Table 6)
Compute the average values of BM_u, BN_u, and BO_u (Table 7).

Step 6. Compute the compromising ranking: From the values of BO_u, EA_3 has the minimum value.
So it has the first rank. The second minimum value is EA_1.

Fig. 3 Pie chart of BIFSES VIKOR

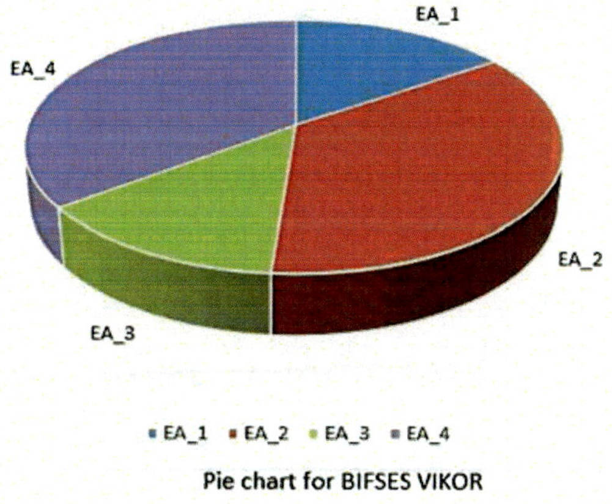

Now, verify the two conditions:
Condition 1: $BO_u(EA_1) - BO_u(EA_3) \geq \frac{1}{d-1}$

$$0.3315 - 0.2889 \geq \frac{1}{4-1},$$
$$0.0426 < 0.33.$$

Condition 1 is not satisfied. Hence, $EA_1, EA_2, \ldots, EA_5$ are compromised solutions. There is no comparative advantage of EA_3 from others.
Condition 2: Since EA_3 is also best ranked by the values of BM_u and BN_u, Condition 2 is satisfied. Hence, it is concluded that fans produced by company EA_3 is the best. Pie chart of BIFSES VIKOR method is given in Fig. 3.

5 Conclusion

In this article, VIKOR method on BIFSES is dealt with. The score function proposed is very effective in finding the score matrix. The BIFSES entropy value is used for calculating the weights of the criteria. The group utility and regret values are found, from which the compromising solution is obtained. Ceiling fans of four different companies are chosen and tested for the qualities speed and energy efficiency. The company that produces the best fan is determined using this method.

Acknowledgments The authors are thankful to the reviewers for their valuable comments, which have greatly helped in improving the quality of the paper.

Conflict of Interest. The authors declare that they have no conflict of interest.

References

1. Zadeh, L.A.: Fuzzy sets. Inf. Control **8**, 338–353 (1965). http://doi.org/10.1016/S0019-9958(65)90241-X
2. Atanassov, K.T.: Intuitionistic fuzzy sets. Fuzzy Sets Syst. **20**, 87–96 (1986). http://doi.org/10.1016/S0165-0114(86)80034-3
3. Zhang, W.R.: Bipolar Fuzzy sets. Proc. FUZZ-IEEE **1**, 835–840 (1998). http://doi.org/10.1109/FUZZY.1998.687599
4. Abdullah, S., Aslam, M., Ullah, K.: Bipolar Fuzzy soft sets and its applications in decision making problem. J. Intell. Fuzzy Syst. **27**, 1–18 (2014). http://doi.org/10.3233/IFS-131031
5. Alkhazaleh, S., Salleh, A.R.: Soft expert sets. Adv. Decis. Sci. **2011**, 1–12 (2011). http://doi.org/10.1155/2011/757868
6. Alkhazaleh, S., Salleh, A.R.: Fuzzy Soft Expert Set and its application. Appl. Math. **05**, 1349–1368 (2014). http://doi.org/10.4236/am.2014.59127
7. Enginoglu, S., Donmez, H.: On soft expert sets. J. New Theory **9**, 69–81 (2015). http://doi.org/0000-0002-7188-9893
8. Geetharamani, G., Sharmiladevi, C., Pandian, J.A.: Fuzzy expert decision set. Int. J. Math. Trends Technol. **20**, 41–46 (2015). http://doi.org/10.14445/22315373/IJMTT-V20P505
9. Seenivasan, M., Ramesh, R.: Heterogeneous queueing system with cost and profit appraises by robust fuzzy ranking approach. In: 4th International Congress on Life, Social and Health Sciences in a Changing World congress proceedings Book, pp. 215–223 (2022)
10. Al-Qudah, Y., Hassan, N.: Bipolar fuzzy Soft Expert Set and its application in decision making. Int. J. Appl. Decis. Sci. **10**, 175–191 (2017). http://doi.org/10.1504/IJADS.2017.084310
11. Chandran, G.S., Salleh, A.R.: Possiblity intuitionistic Fuzzy expert theory and its application decision making. Int. J. Math. Math. Sci. **2015**, 1–11 (2015). http://doi.org/10.1155/2015/314285
12. Tara Sasankal, C., Ravindra, K.: Implementation of VIKOR method for selection of magnesium alloy to suit automotive applications. Int. J. Adv. Sci. Technol. **83**, 49–58 (2015). http://doi.org/10.14257/ijast.2015.83.05
13. Nurmuslimah, S.: Application of VIKOR Method for identifying Amplifier Damage. In: IOP Conference Series: Materials Science and Engineering, vol. 1010, pp. 1–10 (2021). http://doi.org/10.1088/1757-899X/1010/1/012024
14. Opricovic: Fuzzy VIKOR with an application to water resources planning. Expert Syst. Appl. **38**, 12983–12990 (2011). http://doi.org/10.1016/j.eswa.2011.04.097
15. Jiaqi, Y., Naseem, M.H., Zuquan, X., Xueheng, L.: Supplier selection of shipbuilding based on intuitionistic Fuzzy multicriteria decision. Math. Probl. Eng., **2021**, 1–11 (2021). http://doi.org/10.1155/2021/1775053
16. Abu-Faty, G.H., El-Hefnawy, A.N., Kafafy, A.: Calculation of the unknown criteria weights under the neutrosophic TOPSIS and the neutrosophic VIKOR MCGDM problems. Int. J. Sci. Eng. Res. **10**, 1125–1132 (2019)
17. Zhoui, F., Chen, T.-Y.: An integrated multicriteria group decision-making approach for green supplier selection under Pythagorean Fuzzy scenarios. IEEE Access **8**, 165216–165231 (2020). http://doi.org/10.1109/ACCESS.2020.3022377
18. Anitha Shanthi, S., Jayapalan, P.: A Vikor method based on bipolar intuitionistic Fuzzy soft set. Adv. Math. Sci. J. **9**, 1511–1519 (2020). http://doi.org/10.37418/amsj.9.4.5
19. Anitha Shanthi, S., Nishadevi, N., Saranya, M.: Bipolar intuitionistic Fuzzy soft expert sets score functions. Mater. Today Proc. **49**, 1880–1884 (2022). http://doi.org/10.1016/j.matpr.2021.08.076

Efficiency of Eco-friendly Construction Materials in Interval Valued Picture Fuzzy Soft Environment

S. Anita Shanthi and M. Gayathri

Keywords Interval valued picture fuzzy soft set · TODIM Method · Normalized Hamming distance

1 Introduction

Every feature of human reasoning is associated with some sort of ambivalence. Zadeh [1] represented this ambivalence in the form of fuzzy sets. Atanassov [2, 3] and [4] put forward IFS as an extension of FS. Seenivasan et al. [5] designed a robust fuzzy ranking approach. Krohling and De Souza [6] proposed fuzzy TODIM. Zhang et al. [7] Krohling et al [8] and Zhao et al. [9] developed IF TODIM method. Ren et al. [10] extended TODIM method under trapezoidal IF environment. Lu [11] dealt with IVIFN TODIM method. Li et al. [12] and Zhao et al. [13] proposed IVIFS TODIM method. Anita Shanthi et al. [14] proposed BIFS TODIM method. Cuong et al. [15, 16] introduced PF sets to deal with ambiguity in an extensive manner. PF TODIM method was developed by Wei et al. [17] and Jiang et al. [18]. Based on these concepts, TODIM method on IVPFSS is developed.

2 Normalized Hamming Distance on IVPFSS

Definition of IVPFSS is given in [19].

Definition 1 Given a IVPFSS $(P_F, A) = \{x, [\underline{\mu}_{P_{F(e)}}(\delta), \overline{\mu}_{P_{F(e)}}(\delta)], [\underline{\eta}_{P_{F(e)}}(\delta), \overline{\eta}_{P_{F(e)}}(\delta)], [\underline{\nu}_{P_{F(e)}}(\delta), \overline{\nu}_{P_{F(e)}}(\delta)]);$ $\delta \in U, e \in A\}$, $\vartheta_{mn} = 1 - |\underline{\mu} - \underline{\eta} - \underline{\nu} + \overline{\mu} - \overline{\eta} - \overline{\nu}|$, where $\underline{\mu} = \underline{\mu}_{P_{F(e)}}(\delta)$, $\overline{\mu} = \overline{\mu}_{P_{F(e)}}(\delta)$, $\underline{\eta} = \underline{\eta}_{P_{F(e)}}(\delta)$, $\overline{\eta} = \overline{\eta}_{P_{F(e)}}(\delta)$, $\underline{\nu} = \underline{\nu}_{P_{F(e)}}(\delta)$, $\overline{\nu} = \overline{\nu}_{P_{F(e)}}(\delta)$.

S. Anita Shanthi (✉) · M. Gayathri
Department of Mathematics, Annamalai University, Chidambaram, Tamil Nadu, India

H.-H. Leung et al. (eds.), *Recent Developments in Algebra and Analysis*, Trends in Mathematics, https://doi.org/10.1007/978-3-031-37538-5_24

The entropy E_n is

$$E_n = \frac{1}{i}\sum_{m=1}^{i} \vartheta_{mn}, n = 1, 2, \ldots, j.$$

Definition 2 Each criteria is assigned a weight defined as

$$w_n = \frac{1 - E_n}{\sum_{n=1}^{j}(1 - E_n)}, n = 1, 2, \ldots, j.$$

The weight vector $W = (w_1, w_2, \ldots, w_j)$ satisfies $w_1 + w_2 + \ldots + w_j = 1$. Relative weight w_{nk} is

$$w_{nk} = \frac{w_n}{w_k}, \text{ where } w_k = max\ w_n.$$

Definition 3 Let $A = [\underline{\mu}_{P_{F(e)}}(\delta), \overline{\mu}_{P_{F(e)}}(\delta)], [\underline{\eta}_{P_{F(e)}}(\delta), \overline{\eta}_{P_{F(e)}}(\delta)], [\underline{\nu}_{P_{F(e)}}(\delta), \overline{\nu}_{P_{F(e)}}(\delta)]$, be an IVPFSS value. Its score function is

$$SR(P_F, A) = |\underline{\mu}_{P_{F(e)}}(\delta) - \overline{\mu}_{P_{F(e)}}(\delta) + \underline{\eta}_{P_{F(e)}}(\delta) - \overline{\eta}_{P_{F(e)}}(\delta) + \underline{\nu}_{P_{F(e)}}(\delta) - \overline{\nu}_{P_{F(e)}}(\delta)|/3$$

Using this score function, the score matrix denoted by $P_F SR$ is constructed.

Definition 4 Universal set $X = \{\delta_1, \delta_2, \ldots, \delta_i\}$,
parameter set $E = \{e_1, e_2, \ldots, e_j\}$ and $(P_F, A), (P_G, B)$ IVPFSS. NHD between (P_F, A) and (P_G, B) is
$PNH_d((P_F, A), (P_G, B)) =$

$$\frac{1}{6mn}\{\sum_{m=1}^{i}\sum_{n=1}^{j}(|\underline{\mu}_{P_{F(e_i)}}(\delta_j) - \underline{\mu}_{P_{G(e_i)}}(\delta_j)|, |\overline{\mu}_{P_{F(e_i)}}(\delta_j) - \overline{\mu}_{P_{G(\delta_i)}}(x_j)|, |\underline{\eta}_{P_{F(e_i)}}(\delta_j) - \underline{\eta}_{P_{G(e_i)}}(\delta_j)|, |\overline{\eta}_{P_{F(e_i)}}(\delta_j) - \overline{\eta}_{P_{G(e_i)}}(\delta_j)|, |\underline{\nu}_{P_{F(e_i)}}(\delta_j) - \underline{\nu}_{P_{G(e_i)}}(\delta_j)|, |\overline{\nu}_{P_{F(e_i)}}(\delta_j) - \overline{\nu}_{P_{G(e_i)}}(\delta_j)|.$$

Theorem 1 *Distance function PNH_d from $IVPFSS(U)$ to the set of positive real numbers is a metric.*

Proof

(i) $PNH_d((P_F, A), (P_G, B)) > 0$.
(ii) $PNH_d((P_F, A), (P_G, B)) = 0$

$$\Leftrightarrow (\underline{\mu}_{P_{F(e_i)}}(\delta_j) - \underline{\mu}_{P_{G(e_i)}}(\delta_j) + \overline{\mu}_{P_{F(e_i)}}(\delta_j) - \overline{\mu}_{P_{G(e_i)}}(\delta_j)) + (\underline{\eta}_{P_{F(e_i)}}(\delta_j) - \underline{\eta}_{P_{G(e_i)}}(\delta_j) + \overline{\eta}_{P_{F(e_i)}}(\delta_j) - \overline{\eta}_{P_{G(e_i)}}(\delta_j)) + (\underline{\nu}_{P_{F(e_i)}}(\delta_j) - \underline{\nu}_{P_{G(e_i)}}(\delta_j) + \overline{\nu}_{P_{F(e_i)}}(\delta_j) - \overline{\nu}_{P_{G(e_i)}}(\delta_j)) = 0$$

$$\Leftrightarrow \underline{\mu}_{P_{F(e_i)}}(\delta_j) = \underline{\mu}_{P_{G(e_i)}}(\delta_j), \overline{\mu}_{P_{F(e_i)}}(\delta_j) = \overline{\mu}_{P_{G(e_i)}}(\delta_j),$$

$$\underline{\eta}_{P_{F(e_i)}}(\delta_j) = \underline{\eta}_{P_{G(e_i)}}(\delta_j), \overline{\eta}_{P_{F(e_i)}}(\delta_j) = \overline{\eta}_{P_{G(e_i)}}(\delta_j),$$
$$\underline{\nu}_{P_{F(e_i)}}(\delta_j) = \underline{\nu}_{P_{G(e_i)}}(\delta_j), \overline{\nu}_{P_{F(e_i)}}(\delta_j) = \overline{\nu}_{P_{G(e_i)}}(\delta_j).$$
$$\Leftrightarrow (P_F, A) = (P_G, B).$$

(iii) $PNH_d((P_F, A), (P_G, B)) = PNH_d((P_G, B), (P_F, A))$, is obvious.

(iv) $(P_F, A), (P_G, B)$ and (P_H, C) are IVPFSS. Then

$$\begin{aligned}
&(\underline{\mu}_{P_{F(e_i)}}(\delta_j) - \underline{\mu}_{P_{G(e_i)}}(\delta_j) + \overline{\mu}_{P_{F(e_i)}}(\delta_j) - \overline{\mu}_{P_{G(e_i)}}(\delta_j))\\
&+(\underline{\eta}_{P_{F(e_i)}}(\delta_j) - \underline{\eta}_{P_{G(e_i)}}(\delta_j) + \overline{\eta}_{P_{F(e_i)}}(\delta_j) - \overline{\eta}_{P_{G(e_i)}}(\delta_j))\\
&+(\underline{\nu}_{P_{F(e_i)}}(\delta_j) - \underline{\nu}_{P_{G(e_i)}}(\delta_j) + \overline{\nu}_{P_{F(e_i)}}(\delta_j) - \overline{\nu}_{P_{G(e_i)}}(\delta_j))\\
&= (\underline{\mu}_{P_{F(e_i)}}(\delta_j) - \underline{\mu}_{P_{H(e_i)}}(\delta_j) + \underline{\mu}_{P_{H(e_i)}}(\delta_j) - \underline{\mu}_{P_{G(e_i)}}(\delta_j))\\
&\quad +(\overline{\mu}_{P_{F(e_i)}}(\delta_j) - \overline{\mu}_{P_{H(e_i)}}(\delta_j) + \overline{\mu}_{P_{H(e_i)}}(\delta_j) - \overline{\mu}_{P_{G(e_i)}}(\delta_j))\\
&\quad +(\underline{\eta}_{P_{F(e_i)}}(\delta_j) - \underline{\eta}_{P_{H(e_i)}}(\delta_j) + \underline{\eta}_{P_{H(e_i)}}(\delta_j) - \underline{\eta}_{P_{G(e_i)}}(\delta_j))\\
&\quad +(\overline{\eta}_{P_{F(e_i)}}(\delta_j) - \overline{\eta}_{P_{H(e_i)}}(\delta_j) + \overline{\eta}_{P_{H(e_i)}}(\delta_j) - \overline{\eta}_{P_{G(e_i)}}(\delta_j))\\
&\quad +(\underline{\nu}_{P_{F(e_i)}}(\delta_j) - \underline{\nu}_{P_{H(e_i)}}(\delta_j) + \underline{\nu}_{P_{H(e_i)}}(\delta_j) - \underline{\nu}_{P_{G(e_i)}}(\delta_j))\\
&\quad +(\overline{\nu}_{P_{F(e_i)}}(\delta_j) - \overline{\nu}_{P_{H(e_i)}}(\delta_j) + \overline{\nu}_{P_{H(e_i)}}(\delta_j) - \overline{\nu}_{P_{G(e_i)}}(\delta_j))\\
&\le (\underline{\mu}_{P_{F(e_i)}}(\delta_j) - \underline{\mu}_{P_{H(e_i)}}(\delta_j)) + (\underline{\mu}_{P_{H(e_i)}}(\delta_j) - \underline{\mu}_{P_{G(e_i)}}(\delta_j))\\
&\quad +(\overline{\mu}_{P_{F(e_i)}}(\delta_j) - \overline{\mu}_{P_{H(e_i)}}(\delta_j)) + (\overline{\mu}_{P_{H(e_i)}}(\delta_j) - \overline{\mu}_{P_{G(e_i)}}(\delta_j))\\
&\quad +(\underline{\eta}_{P_{F(e_i)}}(\delta_j) - \underline{\eta}_{P_{H(e_i)}}(\delta_j)) + (\underline{\eta}_{P_{H(e_i)}}(\delta_j) - \underline{\eta}_{P_{G(e_i)}}(\delta_j))\\
&\quad +(\overline{\eta}_{P_{F(e_i)}}(\delta_j) - \overline{\eta}_{P_{H(e_i)}}(\delta_j)) + (\overline{\eta}_{P_{H(e_i)}}(\delta_j) - \overline{\eta}_{P_{G(e_i)}}(\delta_j))\\
&\quad +(\underline{\nu}_{P_{F(e_i)}}(\delta_j) - \underline{\nu}_{P_{H(e_i)}}(\delta_j)) + (\underline{\nu}_{P_{H(e_i)}}(\delta_j) - \underline{\nu}_{P_{G(e_i)}}(\delta_j))\\
&\quad +(\overline{\nu}_{P_{F(e_i)}}(\delta_j) - \overline{\nu}_{P_{H(e_i)}}(\delta_j)) + (\overline{\nu}_{P_{H(e_i)}}(\delta_j) - \overline{\nu}_{P_{G(e_i)}}(\delta_j))\\
&= PNH_d((P_F, A), (P_H, C)) + PNH_d((P_H, C), (P_G, B)).
\end{aligned}$$

Hence PNH_d is a metric.

3 TODIM Method for IVPFSS

The alternatives $U = \{r_1, r_2, \ldots, r_m\}$ are to be ranked depending on the parameters $E = \{e_1, e_2, \ldots, e_n\}$. Each alternative r_m is described by a $IVPFSS$ over U.

$$\begin{aligned}
r_m = \{&([\underline{\mu}_{m1}, \overline{\mu}_{m1}], [\underline{\eta}_{m1}, \overline{\eta}_{m1}], [\underline{\nu}_{m1}, \overline{\nu}_{m1}]),\\
&([\underline{\mu}_{m2}, \overline{\mu}_{m2}], [\underline{\eta}_{m2}, \overline{\eta}_{m2}], [\underline{\nu}_{m2}, \overline{\nu}_{m2}]), \ldots,\\
&([\underline{\mu}_{mn}, \overline{\mu}_{mn}], [\underline{\eta}_{mn}, \overline{\eta}_{mn}], [\underline{\nu}_{mn}, \overline{\nu}_{mn}])\},
\end{aligned}$$

$m = 1, 2, \ldots, i, n = 1, 2, \ldots, j$. Taking the data as IVPFSS the best alternative is found.

Definition 5 The IVPFS dominance degree (DD) of an alternative r_m over another alternative r_u depending upon each criteria e_n is:

$P_F\phi(r_m, r_u) = 0, \quad \text{if} \quad r_{mn} = r_{un}$

$$\sqrt{\frac{w_{st} PNH_d(r_{mn}, r_{un})}{\sum_{n=1}^{j} w_{jk}}}, \quad if \quad r_{mn} > r_{un}$$

$$\frac{-1}{P_F\theta}\sqrt{\frac{\sum_{n=1}^{j} w_{jk} PNH_d(r_{mn}, r_{un})}{w_{jk}}}, \quad if \quad r_{mn} < r_{un}$$

Here $PNH_d(r_{mn}, r_{un})$ represents the IVPFSS normalized Hamming distance. The loss factor such that $P_F\theta > 0$. If $r_{mn} > r_{un}$, then $P_F\phi(r_m, r_u)$ represents a gain, and if $r_{mn} < r_{un}$, then $P_F\phi(r_m, r_u)$ indicates a loss.
IVPFS dominance matrix (DM) is:

$$P_F\phi_n = (P_F\phi_n(r_m, r_u)) = \begin{pmatrix} r_1 & 0 & P_F\phi_n(r_1, r_2) & \cdots & P_F\phi_n(r_1, r_m) \\ r_2 & P_F\phi_n(r_2, r_1) & 0 & \cdots & P_F\phi_n(r_2, r_m) \\ \vdots & \vdots & \vdots & \cdots & \vdots \\ r_m & P_F\phi_n(r_m, r_1) & P_F\phi_n(r_m, r_2) & \cdots & 0 \end{pmatrix}$$

Definition 6 IVPFSS overall DD is:
$P_F\delta(r_m, r_u) = \sum_{n=1}^{q} P_F\phi_n(r_m, r_u)$.

Definition 7 IVPFSS global value is:

$$P_F\xi(r_m) = \frac{\sum_{m=1}^{i} P_F\delta(r_m, r_u) - \min_m \sum_{m=1}^{i} P_F\delta(r_m, r_u)}{\max_m \sum_{m=1}^{i} P_F\delta(r_m, r_u) - \min_m \sum_{m=1}^{i} P_F\delta(r_m, r_u)}.$$

Flowchart of IVPFSS TODIM Method is given in Fig. 1.

4 Procedure

The procedure for solving MCDM problem using TODIM method is as follows:

Step 1: IVPFSS decision matrix is constructed.
Step 2: IVPFSS entropy measure (EM_n) of IVPFSS based on parameters e_n is computed by Definition 1.
Step 3: IVPFSS weights w_{jk} is determined by Definition 2.
Step 4: Construct the IVPFSS score matrix by Definition 3.
Step 5: IVPFSS DD $P_F\phi(r_m, r_u)$ is evaluated using Definition 5.
Step 6: IVPFSS overall DD $P_F\delta(r_m, r_u)$ is calculated by using Definition 6.
Step 7: IVPFSS global value is determined by Definition 7.

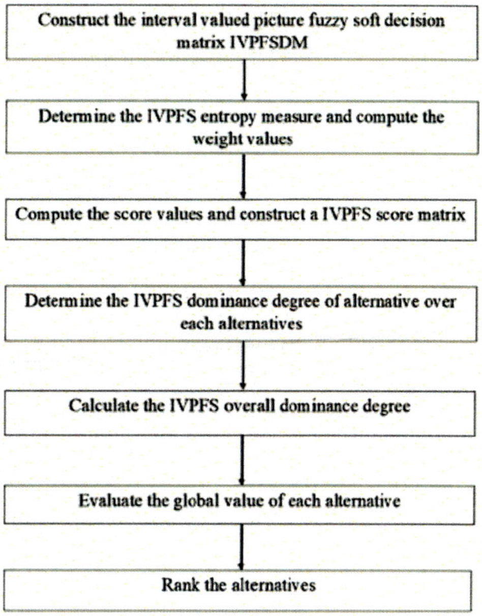

Fig. 1 Flowchart of IVPFSS TODIM method

Step 8: The global values are ranked, and alternative that corresponds to the maximum value is the desirable one.

Example 41 Let $\{\alpha_1, \alpha_2, \alpha_3, \alpha_4, \alpha_5\}$ represent five different types of eco-friendly construction (Fig. 2.) materials: earthen structures; natural fibers and cellulose insulation; slate and stones; Grasscrete, hempcrete, and Ashcrete; and natural clay plaster. The quality of eco-friendly materials is rated based on the parameters $E = \{e_1, e_2, e_3, e_4, e_5, e_6\}$ where e_1 = thermal insulation, e_2 = renewable resources, e_3 = solar orientation, e_4 = indoor air quality, e_5= recyclability resources, and e_6 = stormwater control, using IVPFSS TODIM method.

Step 1. *IVPFS* DM is constructed in Table 1.
Step 2:
The IVPFSS entropy measures EM_n are $EM_1 = 0.392$, $EM_2 = 0.310$, $EM_3 = 0.380$, $EM_4 = 0.114$, $EM_5 = 0.352$, $EM_6 = 0.362$.
Step 3:
Weights W_s are $w_1 = 0.1487$, $w_2 = 0.1687$, $w_3 = 0.1516$, $w_4 = 0.2166$, $w_5 = 0.1584$, $w_6 = 0.1560$.
The relative weights W_{st} are $w_{1t} = 0.6865$, $w_{2t} = 0.7789$, $w_{3t} = 0.6999$, $w_{4t} = 1$, $w_{5t} = 0.7313$, $w_{6t} = 0.7202$.
Step 4:
IVPFSS score matrix is:

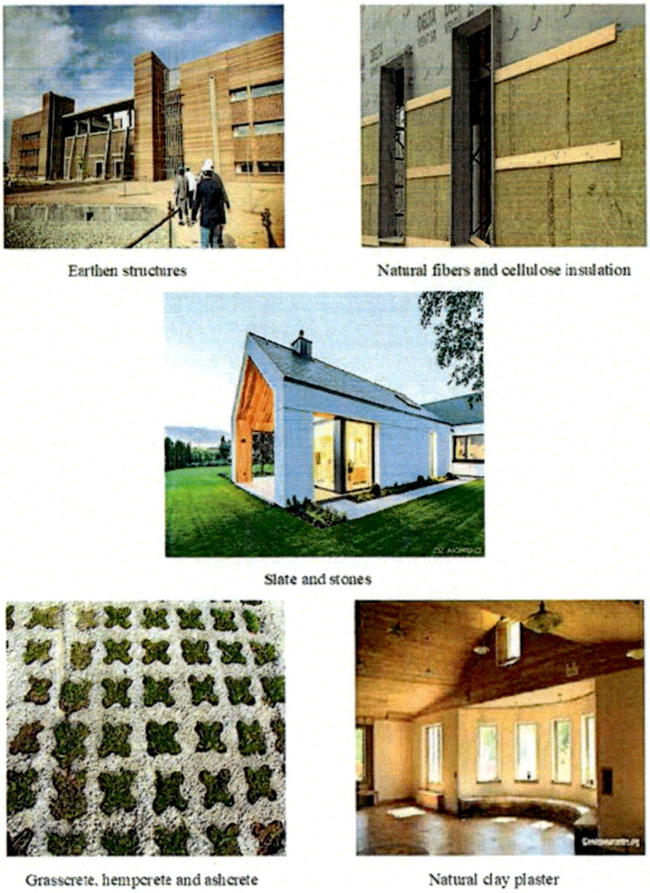

Fig. 2 Constructions of eco-friendly material

$$P_F SR = \begin{pmatrix} 0.173333 & 0.096667 & 0.193333 & 0.153333 & 0.203333 & 0.150000 \\ 0.163333 & 0.100000 & 0.116667 & 0.110000 & 0.116667 & 0.123333 \\ 0.193333 & 0.213333 & 0.180000 & 0.170000 & 0.160000 & 0.183333 \\ 0.190000 & 0.163333 & 0.200000 & 0.126667 & 0.176667 & 0.196667 \\ 0.146667 & 0.250000 & 0.243333 & 0.136667 & 0.130000 & 0.096667 \end{pmatrix}$$

Step 5:

The IVPFSS DD matrices based on the parameter e_n are:

Here $P_F\theta = 1$

Table 1 *IVPFS* decision matrix

U	e_1	e_2
α_1	[00.12,00.20],[00.05,00.37],[00.22,00.34]	[00.09,00.18],[00.25,00.34],[00.30,00.41]
α_2	[00.16,00.24],[00.13,00.32],[00.18,00.40]	[00.16,00.27],[00.40,00.51],[00.06,00.14]
α_3	[00.10,00.31],[00.07,00.23],[00.14,00.35]	[00.07,00.35],[00.20,00.42],[00.03,00.17]
α_4	[00.15,00.25],[00.19,00.30],[00.06,00.42]	[00.11,00.32],[00.24,00.45],[00.15,00.22]
α_5	[00.03,00.17],[00.26,00.38],[00.09,00.27]	[00.04,00.19],[00.12,00.37],[00.08,00.43]
U	e_3	e_4
α_1	[00.14,00.25],[00.08,00.36],[00.21,00.40]	[00.07,00.15],[00.22,00.43],[00.18,00.35]
α_2	[00.19,00.28],[00.11,00.24],[00.30,00.43]	[00.11,00.24],[00.35,00.40],[00.13,00.28]
α_3	[00.07,00.32],[00.16,00.20],[00.10,00.35]	[00.05,00.17],[00.26,00.51],[00.16,00.30]
α_4	[00.03,00.15],[00.27,00.38],[00.04,00.41]	[00.14,00.25],[00.12,00.36],[00.29,00.32]
α_5	[00.12,00.29],[00.05,00.23],[00.17,00.45]	[00.03,00.19],[00.27,00.41],[00.23,00.34]
U	e_5	e_6
α_1	[00.15,00.26],[00.07,00.36],[00.20,00.41]	[00.05,00.13],[00.24,00.45],[00.21,00.37]
α_2	[00.22,00.30],[00.11,00.25],[00.31,00.44]	[00.12,00.26],[00.33,00.40],[00.15,00.31]
α_3	[00.08,00.33],[00.16,00.21],[00.19,00.37]	[00.08,00.29],[00.18,00.36],[00.06,00.22]
α_4	[00.12,00.24],[00.28,00.35],[00.06,00.40]	[00.16,00.42],[00.10,00.27],[00.14,00.30]
α_5	[00.09,00.17],[00.23,00.38],[00.13,00.29]	[00.03,00.11],[00.32,00.43],[00.25,00.35]

$$P_F\phi_1 = \begin{pmatrix} 0 & -0.107614 & 0.024047 & 0.020923 & -0.142040 \\ 0.016002 & 0 & 0.017485 & 0.014079 & -0.161722 \\ -0.119150 & -0.117588 & 0 & -0.129664 & -0.159397 \\ -0.141898 & -0.094682 & 0.019281 & 0 & -0.140708 \\ 0.021121 & 0.024047 & 0.023702 & 0.020923 & 0 \end{pmatrix}$$

Similarly the other values of $P_F\phi$ are calculated.

Step 6: The IVPFSS overall DD of alternative r_m over each other alternative r_u

$$P_F\delta_1 = \begin{pmatrix} 0 & -0.539388 & -0.171355 & -0.137518 & -0.395536 \\ -0.086445 & 0 & 0.121497 & 0.134939 & -0.187564 \\ -0.471487 & -0.737322 & 0 & -0.289293 & -0.502993 \\ -0.564798 & -0.818167 & -0.364881 & 0 & -0.367615 \\ -0.186878 & -0.523318 & -0.168222 & -0.345904 & 0 \end{pmatrix}$$

Similarly, the other values of $P_F\theta$ are calculated.
Step 7: Alternatives are ranked as follows:
From the values of $P_F\xi(r_m)$, the value of alternative α_2 is maximum. So it is chosen as the desirable one.
The global values for different values of $P_F\theta$ are tabulated in Table 2, and the decision graph is shown in Fig. 3.

Table 2 Global values corresponding to $P_F\theta$

U	$P_F\theta = 1$		$P_F\theta = 1.5$		$P_F\theta = 2$	
	$P_F\xi(r_m)$	Ranking	$P_F\xi(r_m)$	Ranking	$P_F\xi(r_m)$	Ranking
α_1	0.415496	3	0.407974	3	0.401304	3
α_2	1	1	1	1	1	1
α_3	0.54515	4	0.044471	4	0.035566	4
α_4	0	5	0	5	0	5
α_5	0.424779	2	0.420031	2	0.415821	2
U	$P_F\theta = 2.5$		$P_F\theta = 3$		$P_F\theta = 3.5$	
	$P_F\xi(r_m)$	Ranking	$P_F\xi(r_m)$	Ranking	$P_F\xi(r_m)$	Ranking
α_1	0.395349	3	0.390003	3	0.385170	3
α_2	1	1	1	1	1	1
α_3	0.027615	4	0.020474	4	0.014024	4
α_4	0	5	0	5	0	5
α_5	0.412062	2	0.408686	2	0.405637	2
U	$P_F\theta = 4$		$P_F\theta = 4.5$		$P_F\theta = 5$	
	$P_F\xi(r_m)$	Ranking	$P_F\xi(r_m)$	Ranking	$P_F\xi(r_m)$	Ranking
α_1	0.380786	3	0.376788	3	0.374413	3
α_2	1	1	1	1	1	1
α_3	0.008170	4	0.002832	4	0	5
α_4	0	5	0	5	0.002049	4
α_5	0.402870	2	0.400343	2	0.399270	2
U	$P_F\theta = 5.5$		$P_F\theta = 6$		$P_F\theta = 6.5$	
	$P_F\xi(r_m)$	Ranking	$P_F\xi(r_m)$	Ranking	$P_F\xi(r_m)$	Ranking
α_1	0.373862	3	0.373359	3	0.372898	3
α_2	1	1	1	1	1	1
α_3	0	5	0	5	0	5
α_4	0.006500	4	0.010569	4	0.014303	4
α_5	0.399840	2	0.400362	2	0.400841	2
U	$P_F\theta = 7$		$P_F\theta = 7.5$		$P_F\theta = 8$	
	$P_F\xi(r_m)$	Ranking	$P_F\xi(r_m)$	Ranking	$P_F\xi(r_m)$	Ranking
α_1	0.372473	3	0.372080	3	0.371716	3
α_2	1	1	1	1	1	1
α_3	0	5	0	5	0	5
α_4	0.017743	4	0.020920	4	0.023865	4
α_5	0.401282	2	0.401690	2	0.402067	2
U	$P_F\theta = 8.5$		$P_F\theta = 9$		$P_F\theta = 9.5$	
	$P_F\xi(r_m)$	Ranking	$P_F\xi(r_m)$	Ranking	$P_F\xi(r_m)$	Ranking
α_1	0.371378	3	0.371063	3	0.370768	3
α_2	1	1	1	1	1	1

(continued)

Table 2 (continued)

U	$P_F\theta = 1$		$P_F\theta = 1.5$		$P_F\theta = 2$	
	$P_F\xi(r_m)$	Ranking	$P_F\xi(r_m)$	Ranking	$P_F\xi(r_m)$	Ranking
α_3	0	5	0	5	0	5
α_4	0.026603	4	0.029153	4	0.031533	4
α_5	0.402419	2	0.402746	2	0.403050	2

U	$P_F\theta = 10$	
	$P_F\xi(r_m)$	Ranking
α_1	0.370493	3
α_2	1	1
α_3	0	5
α_4	0.033764	4
α_5	0.403337	2

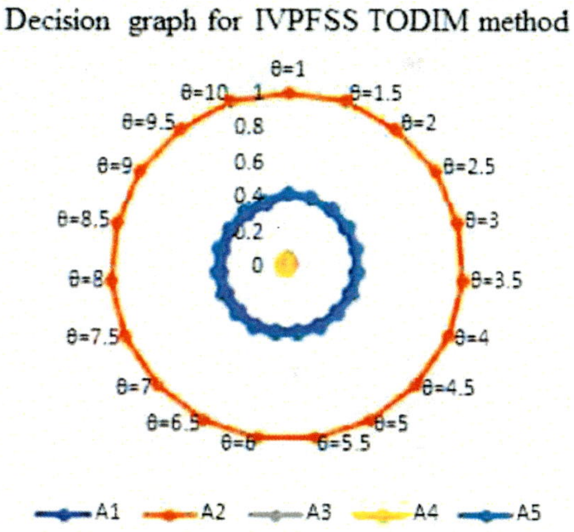

Fig. 3 Decision graph for IVPFSS TODIM method

5 Conclusion

Here TODIM method on IVPFSS is dealt with. Entropy measure on IVPFSS is developed and is used in computing the weights. Normalized Hamming distance is defined, and this is used to find dominance degree matrix. Five types of eco-friendly construction materials are taken, and the quality of these materials is evaluated based on six parameters. It is found that α_2, which corresponds to natural fibers and cellulose insulation, is the best.

Acknowledgments The authors are thankful to the reviewers for their valuable comments, which have greatly helped in improving the quality of the paper.

Conflict of Interest. The authors declare that they have no conflict of interest.

References

1. Zadeh, L.A.: Fuzzy sets. Inf. Control **8**, 338–353 (1965). http://doi.org/10.1016/S0019-9958(65)90421-X
2. Atanassov, K.: Intuitionistic fuzzy sets. Fuzzy Sets Syst. **20**(1), 87–96 (1986). http://doi.org/10.1016/S0165-0114(86)80034-3
3. Atanassov, K.: More on intuitionistic fuzzy sets. Fuzzy Sets Syst. **33**(1), 37–45 (1989). http://doi.org/10.1016/0165-0114(89)90215-7
4. Atanassov, K.: Some operators on intuitionistic fuzzy sets. Notes Intuitionistic Fuzzy Sets **3**(4), 28–33 (1997)
5. Seenivasan, M., Ramesh, R.: Heterogeneous queueing system with cost and profit appraises by robust fuzzy ranking approach. In: 4th International Congress on Life, Social and Health Sciences in a Changing World congress proceedings Book, pp. 215–223 (2022)
6. Krohling, R.A., De Souza, T.T.M.: Combining prospect theory and fuzzy numbers to multi-criteria decision making. Expert Syst. Appl. **39**, 11487–11493 (2012). http://doi.org/10.1016/j.eswa.2012.04.006
7. Zhang, D., Bao, X., Wu, C.: An extended TODIM method based on novel score function and accuracy function under intuitionistic fuzzy environment. Fuzziness Knowledge Based Syst. **27**(6), 905–930 (2019). http://doi.org/10.1142/S0218488519500405
8. Krohling, R.A., Pacheco, A.G.C., Siviero, A.L.T.: IF-TODIM: An intuitionistic fuzzy TODIM to multi-criteria decision making. Knowledge Based Syst. **53**, 142–146 (2013). http://doi.org/10.1016/j.knosys.2013.08.028
9. Zhao, M., Wei, G., Wei, C., Wu, J., Wei, Y.: Extended CPT-TODIM method for interval-valued intuitionistic fuzzy MAGDM and its application to urban ecological risk assessment. J. Intell. Fuzzy Syst. **40**(3), 4091–4106 (2021). http://doi.org/10.3233/JIFS-200534
10. Ren, H., Liu, M., Zhou, H.: Extended TODIM method for MADM problem under trapezoidal intuitionistic fuzzy environment. Int. J. Comput. **14**(2), 1–13 (2019)
11. Lu, J., Wei, C.: TODIM method for performance appraisal on social integration-based rural reconstruction with interval-valued intuitionistic fuzzy information. J. Intell. Fuzzy Syst. **37**(2), 1731–1740 (2019). http://doi.org/10.3233/JIFS-179236
12. Li, Y., Shan, Y., Liu, P.: An extended TODIM method for group decision making with the interval intuitionistic fuzzy sets. Math. Probl. Eng., 1–10 (2015). http://doi.org/10.1155/2015/672140
13. Zhao, M., Wei, G., Wei, C., Wu, J.: Improved TODIM method for intuitionistic fuzzy MAGDM based on cumulative prospect theory and its application on stock investment selection. Int. J. Mach. Learn. Cybern. **12**, 891–901 (2021). http://doi.org/10.1007/s13042-020-01208-1
14. Anita Shanthi, S., Jayapalan, P.: TODIM method based on bipolar intuitionistic fuzzy soft set. J. Phys. **1724**, 1–10 (2021). http://doi.org/10.1088/1742-6596/1724/1/012041
15. Cuong, B.C., Kreinovich, V.: Picture Fuzzy sets - a new concept for computational intelligence problems. In: Proceedings of the Third World Congress on Information and Communication Technologies WIICT, pp. 1–6 (2013). http://doi.org/10.1109/WICT.2013.7113099
16. Cuong, B.C.: Picture Fuzzy sets. J. Comput. Sci. Cybern. **30**(4), 409–420 (2014). http://doi.org/10.15625/1813-9663/30/4/5032
17. Wei, G.: TODIM method for picture fuzzy multiple attribute decision making. Informatica **29**(3), 555–566 (2018). http://doi.org/10.15388/Informatica.2018.181
18. Jiang, Z., Wei, G., Wu, J., Chen, X.: CPT-TODIM method for picture fuzzy multiple attribute group decision making an its application to food enterprise quality credit evaluation. J. Intell. Fuzzy Syst. **40**(5), 10115–10128 (2021). http://doi.org/10.3233/JIFS-202839
19. Anita Shanthi, S., Gayathri, M.: Interval valued picture fuzzy soft set in pattern recognition. Adv. Math. Sci. J. **9**(4), 2019–2029 (2020). http://doi.org/10.37418/amsj.9.4.60

Part II
Analysis

A Subclass of Close-to-Convex Function Involving Srivastava-Tomovski Operator

Elangho Umadevi and Kadhavoor R. Karthikeyan

Keywords Analytic function · Convex function · Starlike function · Close-to-convex function · Mittag-Leffler function · Subordination

MSC Classification 30C45

1 Introduction

Mittag-Leffler function, a special transcendental function has been on the spotlight due to its role in treating problems related to integral and differential equations of fractional order. Refer to Srivastava et al. [1–4] for detailed studies which involved Mittag-Leffler function. Srivastava-Tomovski [5] investigated the properties of the three parameters *Mittag-Leffler function* of the form

$$E_{\rho,\tau}^{\upsilon,\nu}(\xi) = \sum_{n=0}^{\infty} \frac{(\upsilon)_{n\nu}\xi^n}{\Gamma(\rho n + \tau)\, n!}, \quad \xi,\ \rho,\ \tau,\ \upsilon,\ \nu \in \mathbb{C}, Re(\rho) > 0, Re(\nu) > 0, \tag{1}$$

where $\mathbb{C}$ denotes the sets of complex numbers and $(x)_n$ will be used to denote the usual Pochhammer symbol.

Elangho Umadevi and Kadhavoor R. Karthikeyan contributed equally to this work.

E. Umadevi (✉)
Department of Mathematics and Statistics, College of Natural and Health Sciences, Zayed University, Abu Dhabi, UAE
e-mail: z10011@zu.ac.ae

K. R. Karthikeyan
Department of Applied Mathematics and Science, National University of Science & Technology, Muscat, Oman
e-mail: karthikeyan@nu.edu.om

H.-H. Leung et al. (eds.), *Recent Developments in Algebra and Analysis*, Trends in Mathematics, https://doi.org/10.1007/978-3-031-37538-5_25

Let Λ denote the class of functions analytic in $\Theta = \{\xi : |\xi| < 1\}$ having a series expansion

$$\chi(\xi) = \xi + \sum_{n=2}^{\infty} a_n \xi^n. \tag{2}$$

Using *Srivastava-Tomovski generalization of the Mittag-Leffler function* [5], recently Breaz et al. in [6] defined an operator $\mathfrak{J}^m_{\upsilon,\lambda}(\rho, \tau, \upsilon)\chi(\xi) : \Lambda \to \Lambda$ by

$$\mathfrak{J}^m_{\upsilon,\lambda}(\rho, \tau, \upsilon)\chi(\xi) = \xi + \sum_{n=2}^{\infty} [1 - \lambda + \lambda n]^m \frac{\Gamma(\upsilon + n\nu)\Gamma(\rho + \tau)}{\Gamma(\upsilon + \nu)\Gamma(\rho n + \tau)n!} a_n \xi^n. \tag{3}$$

Remark 1 The operator $\mathfrak{J}^m_{\upsilon,\lambda}(\rho, \tau, \upsilon)\chi(\xi)$ was motivated by the studies [7–10].

For $0 \leq \delta < 1$, $\mathcal{S}^*(\delta)$ denote the well-known class of starlike functions of order δ. Gao and Zhou in [11] introduced a class of close-to-convex univalent function denoted by $\mathcal{K}_s$, is the class of all functions $\chi \in \Lambda$ satisfying

$$\operatorname{Re}\left(\frac{\xi^2 \chi'(\xi)}{\kappa(\xi)\kappa(-\xi)}\right) < 0, \quad (\xi \in \Theta) \tag{4}$$

for $\kappa(\xi) \in \mathcal{S}^*(1/2)$. It is well-known that $-\kappa(\xi)\kappa(-\xi)/\xi$ is starlike if $\kappa(\xi) \in \mathcal{S}^*(1/2)$. Hence the class of functions in $\mathcal{K}_s$ is close to convex, which implies that $\chi \in \mathcal{K}_s$ is univalent in Θ. By replacing the right-hand side zero with δ in (4), the class $\mathcal{K}_s$ was extended to $\mathcal{K}_s(\delta)$ by Kowalczyk and Bomba [12].

Recently, Breaz et al. [13] (also see [14]) defined the following function:

$$\Gamma(\mathcal{H}, \mathcal{F}; p; \eta; \Omega) = \frac{[(1 + \mathcal{H})p + \eta(\mathcal{F} - \mathcal{H})]\,\Omega(\xi) + [(1 - \mathcal{H})p - \eta(\mathcal{F} - \mathcal{H})]}{[(\mathcal{F} + 1)\Omega(\xi) + (1 - \mathcal{F})]}, \tag{5}$$

where $-1 \leq \mathcal{F} < \mathcal{H} \leq 1$, $\Omega(\xi) \in \mathcal{P}$ (the class of functions with positive real part) and has an expansion of the form

$$\Omega(\xi) = 1 + R_1 \xi + R_2 \xi^2 + \cdots. \tag{6}$$

Detailed geometrical analysis of $\Gamma(\mathcal{H}, \mathcal{F}; p; \eta; \Omega)$ was discussed by Karthikeyan et al. in [15].

Motivated by Wang and Chen [16] and Karthikeyan et al. [9], we now define the following.

Definition 1 For $\Gamma(\mathcal{H}, \mathcal{F}; p; \eta; \Omega)$ defined as in (5), a function $\chi \in \Lambda$ is said to be in $\mathcal{UCV}^{m,\eta}_{\upsilon,\lambda}(t; \rho, \tau, \upsilon, \mu; \mathcal{H}, \mathcal{F}; \Omega)$ if and only if

$$\frac{t\xi^2\left[\mathfrak{J}^m_{\nu,\lambda}(\rho,\,\tau,\,\upsilon)\chi(\xi)\right]' + \mu t\xi^3\left[\mathfrak{J}^m_{\nu,\lambda}(\rho,\,\tau,\,\upsilon)\chi(\xi)\right]''}{\mathfrak{J}^m_{\nu,\lambda}(\rho,\,\tau,\,\upsilon)\kappa(\xi)\mathfrak{J}^m_{\nu,\lambda}(\rho,\,\tau,\,\upsilon)\kappa(t\xi)} \prec \Gamma(\mathcal{H},\mathcal{F};\,1;\,\eta;\,\Omega), \tag{7}$$

where $\mathfrak{J}^m_{\nu,\lambda}(\rho,\,\tau,\,\upsilon)\kappa(\xi) \in \mathcal{S}^*\left(\frac{1}{2}\right)$, $0 \le \mu \le 1$ and $|t| \le 1,\ t \ne 0$.

Remark 2 For appropriate choice of the parameters, several well-known and new classes can be obtained as special case of $\mathcal{UCV}^{m,\eta}_{\nu,\lambda}(t;\,\rho,\,\tau,\,\upsilon,\,\mu;\,\mathcal{H},\mathcal{F};\,\Omega)$. For example,

1. If $\upsilon = \nu = 1$, $\lambda = 1$, $\rho = \eta = m = 0$, $t = -1$ and $\Omega(\xi) = \frac{1+\xi}{1-\xi}$, then $\mathcal{UCV}^{m,\eta}_{\nu,\lambda}(t;\,\rho,\,\tau,\,\upsilon,\,\mu;\,\mathcal{H},\mathcal{F};\,\Omega)$ reduces to

$$\mathcal{K}_s(\lambda,\mathcal{H},\mathcal{F}) = \left\{\chi \in \Lambda,\ \kappa \in \mathcal{S}^*(1/2);\ \frac{\xi^2\chi'(\xi) + \mu\xi^3\chi''(\xi)}{-\kappa(\xi)\kappa(-\xi)} \prec \frac{1+\mathcal{H}\xi}{1+Yz}\right\},$$

the class $\mathcal{K}_s(\lambda,\mathcal{H},\mathcal{F})$ was introduced by Wang and Chen [16].
2. If $t = -1$, $\upsilon = \nu = 1$, $\lambda = 1$, $\rho = \mu = \eta = m = 0$, $\mathcal{H} = 1$ and $\mathcal{F} = -1$, then $\mathcal{UCV}^{m,\eta}_{\nu,\lambda}(t;\,\rho,\,\tau,\,\upsilon,\,\mu;\,\mathcal{H},\mathcal{F};\,\Omega)$ reduces to

$$\mathcal{K}_s(\Omega) = \left\{\chi \in \Lambda,\ \kappa \in \mathcal{S}^*(1/2);\ \frac{\xi^2\chi'(\xi)}{-\kappa(\xi)\kappa(-\xi)} \prec \Omega(\xi)\right\}.$$

$\mathcal{K}_s(\Omega)$ was investigated by Cho et al. [17].
3. If $\upsilon = \nu = 1$, $\lambda = 1$, $\rho = \mu = m = 0$, $\mathcal{H} = 1$, $\mathcal{F} = -1$ and $\frac{1+\xi}{1-\xi}$, then $\mathcal{UCV}^{m,\eta}_{\nu,\lambda}(t;\,\rho,\,\tau,\,\upsilon,\,\mu;\,\mathcal{H},\mathcal{F};\,\Omega)$ reduces to

$$\mathcal{K}_t(\eta) = \left\{\chi \in \Lambda,\ \kappa \in \mathcal{S}^*(1/2);\ \frac{t\xi^2\chi'(\xi)}{\kappa(\xi)\kappa(t\xi)} \prec \frac{1+(1-2\eta)\xi}{1-\xi}\right\},$$

the class $\mathcal{K}_t(\eta)$ was introduced by Prajapat [18].

2 Main Results

We need the following lemmas to establish our results.

Lemma 1 ([13, Lemma 4]) *Let* $p(\xi) \in \mathcal{P}$ *satisfy the subordination condition*

$$p(\xi) \prec \frac{[(1+\mathcal{H}) + \eta(\mathcal{F}-\mathcal{H})]\,\Omega(\xi) + [(1-\mathcal{H}) - \eta(\mathcal{F}-\mathcal{H})]}{[(\mathcal{F}+1)\Omega(\xi) + (1-\mathcal{F})]},$$

then for $-1 \le \mathcal{F} < \mathcal{H} \le 1$, $0 \le \eta < 1$

$$|p_n| \leq \frac{(\mathcal{H}-\mathcal{F})(1-\eta)R_1}{2}, \quad (n \geq 1). \tag{8}$$

Lemma 2 ([18]) *Let* $\kappa(\xi) = \xi + \sum_{n=2}^{\infty} b_n \xi^n \in \mathcal{S}^*(\frac{1}{2})$, *then* $\frac{\kappa(\xi)\kappa(t\xi)}{t\xi} \in \mathcal{S}^*$.

Using Lemma 2 and proceeding on lines similar to Lemma 2.4 of [9], we get

Lemma 3 ([9]) *Let* $\kappa(\xi) = \xi + \sum_{n=2}^{\infty} b_n \xi^n$ *and* $\mathfrak{J}^m_{\nu,\lambda}(\rho, \tau, \upsilon)\kappa \in \mathcal{S}^*\left(\frac{1}{2}\right)$, *then*

$$G(\xi) = \frac{[\mathfrak{J}^m_{\nu,\lambda}(\rho, \tau, \upsilon)\kappa(\xi)][\mathfrak{J}^m_{\nu,\lambda}(\rho, \tau, \upsilon)\kappa(t\xi)]}{t\xi} = \xi + \sum_{n=2}^{\infty} c_n \xi^n \in \mathcal{S}^*, \tag{9}$$

and $|c_n| \leq n$, *where* $c_n = \left[\varphi_n b_n + t\varphi_{n-1}\varphi_2 b_{n-1} b_2 + \cdots + (t)^{n-1}\varphi_n b_n\right]$ *and*

$$\varphi_n = [1-\lambda+\lambda n]^m \; \frac{\Gamma(\upsilon + n\nu)\Gamma(\rho+\tau)}{\Gamma(\upsilon+\nu)\Gamma(\rho n+\tau)n!}.$$

2.1 Coefficient Estimates

Throughout this chapter, we let

$$\varphi_n = [1-\lambda+\lambda n]^m \; \frac{\Gamma(\upsilon + n\nu)\Gamma(\rho+\tau)}{\Gamma(\upsilon+\nu)\Gamma(\rho n+\tau)n!}$$

and

$$\kappa(\xi) = \frac{\kappa(\xi)\kappa(t\xi)}{t\xi} = \xi + \sum_{n=2}^{\infty} c_n \xi^n,$$

$$\left(c_n = \left[\varphi_n b_n + t\varphi_{n-1}\varphi_2 b_{n-1} b_2 + \cdots + (t)^{n-1}\varphi_n b_n\right]\right).$$

Theorem 1 *Let* $\Omega \in \mathcal{P}$ *be chosen such that*

$$\frac{[(1+\mathcal{H})+\eta(\mathcal{F}-\mathcal{H})]\,\Omega(\xi) + [(1-\mathcal{H})-\eta(\mathcal{F}-\mathcal{H})]}{[(\mathcal{F}+1)\Omega(\xi)+(1-\mathcal{F})]}$$

is convex in Θ. *If* $\chi(\xi) \in \mathcal{UCV}^{m,\eta}_{\nu,\lambda}(t;\ \rho, \tau, \upsilon, \mu;\ \mathcal{H}, \mathcal{F};\ \Omega)$, *then*

$$|\, a_n \,| \leq \frac{1}{n\,[1+\mu(n-1)]\,|\varphi_n|}\left\{|c_n| + \frac{(\mathcal{H}-\mathcal{F})(1-\eta)R_1}{2}\left[1+\sum_{k=2}^{n-1}|c_k|\right]\right\}. \tag{10}$$

Proof Let $\chi(\xi) \in \mathcal{UCV}_{\nu,\lambda}^{m,\eta}(t;\ \rho,\ \tau,\ \upsilon,\ \mu;\ \mathcal{H}, \mathcal{F};\ \Omega)$, then there exists $p(\xi) \in \mathcal{P}$ such that

$$p(\xi)=\frac{\xi\left[\mathfrak{J}_{\nu,\lambda}^{m}(\rho,\ \tau,\ \upsilon)\chi(\xi)\right]'+\mu\xi^2\left[\mathfrak{J}_{\nu,\lambda}^{m}(\rho,\ \tau,\ \upsilon)\chi(\xi)\right]''}{G(\xi)} \prec \Gamma(\mathcal{H}, \mathcal{F};\ 1;\ \eta;\ \Omega), \tag{11}$$

where $\Gamma(\mathcal{H}, \mathcal{F};\ 1;\ \eta;\ \Omega)$ is defined by the equation (5). Clearly $\Gamma(\mathcal{H}, \mathcal{F};\ 1;\ \eta;\ \Omega)$ maps Θ on to a convex domain (assumption). From (11), we have

$$\frac{\xi\left[\mathfrak{J}_{\nu,\lambda}^{m}(\rho,\ \tau,\ \upsilon)\chi(\xi)\right]'+\mu\xi^2\left[\mathfrak{J}_{\nu,\lambda}^{m}(\rho,\ \tau,\ \upsilon)\chi(\xi)\right]''}{G(\xi)} = p(\xi), \quad (p(\xi) \in \mathcal{P})\,. \tag{12}$$

From (12), we get

$$\xi + \sum_{n=2}^{\infty} n\,[1+\mu(n-1)]\,\varphi_n a_n \xi^n = (1+\sum_{n=1}^{\infty} p_n\xi^n)(\xi + \sum_{n=2}^{\infty} c_n\xi^n). \tag{13}$$

Equating the coefficients of ξ^n in (13), we have

$$n\,[1+\mu(n-1)]\,\varphi_n a_n = c_n + p_1c_{n-1} + p_2c_{n-2} + \cdots + p_{n-1}. \tag{14}$$

Using (8) and Lemma 3 , we get

$$\begin{aligned} n\,[1+\mu(n-1)]\,|\varphi_n|\,|a_n| &\le |c_n| + |p_1||c_{n-1}| + |p_2||c_{n-2}| + \cdots + |p_{n-1}| \\ &\le |c_n| + \frac{(\mathcal{H}-\mathcal{F})(1-\eta)R_1}{2}\left[1+\sum_{k=2}^{n-1}|c_k|\right]. \end{aligned} \tag{15}$$

From (15), we get the desired result in (10). □

Corollary 1 *Let $\Omega \in \mathcal{P}$ be chosen in such a manner that*

$$\frac{[(1+\mathcal{H})+\eta(\mathcal{F}-\mathcal{H})]\,\Omega(\xi)+[(1-\mathcal{H})-\eta(\mathcal{F}-\mathcal{H})]}{[(\mathcal{F}+1)\Omega(\xi)+(1-\mathcal{F})]}$$

is convex in Θ. If $\chi(\xi) \in \mathcal{UCV}_{\nu,\lambda}^{m,\eta}(t;\ \rho,\ \tau,\ \upsilon,\ \mu;\ \mathcal{H}, \mathcal{F};\ \Omega)$, then

$$|\,a_n\,| \le \frac{1}{[1+\mu(n-1)]\,|\varphi_n|}\left(1+\frac{(n-1)(\mathcal{H}-\mathcal{F})(1-\eta)R_1}{4}\right).$$

Proof From Lemma 3 $G(\xi) = \frac{[\mathfrak{J}_{\nu,\lambda}^{m}(\rho,\tau,\upsilon)\kappa(\xi)][\mathfrak{J}_{\nu,\lambda}^{m}(\rho,\tau,\upsilon)\kappa(t\xi)]}{t\xi} \in \mathcal{S}^*$, thus $|c_n| \le n$. Now the result follows from (10). □

Setting $\upsilon = \nu = 1$, $\lambda = 1$, $\rho = \mu = m = 0$, $\mathcal{H} = 1$, $\mathcal{F} = -1$ and $\frac{1+\xi}{1-\xi}$ in Corollary 1, we get

Corollary 2 ([18]) *If $\chi(\xi) \in \mathcal{K}_t(\eta)$, then*

$$| a_n | \leq 1 + (n-1)(1-\eta).$$

Remark 3 We note that for an applicable selection of the parameters in Theorem 1, we can get the coefficient inequality obtained by Gao and Zhou [11].

2.2 Fekete-Szegő Problem

In this subsection, we will obtain the Fekete-Szegő inequality for functions belonging to $\mathcal{UCV}^{m,\eta}_{\nu,\lambda}(t;\ \rho,\ \tau,\ \upsilon,\ \mu;\ \mathcal{H}, \mathcal{F};\ \Omega)$

Theorem 2 *If $\chi(\xi) \in \mathcal{UCV}^{m,\eta}_{\nu,\lambda}(t;\ \rho,\ \tau,\ \upsilon,\ \mu;\ \mathcal{H}, \mathcal{F};\ \Omega)$, then for $\vartheta \in \mathbb{C}$, we have*

$$\begin{aligned} | a_3 - \vartheta a_2^2 | \leq & \frac{(\mathcal{H}-\mathcal{F})(1-\eta)|R_1|}{6|\varphi_3|(1+2\mu)} \max\{1, |2L_1 - 1|\} \\ & + \frac{1}{3\,|\varphi_3|\,(1+2\mu)} \max\left\{1, \left|3 - \frac{3\vartheta\varphi_3(1+2\mu)}{(1+\mu)^2\varphi_2^2}\right|\right\} \\ & + \frac{(\mathcal{H}-\mathcal{F})(1-\eta)|R_1|}{4}\left|\frac{3}{4\varphi_3(1+2\mu)} - \frac{\vartheta}{(1+\mu)^2\varphi_2^2}\right|, \end{aligned} \tag{16}$$

where

$$L_1 = \frac{(\mathcal{F}+1)R_1 + 2\left(1 - \frac{R_2}{R_1}\right)}{4} + \frac{3\vartheta(\mathcal{H}-\mathcal{F})(1-\eta)\varphi_3(1+2\mu)R_1}{16(1+\mu)^2\varphi_2^2}.$$

Proof For $p(\xi) \in \mathcal{P}$, we can consider

$$p(\xi) = \frac{1 + w(\xi)}{1 - w(\xi)},$$

where $w(\xi)$ is the Schwartz function. On simple computation, we have

$$\begin{aligned} w(\xi) = \frac{p(\xi) - 1}{p(\xi) + 1} &= \frac{p_1\xi + p_2\xi^2 + p_3\xi^3 + \cdots}{2 + p_1\xi + p_2\xi^2 + p_3\xi^3 + \cdots} \\ &= \frac{1}{2}p_1\xi + \frac{1}{2}\left(p_2 - \frac{1}{2}p_1^2\right)\xi^2 + \frac{1}{2}\left(p_3 - p_1p_2 + \frac{1}{4}p_1^3\right)\xi^3 + \cdots. \end{aligned} \tag{17}$$

Using (17) in (6), we have

$$\Omega(w(\xi)) = 1 + \frac{R_1 p_1}{2}\xi + \frac{R_1}{2}\left[p_2 - \frac{1}{2}\left(1 - \frac{R_2}{R_1}\right)p_1^2\right]\xi^2 + \cdots .$$

As $\chi(\xi) \in \mathcal{UCV}_{\nu,\lambda}^{m,\eta}(t;\ \rho,\ \tau,\ \upsilon,\ \mu;\ \mathcal{H}, \mathcal{F};\ \Omega)$, by (7), we have

$$\frac{\xi\left[\mathfrak{J}_{\nu,\lambda}^{m}(\rho,\ \tau,\ \upsilon)\chi(\xi)\right]' + \mu\xi^2\left[\mathfrak{J}_{\nu,\lambda}^{m}(\rho,\ \tau,\ \upsilon)\chi(\xi)\right]''}{G(\xi)}$$
$$= \frac{[(1+\mathcal{H}) + \eta(\mathcal{F}-\mathcal{H})]\,\Omega[w(\xi)] + [(1-\mathcal{H}) - \eta(\mathcal{F}-\mathcal{H})]}{[(\mathcal{F}+1)\Omega[w(\xi)] + (1-\mathcal{F})]}. \tag{18}$$

From (18), we obtain

$$1 + [2(1+\mu)\varphi_2 a_2 - c_2]\xi + \left[\varphi_3 3(1+2\mu)a_3 - 2(1+\mu)\varphi_3 a_2 c_2 - c_3 + c_2^2\right]\xi^2 + \cdots$$
$$== 1 + \frac{R_1 p_1(\mathcal{H}-\mathcal{F})(1-\eta)}{4}\xi +$$
$$\frac{(\mathcal{H}-\mathcal{F})(1-\eta)R_1}{4}\left[p_2 - p_1^2\left(\frac{(\mathcal{F}+1)R_1 + 2\left(1 - \frac{R_2}{R_1}\right)}{4}\right)\right]\xi^2 + \cdots .$$

Equating the coefficients at ξ and ξ^2 on both sides of the above equation, we get

$$a_2 = \frac{R_1 p_1(\mathcal{H}-\mathcal{F})(1-\eta) + 4c_2}{8\varphi_2(1+\mu)}$$

and

$$a_3 = \frac{(\mathcal{H}-\mathcal{F})(1-\eta)R_1}{12\varphi_3(1+2\mu)}\left[p_2 - p_1^2\left(\frac{(\mathcal{F}+1)R_1 + 2\left(1 - \frac{R_2}{R_1}\right)}{4}\right)\right]$$
$$+ \frac{p_1(\mathcal{H}-\mathcal{F})(1-\eta)R_1 c_2}{12\varphi_3(1+2\mu)} + \frac{c_3}{3\varphi_3(1+2\mu)}.$$

Therefore, we have

$$|\, a_3 - \vartheta a_2^2 \,| = \left|\frac{(\mathcal{H}-\mathcal{F})(1-\eta)R_1}{12\varphi_3(1+2\mu)}\left[p_2 - p_1^2\left(\frac{(\mathcal{F}+1)R_1 + 2\left(1 - \frac{R_2}{R_1}\right)}{4}\right.\right.\right.$$

$$+\frac{3\vartheta(\mathcal{H}-\mathcal{F})(1-\eta)\varphi_3(1+2\mu)R_1}{16(1+\mu)^2\varphi_2^2}\Bigg)\Bigg]$$

$$+\frac{(\mathcal{H}-\mathcal{F})(1-\eta)R_1p_2c_2}{8}\left[\frac{3}{4\varphi_3(1+2\mu)}-\frac{\vartheta}{(1+\mu)^2\varphi_2^2}\right]$$

$$+\frac{1}{3\varphi_3(1+2\mu)}\left[c_3-c_2^2\left(\frac{3\vartheta\varphi_3(1+2\mu)}{4(1+\mu)^2\varphi_2^2}\right)\right]\Bigg| \tag{19}$$

Using Fekete-Szegő inequalities of classes $\mathcal{P}$ (see [19]) and $\mathcal{S}^*$ (see [20]), we can get (16). □

Corollary 3 *If $\chi(\xi) \in X_t(\mathcal{H}, \mathcal{F})$, then for $\vartheta \in \mathbb{C}$, we have*

$$|a_3-\vartheta a_2^2| \le \frac{\mathcal{H}-\mathcal{F}}{3}\max(1, |2\mathcal{L}_2-1|)+\frac{1}{3}\max(1, |3-4\mathcal{L}_3|)$$

$$+2(\mathcal{H}-\mathcal{F})\left|\frac{1}{3}-\frac{\vartheta}{2}\right|,$$

where $\mathcal{L}_2=\frac{1+\mathcal{F}}{2}+\frac{3(\mathcal{H}-\mathcal{F})\vartheta}{8}$, $\mathcal{L}_3=\frac{3\vartheta}{4}$.

2.3 Inclusion Relation

To establish the inclusion relation, we need the following.

Lemma 4 **([21])** *If $-1 \le \mathcal{F}_2 \le \mathcal{F}_1 < \mathcal{H}_1 \le \mathcal{H}_2 \le 1$, then*

$$\frac{1+\mathcal{H}_1\xi}{1+\mathcal{F}_1\xi} \prec \frac{1+\mathcal{H}_2\xi}{1+\mathcal{F}_2\xi}.$$

Theorem 3 *Let $-1 \le \mathcal{F}_2 \le \mathcal{F}_1 < \mathcal{H}_1 \le \mathcal{H}_2 \le 1$, then*

$$\mathcal{UCV}_{\nu,\lambda}^{m,\eta}(t;\ \rho,\ \tau,\ \upsilon,\ \mu;\ \mathcal{H}_1,\mathcal{F}_1;\ \Omega) \subset \mathcal{UCV}_{\nu,\lambda}^{m,\eta}(t;\ \rho,\ \tau,\ \upsilon,\ \mu;\ \mathcal{H}_2,\mathcal{F}_2;\ \Omega).$$

Proof As $\chi(\xi) \in \mathcal{UCV}_{\nu,\lambda}^{m,\eta}(t;\ \rho,\ \tau,\ \upsilon,\ \mu;\ \mathcal{H}_1,\mathcal{F}_1;\ \Omega)$, therefore

$$p(\xi)=\frac{t\xi^2\left[\mathfrak{J}_{\nu,\lambda}^m(\rho,\ \tau,\ \upsilon)\chi(\xi)\right]'+\mu\xi^3\left[\mathfrak{J}_{\nu,\lambda}^m(\rho,\ \tau,\ \upsilon)\chi(\xi)\right]''}{\mathfrak{J}_{\nu,\lambda}^m(\rho,\ \tau,\ \upsilon)\kappa(\xi)\mathfrak{J}_{\nu,\lambda}^m(\rho,\ \tau,\ \upsilon)\kappa(t\xi)}$$

$$\prec \frac{[(1+\mathcal{H}_1)+\eta(\mathcal{F}_1-\mathcal{H}_1)]\,\Omega(\xi)+[(1-\mathcal{H}_1)-\eta(\mathcal{F}_1-\mathcal{H}_1)]}{[(\mathcal{F}_1+1)\Omega(\xi)+(1-\mathcal{F}_1)]}. \tag{20}$$

Since $-1 \leq \mathcal{F}_2 \leq \mathcal{F}_1 < \mathcal{H}_1 < \mathcal{H}_2 \leq 1,$ by Lemma 4, we have

$$p(\xi) \prec \frac{[(1+\mathcal{H}_1)+\eta(\mathcal{F}_1-\mathcal{H}_1)]\,\Omega(\xi)+[(1-\mathcal{H}_1)-\eta(\mathcal{F}_1-\mathcal{H}_1)]}{[(\mathcal{F}_1+1)\Omega(\xi)+(1-\mathcal{F}_1)]}$$
$$\prec \frac{[(1+\mathcal{H}_2)+\eta(\mathcal{F}_2-\mathcal{H}_2)]\,\Omega(\xi)+[(1-\mathcal{H}_2)-\eta(\mathcal{F}_2-\mathcal{H}_2)]}{[(\mathcal{F}_2+1)\Omega(\xi)+(1-\mathcal{F}_2)]}.$$

This yields that $\chi(\xi) \in \mathcal{UCV}_{\nu,\lambda}^{m,\eta}(t;\ \rho,\ \tau,\ \upsilon,\ \mu;\ \mathcal{H}_2, \mathcal{F}_2;\ \Omega)$, and this proves the inclusion relation. □

3 Conclusion

Using Srivastava-Tomovski generalization of the famous Mittag-Leffler function, here we have introduced an differential operator, which generalizes the well-known and new operators. We have studied a subclass of analytic functions in dual with Mittag-Leffler function so that it generalizes or unifies the study of various subclasses of analytic functions. Coefficient inequalities, Fekete-Szegő inequalities, and inclusion relations are the main results of this study.

Conflicts of Interest Both the authors declare that they have no conflict of interest.

Funding This research study received no external funding.

Institutional Review Board Statement Not applicable.

Data Availability No data was used to support this study.

References

1. Srivastava, H.M.: Some parametric and argument variations of the operators of fractional calculus and related special functions and integral transformations. J. Nonlinear Convex Anal. **22**(8), 1501–1520 (2021)
2. Srivastava, H.M., Fernandez, A., Baleanu, D.: Some new fractional-calculus connections between Mittag-Leffler functions. Mathematics **7**(6), (2019). https://doi.org/10.3390/math7060485
3. Srivastava, H.M., Bansal, M., Harjule, P.: A study of fractional integral operators involving a certain generalized multi-index Mittag-Leffler function. Math. Methods Appl. Sci. **41**(16), 6108–6121 (2018). https://doi.org/10.1002/mma.5122
4. Srivastava, H.M.: On an extension of the Mittag-Leffler function. Yokohama Math. J. **16**, 77–88 (1968)
5. Srivastava, H.M., Tomovski, Z.: Fractional calculus with an integral operator containing a generalized Mittag-Leffler function in the kernel. Appl. Math. Comput. 211(1), 198–210 (2009). https://doi.org/10.1016/j.amc.2009.01.055

6. Breaz, D., Karthikeyan, K.R., Umadevi, E., Senguttuvan, A.: Some properties of Bazilevič functions involving srivastava-tomovski operator. Axioms **11**(12), (2022). https://doi.org/10.3390/axioms11120687
7. Cang, Y.-L., Liu, J.-L.: A family of multivalent analytic functions associated with Srivastava-Tomovski generalization of the Mittag-Leffler function. Filomat **32**(13), 4619–4625 (2018). https://doi.org/10.2298/fil1813619c
8. Elhaddad, S., Aldweby, H., Darus, M.: On certain subclasses of analytic functions involving differential operator. Jnānābha **48**, 55–64 (2018)
9. Karthikeyan, K.R., Reddy, K.A., Murugusundaramoorthy, G.: On classes of Janowski functions associated with a conic domain. Ital. J. Pure Appl. Math. **47**, 684–698 (2022)
10. Reddy, K.A., Karthikeyan, K.R., Murugusundaramoorthy, G.: Inequalities for the Taylor coefficients of spiralike functions involving q-differential operator. Eur. J. Pure Appl. Math. **12**(3), 846–856 (2019). https://doi.org/10.29020/nybg.ejpam.v12i3.3429
11. Gao, C., Zhou, S.: On a class of analytic functions related to the starlike functions. Kyungpook Math. J. **45**(1), 123–130 (2005)
12. Kowalczyk, J., Leś-Bomba, E.: On a subclass of close-to-convex functions. Appl. Math. Lett. **23**(10), 1147–1151 (2010). https://doi.org/10.1016/j.aml.2010.03.004
13. Breaz, D., Karthikeyan, K.R., Senguttuvan, A.: Multivalent prestarlike functions with respect to symmetric points. Symmetry **14**(1), 20 (2022). https://doi.org/10.3390/sym14010020
14. Breaz, D., Cotîrlă, L.-I., Umadevi, E., Karthikeyan, K.R.: Properties of meromorphic spiral-like functions associated with symmetric functions. J. Funct. Spaces, **2022**, Article ID: 3444854, 1–10 (2022). https://doi.org/10.1155/2022/3444854
15. Karthikeyan, K.R., Lakshmi, S., Varadharajan, S., Mohankumar, D., Umadevi, E.: Starlike functions of complex order with respect to symmetric points defined using higher order derivatives. Fractal Fraction. **6**(2), 116 (2022). https://doi.org/10.3390/fractalfract6020116
16. Wang, Z.-G., Chen, D.-Z.: On a subclass of close-to-convex functions. Hacet. J. Math. Stat. **38**(2), 95–101 (2009)
17. Cho, N.E., Kwon, O.S., Ravichandran, V.: Coefficient, distortion and growth inequalities for certain close-to-convex functions. J. Inequal. Appl., **2011**, 100 (2011). https://doi.org/10.1186/1029-242X-2011-100
18. Prajapat, J.K.: A new subclass of close-to-convex functions. Surv. Math. Appl. **11**, 11–19 (2016)
19. Keogh, F.R., Merkes, E.P.: A coefficient inequality for certain classes of analytic functions. Proc. Am. Math. Soc. **20**, 8–12 (1969). https://doi.org/10.2307/2035949
20. Koepf, W.: On the Fekete-Szego problem for close-to-convex functions. Proc. Am. Math. Soc. **101**(1), 89–95 (1987). https://doi.org/10.2307/2046556
21. Soni, A., Kant, S.: A new subclass of close-to-convex functions with Fekete-Szego problem. J. Rajasthan Acad. Phys. Sci. **12**(2), 125–138 (2013)

On the Newly Generalized Absolute Summability of an Orthogonal Series with Respect to Hausdorff Matrix

Kalaivani Kamalakkannan and Madhu Venkataraman

Keywords Hausdorff methods · Cesaro method · Regular matrix

1 Introduction

Initially, Moricz and Leindler generalized the concept of absolute summability to generalized summability of an orthogonal series, and they have obtained a condition for the convergence of an series in summability method. Hausdorff matrix includes Cesaro, Riesz, Euler, and many other matrices. In this chapter, we generalize Leindler and Moricz results to Hausdorff matrix.

Definition 1 Let $\{\phi_n\}_{n=1}^{\infty}$ be an orthonormal system in $L_2[0, 1]$. The Hausdorff mean of the series

$$\sum_{k=0}^{\infty} a_k \phi_k \tag{1}$$

is defined by $\sigma_n = \sum\limits_{k=0}^{\infty} b_{n\ k} S_k$, where S_k is k^{th} partial sum of (1) and $(b_{n\ k})_{n,k \in \mathbb{Z}^+}$ is an infinite matrix.

Definition 2 Let γ be a non-negative and non-decreasing function on $[1, \infty)$ and $k \geq 1$. The series (1) is $|A, \gamma|_k$ summable, if

$$\sum_{n=1}^{\infty} \gamma(n)^k n^{k-1} |\sigma_n - \sigma_{n-1}|^k converges. \tag{2}$$

K. Kamalakkannan (✉) · M. Venkataraman
Department of Mathematics, School of Advanced Sciences, Vellore Institute of Technology, Vellore, Tamil Nadu, India
e-mail: kalaivani.kannan@vit.ac.in; madhu.v@vit.ac.in

H.-H. Leung et al. (eds.), *Recent Developments in Algebra and Analysis*, Trends in Mathematics, https://doi.org/10.1007/978-3-031-37538-5_26

Hausdorff Matrix Let Φ be a bounded variation function in $BV[0,1]$. For $n, k \in \mathbb{Z}^+$, we define

$$a_{m\,n} := \begin{cases} \int_0^1 \binom{m}{n} t^n (1-t)^{m-n} d\Phi(t) \text{ if } 0 \le n \le m, \\ 0 \qquad\qquad n > m. \end{cases} \tag{3}$$

Now, we restrict $\Phi(t) = \int_0^1 h(u)\phi_{[0,t]}(u)du$, $h \in L_p[0,1]$ $(1 < p \le 2)$, where ϕ is the characteristic function.

Definition 3 Let $\gamma := \{\gamma(n)\}_{n=1}^{\infty}$ be a set of positive numbers. For $\beta \in \mathbb{R}$, γ is quasi β-power monotone decreasing if $\exists$ constant $M \ge 1$ such that

$$\left(\frac{n}{m}\right)^{\beta} \gamma(n) \le M\gamma(m) \text{ for } m \le n. \tag{4}$$

We denote $\Gamma_\beta = \{\gamma \mid \gamma$ be a non-decreasing on $[1,\infty)$ with the sequence $\{\gamma(n)\}_{n=1}^{\infty}$ is quasi β-power monotone decreasing $\}$.

In 1971, Bolgov [1] obtained for any $\{n_m\}_{m=0}^{\infty}$ be a positive sequence such that $1 \le \gamma \le \frac{n_{m+1}}{n_m} \le \gamma_1$ and $\underline{c} \in \ell_2(\mathbb{N})$ the condition

$$\sum_{m=0}^{\infty} \left\{ \sum_{n=n_m+1}^{n_{m+1}} {n_j}^{1-\frac{2}{q}} |c_n|^2 \right\}^{\frac{k}{2}} < \infty, \tag{5}$$

is sufficient for the series $\sum_{n=1}^{\infty} c_n \psi_n$ absolute Hausdorff summable a.e.

Kantawala [3] gave a sufficient condition $\sum_{m=1}^{\infty} \gamma(2^m)^k 2^{mk} \left\{ \sum_{n=2^m+1}^{2^{m+1}} |c_n|^2 \right\}^{\frac{k}{2}}$ is finite for the series (1) is generalized Euler summable a.e. to the index k.

In 2013, Kalaivani and Youvaraj [2] generalized the result of Kantawala for $\gamma \in \Gamma_\beta$ with $\beta > -\frac{3}{4}$ and gave the necessary condition for the series (1) is generalized absolute Hausdorff summable a.e.

In this chapter, under certain conditions on γ and $1 \le k \le 2$,

1. We extend the result of Bolgov [1] for generalized absolute Hausdorff summability methods.
2. We give sufficient condition in terms of $\{n_m\}_{m=0}^{\infty}$ satisfying certain conditions so that $n_m = 2^m$ is the particular case of theorems obtained in [2, 3].

2 Basic Definition

Definition 4 A set $\gamma := \{\gamma(n)\}_{n=1}^{\infty}$ satisfying the condition, $\exists\ \mu \in \mathbb{N}$ and $K = K(\gamma) \geq 1$ such that

$$2\gamma(n+\mu) \leq \gamma(n) \text{ and } \gamma(n+1) \leq K\gamma(n)\ \ \forall\, n \in \mathbb{N}. \text{ Then } \gamma \text{ is is quasi-}$$

geometrically decreasing sequence.

Theorem 1 *A sequence γ of positve numbers is quasi geometrically decreasing iff*

$$\sum_{n=m}^{\infty} \gamma(n) \leq M\gamma(m) \textit{ for some } M \geq 1,\ \forall\, m \in \mathbb{N}.$$

We refer to [4], for the proof of the theorem.

3 Main Results

Theorem 2 *Let $\{\psi_n\}_{n=0}^{\infty}$ be an orthonormal system in $L_2[0,1]$, $A = (a_{n\,m})$ be a infinite Hausdorff matrix, $k \in [1,2]$ and $\{n^{\frac{k}{2}}\gamma(n)^k\}$ is quasi geometrically decreasing. Then for any $\underline{c} \in \ell_2(\mathbb{Z}^+)$ the condition*

$$\sum_{m=0}^{\infty} \gamma(n_{m+1})^k \left\{ {n_m}^{1-\frac{2}{q}} \sum_{n=n_m+1}^{n_{m+1}} |c_n|^2 \right\}^{\frac{k}{2}} < \infty \tag{6}$$

is sufficient for the orthogonal series $\sum\limits_{n=0}^{\infty} c_n\psi_n$ to be generalized absolute A-summable summable a.e. where $\{n_m\}_{m=0}^{\infty}$ be a positive sequence such that

$$1 \leq \frac{n_{m+1}}{n_m} \leq \gamma_1 \ \ (m = 0, 1, \ldots). \tag{7}$$

Then (6) is necessary if $\{c_n\}_{n=0}^{\infty}$ is absolutely monotonically decreasing.

Proof Let

$$\sigma = \sum_{n=1}^{\infty} \int_a^b \gamma(n)^k n^{k-1} |\sigma_n(x) - \sigma_{n-1}(x)|^k dx$$

$$= \sum_{n=1}^{\infty} \int_a^b \gamma(n)^k n^{k-1} \left| \sum_{u=1}^{n} \frac{u}{n} \left(\int_0^1 a_{n\ u} h(t) dt \right) c_u \phi_u(x) \right|^k dx$$

Choose n_0 as in assumption, we have

$$\sigma \leq \frac{1}{2} \sum_{n=1}^{\infty} \int_a^b \gamma(n)^k n^{k-1} \left| \sum_{u=1}^{n_0} \frac{u}{n} \left(\int_0^1 a_{n\ u} h(t) dt \right) c_u \phi_u(x) \right|^k dx$$

$$+ \frac{1}{2} \sum_{n=n_0+1}^{\infty} \int_a^b \gamma(n)^k n^{k-1} \left| \sum_{u=n_0+1}^{n} \frac{u}{n} \left(\int_0^1 a_{n\ u} h(t) dt \right) c_u \phi_u(x) \right|^k dx$$

$$\leq \sigma_1 + \sigma_2.$$

Let $\mu_k = \max\{1, 2^{k-1}\}$. Then

$$\sigma_1 \leq \frac{\mu_k^{n_0-1}}{2} \sum_{n=1}^{\infty} \gamma(n)^k n^{k-1} \sum_{u=1}^{n_0} \frac{u^k}{n^k} \left(\int_0^1 a_{n\ u} h(t) dt \right)^k |c_u|^k \int_a^b |\phi_u(x)|^k dx$$

$$\leq (b-a)^{1-\frac{k}{2}} \frac{\mu_k^{n_0-1}}{2} \sum_{n=1}^{\infty} \gamma(n)^k n^{k-1} \sum_{u=1}^{n_0} \frac{u^k}{n^k} \left(\int_0^1 a_{n\ u} h(t) dt \right)^k |c_u|^k$$

$$\leq (b-a)^{1-\frac{k}{2}} \frac{\mu_k^{n_0-1}}{2} \sum_{u=1}^{n_0} \sum_{n=u}^{\infty} \gamma(n)^k n^{k-1} \frac{u^k}{n^k} \|h\|_p^k |c_u|^k$$

$$\leq n_0 \|h\|_p^k (b-a)^{1-\frac{k}{2}} \frac{\mu_k^{n_0-1}}{2} \max_{1 \leq u \leq n_0} |c_u|^k \sum_{n=1}^{\infty} \gamma(n)^k n^{k-1}$$

$$\leq n_0 \|h\|_p^k (b-a)^{1-\frac{k}{2}} \frac{\mu_k^{n_0-1}}{2} \max_{1 \leq u \leq n_0} |c_u|^k \gamma(1)^k$$

as $\{\gamma(n)^k n^{\frac{k}{2}}\}_{k=1}^{\infty}$ is quasi geometrically decreasing.

$$\sigma_2 \leq \frac{1}{2} \sum_{n=n_0+1}^{\infty} \int_a^b \gamma(n)^k n^{k-1} \left| \sum_{u=n_0+1}^{n} \frac{u}{n} \left(\int_0^1 a_{n\ u} h(t) dt \right) c_u \phi_u(x) \right|^k dx$$

$$\leq \frac{(b-a)^{1-\frac{k}{2}}}{2}\sum_{n=n_0+1}^{\infty}\gamma(n)^k n^{k-1}\left\{\int_a^b\left|\sum_{u=n_0+1}^{n}\frac{u}{n}\left(\int_0^1 a_{n\ u}h(t)dt\right)c_u\phi_u(x)\right|^2 dx\right\}^{\frac{k}{2}}$$

$$\leq \frac{(b-a)^{1-\frac{k}{2}}}{2}\sum_{n=n_0+1}^{\infty}\gamma(n)^k n^{k-1}\left\{\sum_{u=n_0+1}^{n}\frac{u^2}{n^2}\left(\int_0^1 a_{n\ u}h(t)dt\right)^2|c_u|^2\right\}^{\frac{k}{2}}$$

Let $K_{n\ u}=\frac{u^2}{n^2}\left(\int_0^1 a_{n\ u}h(t)dt\right)^2$. Then

$$\sigma_2 \leq \frac{(b-a)^{1-\frac{k}{2}}}{2}\sum_{n=n_0+1}^{\infty}\gamma(n)^k n^{k-1}\left\{\sum_{u=n_0+1}^{n}K_{n\ u}|c_u|^2\right\}^{\frac{k}{2}}$$

$$\leq \frac{(b-a)^{1-\frac{k}{2}}}{2}\sum_{m=0}^{\infty}\sum_{n=n_m+1}^{n_{m+1}}\gamma(n)^k n^{k-1}\left\{\sum_{j=0}^{m}\sum_{u=n_j+1}^{\min\{n_{j+1},n\}}K_{n\ u}|c_u|^2\right\}^{\frac{k}{2}}$$

$$\leq \frac{(b-a)^{1-\frac{k}{2}}}{2}\sum_{m=0}^{\infty}\left\{\sum_{n=n_m+1}^{n_{m+1}}\sum_{j=0}^{m}\sum_{u=n_j+1}^{\min\{n_{j+1},n\}}K_{n\ u}\gamma(n)^2 n^{\frac{2k-2}{k}}|c_u|^2\right\}^{\frac{k}{2}}(n_{m+1}-n_m)^{1-\frac{k}{2}} \tag{8}$$

$$\leq \frac{(b-a)^{1-\frac{k}{2}}}{2}\sum_{m=0}^{\infty}\left\{\sum_{j=0}^{m}\sum_{u=n_j+1}^{n_{j+1}}\sum_{n=\max\{n_m+1,u\}}^{n_{m+1}}K_{n\ u}\gamma(n)^2 n^{\frac{2k-2}{k}}|c_u|^2\right\}^{\frac{k}{2}}(n_{m+1}-n_m)^{1-\frac{k}{2}}$$

$$\leq \frac{(b-a)^{1-\frac{k}{2}}}{2}\sum_{m=0}^{\infty}\left\{\sum_{j=0}^{m}\sum_{u=n_j+1}^{n_{j+1}}\sum_{n=\max\{n_m+1,u\}}^{n_{m+1}}K_{n\ u}\gamma(n)^2 n_m^{\frac{2}{k}-1}n^{\frac{2k-2}{k}}|c_u|^2\right\}^{\frac{k}{2}}. \tag{9}$$

Then

$$\sum_{n=\max\{n_m+1,u\}}^{n_{m+1}}K_{n\ u}\gamma(n)^2 n^{\frac{2k-2}{k}} \leq \sum_{n=\max\{n_m+1,u\}}^{n_{m+1}}\gamma(n)^2 n^{\frac{2k-2}{k}}\frac{u}{n^{1+\frac{2}{q}}}\|h\|_p^{2-p}\frac{u}{n}\int_0^1 a_{n\ u}(t)|h(t)|^p dt$$

$$\leq \frac{n_{j+1}}{n_m^{\frac{2}{q}+1}}\|h\|_p^{2-p}\int_0^1|h(t)|^p t^u\sum_{n=\max\{n_m+1,u\}}^{n_{m+1}}\gamma(n)^2 n^{\frac{2k-2}{k}}\binom{n-1}{u-1}(1-t)^{n-u}dt$$

$$\leq \frac{n_{j+1}}{n_m^{\frac{2}{q}+1}} \|h\|_p^{2-p} \gamma(n_{m+1})^2 n_{m+1}^{\frac{2k-2}{k}} \int_0^1 |h(t)|^p t^u \sum_{n=u}^{n_{m+1}} \binom{n-1}{u-1} (1-t)^{n-u} dt$$

$$\leq \frac{n_{j+1}}{n_m^{\frac{2}{q}+1}} \|h\|_p^{2-p} \gamma(n_{m+1})^2 n_{m+1}^{\frac{2k-2}{k}} \int_0^1 |h(t)|^p dt$$

$$n_m^{\frac{2}{k}-1} \sum_{n=\max\{n_m+1,u\}}^{n_{m+1}} K_{n\ u} \gamma(n)^2 n^{\frac{2k-2}{k}} \leq \frac{n_{j+1}}{n_m^{\frac{2}{q}}} \|h\|_p^2 \gamma(n_{m+1})^2. \tag{10}$$

Using (10) in (8), we obtain

$$\sigma_2 \leq \frac{(b-a)^{1-\frac{k}{2}}}{2} \sum_{m=0}^{\infty} \left\{ \sum_{j=0}^{m} \sum_{u=n_j+1}^{n_{j+1}} \frac{n_{j+1}}{n_m^{\frac{2}{q}}} |c_u|^2 \|h\|_p^2 \gamma(n_{m+1})^2 \right\}^{\frac{k}{2}}.$$

For $1 \leq k \leq 2$, we have

$$\sigma_2 \leq \frac{(b-a)^{1-\frac{k}{2}}}{2} \sum_{m=0}^{\infty} \sum_{j=0}^{m} \left\{ \sum_{u=n_j+1}^{n_{j+1}} \frac{n_{j+1}}{n_m^{\frac{2}{q}}} |c_u|^2 \|h\|_p^2 \gamma(n_{m+1})^2 \right\}^{\frac{k}{2}}$$

$$\leq \frac{(b-a)^{1-\frac{k}{2}}}{2} \sum_{j=0}^{\infty} \sum_{m=j}^{\infty} \frac{n_{j+1}^{\frac{k}{2}}}{n_m^{\frac{k}{q}}} \|h\|_p^k \gamma(n_{m+1})^k \left\{ \sum_{u=n_j+1}^{n_{j+1}} |c_u|^2 \right\}^{\frac{k}{2}}$$

$$\leq \frac{(b-a)^{1-\frac{k}{2}} \|h\|_p^k}{2} \sum_{j=0}^{\infty} \left\{ \sum_{u=n_j+1}^{n_{j+1}} |c_u|^2 \right\}^{\frac{k}{2}} n_{j+1}^{\frac{k}{2}} \sum_{m=j}^{\infty} \frac{\gamma(n_{m+1})^k}{n_m^{\frac{k}{q}}}$$

$$\leq \frac{(b-a)^{1-\frac{k}{2}} \|h\|_p^k}{2} \sum_{j=0}^{\infty} \left\{ \sum_{u=n_j+1}^{n_{j+1}} |c_u|^2 \right\}^{\frac{k}{2}} \sum_{m=j}^{\infty} \frac{\gamma(n_{m+1})^k}{n_m^{\frac{k}{q}}} n_{m+1}^{\frac{k}{2}}.$$

Then

$$\sigma_2 \leq \frac{(b-a)^{1-\frac{k}{2}} \|h\|_p^k}{2} \sum_{j=0}^{\infty} \left\{ \sum_{u=n_j+1}^{n_{j+1}} |c_u|^2 \right\}^{\frac{k}{2}} n_j^{\frac{k}{q}} \sum_{m=j}^{\infty} \gamma(n_{m+1})^k n_{m+1}^{\frac{k}{2}}$$

$$\leq \frac{(b-a)^{1-\frac{k}{2}} \|h\|_p^k}{2} \sum_{j=0}^{\infty} \left\{ \sum_{u=n_j+1}^{n_{j+1}} |c_u|^2 \right\}^{\frac{k}{2}} \gamma(n_{j+1})^k n_j^{-\frac{k}{q}} n_{j+1}^{\frac{k}{2}}$$

$$\leq \frac{(b-a)^{1-\frac{k}{2}} C\|h\|_p^k}{2} \sum_{j=0}^{\infty} \gamma(n_{j+1})^k \left\{ {n_j}^{1-\frac{2}{q}} \sum_{u=n_j+1}^{n_{j+1}} |c_u|^2 \right\}^{\frac{k}{2}}$$

As we used the property that $\{n^{\frac{k}{2}}\gamma(n)^k\}_{n=1}^{\infty}$ is quasi geometrically decreasing sequence to obtain the above inequality.
For necessity part, the proof obtained by arguments as in Theorem 1 in [2] for the sequence $\{n_m\}_{m=0}^{\infty} = \{2^m\}_{m=0}^{\infty}$.

4 Generalized Hausdorff Matrix

Let $\Phi \in BV[0,1]$. For $\alpha > -1$ and $n, k \in \mathbb{Z}^+$, we define

$$b_{n\,k}^{\alpha} := \begin{cases} \int_0^1 \binom{n+\alpha}{n-k} r^{k+\alpha}(1-r)^{n-k} d\Phi(r) & \text{if } 0 \leq k \leq n, \\ 0 & \text{otherwise.} \end{cases} \tag{11}$$

For $\alpha = 0$, then Hausdorff matrix is the special case of Generalized Hausdorff matrix. In particular $\alpha \in N$, using the techniques in Theorem (1), we obtain the necessary and sufficient condition for summability of a orthogonal series for generalized Hausdorff matrix.

Future Endeavors
Besides $\alpha \in \mathbb{N}$, the natural question arises for $\alpha \in \mathbb{R}$. Also, we could refer the function to obtain the sufficient condition for general sequence $\{n_m\}_{m=0}^{\infty}$ instead of $\{2^m\}_{m=0}^{\infty}$.

Another possible direction for future work is to relax the requirement of the sequence $\{c_n\}_{n=0}^{\infty}$, which is absolutely monotonically decreasing.

References

1. Bolgov, V.A.: Absolute summability of orthogonal series by Hausdorff methods. Mat.Zametki. **1**(5), 485–492 (1971)
2. Kalaivani, K., Youvaraj, G.P.: Generalized absolute Hausdorff summability of orthogonal series. Acta Math. Hungar. **140**(1-2), 169–186 (2013)
3. Kantawala, P.S.: On the generalized absolute summability of orthogonal series. J. Indian Acad. Math. **33**(1), 191–200 (2011)
4. Leindler, L.: On the converse of inequalities of Hardy and Littlewood. Acta Sci. Math. **58**, 191–196 (1993)

Identification and Recognition of Bio-acoustic Events in an Ocean Soundscape Data Using Fourier Analysis

B. Mishachandar, S. Vairamuthu, and B. Selva Rani

Keywords Identification · Fourier analysis · Deep learning · Ocean soundscape

1 Introduction

Marine ecosystems play a vital role in balancing the global ecosystem of the Earth [1]. Every being is directly or indirectly dependent on these massive water bodies for their existence. However, this significance has adversely impacted the ocean dwelling marine habitats especially the large-sized underwater marine species, the marine mammals [2]. The National Centers for Environmental Information has reported intrusion through anthropogenic activities like large-scale shipping, massive fishing, dredging, and the unusual weather or oceanographic events to be the cause for the unusual mortality events reported as mass stranding of marine mammals [3], whereas, this scenario is even more evident in the shallow fronts as their exposure to anthropogenic activities is more vulnerable being very closer to land. Stringent climatic conditions and mostly mislead visual observations through the use of satellite images and surveys don't support conventional visual and aural observatory techniques as proxies for monitoring marine acoustic events. The ease of sound propagation in a water medium favors the use of acoustics in recording massive volumes of real-time soundscape data in an ocean environment [4]. Each year, massive volumes of acoustic recordings are recorded as part of ocean life monitoring projects to observe and assess the acoustic behavior of the bio-acoustic sources in an ocean environment. Traditionally, these animal calls as bio-acoustic data are recorded and inspected manually through visual and aural inspection to identify any acoustic activity. However, this process is highly time and cost incurring, and the possibility of an erroneous detection is substantially high. Besides, identifying individual signals in a soundscape data is manually tedious as

B. Mishachandar · S. Vairamuthu (✉) · B. Selva Rani
School of Computer Science and Engineering, Vellore Institute of Technology, Vellore, Tamil Nadu, India
e-mail: svairamuthu@vit.ac.in

H.-H. Leung et al. (eds.), *Recent Developments in Algebra and Analysis*,
Trends in Mathematics, https://doi.org/10.1007/978-3-031-37538-5_27

they are mostly beyond the human hearing limit, thus increasing the error rate [5]. In this context, the idea of numerically identifying individual acoustic sources in a soundscape data using Fourier analysis aids in precisely identifying the precise number of acoustic events. Post the identification of acoustic events, the proposed deep learning model serves as a recognizer that detects and classifies the bio-acoustic calls from the non-biological signals. This way, a complete identification and recognition system is designed to serve as a relevant choice for monitoring the biodiversity of a given region without expert intelligence.

The idea of perceiving relevant information from these audio sources is to perform sound source detection. In source detection technique, the audio form of data is converted into a two-dimensional spectrogram, and peak finding or fingerprint hashing algorithms are used to define the peaks or fingerprints in the acoustic recording. In this case, the peaks in the acoustic recordings denote the presence of animal activity through vocalizations produced by marine species. Cetacean vocalization ranges from low to high frequencies occupying a wide frequency band with different characteristics. Cetacean vocals are categorized into four major types, namely, clicks, whistles, songs, and pulsed class. Similarly, fish sounds are commonly classified as drums, grunt, and impulse. Frame-wise activity detection in SED is approached using various supervised and unsupervised methods. Some of the most popularly used classifiers include the Gaussian mixture model (GMM), hidden Markov model (HMM), fully convolutional neural networks, recurrent neural network (RNN), and CNN. The choice on the detector depends on the nature of the signal and the application to be developed. Interpreting spectrogram to detect acoustic events requires domain expertise.

Soundscape monitoring helps understand the acoustic ecology of an environment and how different sound sources respond to the dynamically changing ocean [6]. The major sources of a marine soundscape include anthrophony, biophony, and geophony. Increasing ocean-based human activities and changing climatic conditions have rapidly altered the composition of a marine soundscape [7]. This chapter aims to design a recognition system without the need for feature extraction and train the classification model to predict the category of the signal and classify it accordingly. Marine conservatory research aims at achieving a resilient and productive ocean with a small change helping restore the blue world and having a biggest impact in balancing a healthier biodiversity and human well-being.

2 Materials and Methods

2.1 Study Site and Acoustic Recordings

The source of the data used in this chapter is the National Oceanic Atmospheric Administration (NOAA). In association with the QAR Pacific Marine Environmental Laboratory, they have set up twelve mooring sites across the US coastline

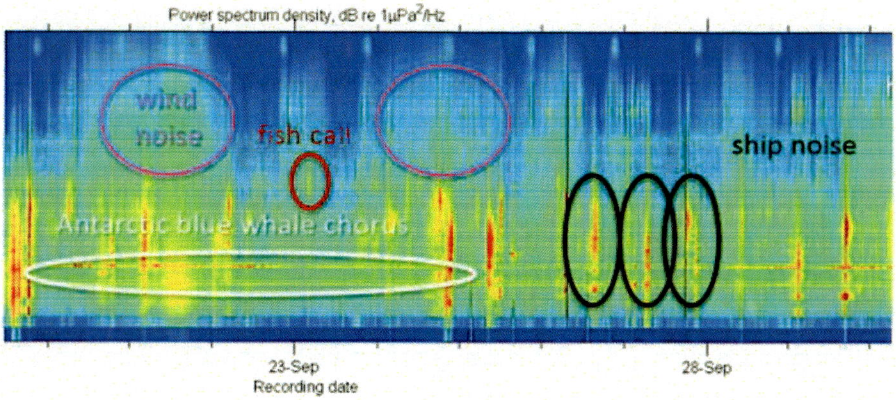

Fig. 1 Annotated long-term spectral average of a marine soundscape of the Perth Canyon [8]

waters. The US Virgin Islands is the 12th deployment site with yearlong ocean acoustic recordings from 2017-05-02T01:20:28 to 2018-05-31T14:15:28. As per the sources state, the passive acoustic recordings comprise predominantly of anthropogenic sources, periodical occurrences of marine species calls, and lower-frequency sounds from geophysical activities or the geophony. Matching to the geographical positioning of this region being very close to the harbor, this site is exposed to an increased level of shipping and vessel activities, which account to the high levels of anthrophony that are observed in the recordings. At the same time, this region is known for its rich biological sources of primarily whales and soniferous fishes. The sources show that the passive acoustic monitoring (PAM) technique was adopted for data collection, collected continuously using a single-channel mooring device, with a sampling rate of 5 kHz with the sensors positioning of −40 m of the ocean depth to cover all frequencies. Figure 1 shows the dataset features in an ocean soundscape data. The biological sounds of over 60 different species of marine mammals are provided by the Watkins Marine Mammal Sound Database [9]. The non-biological sounds from the diverse human activities and the fish and invertebrate sounds are collected from the source "Discovery of sound in the sea," maintained by the University of Rhode Island [10].

3 Methodology

3.1 Identification of Acoustic Events Using Fourier Analysis

The identification of acoustic events in an ocean soundscape data using Fourier analysis forms the vital contribution of this chapter. It is based on the concept that a waveform or a time domain signal is a sum of series of sinusoids of various frequencies, amplitudes, and phases. Fourier analysis is the series of sine waves that decompose a periodic or aperiodic signal to individual sine wave components in its frequency domain of a single-sided Fourier spectrum. In this context, Fourier analysis is used to represent the individual frequency components in an ocean soundscape audio signal that in a way numerically represents the local biodiversity as individual bio-acoustic calls of a marine soundscape using FFT. For an aperiodic time, function x(t) with a period T. The Fourier coefficients X(n) are defined by

$$r = \frac{1}{T}\int_{\frac{-T}{2}}^{\frac{T}{2}} x(t)e^{-jw_o nt} \tag{1}$$

where,

$$w_o = \frac{2\pi}{T} \tag{2}$$

using the coefficients x(m), x(t) can be represented as,

$$x(t) = \sum_{n=-\infty}^{\infty} x(n)e^{jw_o mt} = \sum_{n=-\infty}^{\infty} x(n)e^{j\frac{2\pi}{T}mt} \tag{3}$$

It takes into the property of the symmetry in sine waves to reconstruct the individual bio-acoustic signals. The number of frequencies in a Fourier spectrum equals to the number of individual samples in its raw waveform. Figure 2 shows the spectral peaks of two individual acoustic sources fetched using Fourier analysis on a single sided Fourier spectrum.

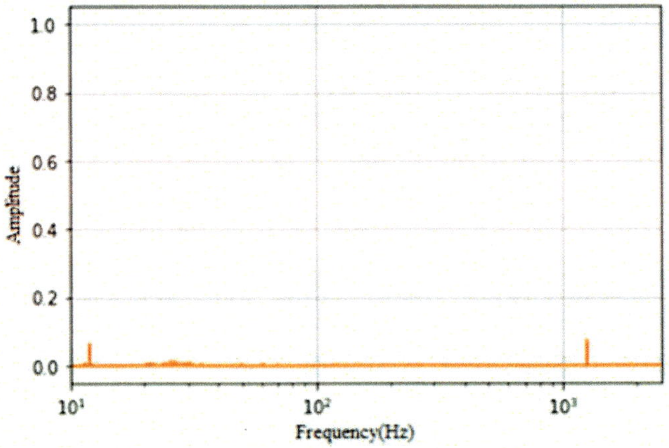

Fig. 2 Spectral content of an ocean soundscape signal with two minor frequencies at 0.1 and 1 kHz

$$\Delta f = \frac{1}{T} = \frac{F_s}{N} \tag{4}$$

where N is the length of the signal and FS is the sample rate of the raw waveform frequency. Frequency ranges of 0 Hz to the highest frequency of Nyquist frequency can be reconstructed using FFT. After the identification of individual acoustic sources from the soundscape data, the numbers of sources are identified, and they are recognized and classified as biological or non-biological sources.

3.2 Recognition of Acoustic Events Using Deep Learning

3.3 Visual Representation of Acoustic Events

The technique of converting an audio signal in its time domain to 2D image, like spectrograms, aids experts in performing visual inspection to identify acoustic events, precisely bio-acoustic events [11]. Spectrograms are 2D imagelike representation of a time domain signal that varies across time. It is extensively applied on voiced signals to extract meaningful features to perform speech recognition, image identification, gesture and palm recognition, and many more applications of audio signal processing. Figure 3 presents the 1D time domain and 2D spectrogram representation extracted using short-term Fourier transform (STFT). The spectrogram shows the prevalence of acoustic events occupying across the entire frequency spectrum. In the case of a non-biological spectrogram, the abiotic acoustic events show more prevalence in the higher frequencies of the spectrum. A short clip of the overall audio data is used to generate the spectrogram.

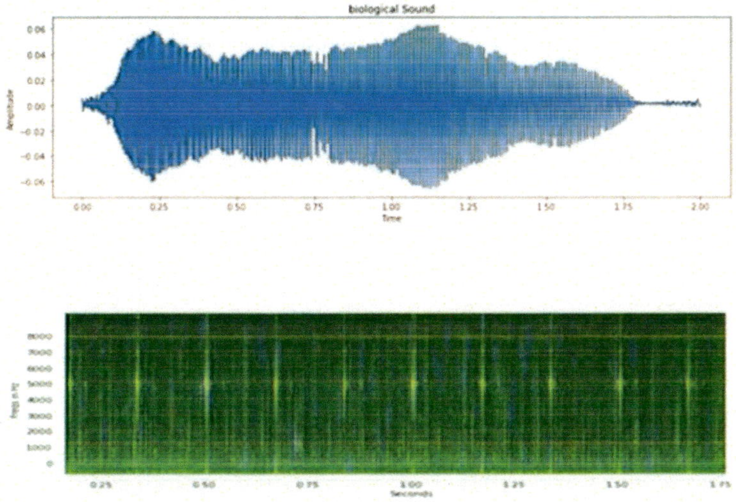

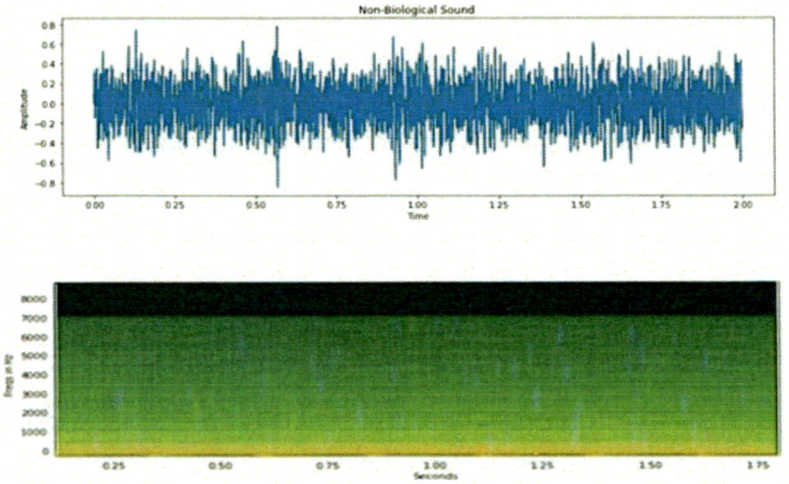

Fig. 3 From the top, time domain and spectrogram representation of the biological and the non-biological sources

3.4 Training, Testing, and Validation Data Split

A preliminary study was performed to collect the list of marine species from the region of study, and their corresponding audio data was collected to constitute the training and testing data. Audio information of all the possible marine life in this region was collected to improve the robustness of the model in detecting the sources in the validation data, which is the source-separated soundscape data. The training samples consisted of spectrograms with corresponding labels and mapping generated and stored in a JSON file. An 80-20-20 split of training, testing, and validation data was adopted in the model preparation. The two categories of data, namely, biological and non-biological sounds, were equally sized and labeled to constitute the training and the testing data. The model was validated using the denoised soundscape data of the US Virgin Islands to classify it to be either biological or a non-biological. The long durational denoised data was clipped into short segments of audio, with 2s length each. This way, the separated data is classified to either of the two classes to identify the predominant source of the separated long durational data. This way, the need for profound feature extraction techniques is evaded using the proposed model to identify target signals in the audio data amidst the effect of other ocean noise sources.

3.5 Model Creation

Two deep learning models, namely, convolutional neural networks (CNN) and recurrent neural networks (RNN-LSTM), were the preferred choice for the model creation. The choice on the model was made based on the nature of the data, which is a long durational sequential labeled data clipped into short segmental frames. 2D CNN for learning is more robust as the model is capable of recognizing feeble animal sounds as contours in a spectrogram. Model structuring is designed to learn more features from the time domain by stacking three consecutive convolutional layers followed by a max pooling to pull down the loss of prominent features. The filter size of the model is progressively increased, and padding is used to preserve the input matrix dimension. The ReLu activation function is used to avoid the vanishing gradient problem that facilitates a faster learning process. A minimalistic model architecture is achieved with this model design.

On a general note, CNN is the preferred choice of deep learning model when dealing with 2D image data. The 2D image form, an audio signal known as spectrogram, is generated from time-domain signals using various feature extraction techniques like Fourier transform (FT), short-term Fourier transform (STFT), wavelet transform (WT), spectral centroid, spectral bandwidth, and zero-crossing rate. When compared to other conventional convolutional models, the LSTM model is a type of recurrent neural network (RNN) where the learning is performed sequentially, and it is the preferred choice of model in complex recognition problems. The proposed classification model is fed with sequential long durational passive acoustic data collected for 3 months, recorded on the basis of 4 hours per day summing up to 1200 hours. The long durational data is clipped to definite-size audio samples definitive to the length of the species call. The clipped biological and non-biological calls are identified and classified to either of the two classes. Conventional CNN and RNN models are used to access the best performing model that identifies the sources with almost similar acoustic structure. The model architecture is kept simple to reduce the complexity and to achieve far more precise results compared to other less complicated traditional machine learning algorithms like SVM and other conventional clustering techniques. The entire model is scripted in Python 3.6 and implemented on Spyder 4.4 IDE with TensorFlow 2.0.

4 Results and Discussion

The spectrograms generated from the image-based data are used for training the model of denoised soundscape data consisting of marine species vocalization and anthropogenic noise separated in the audio source separation is clipped and used in model validation. The training and validation loss of the model is controlled through the model structuring, where dropouts and flattening layers are combined with the convolutional layers to balance the model architecture. The model's performance was evaluated with varying epoch sizes like 10, 25, and 50, to check its effect on the overall model accuracy. The recognition model was developed on Windows 10, 64bit PC, with Intel(R) Core (TM) i5-7200U CPU, and NVIDIA GeForce 304 GTX 1080 GPU and TensorFlow 2.2. The model processing was performed using a GPU to increase the computation speed with large volumes of data. The complexity of the model is quite reduced as the model is expected to recognize an entire category of the audio source rather than individual sources. Recognizing individual sources in an audio recording would require learning from an overcomplete dictionary, which is tedious. The proposed model is evaluated using the accuracy metric in which the 2D CNN model achieves a validation loss of 0.25 and an accuracy rate of 94.2%. The binary classification accuracy shows the adaptability of a convolutional model that suits best in handling image data and also performs best when dealing with audio-based data; however, the performance shows an inconsistent classification with steady drops in the accuracy value and varying epochs. The RNN-LSTM model shows greater adaptability toward the continuous data and fetch and accuracy of 96% and a validation loss of 0.07.

On comparing the models, the 2D CNN shows an upper hand compared to the RNN-LSTM in terms of handling structures in a non-voiced data in audio recognition. However, this study also proves the efficient applicability of simple deep learning models in performing complex audio recognition and classification tasks easily. A fixed dictionary of the sources in a deep ocean is still undefined as the evolving marine world unwraps new species each day. On this note, the results of the binary classifier in this study prove the proposed model that identifies the ocean life as a significant contribution in this regard. Figure 4 shows the performance of Fig. 4a CNN Fig. 4b RNN-LSTM in a graph plotted accuracy against the number of epochs.

The graphs for model loss in the training and the validation process in Figure 2 show 2D CNN-based binary classification to exhibit the model's training and validation process with the loss of the model reducing significantly with varying epochs and converging to zero at the last few epochs, proving the model's adaptability in handling new data. An important inference from a steady drop in the RNN model's accuracy is observed at varied epochs, proving its adaptability in handling non-voiced structures in a spectrogram to be less efficient as the vocalization pattern as whistle contours do not possess a definite structure as in a usual 2D image. On a precise note, the designed automated system recognizes the bio-acoustic events and is accessed on the basis of the validation accuracy. Efficient conventional deep learning models like convolutional neural networks (CNN) and recurrent neural networks (RNN-LSTM) were put to test their efficiency in handling the real-time passive acoustic ocean soundscape data. Unlike other popular audio AI domains like speech and music, the applicability of AI on ocean acoustics is nascent. Research efforts in this regard are expected to grow in the future for its significance in marine life conservation.

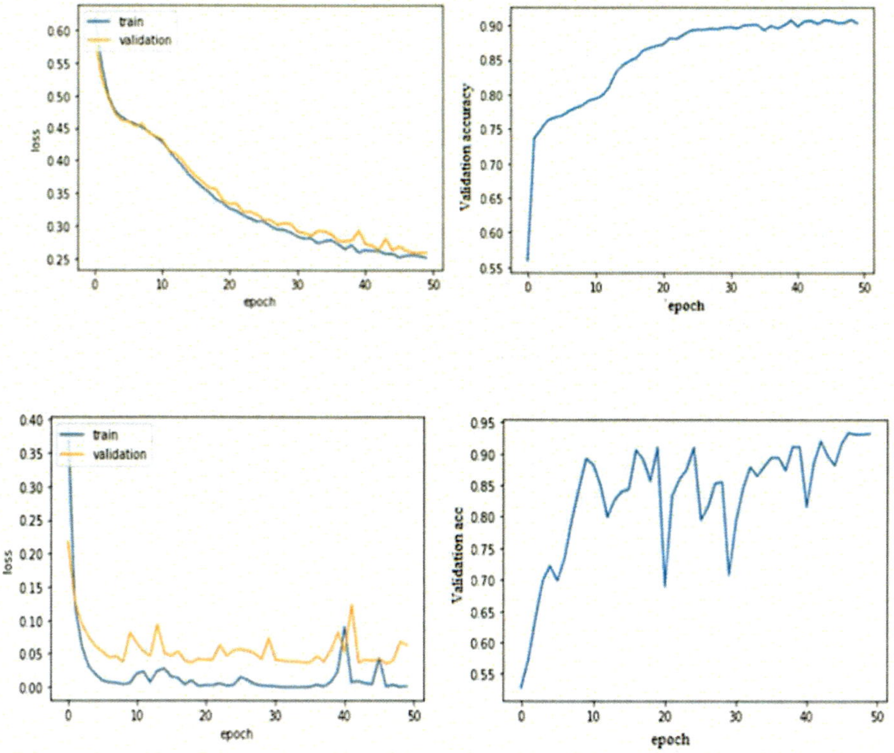

Fig. 4 From the top, time domain and spectrogram representation of the biological and the non-biological sources

5 Conclusion

The conservation of rapidly depleting marine ecosystems and its habitat has necessitated the need for marine life monitoring and conservatory policies with preventive measures. In this regard, artificial intelligence, especially deep learning, has gained great significance in the design and development of robust automated systems to interpret and perform actions with far more precision and intelligence than humans. This chapter is one such effort to effectively recognize the number of sources and classify the bio-acoustic events in an ocean soundscape data using CNN. The model's performance was tested on its ability in effectively handling biological and non-biological calls that possess nearly similar acoustic patterns in their 2D image form. The results shows CNN model, a conventional choice of model in handling 2D image-based data, also efficiently handles sequential non-voiced data better compared to RNN model. As part of future work, we intend to identify the specific bio-acoustic source in the ocean soundscape data using a hybrid deep learning model and apply transfer learning to extend to new species with limited data. The ocean is a vital source of life sustenance on this planet. A little contribution in this regard can help revive the massive blue world and protect it in the long run.

References

1. Wall, C.C., Haver, S.M., Hatch, L.T., Miksis-Olds, J., Bochenek, R., Dziak, R.P., Gedamke, J.: The next wave of passive acoustic data management: how centralized access can enhance science. Frontiers Marine Sci. **8**, 703682 (2021)
2. Sattar, F., Cullis-Suzuki, S., Jin, F.: Identification of fish vocalizations from ocean acoustic data. Appl. Acoust. **110**, 248–255 (2016)
3. Williams, R., Wright, A.J., Ashe, E., Blight, L.K., Bruintjes, R., Canessa, R., Wale, M.A.: Impacts of anthropogenic noise on marine life: publication patterns, new discoveries, and future directions in research and management. Ocean Coastal Manag. **115**, 17–24 (2015)
4. Zou, Z., Badiey, M.: Effects of wind speed on shallow-water broadband acoustic transmission. IEEE J. Ocean Eng. **43**, 1–13 (2017). https://doi.org/10.1109/JOE.2017.2737049
5. Mishachandar, B., Vairamuthu, S.: An underwater cognitive acoustic network strategy for efficient spectrum utilization. Appl. Acoust. **175**, 107861 (2021). https://doi.org/10.1016/j.apacoust.2020.107861
6. Nanaware, S., Shastri, R., Joshi, Y., Das, A.: Passive acoustic detection and classification of marine mammal vocalizations. In: 2014 International Conference on Communication and Signal Processing, pp. 493–497. https://doi.org/10.1109/ICCSP.2014.6949891
7. Mishachandar, B., Vairamuthu, S.: Diverse ocean noise classification using deep learning. Appl. Acoust. **181**, 108141 (2021). https://doi.org/10.1016/j.apacoust.2021.108141
8. Erbe, C.,Verma, A., McCauley, R.,Gavrilov, A., Parnum, I.: The marine soundscape of the Perth Canyon. Progr. Oceanogr. **137**, 38–51 (2015)
9. Sayigh, L., Daher, M.A., Allen, J., Gordon, H., Joyce, K., Stuhlmann, C., Tyack, P.: The Watkins marine mammal sound database: An online, freely accessible resource. In: Proceedings of Meetings on Acoustics 4ENAL, vol. 27, no. 1, p. 040013. Acoustical Society of America, Melville (2016)
10. Vigness-Raposa, K. J., Scowcroft, G., Miller, J.H., Ketten, D.: Discovery of sound in the Sea: An online resource. In: The Effects of Noise on Aquatic Life, pp. 135–138. Springer, Berlin (2012)
11. Mishachandar, B.,Vairamuthu, S.: A review on underwater acoustic sensor networks: perspective of internet of things. Int. J. Innovative Technol. Explor. Eng. **8**(6), 1603–1615 (2019).

Neutrosophic Nano M Continuous Mappings via Neutrosophic Nano-topological Spaces

K. Saraswathi, A. Vadivel, S. Tamilselvan, and C. John Sundar

Keywords $N_s eu\mathcal{N}o$ set · $N_s eu\mathcal{N}Mo$ set · $N_s eu\mathcal{N}MCts$ · $N_s eu\mathcal{N}Mq$-nbhd

1 Introduction

Zadeh [31] was the first one to introduce fuzzy set and its topological spaces by Chang [4]. Atanassov [2, 3] introduced intuitionistic fuzzy sets and its topological spaces by Coker [5]. In 1995, Smarandache [18, 19] introduced neutrosophic logic and its topological spaces by Salama et al. [17] and their applications by several authors [1, 6, 11, 14, 24, 27]. Pawlak [15] was the first one to introduced the rough set theory. In 2013, Lellis Thivagar [9] presented nano-topology and nano-topological spaces.

Pankajam [12] introduced δ-open sets in nano-topological space and Vadivel et al. [22, 23, 28–30] in a neutrosophic topological space. El-Maghrabi and Al-Juhani [7] introduced M-open sets in topological spaces and in nano-topological spaces by Padma et al. [13].

Recently, a novel idea of neutrosophic nano-topology was investigated by Lellis Thivagar et al. [10]. Recently, Thangammal et al. [20, 21] and Kalaiyarasan et al. [8] introduced some open sets in fuzzy nano-topological spaces, and Vadivel et al.

K. Saraswathi
Department of Mathematics, Thiruvalluvar Government Arts College, Rasipuram, Namakkal, India

A. Vadivel (✉)
Department of Mathematics, Arignar Anna Government Arts College, Namakkal, India

Department of Mathematics, Annamalai University, Annamalai Nagar, Chidambaram, India

S. Tamilselvan
Engineering Mathematics, Annamalai University, Annamalai Nagar, Chidambaram, India

C. J. Sundar
Department of Mathematics, Sri Venkateshwaraa College of Engineering and Technology, Puducherry, India

H.-H. Leung et al. (eds.), *Recent Developments in Algebra and Analysis*, Trends in Mathematics, https://doi.org/10.1007/978-3-031-37538-5_28

[25, 26] investigated δ-open sets and M-open sets in neutrosophic nano-topological spaces.

2 Preliminaries

The basic definitions used in this article are given and studied in these articles [10, 13, 16, 25, 26].

3 Neutrosophic Nano M Continuous Functions

Throughout this section, let a function $h : (U_1, \tau_N(F_1)) \rightarrow (U_2, \tau_N(F_2))$ be h.

Definition 1 A function h is said to be neutrosophic nano (resp. θ, θ semi, δ pre & M) continuous (briefly, $N_seu\mathcal{N}Cts$ (resp. $N_seu\mathcal{N}\theta Cts$, $N_seu\mathcal{N}\theta SCts$, $N_seu\mathcal{N}\delta\mathcal{P}Cts$ & $N_seu\mathcal{N}MCts$)), if for each $N_seu\mathcal{N}o$ set S of U_2, the set $h^{-1}(S)$ is $N_seu\mathcal{N}o$ (resp. $N_seu\mathcal{N}\theta o$, $N_seu\mathcal{N}\theta So$, $N_seu\mathcal{N}\delta\mathcal{P}o$ & $N_seu\mathcal{N}Mo$) sets of U_1.

Theorem 1 *Let h be a mapping. Then, every*

(i) $N_seu\mathcal{N}\theta Cts$ *is* $N_seu\mathcal{N}Cts$.
(ii) $N_seu\mathcal{N}Cts$ *is* $N_seu\mathcal{N}\delta\mathcal{P}Cts$.
(iii) $N_seu\mathcal{N}\theta Cts$ *is* $N_seu\mathcal{N}\theta SCts$.
(iv) $N_seu\mathcal{N}\theta SCts$ *is* $N_seu\mathcal{N}MCts$.
(v) $N_seu\mathcal{N}\delta\mathcal{P}Cts$ *is* $N_seu\mathcal{N}MCts$.
(vi) Every $N_seu\mathcal{N}\delta\mathcal{P}Cts$ *is a* $N_seu\mathcal{N}eCts$.
(vii) Every $N_seu\mathcal{N}MCts$ *is a* $N_seu\mathcal{N}eCts$.

Proof

(i) Let h be $N_seu\mathcal{N}\theta Cts$ and L is a $N_seu\mathcal{N}o$ in U_2. Then $h^{-1}(L)$ is $N_seu\mathcal{N}\theta o$ in U_1. Since every $N_seu\mathcal{N}\theta o$ is $N_seu\mathcal{N}o$, $h^{-1}(L)$ is $N_seu\mathcal{N}o$ in U_1. Therefore h is $N_seu\mathcal{N}Cts$.
(ii) Let h be $N_seu\mathcal{N}Cts$ and L is a $N_seu\mathcal{N}o$ in U_2. Then $h^{-1}(L)$ is $N_seu\mathcal{N}o$ in U_1. Since every $N_seu\mathcal{N}o$ is $N_seu\mathcal{N}\delta\mathcal{P}o$, $h^{-1}(L)$ is $N_seu\mathcal{N}\delta\mathcal{P}o$ in U_1. Therefore h is $N_seu\mathcal{N}\delta\mathcal{P}Cts$.
(iii) Let h be $N_seu\mathcal{N}\theta Cts$ and L is a $N_seu\mathcal{N}o$ in U_2. Then $h^{-1}(L)$ is $N_seu\mathcal{N}\theta o$ in U_1. Since every $N_seu\mathcal{N}\theta o$ is $N_seu\mathcal{N}\theta So$, $h^{-1}(L)$ is $N_seu\mathcal{N}\theta So$ in U_1. Therefore h is $N_seu\mathcal{N}\theta SCts$.
(iv) Let h be $N_seu\mathcal{N}\theta SCts$ and L is a $N_seu\mathcal{N}o$ in U_2. Then $h^{-1}(L)$ is $N_seu\mathcal{N}\theta So$ in U_1. Since every $N_seu\mathcal{N}\theta So$ is $N_seu\mathcal{N}Mo$, $h^{-1}(L)$ is $N_seu\mathcal{N}Mo$ in U_1. Therefore h is $N_seu\mathcal{N}MCts$.

(v) Let h be $N_s eu\mathcal{N}\delta\mathcal{P}Cts$ and L is a $N_s eu\mathcal{N}o$ in U_2. Then $h^{-1}(L)$ is $N_s eu\mathcal{N}\delta\mathcal{P}o$ in U_1. Since every $N_s eu\mathcal{N}\delta\mathcal{P}o$ is $N_s eu\mathcal{N}Mo$, $h^{-1}(L)$ is $N_s eu\mathcal{N}Mo$ in U_1. Therefore h is $N_s eu\mathcal{N}MCts$.

(vi) Let h be $N_s eu\mathcal{N}\delta\mathcal{P}Cts$ and L is a $N_s eu\mathcal{N}o$ in U_2. Then $h^{-1}(L)$ is $N_s eu\mathcal{N}\delta\mathcal{P}o$ in U_1. Since every $N_s eu\mathcal{N}\delta\mathcal{P}o$ is $N_s eu\mathcal{N}eo$, $h^{-1}(L)$ is $N_s eu\mathcal{N}eo$ in U_1. Therefore h is $N_s eu\mathcal{N}eCts$.

(vii) Let h be $N_s eu\mathcal{N}MCts$ and L is a $N_s eu\mathcal{N}o$ in U_2. Then $h^{-1}(L)$ is $N_s eu\mathcal{N}Mo$ in U_1. Since every $N_s eu\mathcal{N}Mo$ is $N_s eu\mathcal{N}eo$, $h^{-1}(L)$ is $N_s eu\mathcal{N}eo$ in U_1. Therefore h is $N_s eu\mathcal{N}eCts$.

□

The converse of Theorem 1 need not be true.

Example 1 Assume $U = \{s_1, s_2, s_3, s_4\}$ and $U/R = \{\{s_1, s_4\}, \{s_2\}, \{s_3\}\}$. Let $S_o = \left\{\left\langle\frac{s_1}{0.2,0.5,0.8}\right\rangle, \left\langle\frac{s_2}{0.3,0.5,0.7}\right\rangle, \left\langle\frac{s_3}{0.4,0.5,0.6}\right\rangle, \left\langle\frac{s_4}{0.1,0.5,0.9}\right\rangle\right\}$ be a $N_s eu\ subs$ of U.

$\underline{N_s eu\mathcal{N}}(S_o) = \left\{\left\langle\frac{s_1,s_4}{0.1,0.5,0.9}\right\rangle, \left\langle\frac{s_2}{0.3,0.5,0.7}\right\rangle, \left\langle\frac{s_3}{0.4,0.5,0.6}\right\rangle\right\}$,

$\overline{N_s eu\mathcal{N}}(S_o) = \left\{\left\langle\frac{s_1,s_4}{0.2,0.5,0.8}\right\rangle, \left\langle\frac{s_2}{0.3,0.5,0.7}\right\rangle, \left\langle\frac{s_3}{0.4,0.5,0.6}\right\rangle\right\}$,

$B_{N_s eu\mathcal{N}}(S_o) = \left\{\left\langle\frac{s_1,s_4}{0.2,0.5,0.8}\right\rangle, \left\langle\frac{s_2}{0.3,0.5,0.7}\right\rangle, \left\langle\frac{s_3}{0.4,0.5,0.6}\right\rangle\right\}$.

Thus $\tau_N(S_o) = \{0_N, 1_N, \underline{N_s eu\mathcal{N}}(S_o), \overline{N_s eu\mathcal{N}}(S_o) = B_{N_s eu\mathcal{N}}(S_o)\}$.

Then $h : (U, \tau_N(F)) \to (U, \tau_N(F))$ is an identity function, the set $A_o = \left\{\left\langle\frac{s_1,s_4}{0.1,0.5,0.9}\right\rangle, \left\langle\frac{s_2}{0.3,0.5,0.7}\right\rangle, \left\langle\frac{s_3}{0.4,0.5,0.6}\right\rangle\right\}$ is $N_s eu\mathcal{N}Cts$ but not $N_s eu\mathcal{N}\theta Cts$.

Since, A_o is a $N_s eu\mathcal{N}o$ set in U but $h^{-1}(A_o)$ is not $N_s eu\mathcal{N}\theta o$ set in U.

Example 2 Assume $U = \{s_1, s_2, s_3, s_4\}$ and $U/R = \{\{s_1, s_4\}, \{s_2\}, \{s_3\}\}$. Let $S_o = \left\{\left\langle\frac{s_1}{0.2,0.5,0.8}\right\rangle, \left\langle\frac{s_2}{0.3,0.5,0.7}\right\rangle, \left\langle\frac{s_3}{0.4,0.5,0.6}\right\rangle, \left\langle\frac{s_4}{0.1,0.5,0.9}\right\rangle\right\}$ be a $N_s eu\ subs$ of U.

$\underline{N_s eu\mathcal{N}}(S_o) = \left\{\left\langle\frac{s_1,s_4}{0.1,0.5,0.9}\right\rangle, \left\langle\frac{s_2}{0.3,0.5,0.7}\right\rangle, \left\langle\frac{s_3}{0.4,0.5,0.6}\right\rangle\right\}$,

$\overline{N_s eu\mathcal{N}}(S_o) = \left\{\left\langle\frac{s_1,s_4}{0.2,0.5,0.8}\right\rangle, \left\langle\frac{s_2}{0.3,0.5,0.7}\right\rangle, \left\langle\frac{s_3}{0.4,0.5,0.6}\right\rangle\right\}$,

$B_{N_s eu\mathcal{N}}(S_o) = \left\{\left\langle\frac{s_1,s_4}{0.2,0.5,0.8}\right\rangle, \left\langle\frac{s_2}{0.3,0.5,0.7}\right\rangle, \left\langle\frac{s_3}{0.4,0.5,0.6}\right\rangle\right\}$.

Thus $\tau_N(S_o) = \{0_N, 1_N, \underline{N_s eu\mathcal{N}}(S_o), \overline{N_s eu\mathcal{N}}(S_o) = B_{N_s eu\mathcal{N}}(S_o)\}$.

Also, $V = \{t_1, t_2, t_3, t_4\}$ and $V/R = \{\{t_1, t_4\}, \{t_2\}, \{t_3\}\}$.

Let $T_o = \left\{\left\langle\frac{t_1}{0.1,0.5,0.9}\right\rangle, \left\langle\frac{t_2}{0.1,0.5,0.9}\right\rangle, \left\langle\frac{t_3}{0.4,0.5,0.6}\right\rangle, \left\langle\frac{t_4}{0.1,0.5,0.9}\right\rangle\right\}$ be a $N_s eu\ subs$ of V.

$\underline{N_s eu\mathcal{N}}(T_o) = \left\{\left\langle\frac{t_1,t_4}{0.1,0.5,0.9}\right\rangle, \left\langle\frac{t_2}{0.1,0.5,0.9}\right\rangle, \left\langle\frac{t_3}{0.4,0.5,0.6}\right\rangle\right\} = \overline{N_s eu\mathcal{N}}(T_o) = B_{N_s eu\mathcal{N}}(T_o)$.

Thus $\sigma_N(T_o) = \{0_N, 1_N, \underline{N_s eu\mathcal{N}}(T_o) = \overline{N_s eu\mathcal{N}}(T_o) = B_{N_s eu\mathcal{N}}(T_o)\}$.

Then $h : (U, \tau_N(F)) \to (V, \sigma_N(F))$ is an identity function, the set $A_o = \left\{\left\langle\frac{t_1,t_4}{0.1,0.5,0.9}\right\rangle, \left\langle\frac{t_2}{0.1,0.5,0.9}\right\rangle, \left\langle\frac{t_3}{0.4,0.5,0.6}\right\rangle\right\}$ is $N_s eu\mathcal{N}\delta\mathcal{P}Cts$ (resp. $N_s eu\mathcal{N}MCts$) but not $N_s eu\mathcal{N}Cts$ (resp. $N_s eu\mathcal{N}\theta SCts$). Since, A_o is a $N_s eu\mathcal{N}o$ set in V but $h^{-1}(A_o)$ is not $N_s eu\mathcal{N}o$ (resp. $N_s eu\mathcal{N}\theta So$) set set in U.

Example 3 Assume $U = \{s_1, s_2, s_3, s_4\}$ and $U/R = \{\{s_1, s_4\}, \{s_2\}, \{s_3\}\}$. Let $S_o = \left\{\left\langle\frac{s_1}{0.2,0.5,0.8}\right\rangle, \left\langle\frac{s_2}{0.3,0.5,0.7}\right\rangle, \left\langle\frac{s_3}{0.4,0.5,0.6}\right\rangle, \left\langle\frac{s_4}{0.1,0.5,0.9}\right\rangle\right\}$ be a $N_s eu$ *subs* of U. $\underline{N_s eu\mathcal{N}}(S_o) = \left\{\left\langle\frac{s_1,s_4}{0.1,0.5,0.9}\right\rangle, \left\langle\frac{s_2}{0.3,0.5,0.7}\right\rangle, \left\langle\frac{s_3}{0.4,0.5,0.6}\right\rangle\right\}$,

$\overline{N_s eu\mathcal{N}}(S_o) = \left\{\left\langle\frac{s_1,s_4}{0.2,0.5,0.8}\right\rangle, \left\langle\frac{s_2}{0.3,0.5,0.7}\right\rangle, \left\langle\frac{s_3}{0.4,0.5,0.6}\right\rangle\right\}$,

$B_{N_s eu\mathcal{N}}(S_o) = \left\{\left\langle\frac{s_1,s_4}{0.2,0.5,0.8}\right\rangle, \left\langle\frac{s_2}{0.3,0.5,0.7}\right\rangle, \left\langle\frac{s_3}{0.4,0.5,0.6}\right\rangle\right\}$.

Thus $\tau_N(S_o) = \{0_N, 1_N, \underline{N_s eu\mathcal{N}}(S_o), \overline{N_s eu\mathcal{N}}(S_o) = B_{N_s eu\mathcal{N}}(S_o)\}$.

Also, $V = \{t_1, t_2, t_3, t_4\}$ and $V/R = \{\{t_1, t_4\}, \{t_2\}, \{t_3\}\}$.

Let $T_o = \left\{\left\langle\frac{t_1}{0.8,0.5,0.2}\right\rangle, \left\langle\frac{t_2}{0.7,0.5,0.3}\right\rangle, \left\langle\frac{t_3}{0.6,0.5,0.4}\right\rangle, \left\langle\frac{t_4}{0.8,0.5,0.2}\right\rangle\right\}$ be a $N_s eu$ *subs* of V.

$\underline{N_s eu\mathcal{N}}(T_o) = \left\{\left\langle\frac{t_1,t_4}{0.8,0.5,0.2}\right\rangle, \left\langle\frac{t_2}{0.7,0.5,0.3}\right\rangle, \left\langle\frac{t_3}{0.6,0.5,0.4}\right\rangle\right\} = \overline{N_s eu\mathcal{N}}(T_o)$,

$B_{N_s eu\mathcal{N}}(T_o) = \left\{\left\langle\frac{t_1,t_4}{0.2,0.5,0.8}\right\rangle, \left\langle\frac{t_2}{0.3,0.5,0.7}\right\rangle, \left\langle\frac{t_3}{0.4,0.5,0.6}\right\rangle\right\}$.

Thus $\sigma_N(T_o) = \{0_N, 1_N, \underline{N_s eu\mathcal{N}}(T_o) = \overline{N_s eu\mathcal{N}}(T_o), B_{N_s eu\mathcal{N}}(T_o)\}$.

Then $h : (U, \tau_N(F)) \rightarrow (V, \sigma_N(F))$ is an identity function, the set $B_o = \left\{\left\langle\frac{t_1,t_4}{0.8,0.5,0.2}\right\rangle, \left\langle\frac{t_2}{0.7,0.5,0.3}\right\rangle, \left\langle\frac{t_3}{0.6,0.5,0.4}\right\rangle\right\}$ is $N_s eu\mathcal{N}\theta SCts$ (resp. $N_s eu\mathcal{N}M\ Cts$ and $N_s eu\mathcal{N}eCts$) but not $N_s eu\mathcal{N}\theta Cts$ (resp. $N_s eu\mathcal{N}\delta\mathcal{P}Cts$ and $N_s eu\mathcal{N}\ \delta\mathcal{P}Cts$). Since, B_o is a $N_s eu\mathcal{N}o$ set in V but $h^{-1}(B_o)$ is not $N_s eu\mathcal{N}\theta o$ (resp. $N_s eu\mathcal{N}\delta\mathcal{P}o$ and $N_s eu\mathcal{N}\delta\mathcal{P}o$) set in U.

Remark 1 Figure 1 shows the relationships of $N_s eu\mathcal{N}MCts$ mappings in $N_s eu\mathcal{N}ts$.

Theorem 2 *A function h is* $N_s eu\mathcal{N}MCts$ *iff the inverse image of every* $N_s eu\ \mathcal{N}c$ *in* U_2 *is* $N_s eu\mathcal{N}Mc$ *in* U_1.

Proof Let h be $N_s eu\mathcal{N}MCts$ & O_o is $N_s eu\mathcal{N}o$ in U_2. (i.e.) $U_2 - O_o$ is $N_s eu\mathcal{N}o$ in U_2. Since h is $N_s eu\mathcal{N}MCts$, $h^{-1}(U_2 - O_o)$ is $N_s eu\mathcal{N}Mo$ in U_1. (i.e.) $U_1 - h^{-1}(O_o)$ is $N_s eu\mathcal{N}Mo$ in U_1. Therefore $h^{-1}(O_o)$ is $N_s eu\mathcal{N}Mc$ in U_1.

Conversely, let the inverse image of every $N_s eu\mathcal{N}c$ be $N_s eu\mathcal{N}Mc$. Let C be $N_s eu\mathcal{N}o$ in U_2. Then $U_2 - C$ is $N_s eu\mathcal{N}c$ in U_2, implies $h^{-1}(U_2 - C)$ is $N_s eu\mathcal{N}Mc$ in U_1. (i.e.) $U_1 - h^{-1}(C)$ is $N_s eu\mathcal{N}Mc$ in U_1. Therefore $h^{-1}(C)$ is $N_s eu\mathcal{N}Mo$ in U_1. Thus, the inverse image of every $N_s eu\mathcal{N}o$ in U_2 is $N_s eu\mathcal{N}Mo$ in U_1. Hence, h is $N_s eu\mathcal{N}MCts$ on U_1. □

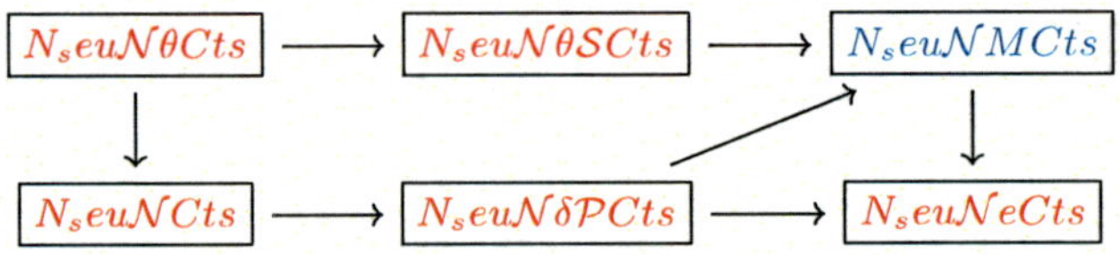

Fig. 1 $N_s eu\mathcal{N}MCts$ mappings in $N_s eu\mathcal{N}ts$

Theorem 3 *A function h is $N_s eu\mathcal{N}MCts$ iff $h(N_s eu\mathcal{N}Mcl(M)) \subseteq N_s eu\mathcal{N}cl(h(M))$ for every subset M of U_1.*

Remark 2 A function h is $N_s eu\mathcal{N}MCts$ then $h(N_s eu\mathcal{N}Mcl(K))$ is not necessarily equal to $N_s eu\mathcal{N}cl(h(K))$ where $K \in I^{U_1}$.

Example 4 In Example 1, $h : (U, \tau_N(F)) \rightarrow (U, \tau_N(F))$ is $N_s eu\mathcal{N}MCts$. Let $A_o = \left\{\left\langle\frac{s_1,s_4}{0.1,0.5,0.9}\right\rangle, \left\langle\frac{s_2}{0.2,0.5,0.8}\right\rangle, \left\langle\frac{s_3}{0.1,0.5,0.9}\right\rangle\right\} \in U$. Then $N_s eu\mathcal{N}Mcl(A_o) = h(N_s eu\mathcal{N}Mcl(A_o)) = h(A_o) = A_o$.

But

$$N_s eu\mathcal{N}cl(h(A_o)) = N_s eu\mathcal{N}cl(A_o)$$
$$= \left\{\left\langle\frac{s_1, s_4}{0.8, 0.5, 0.2}\right\rangle, \left\langle\frac{s_2}{0.7, 0.5, 0.3}\right\rangle, \left\langle\frac{s_3}{0.6, 0.5, 0.4}\right\rangle\right\}.$$

Thus $h(N_s eu\mathcal{N}Mcl(A_o)) \neq N_s eu\mathcal{N}cl(h(A_o))$.

Theorem 4 *A function h is $N_s eu\mathcal{N}MCts$ iff $N_s eu\mathcal{N}Mcl(h^{-1}(S)) \subseteq h^{-1}(N_s eu\mathcal{N}cl(S))$ for every subset S of U_2.*

Proof If h is $N_s eu\mathcal{N}MCts$ & $S \subseteq U_2$. $N_s eu\mathcal{N}cl(S)$ is $N_s eu\mathcal{N}c$ in U_2. Thus, $h^{-1}(N_s eu\mathcal{N}cl(S))$ is $N_s eu\mathcal{N}Mc$ in U_1. Therefore $N_s eu\mathcal{N}Mcl(h^{-1}(N_s eu\mathcal{N}cl(S))) = h^{-1}(N_s eu\mathcal{N}cl(S))$. Since $S \subseteq N_s eu\mathcal{N}cl(S)$, $h^{-1}(S) \subset h^{-1}(N_s eu\mathcal{N}\ cl(S))$. Therefore $N_s eu\mathcal{N}Mcl(h^{-1}(S)) \subset N_s eu\mathcal{N}Mcl(h^{-1}(N_s eu\mathcal{N}cl(S))) = h^{-1}(N_s eu\mathcal{N}cl(S))$. That is $N_s eu\mathcal{N}Mcl(h^{-1}(S)) \subseteq h^{-1}(N_s eu\mathcal{N}cl(S))$.

Conversely, let $N_s eu\mathcal{N}Mcl(h^{-1}(S)) \subseteq h^{-1}(N_s eu\mathcal{N}cl(S))$ for every subset S of U_2. If S is $N_s eu\mathcal{N}c$ in U_2, then $N_s eu\mathcal{N}cl(S) = S$. By assumption, $N_s eu\mathcal{N}Mcl(h^{-1}(S)) \subseteq h^{-1}(N_s eu\mathcal{N}cl(S)) = h^{-1}(S)$. Thus $N_s eu\mathcal{N}Mcl(h^{-1}(S)) \subseteq h^{-1}(S)$. But $h^{-1}(S) \subseteq N_s eu\mathcal{N}Mcl(h^{-1}(S))$. Therefore $N_s eu\mathcal{N}M\ cl(h^{-1}(S)) = h^{-1}(S)$. Hence $h^{-1}(S)$ is $N_s eu\mathcal{N}Mc$ in U_1, for every $N_s eu\mathcal{N}c$ set S in U_2. Therefore h is $N_s eu\mathcal{N}MCts$ on U_1. □

Remark 3 A function h is $N_s eu\mathcal{N}MCts$ then $N_s eu\mathcal{N}Mcl(h^{-1}(L))$ is not necessarily equal to $h^{-1}(N_s eu\mathcal{N}cl(L))$ where $L \in I^{U_2}$.

Example 5 In Example 1, $h : (U, \tau_N(F)) \rightarrow (U, \tau_N(F))$ is $N_s eu\mathcal{N}MCts$. Let $B_o = \left\{\left\langle\frac{s_1,s_4}{0.1,0.5,0.9}\right\rangle, \left\langle\frac{s_2}{0.2,0.5,0.8}\right\rangle, \left\langle\frac{s_3}{0.1,0.5,0.9}\right\rangle\right\} \in U$.

Then $N_s eu\mathcal{N}Mclh^{-1}(B_o) = N_s eu\mathcal{N}Mcl(B_o) = B_o$. But

$$h^{-1}(N_s eu\mathcal{N}cl(B_o)) = h^{-1}(\left\{\left\langle\frac{s_1, s_4}{0.8, 0.5, 0.2}\right\rangle, \left\langle\frac{s_2}{0.7, 0.5, 0.3}\right\rangle, \left\langle\frac{s_3}{0.6, 0.5, 0.4}\right\rangle\right\})$$
$$= \left\{\left\langle\frac{s_1, s_4}{0.8, 0.5, 0.2}\right\rangle, \left\langle\frac{s_2}{0.7, 0.5, 0.3}\right\rangle, \left\langle\frac{s_3}{0.6, 0.5, 0.4}\right\rangle\right\}.$$

Thus $N_s eu\mathcal{N}Mcl(h^{-1}(B_o)) \neq h^{-1}(N_s eu\mathcal{N}cl(B_o))$.

Theorem 5 *A function h is $N_s euNMCts$ iff $h^{-1}(N_s euNint(S)) \subseteq N_s euNM$ $int(h^{-1}(S))$ for every subset S of U_2.*

Proof If h is $N_s euNMCts$ and $S \subseteq U_2$. $N_s euNint(S)$ is $N_s euNo$ in U_2 and hence, $h^{-1}(N_s euNint(S))$ is $N_s euNMo$ in U_1. Therefore $N_s euNMint(h^{-1}(N_s euNint(S))) = h^{-1}(N_s euNint(S))$. So, $N_s euNint(S) \subseteq S$, implies $h^{-1}(N_s eu\ Nint(S)) \subseteq h^{-1}(S)$. Then, $N_s euNMint(h^{-1}(N_s euNint\ (S))) \subseteq N_s euNMint(h^{-1}(S))$. That is $h^{-1}(N_s euNint(S)) \subseteq N_s euNMint(h^{-1}(S))$.

Conversely, let $h^{-1}(N_s euNint(S)) \subseteq N_s euNMint(h^{-1}(S))$ for every subset S of U_2. If S is $N_s euNo$ in U_2, then $N_s euNint(S) = S$. Based on our assumption, $h^{-1}(N_s euNint(S)) \subseteq N_s euNMint(h^{-1}(S))$. Thus $h^{-1}(S) \subseteq N_s euNMint(h^{-1}(S))$. But $N_s euNMint(h^{-1}(S)) \subseteq h^{-1}(S)$. Hence, $N_s eu\ NMint(h^{-1}(S)) = h^{-1}(S)$. That is, $h^{-1}(S)$ is $N_s euNMo$ in U_1, for every $N_s euNo$ set S in U_2. Therefore h is $N_s euNMCts$ on U_1. □

Remark 4 A function h is $N_s euNMCts$ then $h^{-1}(N_s euNint(L))$ is not necessarily equal to $N_s euNMint(h^{-1}(L))$ where $L \in I^{U_2}$.

Example 6 In Example 1, $h : (U, \tau_N(F)) \to (U, \tau_N(F))$ is $N_s euNMCts$. Let $B_o = \left\{\left\langle \frac{s_1, s_4}{0.9, 0.5, 0.1} \right\rangle, \left\langle \frac{s_2}{0.8, 0.5, 0.2} \right\rangle, \left\langle \frac{s_3}{0.9, 0.5, 0.1} \right\rangle\right\} \in U$.

Then $N_s euNMint(h^{-1}(B_o)) = N_s euNMint(B_o) = B_o$.

But $h^{-1}(N_s euNint(B_o)) = h^{-1}\left(\left\{\left\langle \frac{s_1, s_4}{0.2, 0.5, 0.8} \right\rangle, \left\langle \frac{s_2}{0.3, 0.5, 0.7} \right\rangle, \left\langle \frac{s_3}{0.4, 0.5, 0.6} \right\rangle\right\}\right) = \left\{\left\langle \frac{s_1, s_4}{0.2, 0.5, 0.8} \right\rangle, \left\langle \frac{s_2}{0.3, 0.5, 0.7} \right\rangle, \left\langle \frac{s_3}{0.4, 0.5, 0.6} \right\rangle\right\}$.

Thus $N_s euNMint(h^{-1}(B_o)) \neq h^{-1}(N_s euNint(B_o))$.

Theorem 6 *Let h be a function. Then, h is a $N_s euNMCts$ function iff $N_s euN$ $Mcl(h^{-1}(S_o)) \subseteq h^{-1}(N_s euNMcl(S_o))$ for all $N_s eu$ subs S_o in U_2.*

Proof Let S_o be any $N_s eu\ subs$ in U_2 and h be a $N_s euNMCts$ function. From Theorem 4 (i), $h^{-1}(S_o) \subseteq h^{-1}(N_s euNMcl(S_o))$. Then, $N_s euNMcl(h^{-1}(S_o)) \subseteq N_s euNMcl(h^{-1}(N_s euNMcl(S_o)))$. Since $N_s euNMcl(S_o)$ is $N_s euNMc$ set in U_2, by Theorem 4, $h^{-1}(N_s euNMcl(S_o))$ is $N_s euNMc$ set in U_1. Thus, $N_s euNM$ $cl(h^{-1}(S_o)) \subseteq N_s euNMcl(h^{-1}(N_s euNMcl(S_o))) = h^{-1}(N_s euNM\ cl(S_o))$.

Conversely, $N_s euNMcl(h^{-1}(S_o)) \subseteq h^{-1}(N_s euNMcl(S_o))$ for all $N_s eu\ subs$ S_o in U_2. Let F_o be a $N_s euNc$ set in U_2. Since every $N_s euNc$ set is $N_s euNMc$ set, $N_s euNMcl(h^{-1}(F_o)) \subseteq h^{-1}(N_s euNMcl(F_o)) = h^{-1}(F_o)$. From Theorem 4, h is a $N_s euNMCts$ function. □

Theorem 7 *Let h be a bijective function. Then h is $N_s euNMCts$ iff $N_s euNM$ $int(h(S_o)) \subseteq h(N_s euNMint(S_o))$ for all $N_s eu$ subs S_o in U_1.*

Proof Let S_o be any $N_s eu\ subs$ in U_1 and h be a bijective and $N_s euNMCts$ function. Let $h(S_o) = T_o$. From Theorem 5 (i), $h^{-1}(N_s euNMint(T_o)) \subseteq h^{-1}(T_o)$. Since h is an injective function, $h^{-1}(T_o) = S_o$, so that $h^{-1}(N_s euNM\ int(T_o)) \subseteq S_o$. Therefore, $N_s euNMint(h^{-1}(N_s euNMint(T_o))) \subseteq N_s euNM\ int(S_o)$. Since h is $N_s euNMCts$, $h^{-1}(N_s euNMint(T_o))$ is $N_s euNMo$ set in U_1 and $h^{-1}(N_s eu$

$\mathcal{N}Mint(T_o)) \subseteq N_s eu\mathcal{N}Mint(S_o), h(h^{-1}(N_s eu\mathcal{N}Mint(T_o))) \subseteq h(N_s eu\mathcal{N}Mint(S_o))$. Hence, $N_s eu\mathcal{N}Mint(h(S_o)) \subseteq h(N_s eu\mathcal{N}Mint(S_o))$.

Conversely, $N_s eu\mathcal{N}Mint(h(S_o)) \subseteq h(N_s eu\mathcal{N}Mint(S_o))$ for all $N_s eu\ subs$ S_o in U_1. Let V be a $N_s eu\mathcal{N}o$ set in U_2. Then V is $N_s eu\mathcal{N}Mo$ in U_2. Since h is surjective and Theorem 5 (ii), $V = N_s eu\mathcal{N}Mint(V) = N_s eu\mathcal{N}Mint(h(h^{-1}(V))) \subseteq h(N_s eu\mathcal{N}Mint(h^{-1}(V)))$. It follows that, $h^{-1}(V) \subseteq N_s eu\mathcal{N}Mint(h^{-1}(V))$. Hence, $h^{-1}(V)$ is $N_s eu\mathcal{N}Mo$ set in U_1. Hence by Definition 1, h is a $N_s eu\mathcal{N}MCts$ function. □

Definition 2 For any two $N_s eu\ subs$'s S and T, then SqT means that S is neutrosophic nano-quasi-coincident with T if $\exists\ s \in U \ni S(s) + T(s) > 1$. That is, $\{\langle s, \mu_S(s) + \mu_T(s), \sigma_S(s) + \sigma_T(s), \nu_S(s) + \nu_T(s)\rangle : s \in U\} > 1$.

If S is not neutrosophic nano-quasi-coincident with T, then we write $S \not{q} T$.

Definition 3 Let S and T be any two $N_s eu\ subs$'s of a $N_s eu\mathcal{N}ts$'s. Then S is neutrosophic nano q-neighbourhood with T if $\exists$ a $N_s eu\mathcal{N}o$ set O with $SqO \subseteq T$.

Definition 4 A $N_s eu\ subs\ S$ in a $N_s eu\mathcal{N}ts$ $(U, \tau_N(F))$ is called a neutrosophic nano M q-neighborhood (briefly, $N_s eu\mathcal{N}Mq$-$nbhd$) of a fuzzy point $x_{(u,v,w)}$ if $\exists$ a $N_s eu\mathcal{N}Mo$ set V in $(U, \tau_N(F)) \ni x_{(u,v,w)}qV \subseteq S$.

Proposition 1 *Let h be a $N_s eu\mathcal{N}ts$. Then the following assertions are equivalent:*

(i) h is $N_s eu\mathcal{N}MCts$.
(ii) For each fuzzy point $x_{(u,v,w)} \in U_1$ and every $N_s eu\mathcal{N}Mq$-nbhd S of $h(x_{(u,v,w)})$, $\exists$ a $N_s eu\mathcal{N}Mo$ set T in $U_1 \ni x_{(u,v,w)} \in T \subseteq h^{-1}(S)$.
(iii) For each fuzzy point $x_{(u,v,w)} \in U_1$ and every $N_s eu\mathcal{N}Mq$-nbhd S of $h(x_{(u,v,w)})$, $\exists$ a $N_s eu\mathcal{N}Mo$ set T in $U_1 \ni x_{(u,v,w)} \in T$ and $h(T) \subseteq S$.

Remark 5 The composition of two $N_s eu\mathcal{N}MCts$ functions need not be $N_s eu\mathcal{N}MCts$ as seen from the following example.

Example 7 Assume $U = \{s_1, s_2, s_3, s_4\}$ and $U/R = \{\{s_1, s_4\}, \{s_2\}, \{s_3\}\}$. Let $S_o = \left\{\left\langle\frac{s_1}{0.2,0.5,0.8}\right\rangle, \left\langle\frac{s_2}{0.3,0.5,0.7}\right\rangle, \left\langle\frac{s_3}{0.4,0.5,0.6}\right\rangle, \left\langle\frac{s_4}{0.1,0.5,0.9}\right\rangle\right\}$ be a $N_s eu\ subs$ of U.

$\underline{N_s eu\mathcal{N}}(S_o) = \left\{\left\langle\frac{s_1,s_4}{0.1,0.5,0.9}\right\rangle, \left\langle\frac{s_2}{0.3,0.5,0.7}\right\rangle, \left\langle\frac{s_3}{0.4,0.5,0.6}\right\rangle\right\}$,

$\overline{N_s eu\mathcal{N}}(S_o) = \left\{\left\langle\frac{s_1,s_4}{0.2,0.5,0.8}\right\rangle, \left\langle\frac{s_2}{0.3,0.5,0.7}\right\rangle, \left\langle\frac{s_3}{0.4,0.5,0.6}\right\rangle\right\}$,

$B_{N_s eu\mathcal{N}}(S_o) = \left\{\left\langle\frac{s_1,s_4}{0.2,0.5,0.8}\right\rangle, \left\langle\frac{s_2}{0.3,0.5,0.7}\right\rangle, \left\langle\frac{s_3}{0.4,0.5,0.6}\right\rangle\right\}$.

Thus $\tau_N(S_o) = \{0_N, 1_N, \underline{N_s eu\mathcal{N}}(S_o), \overline{N_s eu\mathcal{N}}(S_o) = B_{N_s eu\mathcal{N}}(S_o)\}$.

Also, $V = \{t_1, t_2, t_3, t_4\}$ and $V/R = \{\{t_1, t_4\}, \{t_2\}, \{t_3\}\}$.

Let $T_o = \left\{\left\langle\frac{t_1}{0.1,0.5,0.9}\right\rangle, \left\langle\frac{t_2}{0.2,0.5,0.8}\right\rangle, \left\langle\frac{t_3}{0.1,0.5,0.9}\right\rangle, \left\langle\frac{t_4}{0.1,0.5,0.9}\right\rangle\right\}$ be a $N_s eu\ subs$ of V.

$\underline{N_s eu\mathcal{N}}(T_o) = \left\{\left\langle\frac{t_1,t_4}{0.1,0.5,0.9}\right\rangle, \left\langle\frac{t_2}{0.2,0.5,0.8}\right\rangle, \left\langle\frac{t_3}{0.1,0.5,0.9}\right\rangle\right\} = \overline{N_s eu\mathcal{N}}(T_o) = B_{N_s eu\mathcal{N}}(T_o)$.

Thus $\sigma_N(T_o) = \{0_N, 1_N, \underline{N_s eu\mathcal{N}}(T_o) = \overline{N_s eu\mathcal{N}}(T_o) = B_{N_s eu\mathcal{N}}(T_o)\}$.

Also, $W = \{r_1, r_2, r_3, r_4\}$ and $W/R = \{\{r_1, r_4\}, \{r_2\}, \{r_3\}\}$.

Let $R_o = \left\{\left\langle\frac{r_1}{0.1,0.5,0.9}\right\rangle, \left\langle\frac{r_2}{0.0,0.0,1.0}\right\rangle, \left\langle\frac{r_3}{0.6,0.5,0.4}\right\rangle, \left\langle\frac{r_4}{0.1,0.5,0.9}\right\rangle\right\}$ be a $N_s eu$ *subs* of W.

$\underline{N_s eu\mathcal{N}}(R_o) = \left\{\left\langle\frac{r_1, r_4}{0.1,0.5,0.9}\right\rangle, \left\langle\frac{r_2}{0.0,0.0,1.0}\right\rangle, \left\langle\frac{r_3}{0.6,0.5,0.4}\right\rangle\right\} = \overline{N_s eu\mathcal{N}}(R_o) = B_{N_s eu\mathcal{N}}(R_o)$.

Thus $\rho_N(R_o) = \{0_N, 1_N, \underline{N_s eu\mathcal{N}}(R_o) = \overline{N_s eu\mathcal{N}}(R_o) = B_{N_s eu\mathcal{N}}(R_o)\}$.

Then $h_1 : (U, \tau_N(F)) \to (V, \sigma_N(F))$ and $h_2 : (V, \sigma_N(F)) \to (W, \rho_N(F))$ are $N_s eu\mathcal{N}MCts$ but $(h_2 \circ h_1)$ is not $N_s eu\mathcal{N}MCts$.

Since, $B_o = \left\{\left\langle\frac{r_1, r_4}{0.1,0.5,0.9}\right\rangle, \left\langle\frac{r_2}{0.0,0.0,1.0}\right\rangle, \left\langle\frac{r_3}{0.6,0.5,0.4}\right\rangle\right\}$ is $N_s eu\mathcal{N}o$ set in W but $(h_2 \circ h_1)^{-1}(B_o)$ is not $N_s eu\mathcal{N}Mo$ set in U.

Theorem 8 *Let $h_1 : (U_1, \tau_N(F_1)) \to (U_2, \tau_N(F_2))$ and $h_2: (U_2, \tau_N(F_2)) \to (U_3, \tau_N(F_3))$ be any two functions. If h_1 is a $N_s eu\mathcal{N}MCts$ & h_2 is $N_s eu\mathcal{N}Cts$ functions, then $h_2 \circ h_1$ is $N_s eu\mathcal{N}MCts$.*

Proof Let C be any $N_s eu\mathcal{N}c$ in U_3. As h_2 is $N_s eu\mathcal{N}Cts$, $h_2^{-1}(C)$ is $N_s eu\mathcal{N}c$ in U_2. Since h_1 is $N_s eu\mathcal{N}MCts$, implies $h_1^{-1}(h_2^{-1}(C)) = (h_2 \circ h_1)^{-1}(C)$ is $N_s eu\mathcal{N}Mc$ in U_1. Therefore $h_2 \circ h_1$ is $N_s eu\mathcal{N}MCts$. □

References

1. Al-Hamido, R.K.: On neutrosophic crisp supra bi-topological spaces. International Journal of Neutrosophic Sciences. 4 (1), 36-46 (2020).
2. Atanassov, K., Stoeva, S.: Intuitionistic fuzzy sets. In Proceedings of the Polish Symposium on Interval and Fuzzy Mathematics, Poznan, Poland, 23-26 (1983).
3. Atanassov, K.: Intuitionistic fuzzy sets. Fuzzy Sets and Systems. 20, 87-96 (1986).
4. Chang, C.L.: Fuzzy topological spaces. J. Math. Anal. Appl. 24, 182-190 (1968).
5. Coker, D.: An introduction to intuitionistic fuzzy topological spaces. Fuzzy sets and systems. 88, 81-89 (1997).
6. Das, S.K., Edalatpanah, S.A.: A new ranking function of triangular neutrosophic number and its application in inter programming. International Journal of Neutrosophic Sciences. 4 (2), 82-92 (2020).
7. El-Maghrabi, A.I., Al-Juhani, M.A.: M-open sets in topological spaces. Pioneer J. Math. Sci. 4 (2), 213–230 (2011).
8. Kalaiyarasan, V., Tamilselvan, S., Vadivel, A., Sundar, C.J.: Normal spaces associated with fuzzy nano M-open sets and its application. Journal of Mathematics and Computer Science. 29 (2), 156-166 (2023).
9. Lellis Thivagar, M., Richard, C.: On nano forms of weekly open sets. International journal of mathematics and statistics invention. 1 (1), 31-37 (2013).
10. Lellis Thivagar, M., Jafari, S., Sutha Devi, V., Antonysamy, V.: A novel approach to nano topology via neutrosophic sets. Neutrosophic Sets and Systems. 20, 86-94 (2018).
11. Mullai, M., Sangeetha, K., Surya, R., Madhan Kumar, G., Jayabalan, R., Broumi, S.: A single valued neutrosophic inventory model with neutrosophic random variable. International Journal of Neutrosophic Sciences. 1 (2), 52-63 (2020).
12. Pankajam, V., Kavitha, K.: δ-open sets and δ-nano continuity in δ-nano topological spaces. International Journal of Innovative Science and Research Technology. 2 (12), 110-118 (2017).
13. Padma, A., Saraswathi, M., Vadivel, A., Saravanakumar, G.: New Notions of Nano M-open Sets. Malaya Journal of Matematik. S (1), 656-660 (2019).

14. Parimala, M., Jeevitha, R., Jafari, S., Samarandache, F., Karthika, M.: Neutrosophic nano $\alpha\psi$-closed sets in neutrosophic nano topological spaces. Journal of advance Research in Dynamical and Control Systems. 10, 523-531 (2018).
15. Pawlak, Z.: Rough sets. International Journal of Computer and Information Sciences. 11, 341-356 (1982).
16. Salama, A.A., Alblowi, S.A.: Generalized neutrosophic set and generalized neutrosophic topological spaces. Journal of Computer Sci. Engineering. 2 (7), 129-132 (2012).
17. Salama, A.A., Alblowi, S.A.: Neutrosophic Set and Neutrosophic Topological Spaces. IOSR Journal of Mathematics. 3 (4), 31-35 (2012).
18. Smarandache, F.: A Unifying field in logics: neutrosophic logic. neutrosophy, neutrosophic set, neutrosophic probability. (Rehoboth: American Research Press), (1999).
19. Smarandache, F.: Neutrosophy and neutrosophic logic. 1st Int. Conf. on Neutrosophy, Neutrosophic Logic, Set, Probability and Statistics, University of New Mexico, Gallup NM, 87301, USA, (2002).
20. Thangammal, R., Saraswathi, M., Vadivel, A., Sundar, C.J.: Fuzzy nano Z -open sets in fuzzy nano topological spaces. Journal of Linear and Topological Algebra. 11 (01), 27-38 (2022).
21. Thangammal, R., Saraswathi, M., Vadivel, A., Noeiaghdam, S., Sundar, C.J., Govindan, V., Iampan, A.: Fuzzy Nano Z-locally Closed Sets, Extremally Disconnected Spaces, Normal Spaces, and Their Application. Advances in Fuzzy Systems. 2022, 3364170 (2022).
22. Vadivel, A., Sundar, C.J.: Neutrosophic δ-Open Maps and Neutrosophic δ-Closed Maps. International Journal of Neutrosophic Science. 13 (2), 66-74 (2021).
23. Vadivel, A., Sundar, C.J.: New Operators Using Neutrosophic δ-Open Set. Journal of Neutrosophic and Fuzzy Systems. 1 (2), 61-70 (2021).
24. Vadivel, A., Sundar, C.J.: Application of Neutrosophic Sets Based on Mobile Network Using Neutrosophic Functions. Emerging Trends in Industry 4.0, 1-8 (2021).
25. Vadivel, A., Sundar, C.J., Kirubadevi, K., Tamilselvan, S.: More on Neutrosophic Nano Open Sets. International Journal of Neutrosophic Science. 18 (4), 204-222 (2022).
26. Vadivel, A., Sundar, C.J., Saraswathi, K., Tamilselvan, S.: Neutrosophic Nano M Open Sets. International Journal of Neutrosophic Science. 19 (1), 132-147 (2022).
27. Vadivel, A., Moogambigai, N., Tamilselvan, S., Thangaraja, P.: Application of Neutrosophic Sets Based on Neutrosophic Score Function in Material Selection. 2022 First International Conference on Electrical, Electronics, Information and Communication Technologies, 1-5 (2022).
28. Vadivel, A., Seenivasan, M., Sundar, C.J.: An Introduction to δ-open sets in a Neutrosophic Topological Spaces. Journal of Physics: Conference Series. 1724, 012011 (2021).
29. Vadivel, A., Thangaraja, P., Sundar, C.J.: Neutrosophic e-continuous maps and neutrosophic e-irresolute maps. Turkish Journal of Computer and Mathematics Education. 12 (1S), 369-375 (2021).
30. Vadivel, A., Thangaraja, P., Sundar, C.J.: Neutrosophic e-Open Maps, Neutrosophic e-Closed Maps and Neutrosophic e-Homeomorphisms in Neutrosophic Topological Spaces. AIP Conference Proceedings. 2364, 020016 (2021).
31. Zadeh, L.A.: Fuzzy sets. Information and control. 8, 338-353 (1965).

Summability in Measure of Two-Dimensional Walsh-Fourier Series

Ushangi Goginava

Keywords Two-dimensional Walsh-Fourier series · Summability Cesáro means with varying parameters · Convergence in measure

2000 Mathematics Subject Classification 42C10

1 Introduction

We shall denote the set of non-negative integers by $\mathbb{N}$. The Walsh(-Paley) system $(w_n : n \in \mathbb{N})$ was introduced by Paley in 1932. Let $S_n(f)$ denote the nth partial sum of one-dimensional Walsh-Fourier series, and let $S_{n,m}(f)$ denote the rectangular partial sums of two-dimensional Walsh-Fourier series. It is well-known that the partial sums $S_n(f)$ of every integrable function are convergent in measure to the function f. A similar theorem does not hold for quadratic sums $S_{n,n}(f)$ of two-dimensional Walsh-Fourier series. In particular, it is known that [6, 7, 9, 11, 12] there exists such an integrable function whose square partial sums $S_{n,n}(f)$ are divergent in measure.

The $(C; \alpha_n, \beta_m)$ means of two-dimensional Walsh-Fourier series are defined as follows:

$$\sigma_{n,m}^{(\alpha_n,\beta_m)}(f, x, y) = \frac{1}{A_{n-1}^{\alpha_n} A_{m-1}^{\beta_m}} \sum_{i=1}^{n} \sum_{j=1}^{m} A_{n-i}^{\alpha_n-1} A_{m-j}^{\beta_m-1} S_{i,j}(f, x, y),$$

The author is very thankful to United Arab Emirates University (UAEU) for the Start-up Grant 12S100.

U. Goginava (✉)
Department of Mathematical Sciences, United Arab Emirates University, Al Ain, Abu Dhabi, UAE
e-mail: ugoginava@uaeu.ac.ae

H.-H. Leung et al. (eds.), *Recent Developments in Algebra and Analysis*, Trends in Mathematics, https://doi.org/10.1007/978-3-031-37538-5_29

where

$$A_n^{\alpha_n} := \frac{(1+\alpha_n)\dots(n+\alpha_n)}{n!}$$

for any $n \in \mathbb{N}, \alpha_n \neq -1, -2, \dots$.

In the case when the functions α_n and β_m are positive constant numbers $\alpha_n = \alpha$ and $\beta_m = \beta$, then $(C; \alpha_n, \beta_m)$ means coincide with the well-known $(C; \alpha, \beta)$ means, and it is well-known that for each integrable function $(C; \alpha, \beta)$ means are convergent in L_1−norm and, therefore, in measure.

The presented paper will discuss the case when $\alpha_n, \beta_n \in (0, 1)$ and $\alpha_n, \beta_n \to 0$ as $n \to \infty$.

We Can Formulate the Following Questions *Let* α_n , $\beta_n \in (0, 1)$ and *tend to zero as* $n \to \infty$. *Does the* $L_1(\mathbb{I}^2)$*-class provide convergence in measure of* $\sigma_{n,m}^{(\alpha_n,\beta_m)}(f)$?

In the present chapter, the main aim is to establish the necessary and sufficient conditions for $\{\alpha_n\}$ and $\{\beta_m\}$ in order (C, α_n, β_m) means with varying parameters of two-dimensional Walsh-Fourier series to be convergent in measure.

2 Definitions and Notation

Set $\mathbb{I} := [0, 1)$ and $\mathbb{I}^2 := [0, 1) \times [0, 1)$. We denote by $L_0 = L_0(\mathbb{I}^2)$ the Lebesgue space of functions that are measurable and finite almost everywhere on $\mathbb{I}^2$; $\boldsymbol{\mu}(A)$ is the two-dimensional Lebesgue measure of the set $A \subset \mathbb{I}^2$. The one-dimensional Lebesgue measure of the set $A \subset \mathbb{I}$ will be denoted by $\mu(A)$. We denote by $L_1(\mathbb{I}^2)$ the class of all measurable functions f satisfying $\|f\|_1 := \int_{\mathbb{I}^2} |f| < \infty$. For the functions of two variables and for the set $E \subset \mathbb{I}^2$, the following notions will be used:

$$\boldsymbol{\mu}_1(E) := \int_{\mathbb{I}} \chi_E(x, y)\, dx \text{ for a.e. } y \in \mathbb{I} \tag{1}$$

and

$$\boldsymbol{\mu}_2(E) := \int_{\mathbb{I}} \chi_E(x, y)\, dy \text{ for a.e. } x \in \mathbb{I}, \tag{2}$$

where χ_E is characteristic function of the set E.

The (C, α_n) kernel is defined by

$$K_n^{\alpha_n} = \frac{1}{A_{n-1}^{\alpha}} \sum_{j=1}^{n} A_{n-j}^{\alpha_n-1} D_j.$$

It is evident that

$$\sigma_{n,m}^{(\alpha_n,\beta_m)}(f,x,y) = f * \left(K_n^{\alpha_n} \otimes K_m^{\beta_m}\right)(x,y) \tag{3}$$
$$= \sigma_m^{\beta_m}\left(\sigma_n^{\alpha_n}(f);x,y\right) = \sigma_n^{\alpha_n}\left(\sigma_m^{\beta_m}(f);x,y\right),$$

where $\sigma_n^{\alpha_n}(f) := f * K_n^{\alpha_n}$ is the (C,α_n) means of one-dimensional Walsh-Fourier series of the function $f \in L_1(\mathbb{I})$.

3 Main Results

We assume that $\alpha_n, \beta_m \in (0,1)$ and $\lim_{n\to\infty} \alpha_n = \alpha, \quad \lim_{n\to\infty} \beta_n = \beta$.

Theorem 1 *Let $f \in L_1(\mathbb{I}^2)$ and $\alpha + \beta > 0$. Then*

(a)

$$\mu\left\{(x,y) \in \mathbb{I}^2 : \left|\sigma_{n,m}^{(\alpha_n,\beta_m)}(f,x,y)\right| > \lambda\right\} \le \frac{c}{\lambda}\int_{\mathbb{I}^2} |f(x,y)|\, dxdy;$$

(b)

$$\sigma_{n,m}^{(\alpha_n,\beta_m)}(f) \to f \quad \text{in measure on } \mathbb{I}^2, \text{ as } n,m \to \infty.$$

Theorem 2 *Let*

$$\alpha + \beta = 0. \tag{4}$$

The set of the functions from the space $L_1(\mathbb{I}^2)$ with sequence $\sigma_{n,n}^{(\alpha_n,\beta_n)}(f)$ convergent in measure on $\mathbb{I}^2$ is of first Baire category in $L_1(\mathbb{I}^2)$.

4 Proof of the Theorems

Proof of Theorem 1 Since $\alpha + \beta > 0$ at least one α and β is not equal to zero. Without loss of generality, we can assume that $\beta > 0$. Set

$$\Omega := \left\{(x,y) : \left|\sigma_{n,m}^{(\alpha_n,\beta_m)}(f,x,y)\right| > \lambda\right\}.$$

Since (see [3])

$$\lambda\mu\left(\left\{x \in \mathbb{I} : \left|\sigma_n^{\alpha_n}(f,x)\right| > \lambda\right\}\right) \le c\,\|f\|_1, \quad f \in L_1(\mathbb{I}) \text{ and } \alpha_n \in (0,1),$$

by Fubin's theorem, we can write (see (3))

$$\mu(\Omega) = \int_{\mathbb{I}^2} \chi_\Omega(x, y)\, dxdy = \int_{\mathbb{I}} \left(\int_{\mathbb{I}} \chi_\Omega(x, y)\, dx \right) dy \tag{5}$$

$$= \int_{\mathbb{I}} \mu_1 \left\{ x : \left| \sigma_n^{\alpha_n} \left(\sigma_m^{\beta_m}(f), x, y \right) \right| > \lambda \right\} dy$$

$$\leq \int_{\mathbb{I}} \frac{c \left\| \sigma_m^{\beta_m}(f; \cdot, y) \right\|_1}{\lambda} dy = \int_{\mathbb{I}} \frac{c \left\| \sigma_m^{\beta_m}(f; x, \cdot) \right\|_1}{\lambda} dx.$$

Since $\beta > 0$, using estimations (see [4])

$$\left\| K_n^{\beta_n} \right\|_1 \sim \frac{1}{n^{\beta_n}} \sum_{k=0}^{|n|} \left| \varepsilon_k(n) - \varepsilon_{k+1}(n) \right| 2^{k\beta_n}$$

we conclude that

$$\sup_m \left\| K_m^{\beta_m} \right\|_1 \leq c < \infty.$$

On the other hand, $f(x, \cdot) \in L_1(\mathbb{I})$, for a. e. $x \in \mathbb{I}$ and $f \in L_1(\mathbb{I}^2)$ and $\left\| \sigma_m^{\beta_m}(f; x, \cdot) \right\|_1 \leq \| f(x, \cdot) \|_1 \sup_m \left\| K_m^{\beta_m} \right\|_1 \leq c \| f(x, \cdot) \|_1$. Hence, from (5) we have $\mu(\Omega) \leq \frac{c}{\lambda} \int_{\mathbb{I}^2} |f(x, y)|\, dxdy$. By density of polynomials and by virtue of standart arguments (see [13], ch. VII), we can obtain validity of (b) of Theorem 1. □

Proof of Theorem 2 *It is proved in [3] that*

$$K_n^{\alpha_n} = \frac{1}{A_{n-1}^{\alpha_n}} \sum_{s=0}^{|n|} \varepsilon_s(n) w_{n^{(s)}-1} \sum_{j=1}^{2^s-1} A_{n^{(s-1)}+j}^{\alpha_n-2} j K_j \tag{6}$$

$$-\frac{1}{A_{n-1}^{\alpha_n}} \sum_{s=0}^{|n|} \varepsilon_s(n) w_{n^{(s)}-1} A_{n^{(s)}-1}^{\alpha_n-1} 2^s K_{2^s}$$

$$+\frac{1}{A_{n-1}^{\alpha_n}} \sum_{s=0}^{|n|} \varepsilon_s(n) w_{n^{(s)}-1} A_{n^{(s)}-1}^{\alpha_n} D_{2^s}$$

$$=: T_n^{(1)}(\alpha_n) + T_n^{(2)}(\alpha_n) + T_n^{(3)}(\alpha_n),$$

where

$$n^{(s)} := \sum_{j=s}^{\infty} \varepsilon_j(n) 2^j, n_{(s)} = n - n^{(s+1)} = \sum_{j=0}^{s} \varepsilon_j(n) 2^j.$$

Then the operator $\sigma_{n,m}^{(\alpha_n,\beta_m)}(f)$ can be represented as follows:

$$\begin{aligned}&\sigma_{n,m}^{(\alpha_n,\beta_m)}(f,x,y)\\&=\sum_{i,j=1}^{3} f * \left(T_n^{(i)}(\alpha_n) \otimes T_m^{(j)}(\beta_m)\right)(x,y) = \sum_{i,j=1}^{3} J(i,j).\end{aligned}$$

Since (see [10, p. 46]) $\sup_j \|K_j\|_1 < \infty$, we have

$$\begin{aligned}&\left\|T_n^{(1)}(\alpha_n)\right\|_1\\&\leq \frac{1}{A_{n-1}^{\alpha_n}} \sum_{s=0}^{|n|} \varepsilon_s(n) \sum_{j=1}^{2^s-1} \left|A_{n_{(s-1)}+j}^{\alpha_n-2}\right| j \|K_j\|_1\\&\leq c\frac{1}{A_{n-1}^{\alpha_n}} \sum_{s=0}^{|n|} \varepsilon_s(n) \sum_{j=1}^{2^s-1} \frac{|\alpha_n-1|}{\alpha_n-1+n_{(s-1)}+j} A_{n_{(s-1)}+j}^{\alpha_n-1} j\\&= c\frac{1}{A_{n-1}^{\alpha_n}} \sum_{s=0}^{|n|} \varepsilon_s(n) \left(A_{n_{(s)}}^{\alpha_n} - A_{n_{(s-1)}}^{\alpha_n}\right) \leq c < \infty.\end{aligned}$$

Hence,

$$\begin{aligned}&\|J(1,1)\|_1\\&\leq c\|f\|_1 \left\|T_n^{(1)}(\alpha_n) \otimes T_m^{(1)}(\beta_m)\right\|_1\\&\leq c\|f\|_1.\end{aligned} \tag{7}$$

Analogously, we can prove that

$$\|J(i,j)\|_1 \leq c\|f\|_1, (i,j) \in \{(1,2),(2,1),(2,2)\}. \tag{8}$$

Since

$$T_n^{(3)}(\alpha_n) = \frac{w_n}{A_{n-1}^{\alpha_n}} \sum_{s=0}^{|n|} \varepsilon_s(n) A_{n_{(s)}-1}^{\alpha_n} \left(D_{2^{s+1}} - D_{2^s}\right),$$

we get

$$
\begin{aligned}
& f * T_n^{(3)} (\alpha_n) \\
& = f * \left(\frac{w_n}{A_{n-1}^{\alpha_n}} \sum_{s=0}^{|n|} \varepsilon_s (n) A_{n_{(s)}-1}^{\alpha_n} \left(D_{2^{s+1}} - D_{2^s} \right) \right) \\
& = w_n \sum_{s=0}^{|n|} \varepsilon_s (n) \frac{A_{n_{(s)}-1}^{\alpha_n}}{A_{n-1}^{\alpha_n}} \left(S_{2^{s+1}} (f w_n) - S_{2^s} (f w_n) \right)
\end{aligned}
$$

Since

$$
\sup_k \left(\varepsilon_k (n) \frac{A_{n_{(k)}-1}^{\alpha_n}}{A_{n-1}^{\alpha_n}} \right) \leq 1,
$$

we conclude that (see [10, p. 97])

$$
\lambda \mu \left(\left\{ \left| f * T_n^{(3)} (\alpha_n) \right| > \lambda \right\} \right) \leq c \| f \|_1 .
$$

If we use the same way that was used to prove Theorem 1, we get that

$$
\mu \left(\{ |J (i, j)| > \lambda \} \right) \leq \frac{c}{\lambda} \int_{\mathbb{I}^2} |f| , (i, j) \in \{ (1, 3) , (2, 3) , (3, 1) , (3, 2) \} . \tag{9}
$$

Set

$$
n_A := 2^{2A} + 2^{2A-2} + \cdots + 2^2 + 2^0, \quad \delta_A := \max \left\{ \alpha_{n_A}, \beta_{n_A} \right\}
$$

and define

$$
f_A (x, y) := D_{2^{2A+1}} (x) D_{2^{2A+1}} (y) ,
$$

where $D_{2^n} := \sum_{k=0}^{2^n-1} w_k$. First, we assume that $\overline{\lim_{A \to \infty}} (A \delta_A) \geq 1$. Then there exists a sequence $\{ A_k : k \in \mathbb{N} \}$ such that $\delta_{A_k} > \frac{2}{3A_k}$. Set$\lambda_{A_k} := 4A_k - \frac{1}{\delta_{A_k}}$.Then it is easy to see that $\delta_{A_k} \to 0$ as $k \to \infty$ and $\sup_A \lambda_{A_k} = \infty$. Without loss of generality, we can assume that $A_k = A$ and $\lambda_A \to \infty$ as $A \to \infty$. Now, we discuss the operator $f_A * T_{n_A}^{(3)} \left(\alpha_{n_A} \right) \otimes T_{n_A}^{(3)} \left(\beta_{n_A} \right)$. We have

$$
\left| f_A * \left(T_{n_A}^{(3)} \left(\alpha_{n_A} \right) \otimes T_{n_A}^{(3)} \left(\beta_{n_A} \right) \right) \right| \tag{10}
$$

$$= \left| \frac{1}{A_{n_A-1}^{\alpha_{n_A}}} \sum_{s=0}^{2A} \varepsilon_s(n_A) A_{(n_A)_{(s)}-1}^{\alpha_{n_A}} \left(D_{2^{s+1}} - D_{2^s}\right) \right|$$

$$\times \left| \frac{1}{A_{n_A-1}^{\beta_{n_A}}} \sum_{s=0}^{2A} \varepsilon_s(n_A) A_{(n_A)_{(s)}-1}^{\beta_{n_A}} \left(D_{2^{s+1}} - D_{2^s}\right) \right|.$$

We assume that $x \in I_a \backslash I_{a+1}$, $a = 0, 1, \ldots, 2A-1$. Since (see [10, p. 7])

$$D_{2^s} = 2^s \chi_{[0,2^s)} \tag{11}$$

from (10) we obtain

$$\frac{1}{A_{n_A-1}^{\alpha_{n_A}}} \sum_{s=0}^{2A} \varepsilon_s(n_A) A_{(n_A)_{(s)}-1}^{\alpha_{n_A}} \left(D_{2^{s+1}}(x) - D_{2^s}(x)\right)$$

$$= \frac{1}{A_{n_A-1}^{\alpha_{n_A}}} \sum_{s=0}^{a-1} \varepsilon_s(n_A) A_{(n_A)_{(s)}-1}^{\alpha_{n_A}} 2^s - \varepsilon_a(n_A) A_{(n_A)_{(a)}-1}^{\alpha_{n_A}} 2^a.$$

Since $|\varepsilon_{a-1}(n_A) - \varepsilon_a(n_A)| = 1$, two cases are possible: (a) $\varepsilon_a(n_A) = 0$ and $\varepsilon_{a-1}(n_A) = 1$; (b) $\varepsilon_a(n_A) = 1$ and $\varepsilon_{a-1}(n_A) = 0$. First we consider case (a). Since (see [1]) $A_k^{\alpha_n} \sim k^{\alpha_n}$, when $0 < \alpha_n \le 1$ from (11) we have

$$\left| \frac{1}{A_{n_A-1}^{\alpha_{n_A}}} \sum_{s=0}^{2A} \varepsilon_s(n_A) A_{(n_A)_{(s)}-1}^{\alpha_{n_A}} \left(D_{2^{s+1}}(x) - D_{2^s}(x)\right) \right| \tag{12}$$

$$\ge \frac{A_{(n_A)_{(a-1)}-1}^{\alpha_{n_A}} 2^{a-1}}{A_{n_A-1}^{\alpha_{n_A}}} \ge \frac{c 2^{a(1+\alpha_A)}}{2^{2A\alpha_A}} \ge \frac{c 2^{a(1+\delta_A)}}{2^{2A\delta_A}}.$$

Now, we consider case (b), and we have

$$\left| \frac{1}{A_{n_A-1}^{\alpha_{n_A}}} \sum_{s=0}^{2A} \varepsilon_s(n_A) A_{(n_A)_{(s)}-1}^{\alpha_{n_A}} \left(D_{2^{s+1}}(x) - D_{2^s}(x)\right) \right|$$

$$\ge \frac{1}{A_{n_A-1}^{\alpha_{n_A}}} \left| A_{2^a-1}^{\alpha_{n_A}} 2^a - \sum_{s=0}^{a-2} A_{2^{s+1}-1}^{\alpha_{n_A}} 2^s \right|.$$

Since

$$\sum_{s=1}^{a-1} A_{2^s-1}^{\alpha_{n_A}} 2^{s-1} \le \frac{1}{2} \sum_{s=0}^{a-2} \sum_{l=2^s}^{2^{s+1}-1} A_l^{\alpha_{n_A}} = \frac{1}{2} \sum_{l=1}^{2^{a-1}-1} A_l^{\alpha_{n_A}} \le \frac{1}{2} A_{2^{a-1}-1}^{\alpha_{n_A}+1}$$

and

$$A_{2^{a-1}-1}^{\alpha_{n_A}+1} = \frac{\alpha_{n_A} + 2^{a-1}}{\alpha_{n_A}+1} A_{2^{a-1}-1}^{\alpha_{n_A}} \leq 2^{a-1} A_{2^{a-1}-1}^{\alpha_{n_A}},$$

we obtain,

$$\left| \frac{1}{A_{n_A-1}^{\alpha_{n_A}}} \sum_{s=0}^{2A} \varepsilon_s (n_A) A_{(n_A)(s)-1}^{\alpha_{n_A}} \left(D_{2^{s+1}}(x) - D_{2^s}(x) \right) \right| \tag{13}$$

$$\geq \frac{c 2^{a(1+\alpha_{n_A})}}{2^{2A\alpha_{n_A}}} \geq \frac{c 2^{a(1+\delta_A)}}{2^{2A\delta_A}}.$$

Let $(x, y) \in (I_a \backslash I_{a+1}) \times (I_b \backslash I_{b+1})$ for some $(a, b) \in \{0, 1, \ldots, 2A-1\} \times \{0, 1, \ldots, 2A-1\}$. Then combining (10), (12), and (13), we have the following lower estimation:

$$\left| f_A * \left(T_{n_A}^{(3)} (\alpha_{n_A}) \otimes T_{n_A}^{(3)} (\beta_{n_A}) \right) \right|$$

$$\geq \frac{c 2^{(a+b)(\delta_A+1)}}{2^{4A\delta_A}} \geq c 2^{a+b},$$

when $2A - 1/\delta_A \leq a, b \leq 2A$. Consequently,

$$\mu \left(\left\{ (x, y) : \left| f_A * \left(T_{n_A}^{(3)} (\alpha_{n_A}) \otimes T_{n_A}^{(3)} (\beta_{n_A}) \right) \right| > 2^{\lambda_A} \right\} \right) \tag{14}$$

$$\geq c \sum_{2A-1/\delta_A \leq a < 2A} \sum_{4A-1/\delta_A - a \leq b \leq 2A} \frac{1}{2^{a+b}} \geq \frac{c 1/\delta_A}{2^{\lambda_A}}.$$

Combining (6), (7), (8), (9), and (14), we conclude that

$$\mu \left(\left\{ (x, y) : \left| f_A * \left(K_{n_A}^{\alpha_{n_A}} \otimes K_{n_A}^{\beta_{n_A}} \right) (x, y) \right| > 2^{\lambda_A} \right\} \right) \geq c \frac{1/\delta_A}{2^{\lambda_A}}. \tag{15}$$

Set

$$Q_A := \left\{ (x, y) : \left| f_A * \left(K_{n_A}^{\alpha_{n_A}} \otimes K_{n_A}^{\beta_{n_A}} \right) (x, y) \right| > 2^{\lambda_A} \right\}.$$

Now, we prove the following: There exists $(x_1, y_1), \ldots, (x_{l(A)}, y_{l(A)}) \in \mathbb{I}^2, l(A) := [2^{\lambda_A} \delta_A] + 1$, such that

$$\left| \bigcup_{j=1}^{l(A)} \left(Q_A \dotplus (x_j, y_j) \right) \right| \geq c > 0. \tag{16}$$

On the other hand, the existence of such pairs $(x_1, y_1), \ldots, (x_{l(A)}, y_{l(A)}) \in \mathbb{I}^2$ for which inequality (16) will take place for $c = 1/2$ was proved in the paper [8]. Then using Stein's method from (16), we can show that ([2], pp. 7-12) there exists $t_0 \in \mathbb{I}$, such that

$$\mu\left\{(x, y) \in \mathbb{I}^2 : \left|\sum_{j=1}^{l(A)} r_j(t_0) \left(f_A * \left(K_{n_A}^{\alpha_{n_A}} \otimes K_{n_A}^{\beta_{n_A}}\right)(x \dotplus x_j, y \dotplus y_j)\right)\right| > 2^{\lambda_A}\right\}$$
$$\geq \frac{1}{8}. \tag{17}$$

Set

$$G_A(x, y) = \frac{1}{l(A)} \sum_{j=1}^{l(A)} r_j(t_0) f_A(x \dotplus x_j, y \dotplus y_j).$$

It is easy to show that

$$\|G_A\|_1 \leq 1$$

and

$$\mu\left\{(x, y) \in \mathbb{I}^2 : \left|G_A * \left(K_{n_A}^{\alpha_{n_A}} \otimes K_{n_A}^{\beta_{n_A}}\right)\right| > 2^{\lambda_A}\right\} \geq \frac{1}{8}. \tag{18}$$

The theorem can be proved from inequality (18) and applying the method that was used in the paper (see [5]).

Finally, let us consider the case when $\delta_A < \frac{1}{3A}$, $A \geq A_0 \in N$. Then from

$$\left|f_A * \left(T_{n_A}^{(3)}(\alpha_{n_A}) \otimes T_{n_A}^{(3)}(\beta_{n_A})\right)\right|$$
$$\geq c2^{a+b}, (x, y) \in (I_a \backslash I_{a+1}) \times (I_b \backslash I_{b+1}), a, b = 0, 1, \ldots, 2A - 1.$$

Consequently, we can write

$$\mu\left(\left\{(x, y) : \left|f_A * \left(T_{n_A}^{(3)}(\alpha_{n_A}) \otimes T_{n_A}^{(3)}(\beta_{n_A})\right)\right| > 2^{3A}\right\}\right)$$
$$\geq c \sum_{a=0}^{3A} \sum_{b=3A-a}^{3A} \frac{1}{2^{a+b}} \geq \frac{cA}{2^{3A}},$$

and in the same way as the above, the proof of Theorem 2 will be completed. □

References

1. Akhobadze, T.: Lebesgue constants of generalized Cesáro (C, α_n)-means of trigonometric series. Bull. Georgian Natl. Acad. Sci. **175**(2), 24–26 (2007)
2. Garsia, A.M.: Topics in Almost Everywhere Convergence. Lectures in Advanced Mathematics, no. 4. Markham Publishing, Chicago (1970)
3. Gát, G., Goginava, U.: Almost everywhere convergence and divergence of cesáro means with varying parameters of Walsh-Fourier series. Arab. J. Math. **11**(2), 241–259 (2022)
4. Gát, G., Goginava, U.: Cesàro means with varying parameters of Walsh-Fourier series. Period. Math. Hungar. **87**(1), 57–74 (2023)
5. Gát, G., Goginava, U., Tkebuchava, G.: Convergence in measure of logarithmic means of quadratical partial sums of double Walsh-Fourier series. J. Math. Anal. Appl. **323**(1), 535–549 (2006)
6. Getsadze, R.D.: Divergence in measure of general multiple orthogonal Fourier series. Dokl. Akad. Nauk SSSR **306**(1), 24–25 (1989)
7. Getsadze, R.D.: Divergence in measure of general multiple orthogonal Fourier series. volume 190, pages 75–87. 1989. Translated in Proceedings of the Steklov Institute of Mathematics, vol. 1, pp. 77–90 (1992). Theory of functions (Russian). Amberd (1987)
8. Goginava, U.: Subsequences of spherical sums of double Walsh-Fourier series. Nonlinear Stud. **26**(4), 821–830 (2019)
9. Konyagin, S.V.: On a subsequence of Fourier-Walsh partial sums. Mat. Zametki **54**(4), 69–75, 158 (1993)
10. Schipp, F., Wade, W.R., Simon, P.: Walsh Series. Adam Hilger Ltd, Bristol (1990). An introduction to dyadic harmonic analysis. With the Collaboration of J. Pál
11. Tkebuchava, G.E.: On multiple Fourier, Fourier-Walsh and Fourier-Haar series in nonreflexive separable Orlicz spaces. Soobshch. Akad. Nauk Gruzii **149**(2), 199–202 (1994/1995)
12. Tkebuchava, G.: Subsequences of partial sums of multiple Fourier and Fourier-Walsh series. Bull. Georgian Acad. Sci. **169**(2), 252–253 (2004)
13. Zygmund, A.: Trigonometric Series. Vol. I, II. Cambridge Mathematical Library, 3rd edn. Cambridge University Press, Cambridge (2002). With a foreword by Robert A. Fefferman

Some Topological Operators Using Neutrosophic Nano *M* Open Sets

K. Saraswathi, A. Vadivel, S. Tamilselvan, and C. John Sundar

Keywords $N_seu\mathcal{N}Mo$ set · $N_seu\mathcal{N}MFr$ · $N_seu\mathcal{N}MBr$ · $N_seu\mathcal{N}MExt$

1 Introduction

Zadeh [32] was the first one to introduce fuzzy set and its topological spaces by Chang [4]. Atanassov [2, 3] introduced intuitionistic fuzzy sets and its topological spaces by Coker [5]. In 1995, Smarandache [18, 19] introduced neutrosophic logic and its topological spaces by Salama et al. [17] and their applications by several authors [1, 6, 11, 14, 24, 30]. Pawlak [15] was the first one to introduce the rough set theory. In 2013, Lellis Thivagar [9] presented nano-topology and nano-topological spaces.

Pankajam [13] introduced δ-open sets in nano-topological space and Vadivel et al. [22, 23, 25–27] in a neutrosophic topological space. El-Maghrabi and Al-Juhani [7] introduced M-open sets in topological spaces and in nano-topological spaces by Padma et al. [12].

Recently, a novel idea of neutrosophic nano-topology was investigated by Lellis Thivagar et al. [10]. Recently, Thangammal et al. [20, 21] and Kalaiyarasan et al. [8] introduced some open sets in fuzzy nano-topological spaces, and Vadivel et al.

K. Saraswathi
Department of Mathematics, Thiruvalluvar Government Arts College, Rasipuram, Namakkal, India

A. Vadivel (✉)
Department of Mathematics, Arignar Anna Government Arts College, Namakkal, India

Department of Mathematics, Annamalai University, Annamalai Nagar, Chidambaram, India

S. Tamilselvan
Engineering Mathematics, Annamalai University, Annamalai Nagar, Chidambaram, India

C. J. Sundar
Department of Mathematics, Sri Venkateshwaraa College of Engineering and Technology, Puducherry, India

H.-H. Leung et al. (eds.), *Recent Developments in Algebra and Analysis*,
Trends in Mathematics, https://doi.org/10.1007/978-3-031-37538-5_30

[28, 29] investigated δ-open sets and M-open sets in neutrosophic nano-topological spaces. In 2020, Vijayalakshmi and Mookambika [31] introduced neutrosophic nano-frontier in neutrosophic nano-topological spaces.

2 Preliminaries

The basic definitions used in this article are defined and studied in these articles [10, 12, 16, 28, 29].

3 Neutrosophic Nano M Frontier

In this section, neutrosophic nano M-frontier operators are introduced and their properties discussed.

Throughout this section, let $(U, \tau_N(F))$ be a $N_seu\mathcal{N}ts$ and S_o be a $N_seu\ subs$ of $(U, \tau_N(F))$.

Definition 1 The neutrosophic nano (resp. θ, θ semi, δ pre & M) frontier of a $N_seu\ subsS_o$ were denoted by $N_seu\mathcal{N}Fr(S_o)$ (resp. $N_seu\mathcal{N}\theta Fr(S_o)$, $N_seu\mathcal{N}\theta SFr(S_o)$, $N_seu\mathcal{N}\delta\mathcal{P}Fr(S_o)$&$N_seu\mathcal{N}MFr(S_o)$) and defined by $N_seu\mathcal{N}Fr(S_o) = N_seu\mathcal{N}cl(S_o) \cap N_seu\mathcal{N}cl(S_o{}^c)$ (resp. $N_seu\mathcal{N}\theta Fr(S_o) = N_seu\mathcal{N}\theta cl(S_o) \cap N_seu\mathcal{N}\theta cl(S_o{}^c)$, $N_seu\mathcal{N}\theta SFr(S_o) = N_seu\mathcal{N}\theta Scl(S_o) \cap N_seu\mathcal{N}\theta Scl(S_o{}^c)$, $N_seu\mathcal{N}\delta\mathcal{P}Fr(S_o) = N_seu\mathcal{N}\delta\mathcal{P}cl(S_o) \cap N_seu\mathcal{N}\delta\mathcal{P}cl(S_o{}^c)$&$N_seu\mathcal{N}MFr(S_o) = N_seu\mathcal{N}Mcl(S_o) \cap N_seu\mathcal{N}Mcl(S_o{}^c)$).

Example 1 Assume $U = \{s_1, s_2, s_3, s_4\}$ and $U/R = \{\{s_1, s_4\}, \{s_2\}, \{s_3\}\}$. Let $S_o = \left\{\left\langle\frac{s_1}{0.2,0.5,0.8}\right\rangle, \left\langle\frac{s_2}{0.3,0.5,0.7}\right\rangle, \left\langle\frac{s_3}{0.4,0.5,0.6}\right\rangle, \left\langle\frac{s_4}{0.1,0.5,0.9}\right\rangle\right\}$ be a $N_seu\ subs$ of U.

$\underline{N_seu\mathcal{N}}(S_o) = \left\{\left\langle\frac{s_1,s_4}{0.1,0.5,0.9}\right\rangle, \left\langle\frac{s_2}{0.3,0.5,0.7}\right\rangle, \left\langle\frac{s_3}{0.4,0.5,0.6}\right\rangle\right\}$,

$\overline{N_seu\mathcal{N}}(S_o) = \left\{\left\langle\frac{s_1,s_4}{0.2,0.5,0.8}\right\rangle, \left\langle\frac{s_2}{0.3,0.5,0.7}\right\rangle, \left\langle\frac{s_3}{0.4,0.5,0.6}\right\rangle\right\}$,

$B_{N_seu\mathcal{N}}(S_o) = \left\{\left\langle\frac{s_1,s_4}{0.2,0.5,0.8}\right\rangle, \left\langle\frac{s_2}{0.3,0.5,0.7}\right\rangle, \left\langle\frac{s_3}{0.4,0.5,0.6}\right\rangle\right\}$.

Thus $\tau_N(S_o) = \{0_N, 1_N, \underline{N_seu\mathcal{N}}(S_o), \overline{N_seu\mathcal{N}}(S_o) = B_{N_seu\mathcal{N}}(S_o)\}$.

Then $A = \left\{\left\langle\frac{s_1,s_4}{0.1,0.5,0.9}\right\rangle, \left\langle\frac{s_2}{0.2,0.5,0.8}\right\rangle, \left\langle\frac{s_3}{0.1,0.5,0.9}\right\rangle\right\}$,

$N_seu\mathcal{N}Mcl(A) = \left\{\left\langle\frac{s_1,s_4}{0.1,0.5,0.9}\right\rangle, \left\langle\frac{s_2}{0.2,0.5,0.8}\right\rangle, \left\langle\frac{s_3}{0.1,0.5,0.9}\right\rangle\right\}$,

$N_seu\mathcal{N}Mcl(A^c) = \left\{\left\langle\frac{s_1,s_4}{0.9,0.5,0.1}\right\rangle, \left\langle\frac{s_2}{0.8,0.5,0.2}\right\rangle, \left\langle\frac{s_3}{0.9,0.5,0.1}\right\rangle\right\}$.

$N_seu\mathcal{N}MFr(A) = \left\{\left\langle\frac{s_1,s_4}{0.1,0.5,0.9}\right\rangle, \left\langle\frac{s_2}{0.2,0.5,0.8}\right\rangle, \left\langle\frac{s_3}{0.1,0.5,0.9}\right\rangle\right\}$.

Remark 1 For a $N_seu\ subsA$ of U, $N_seu\mathcal{N}MFr(A)$ is $N_seu\mathcal{N}Mc$.

Theorem 1 *For a* $N_s eu$ *subs* S_o *in* $N_s eu\mathcal{N}ts(U, \tau_N(F))$, $N_s eu\mathcal{N}MFr(S_o) = N_s eu\mathcal{N}MFr(S_o^c)$.

Proof Let S_o be a $N_s eu$ *subs* in $N_s eu\mathcal{N}ts(U, \tau_N(F))$. Then by Definition 1, $N_s eu\mathcal{N}MFr(S_o) = N_s eu\mathcal{N}Mcl(S_o) \cap N_s eu\mathcal{N}Mcl(S_o^c) = N_s eu\mathcal{N}Mcl(S_o^c) \cap N_s eu\mathcal{N}Mcl(S_o) = N_s eu\mathcal{N}Mcl(S_o^c) \cap (N_s eu\mathcal{N}Mcl(S_o^c)^c)$. Again by Definition 1, this is equal to $N_s eu\mathcal{N}MFr(S_o^c)$. Hence $N_s eu\mathcal{N}MFr(S_o) = N_s eu\mathcal{N}MFr(S_o^c)$. □

Theorem 2 *For,* $N_s eu\mathcal{N}MFr(S_o) = N_s eu\mathcal{N}Mcl(S_o) - N_s eu\mathcal{N}Mint(S_o)$.

Proof Let S_o be in $(U, \tau_N(F))$. By Theorem 4.2. (i) in [29], $(N_s eu\mathcal{N}Mcl(S_o^c))^c = N_s eu\mathcal{N}Mint(S_o)$ and by Definition 1, $N_s eu\mathcal{N}MFr(S_o) = N_s eu\mathcal{N}Mcl(S_o) \cap (N_s eu\mathcal{N}Mcl(S_o^c)) = N_s eu\mathcal{N}Mcl(S_o) \cap (N_s eu\mathcal{N}Mint(S_o^c))^c$. By using $S_o - T_o = S_o \cap {T_o}^c$, $N_s eu\mathcal{N}MFr(S_o) = N_s eu\mathcal{N}Mcl(S_o) - N_s eu\mathcal{N}Mint(S_o)$. Hence $N_s eu\mathcal{N}MFr(S_o) = N_s eu\mathcal{N}Mcl(S_o) - N_s eu\mathcal{N}Mint(S_o)$. □

Theorem 3 *A* $N_s eu$ *subs* S_o *is* $N_s eu\mathcal{N}Mc$ *set in* U *iff* $N_s eu\mathcal{N}MFr(S_o) \subseteq S_o$.

Proof Let S_o be a $N_s eu\mathcal{N}Mc$ in the $N_s eu\mathcal{N}ts(U, \tau_N(F))$. Then by Definition 1, $N_s eu\mathcal{N}MFr(S_o) = N_s eu\mathcal{N}Mcl(S_o) \cap N_s eu\mathcal{N}Mcl(S_o^c) \subseteq N_s eu\mathcal{N}Mcl(S_o)$. By using Theorem 4.2. (ii) in [29], $N_s eu\mathcal{N}Mcl(S_o) = S_o$. Hence $N_s eu\mathcal{N}MFr(S_o) \subseteq S_o$, if S_o is $N_s eu\mathcal{N}Mc$ in U.

Conversely, Assume that, $N_s eu\mathcal{N}MFr(S_o) \subseteq S_o$. Then $N_s eu\mathcal{N}Mcl(S_o) - N_s eu\mathcal{N}Mint(S_o) \subseteq S_o$. Since $N_s eu\mathcal{N}Mint(S_o) \subseteq S_o$, we conclude that $N_s eu\mathcal{N}Mcl(S_o) = S_o$, and hence S_o is $N_s eu\mathcal{N}Mc$. □

Theorem 4 *If* S_o *is a* $N_s eu\mathcal{N}Mo$ *set in* U*, then* $N_s eu\mathcal{N}MFr(S_o) \subseteq S_o^c$.

Proof Let S_o be a $N_s eu\mathcal{N}Mo$ set in the $N_s eu\mathcal{N}ts(U, \tau_N(F))$. By Definition of $N_s eu\mathcal{N}Mo$ set, S_o^c is $N_s eu\mathcal{N}Mc$ in U. By Theorem 3, $N_s eu\mathcal{N}MFr(S_o^c) \subseteq S_o^c$ and by Theorem 3, we get $N_s eu\mathcal{N}MFr(S_o) \subseteq S_o^c$. □

The converse of Theorem 4 is not true as shown by the following example.

Example 2 In Example 1, let $A = \left\{\left\langle\frac{s_1, s_4}{0.2, 0.5, 0.8}\right\rangle, \left\langle\frac{s_2}{0.2, 0.5, 0.8}\right\rangle, \left\langle\frac{s_3}{0.5, 0.5, 0.5}\right\rangle\right\}$,

$N_s eu\mathcal{N}Mcl(A) = \left\{\left\langle\frac{s_1, s_4}{0.2, 0.5, 0.8}\right\rangle, \left\langle\frac{s_2}{0.2, 0.5, 0.8}\right\rangle, \left\langle\frac{s_3}{0.5, 0.5, 0.5}\right\rangle\right\}$,

$N_s eu\mathcal{N}Mcl(A^c) = \left\{\left\langle\frac{s_1, s_4}{0.8, 0.5, 0.2}\right\rangle, \left\langle\frac{s_2}{0.8, 0.5, 0.2}\right\rangle, \left\langle\frac{s_3}{0.6, 0.5, 0.4}\right\rangle\right\}$.

$N_s eu\mathcal{N}MFr(A) = \left\{\left\langle\frac{s_1, s_4}{0.2, 0.5, 0.8}\right\rangle, \left\langle\frac{s_2}{0.2, 0.5, 0.8}\right\rangle, \left\langle\frac{s_3}{0.5, 0.5, 0.5}\right\rangle\right\}$.

$A^c = \left\{\left\langle\frac{s_1, s_4}{0.8, 0.5, 0.2}\right\rangle, \left\langle\frac{s_2}{0.8, 0.5, 0.2}\right\rangle, \left\langle\frac{s_3}{0.5, 0.5, 0.5}\right\rangle\right\}$.

$N_s eu\mathcal{N}MFr(A) \subseteq A^c$. But A is not $N_s eu\mathcal{N}Mo$ set.

Theorem 5 *Let* $S_o \subseteq T_o$ *and* T_o *be any* $N_s eu\mathcal{N}Mc$ *set in* U*. Then* $N_s eu\mathcal{N}MFr(S_o) \subseteq T_o$.

Proof By Theorem 4.9. (iv) in [29], $S_o \subseteq T_o$, $N_s eu\mathcal{N}Mcl(S_o) \subseteq N_s eu\mathcal{N}Mcl(T_o)$. By Definition 1, $N_s eu\mathcal{N}MFr(S_o) = N_s eu\mathcal{N}Mcl(S_o) \cap N_s eu\mathcal{N}Mcl({S_o}^c) \subseteq$

$N_s eu\mathcal{N}Mcl(T_o) \cap N_s eu\mathcal{N}Mcl(S_o{}^c) \subseteq N_s eu\mathcal{N}Mcl(T_o)$. Then by Theorem 4.9. (iv) in [29], this is equal to T_o. Hence $N_s eu\mathcal{N}MFr(S_o) \subseteq T_o$. □

Theorem 6 *For S_o, $(N_s eu\mathcal{N}MFr(S_o))^c = N_s eu\mathcal{N}Mint(S_o) \cup N_s eu\mathcal{N}Mint(S_o{}^c)$.*

Proof Let S_o be in $(U, \tau_N(F))$.

Then by Definition 1, $(N_s eu\mathcal{N}MFr(S_o))^c = (N_s eu\mathcal{N}Mcl(S_o) \cap N_s eu\mathcal{N}Mcl(S_o{}^c))^c = ((N_s eu\mathcal{N}Mcl(S_o))^c \cup (N_s eu\mathcal{N}Mcl(S_o{}^c))^c$. By Theorem 4.2. (i) in [29], which is equal to $N_s eu\mathcal{N}Mint(S_o{}^c) \cup N_s eu\mathcal{N}Mint(S_o)$. Hence $(N_s eu\mathcal{N}MFr(S_o))^c = N_s eu\mathcal{N}Mint(S_o) \cup N_s eu\mathcal{N}Mint(S_o{}^c)$. □

Theorem 7 *For a $N_s eu$ subs S_o in the $N_s eu\mathcal{N}ts(U, \tau_N(F))$, then $N_s eu\mathcal{N}MFr(S_o) \subseteq N_s eu\mathcal{N}Fr(S_o)$.*

Proof Let S_o be in $(U, \tau_N(F))$. Then by Proposition 4.12. (4) in [29], $N_s eu\mathcal{N}Mcl(S_o) \subseteq N_s eu\mathcal{N}cl(S_o)$ and $N_s eu\mathcal{N}Mcl(S_o{}^c) \subseteq N_s eu\mathcal{N}cl(S_o{}^c)$. By Definition 1, $N_s eu\mathcal{N}MFr(S_o) = N_s eu\mathcal{N}Mcl(S_o) \cap N_s eu\mathcal{N}Mcl(S_o{}^c) \subseteq N_s eu\mathcal{N}cl(S_o) \cap N_s eu\mathcal{N}cl(S_o{}^c)$, this is equal to $N_s eu\mathcal{N}Fr(S_o)$. Hence $N_s eu\mathcal{N}MFr(S_o) \subseteq N_s eu\mathcal{N}Fr(S_o)$. □

Theorem 8 *For a $N_s eu$ subs S_o in the $N_s eu\mathcal{N}ts(U, \tau_N(F))$, $N_s eu\mathcal{N}Mcl(N_s eu\mathcal{N}MFr(S_o)) \subseteq N_s eu\mathcal{N}MFr(S_o)$.*

Proof Let S_o be the $N_s eu$ *subs* in the $N_s eu\mathcal{N}ts(U, \tau_N(F))$. Then by Definition 1, $N_s eu\mathcal{N}Mcl(N_s eu\mathcal{N}MFr(S_o)) = N_s eu\mathcal{N}Mcl(N_s eu\mathcal{N}Mcl(S_o) \cap (N_s eu\mathcal{N}Mcl(S_o{}^c))) \subseteq (N_s eu\mathcal{N}Mcl(N_s eu\mathcal{N}Mcl(S_o))) \cap (N_s eu\mathcal{N}Mcl(N_s eu\mathcal{N}Mcl(S_o{}^c)))$. By Theorem 4.9. (iii) in [29], $N_s eu\mathcal{N}Mcl(N_s eu\mathcal{N}MFr(S_o)) = N_s eu\mathcal{N}Mcl(S_o) \cap (N_s eu\mathcal{N}Mcl(S_o{}^c))$. By Definition 1, this is equal to $N_s eu\mathcal{N}MFr(S_o)$. □

Theorem 9 *For a $N_s eu$ subs S_o in the $N_s eu\mathcal{N}ts(U, \tau_N(F))$, $N_s eu\mathcal{N}MFr(N_s eu\mathcal{N}Mint(S_o)) \subseteq N_s eu\mathcal{N}MFr(S_o)$.*

Theorem 10 *For a $N_s eu$ subs S_o in the $N_s eu\mathcal{N}ts(U, \tau_N(F))$, then $N_s eu\mathcal{N}MFr(N_s eu\mathcal{N}Mcl(S_o)) \subseteq N_s eu\mathcal{N}MFr(S_o)$.*

Theorem 11 *For S_o, $N_s eu\mathcal{N}Mint(S_o) \subseteq S_o - N_s eu\mathcal{N}MFr(S_o)$.*

Remark 2 In $N_s eu\mathcal{N}t$, the following conditions do not hold in general:

(i) $N_s eu\mathcal{N}MFr(S_o) \cap N_s eu\mathcal{N}Mint(S_o) = 0_N$,
(ii) $N_s eu\mathcal{N}Mint(S_o) \cup N_s eu\mathcal{N}MFr(S_o) = N_s eu\mathcal{N}Mcl(S_o)$,
(iii) $N_s eu\mathcal{N}Mint(S_o) \cup N_s eu\mathcal{N}Mint(S_o{}^c) \cup N_s eu\mathcal{N}MFr(S_o) = 1_N$.

Example 3 In Example 1, let $A = \left\{\left\langle\frac{s_1, s_4}{0.2, 0.5, 0.8}\right\rangle, \left\langle\frac{s_2}{0.4, 0.5, 0.6}\right\rangle, \left\langle\frac{s_3}{0.4, 0.5, 0.6}\right\rangle\right\}$, then

$N_s eu\mathcal{N}Mcl(A) = \left\{\left\langle\frac{s_1, s_4}{0.8, 0.5, 0.2}\right\rangle, \left\langle\frac{s_2}{0.7, 0.5, 0.3}\right\rangle, \left\langle\frac{s_3}{0.6, 0.5, 0.4}\right\rangle\right\}$,

$N_s eu\mathcal{N}Mcl(A^c) = \left\{\left\langle\frac{s_1, s_4}{0.8, 0.5, 0.2}\right\rangle, \left\langle\frac{s_2}{0.7, 0.5, 0.3}\right\rangle, \left\langle\frac{s_3}{0.6, 0.5, 0.4}\right\rangle\right\}$.

$N_s eu\mathcal{N}MFr(A) = \left\{\left\langle\frac{s_1, s_4}{0.8, 0.5, 0.2}\right\rangle, \left\langle\frac{s_2}{0.7, 0.5, 0.3}\right\rangle, \left\langle\frac{s_3}{0.6, 0.5, 0.4}\right\rangle\right\}$.

$N_s eu\mathcal{N}Mint(A) = \left\{\left\langle\frac{s_1,s_4}{0.2,0.5,0.8}\right\rangle, \left\langle\frac{s_2}{0.3,0.5,0.7}\right\rangle, \left\langle\frac{s_3}{0.4,0.5,0.6}\right\rangle\right\}$,
$N_s eu\mathcal{N}Mint(A^c) = \left\{\left\langle\frac{s_1,s_4}{0.2,0.5,0.8}\right\rangle, \left\langle\frac{s_2}{0.3,0.5,0.7}\right\rangle, \left\langle\frac{s_3}{0.4,0.5,0.6}\right\rangle\right\}$.

(i) $N_s eu\mathcal{N}MFr(A) \cap N_s eu\mathcal{N}Mint(A) \neq 0_N$.
(ii) $N_s eu\mathcal{N}Mint(A) \cup N_s eu\mathcal{N}Mint(A^c) \cup N_s eu\mathcal{N}MFr(A) \neq 1_N$.

Example 4 In Example 1, let $A = \left\{\left\langle\frac{s_1,s_4}{0.3,0.5,0.7}\right\rangle, \left\langle\frac{s_2}{0.3,0.5,0.7}\right\rangle, \left\langle\frac{s_3}{0.7,0.5,0.3}\right\rangle\right\}$,
$N_s eu\mathcal{N}Mcl(A) = \left\{\left\langle\frac{s_1,s_4}{0.8,0.5,0.2}\right\rangle, \left\langle\frac{s_2}{0.7,0.5,0.3}\right\rangle, \left\langle\frac{s_3}{0.7,0.5,0.3}\right\rangle\right\}$,
$N_s eu\mathcal{N}Mcl(A^c) = \left\{\left\langle\frac{s_1,s_4}{0.7,0.5,0.3}\right\rangle, \left\langle\frac{s_2}{0.7,0.5,0.3}\right\rangle, \left\langle\frac{s_3}{0.3,0.5,0.7}\right\rangle\right\}$.
$N_s eu\mathcal{N}MFr(A) = \left\{\left\langle\frac{s_1,s_4}{0.7,0.5,0.3}\right\rangle, \left\langle\frac{s_2}{0.7,0.5,0.3}\right\rangle, \left\langle\frac{s_3}{0.3,0.5,0.7}\right\rangle\right\}$.
$N_s eu\mathcal{N}Mint(A) = \left\{\left\langle\frac{s_1,s_4}{0.3,0.5,0.7}\right\rangle, \left\langle\frac{s_2}{0.3,0.5,0.7}\right\rangle, \left\langle\frac{s_3}{0.7,0.5,0.3}\right\rangle\right\}$.
$N_s eu\mathcal{N}Mint(A) \cup N_s eu\mathcal{N}MFr(A) \neq N_s eu\mathcal{N}Mcl(A)$.

Theorem 12 *Let $S_o\&T_o$ be $N_s eu$ subs's in the $N_s eu\mathcal{N}ts(U, \tau_N(F))$. Then $N_s eu\mathcal{N}MFr(S_o \cup T_o) \subseteq N_s eu\mathcal{N}MFr(S_o) \cup N_s eu\mathcal{N}MFr(T_o)$.*

Theorem 13 *For any $N_s eu$ subs's S_o and T_o in the $N_s eu\mathcal{N}ts(U, \tau_N(F))$, $N_s eu\mathcal{N}MFr(S_o \cap T_o) \subseteq (N_s eu\mathcal{N}MFr(S_o) \cap (N_s eu\mathcal{N}Mcl(T_o))) \cup (N_s eu\mathcal{N}MFr(T_o) \cap N_s eu\mathcal{N}Mcl(S_o))$.*

Corollary 1 *For any $N_s eu$ subs's S_o and T_o in the $N_s eu\mathcal{N}ts(U, \tau_N(F))$, $N_s eu\mathcal{N}MFr(S_o \cap T_o) \subseteq N_s eu\mathcal{N}MFr(S_o) \cup N_s eu\mathcal{N}MFr(T_o)$.*

Theorem 14 *For any $N_s eu$ subsS_o in the $N_s eu\mathcal{N}ts(U, \tau_N(F))$,*

(i) $N_s eu\mathcal{N}MFr(N_s eu\mathcal{N}MFr(S_o)) \subseteq N_s eu\mathcal{N}MFr(S_o)$,
(ii) $N_s eu\mathcal{N}MFr(N_s eu\mathcal{N}MFr(N_s eu\mathcal{N}MFr(S_o))) \subseteq N_s eu\mathcal{N}MFr(N_s eu\mathcal{N}MFr(S_o))$.

Remark 3 Theorems 1–14, Remarks 1 & 2 and Corollary 1 holds for $N_s eu\mathcal{N}o$, $N_s eu\mathcal{N}\theta o$, $N_s eu\mathcal{N}\theta\mathcal{S}o\&N_s eu\mathcal{N}\delta\mathcal{P}o$ sets.

4 Neutrosophic Nano M Border and Neutrosophic Nano M Exterior

In this section, neutrosophic nano M border and neutrosophic nano M exterior operators are introduced and their properties discussed.

Definition 2 The neutrosophic nano (resp. θ, $\theta\mathcal{S}$, $\delta\mathcal{P}\&M$) border of S_o (briefly, $N_s eu\mathcal{N}Br(S_o)$ (resp. $N_s eu\mathcal{N}\theta Br(S_o)$, $N_s eu\mathcal{N}\theta\mathcal{S}Br(S_o)$, $N_s eu\mathcal{N}\delta\mathcal{P}Br(S_o)\&N_s eu\mathcal{N}MBr(S_o)$)) is defined by $N_s eu\mathcal{N}Br(S_o) = S_o - N_s eu\mathcal{N}int(S_o)$ (resp. $N_s eu\mathcal{N}\theta Br(S_o) = S_o - N_s eu\mathcal{N}\theta int(S_o)$, $N_s eu\mathcal{N}\theta\mathcal{S}Br(S_o) = S_o -$

$N_s eu\mathcal{N}\theta Sint(S_o)$, $N_s eu\mathcal{N}\delta\mathcal{P}Br(S_o) = S_o - N_s eu\mathcal{N}\delta\mathcal{P}int(S_o) \& N_s eu\mathcal{N}MBr(S_o) = S_o - N_s eu\mathcal{N}Mint(S_o))$.

Example 5 In Example 1, let $A = \left\{\left\langle\frac{s_1,s_4}{0.2,0.5,0.8}\right\rangle, \left\langle\frac{s_2}{0.4,0.5,0.6}\right\rangle, \left\langle\frac{s_3}{0.4,0.5,0.6}\right\rangle\right\}$,

$N_s eu\mathcal{N}Mint(A) = \left\{\left\langle\frac{s_1,s_4}{0.2,0.5,0.8}\right\rangle, \left\langle\frac{s_2}{0.3,0.5,0.7}\right\rangle, \left\langle\frac{s_3}{0.4,0.5,0.6}\right\rangle\right\}$.

$N_s eu\mathcal{N}MBr(A) = \left\{\left\langle\frac{s_1,s_4}{0.2,0.5,0.8}\right\rangle, \left\langle\frac{s_2}{0.4,0.5,0.6}\right\rangle, \left\langle\frac{s_3}{0.4,0.5,0.6}\right\rangle\right\}$.

Theorem 15 *If a $N_s eu$ subsS_o of $N_s eu\mathcal{N}ts(U, \tau_N(F))$ is $N_s eu\mathcal{N}Mc$, then $N_s eu\mathcal{N}MBr(S_o) = N_s eu\mathcal{N}MFr(S_o)$.*

Proof Let S_o be a $N_s eu\mathcal{N}Mc$ subset of U. Then by Theorem 4.2. (ii) in [29], $N_s eu\mathcal{N}Mcl(S_o) = S_o$. Now, $N_s eu\mathcal{N}MFr(S_o) = N_s eu\mathcal{N}Mcl(S_o) - N_s eu\mathcal{N}Mint(S_o) = S_o - N_s eu\mathcal{N}Mint(S_o) = N_s eu\mathcal{N}MBr(S_o)$. □

Theorem 16 *For a $N_s eu$ subsS_o of $N_s eu\mathcal{N}ts(U, \tau_N(F))$, $S_o = N_s eu\mathcal{N}Mint(S_o) \cup N_s eu\mathcal{N}MBr(S_o)$.*

Proof Let $x_{(u,v,w)} \in S_o$. If $x_{(u,v,w)} \in N_s eu\mathcal{N}Mint(S_o)$, then the result is obvious. If $x_{(u,v,w)} \notin N_s eu\mathcal{N}Mint(S_o)$, then by the definition of $N_s eu\mathcal{N}MBr(S_o)$, $x_{(u,v,w)} \in N_s eu\mathcal{N}MBr(S_o)$. Hence $x_{(u,v,w)} \in N_s eu\mathcal{N}Mint(S_o) \cup N_s eu\mathcal{N}MBr(S_o)$ and so $S_o \subseteq N_s eu\mathcal{N}Mint(S_o) \cup N_s eu\mathcal{N}MBr(S_o)$. On the other hand, since $N_s eu\mathcal{N}Mint(S_o) \subseteq S_o$ & $N_s eu\mathcal{N}MBr(S_o) \subseteq S_o$, we have $N_s eu\mathcal{N}Mint(S_o) \cup N_s eu\mathcal{N}MBr(S_o) \subseteq S_o$. □

Theorem 17 *For S_o, $N_s eu\mathcal{N}Mint(S_o) \cap N_s eu\mathcal{N}MBr(S_o) = 0_N$.*

Proof Suppose $N_s eu\mathcal{N}Mint(S_o) \cap N_s eu\mathcal{N}MBr(S_o) \neq 0_N$. Let $x_{(u,v,w)} \in N_s eu\mathcal{N}Mint(S_o) \cap N_s eu\mathcal{N}MBr(S_o)$. Then $x_{(u,v,w)} \in N_s eu\mathcal{N}Mint(S_o)$ and $x_{(u,v,w)} \in N_s eu\mathcal{N}MBr(S_o)$. Since $N_s eu\mathcal{N}MBr(S_o) = S_o - N_s eu\mathcal{N}Mint(S_o)$, then $x_{(u,v,w)} \in S_o$. But $x_{(u,v,w)} \in N_s eu\mathcal{N}Mint(S_o)$, $x_{(u,v,w)} \in S_o$. There is a contradiction. Hence $N_s eu\mathcal{N}Mint(S_o) \cap N_s eu\mathcal{N}MBr(S_o) = 0_N$. □

Theorem 18 *For a $N_s eu$ subsS_o of $N_s eu\mathcal{N}ts(U, \tau_N(F))$, S_o is a $N_s eu\mathcal{N}Mo$ set if and only if $N_s eu\mathcal{N}MBr(S_o) = 0_N$.*

Proof **Necessity:** Suppose S_o is $N_s eu\mathcal{N}Mo$. Then by Theorem 4.2. (ii) in [29], $N_s eu\mathcal{N}Mint(S_o) = S_o$. Now, $N_s eu\mathcal{N}MBr(S_o) = S_o - N_s eu\mathcal{N}Mint(S_o) = S_o - S_o = 0_N$.

Sufficiency: Suppose $N_s eu\mathcal{N}MBr(S_o) = 0_N$. This implies, $S_o - N_s eu\mathcal{N}Mint(S_o) = 0_N$. Therefore, $S_o = N_s eu\mathcal{N}Mint(S_o)$, and hence S_o is $N_s eu\mathcal{N}Mo$. □

Corollary 2 *For a $N_s eu\mathcal{N}ts$, $N_s eu\mathcal{N}MBr(0_N) = 0_N$ & $N_s eu\mathcal{N}MBr(1_N) = 0_N$.*

Proof Since $0_N \& 1_N$ are $N_s eu\mathcal{N}Mo$, by Theorem 18, $N_s eu\mathcal{N}MBr(0_N) = 0_N$ and $N_s eu\mathcal{N}MBr(1_N) = 0_N$. □

Theorem 19 *For S_o, $N_s eu\mathcal{N}MBr(N_s eu\mathcal{N}Mint(S_o)) = 0_N$.*

Proof By the definition of $N_s eu\mathcal{N}MBr$, $N_s eu\mathcal{N}MBr(N_s eu\mathcal{N}Mint(S_o)) = N_s eu\mathcal{N}Mint(S_o) - N_s eu\mathcal{N}Mint(N_s eu\mathcal{N}Mint(S_o))$. By Theorem 4.6. (iv)

in [29], $N_seu\mathcal{N}Mint(N_seu\mathcal{N}Mint(S_o)) = N_seu\mathcal{N}Mint(S_o)$, and hence $N_seu\mathcal{N}MBr(N_seu\mathcal{N}Mint(S_o)) = 0_N$. □

Theorem 20 *For S_o, $N_seu\mathcal{N}Mint(N_seu\mathcal{N}MBr(S_o)) = 0_N$.*

Proof Let $x_{(u,v,w)} \in N_seu\mathcal{N}Mint(N_seu\mathcal{N}MBr(S_o))$. Since $N_seu\mathcal{N}MBr(S_o) \subseteq S_o$, by Theorem 4.2. (i), $N_seu\mathcal{N}Mint(N_seu\mathcal{N}MBr(S_o)) \subseteq N_seu\mathcal{N}Mint(S_o)$. Hence $x_{(u,v,w)} \in N_seu\mathcal{N}Mint(S_o)$. Since $N_seu\mathcal{N}Mint(N_seu\mathcal{N}MBr(S_o)) \subseteq N_seu\mathcal{N}MBr(S_o)$, $x_{(u,v,w)} \in N_seu\mathcal{N}MBr(S_o)$. Therefore, $x_{(u,v,w)} \in N_seu\mathcal{N}Mint(S_o) \cap N_seu\mathcal{N}MBr(S_o)$, $x_{(u,v,w)} = 0_N$. □

Theorem 21 *For S_o, $N_seu\mathcal{N}MBr(N_seu\mathcal{N}MBr(S_o)) = N_seu\mathcal{N}MBr(S_o)$.*

Proof By the definition of $N_seu\mathcal{N}MBr$, $N_seu\mathcal{N}MBr(N_seu\mathcal{N}MBr(S_o)) = N_seu\mathcal{N}MBr(S_o) - N_seu\mathcal{N}Mint(N_seu\mathcal{N}MBr(S_o))$. By Theorem 20, $N_seu\mathcal{N}Mint(N_seu\mathcal{N}MBr(S_o)) = 0_N$, and hence $N_seu\mathcal{N}MBr(N_seu\mathcal{N}MBr(S_o)) = N_seu\mathcal{N}MBr(S_o)$. □

Theorem 22 *For S_o, the following statements are equivalent:*

(i) S_o *is* $N_seu\mathcal{N}Mo$.
(ii) $S_o = N_seu\mathcal{N}Mint(S_o)$.
(iii) $N_seu\mathcal{N}MBr(S_o) = 0_N$.

Proof (i) $\rightarrow$ (ii) Obvious from Theorem 4.6. (ii) in [29].

(ii) $\rightarrow$ (iii). Suppose that $S_o = N_seu\mathcal{N}Mint(S_o)$. Then by definition of $N_seu\mathcal{N}Mint$, $N_seu\mathcal{N}MBr(S_o) = N_seu\mathcal{N}Mint(S_o) - N_seu\mathcal{N}Mint(S_o) = 0_N$.

(iii) $\rightarrow$ (i). Let $N_seu\mathcal{N}MBr(S_o) = 0_N$. Then by Definition 2, $S_o - N_seu\mathcal{N}Mint(S_o) = 0_N$, and hence $S_o = N_seu\mathcal{N}Mint(S_o)$. □

Theorem 23 *Let S_o be a N_seu subs of $N_seu\mathcal{N}ts(U, \tau_N(F))$. Then, $N_seu\mathcal{N}MBr(S_o) = S_o \cap N_seu\mathcal{N}Mcl(S_o^c)$.*

Proof Since $N_seu\mathcal{N}MBr(S_o) = S_o - N_seu\mathcal{N}Mint(S_o)$ and by Theorem 4.2. (i) in [29], $N_seu\mathcal{N}MBr(S_o) = S_o - (N_seu\mathcal{N}Mcl(S_o^c))^c = S_o \cap (N_seu\mathcal{N}Mcl(S_o^c)^c) = S_o \cap N_seu\mathcal{N}Mcl(S_o^c)$. □

Theorem 24 *For S_o, $N_seu\mathcal{N}MBr(S_o) \subseteq N_seu\mathcal{N}MFr(S_o)$.*

Proof Since $S_o \subseteq N_seu\mathcal{N}Mcl(S_o)$, $S_o - N_seu\mathcal{N}Mint(S_o) \subseteq N_seu\mathcal{N}Mcl(S_o) - N_seu\mathcal{N}Mint(S_o)$. That implies, $N_seu\mathcal{N}MBr(S_o) \subseteq N_seu\mathcal{N}MFr(S_o)$. The proof of the others are similar. □

Remark 4 Theorems 15, 16, 17, 18, 19, 20, 21, 22, 23, 24 & Corollary 2 holds for $N_seu\mathcal{N}o$, $N_seu\mathcal{N}\theta o$, $N_seu\mathcal{N}\theta So \& N_seu\mathcal{N}\delta\mathcal{P}o$ sets.

Definition 3 The neutrosophic nano (resp. θ, θS, $\delta\mathcal{P}\&M$) interior of ${S_o}^c$ is called the neutrosophic nano (resp. θ, θS, $\delta\mathcal{P}\&M$) exterior of S_o (briefly, $N_seu\mathcal{N}Ext(S_o)$ (resp. $N_seu\mathcal{N}\theta Ext(S_o)$, $N_seu\mathcal{N}\theta SExt(S_o)$, $N_seu\mathcal{N}\delta\mathcal{P}Ext(S_o) \& N_seu\mathcal{N}MExt(S_o)$)). That is, $N_seu\mathcal{N}Ext(S_o) = N_seu\mathcal{N}int({S_o}^c)$ (resp. $N_seu\mathcal{N}\theta Ext(S_o) =$

$N_seu\mathcal{N}\theta int(S_o{}^c)$, $N_seu\mathcal{N}\theta SExt(S_o) = N_seu\mathcal{N}\theta Sint(S_o{}^c)$, $N_seu\mathcal{N}\delta\mathcal{P}Ext(S_o) = N_seu\mathcal{N}\delta\mathcal{P}int(S_o{}^c)\&N_seu\mathcal{N}MExt(S_o) = N_seu\mathcal{N}Mint(S_o{}^c)$).

Example 6 In Example 1, let $A = \left\{\left\langle\frac{s_1,s_4}{0.2,0.5,0.8}\right\rangle,\left\langle\frac{s_2}{0.4,0.5,0.6}\right\rangle,\left\langle\frac{s_3}{0.4,0.5,0.6}\right\rangle\right\}$,

$N_seu\mathcal{N}Mint(A^c) = \left\{\left\langle\frac{s_1,s_4}{0.2,0.5,0.8}\right\rangle,\left\langle\frac{s_2}{0.3,0.5,0.7}\right\rangle,\left\langle\frac{s_3}{0.4,0.5,0.6}\right\rangle\right\}$.

$N_seu\mathcal{N}MExt(A) = \left\{\left\langle\frac{s_1,s_4}{0.2,0.5,0.8}\right\rangle,\left\langle\frac{s_2}{0.3,0.5,0.7}\right\rangle,\left\langle\frac{s_3}{0.4,0.5,0.6}\right\rangle\right\}$.

Theorem 25 *For a N_seu subs S_o of $N_seu\mathcal{N}ts(U,\tau_N(F))$, $N_seu\mathcal{N}MExt(S_o) = (N_seu\mathcal{N}Mcl(S_o))^c$.*

Proof We know that, $1_N - N_seu\mathcal{N}Mcl(S_o) = N_seu\mathcal{N}Mint(S_o^c)$, then $N_seu\mathcal{N}MExt(S_o) = N_seu\mathcal{N}Mint(S_o^c) = (N_seu\mathcal{N}Mcl(S_o))^c$. □

Theorem 26 *For a N_seu subs S_o of $N_seu\mathcal{N}ts(U,\tau_N(F))$, $N_seu\mathcal{N}MExt(N_seu\mathcal{N}MExt(S_o)) = N_seu\mathcal{N}Mint(N_seu\mathcal{N}Mcl(S_o)) \supseteq N_seu\mathcal{N}Mint(S_o)$.*

Proof Now, $N_seu\mathcal{N}MExt(N_seu\mathcal{N}MExt(S_o)) = N_seu\mathcal{N}MExt(N_seu\mathcal{N}Mint(S_o^c)) = N_seu\mathcal{N}Mint((N_seu\mathcal{N}Mint(S_o^c))^c) = N_seu\mathcal{N}Mint(N_seu\mathcal{N}Mcl(S_o)) \supseteq N_seu\mathcal{N}Mint(S_o)$. □

Theorem 27 *For a N_seu subs S_o of $N_seu\mathcal{N}ts(U,\tau_N(F))$, If $S_o \subseteq T_o$, then $N_seu\mathcal{N}MExt(T_o) \subseteq N_seu\mathcal{N}MExt(S_o)$.*

Proof Suppose $S_o \subseteq T_o$. Now, $N_seu\mathcal{N}MExt(T_o) = N_seu\mathcal{N}Mint(T_o^c) \subseteq N_seu\mathcal{N}Mint(S_o^c) = N_seu\mathcal{N}MExt(S_o)$. □

Theorem 28 *For a N_seu subs S_o of $N_seu\mathcal{N}ts(U,\tau_N(F))$, $N_seu\mathcal{N}MExt(1_N) = 0_N\&N_seu\mathcal{N}MExt(0_N) = 1_N$.*

Proof Now, $N_seu\mathcal{N}MExt(1_N) = N_seu\mathcal{N}Mint((1_N)^c) = N_seu\mathcal{N}Mint(0_N)\&N_seu\mathcal{N}MExt(0_N) = N_seu\mathcal{N}Mint((0_N)^c) = N_seu\mathcal{N}Mint(1_N)$. Since $0_N\&1_N$ are $N_seu\mathcal{N}Mo$ sets, then $N_seu\mathcal{N}Mint(0_N) = 0_N\&N_seu\mathcal{N}Mint(1_N) = 1_N$. Hence $N_seu\mathcal{N}MExt(0_N) = 1_N\&N_seu\mathcal{N}MExt(1_N) = 0_N$. □

Theorem 29 *For a N_seu subs S_o of $N_seu\mathcal{N}ts(U,\tau_N(F))$, $N_seu\mathcal{N}MExt(S_o) = N_seu\mathcal{N}MExt((N_seu\mathcal{N}MExt(S_o))^c)$.*

Proof Now, $N_seu\mathcal{N}MExt((N_seu\mathcal{N}MExt(S_o))^c) = N_seu\mathcal{N}MExt((N_seu\mathcal{N}Mint(S_o^c))^c) = N_seu\mathcal{N}Mint((((N_seu\mathcal{N}Mint(S_o^c))^c))^c) = N_seu\mathcal{N}Mint(N_seu\mathcal{N}Mint(S_o^c)) = N_seu\mathcal{N}Mint(S_o^c) = N_seu\mathcal{N}MExt(S_o)$. □

Theorem 30 *For a N_seu subs's $S_o\&T_o$ of $N_seu\mathcal{N}ts(U,\tau_N(F))$, the followings are valid.*

(i) $N_seu\mathcal{N}MExt(S_o \cup T_o) \subseteq N_seu\mathcal{N}MExt(S_o) \cap N_seu\mathcal{N}MExt(T_o)$.
(ii) $N_seu\mathcal{N}MExt(S_o \cap T_o) \supseteq N_seu\mathcal{N}MExt(S_o) \cup N_seu\mathcal{N}MExt(T_o)$.

Remark 5 Theorems 25, 26, 27, 28, 29 & 30 holds for $N_seu\mathcal{N}o$, $N_seu\mathcal{N}\theta o$, $N_seu\mathcal{N}\theta So\&N_seu\mathcal{N}\delta\mathcal{P}o$ sets.

References

1. Al-Hamido, R.K.: On neutrosophic crisp supra bi-topological spaces. Int. J. Neutrosoph. Sci. **4**(1), 36–46 (2020)
2. Atanassov, K.: Intuitionistic fuzzy sets. Fuzzy Sets Syst. **20**, 87–96 (1986)
3. Atanassov, K., Stoeva, S.: Intuitionistic fuzzy sets. In: Proceedings of the Polish Symposium on Interval and Fuzzy Mathematics, Poznan, Poland, pp. 23–26 (1983)
4. Chang, C.L.: Fuzzy topological spaces. J. Math. Anal. Appl. **24**, 182–190 (1968)
5. Coker, D.: An introduction to intuitionistic fuzzy topological spaces. Fuzzy Sets Syst. **88**, 81–89 (1997)
6. Das, S.K., Edalatpanah, S.A.: A new ranking function of triangular neutrosophic number and its application in inter programming. Int. J. Neutrosoph. Sci. **4**(2), 82–92 (2020)
7. El-Maghrabi, A.I., Al-Juhani, M.A.: M-open sets in topological spaces. Pioneer J. Math. Sci. **4**(2), 213–230 (2011)
8. Kalaiyarasan, V., Tamilselvan, S., Vadivel, A., Sundar, C.J.: Normal spaces associated with fuzzy nano M-open sets and its application. J. Math. Comput. Sci. **29**(2), 156–166 (2023)
9. Lellis Thivagar, M., Richard, C.: On nano forms of weekly open sets. Int. J. Math. Statist. Invent. **1**(1), 31–37 (2013)
10. Lellis Thivagar, M., Jafari, S., Sutha Devi, V., Antonysamy, V.: A novel approach to nano topology via neutrosophic sets. Neutrosoph. Sets Syst. **20**, 86–94 (2018)
11. Mullai, M., Sangeetha, K., Surya, R., Madhan Kumar, G., Jayabalan, R., Broumi, S.: A single valued neutrosophic inventory model with neutrosophic random variable. Int. J. Neutrosoph. Sci. **1**(2), 52–63 (2020)
12. Padma, A., Saraswathi, M., Vadivel, A., Saravanakumar, G.: New notions of nano M-open sets. Malaya J. Matematik. **S**(1), 656–660 (2019)
13. Pankajam, V., Kavitha, K.: δ-open sets and δ-nano continuity in δ-nano topological spaces. Int. J. Innovat. Sci. Res. Technol. **2**(12), 110–118 (2017)
14. Parimala, M., Jeevitha, R., Jafari, S., Samarandache, F., Karthika, M.: Neutrosophic nano $\alpha\psi$-closed sets in neutrosophic nano topological spaces. J. Adv. Res. Dyn. Control Syst. **10**, 523–531 (2018)
15. Pawlak, Z.: Rough sets. Int. J. Comput. Informat. Sci. **11**, 341–356 (1982)
16. Salama, A.A., Alblowi, S.A.: Generalized neutrosophic set and generalized neutrosophic topological spaces. J. Comput. Sci. Eng. **2**(7), 129–132 (2012)
17. Salama, A.A., Alblowi, S.A.: Neutrosophic set and neutrosophic topological spaces. IOSR J. Math. **3**(4), 31–35 (2012)
18. Smarandache, F.: A Unifying Field in Logics: Neutrosophic Logic. Neutrosophy, Neutrosophic Set, Neutrosophic Probability. American Research Press, Rehoboth (1999)
19. Smarandache, F.: Neutrosophy and neutrosophic logic. In: First International Conference on Neutrosophy, Neutrosophic Logic, Set, Probability and Statistics, University of New Mexico, Gallup NM, USA (2002).
20. Thangammal, R., Saraswathi, M., Vadivel, A., Sundar, C.J.: Fuzzy nano Z-open sets in fuzzy nano topological spaces. J. Linear Topolog. Algebra. **11**(01), 27–38 (2022)
21. Thangammal, R., Saraswathi, M., Vadivel, A., Noeiaghdam, S., Sundar, C.J., Govindan, V., Iampan, A.: Fuzzy nano Z-locally closed sets, extremally disconnected spaces, normal spaces, and their application. Adv. Fuzzy Syst. **2022**, 3364170 (2022)
22. Vadivel, A., Sundar, C.J.: Neutrosophic δ-open maps and neutrosophic δ-closed maps. Int. J. Neutrosoph. Sci. **13**(2), 66–74 (2021)
23. Vadivel, A., Sundar, C.J.: New operators using neutrosophic δ-open set. J. Neutrosoph. Fuzzy Syst. **1**(2), 61–70 (2021)
24. Vadivel, A., Sundar, C.J.: Application of neutrosophic sets based on mobile network using neutrosophic functions. Emerg. Trends Ind. **4**, 1–8 (2021)
25. Vadivel, A., Seenivasan, M., Sundar, C.J.: An introduction to δ-open sets in a neutrosophic topological spaces. J. Phys. Confer. Ser. **1724**, 012011 (2021)

26. Vadivel, A., Thangaraja, P., Sundar, C.J.: Neutrosophic e-continuous maps and neutrosophic e-irresolute maps. Turkish J. Comput. Math. Educat. **12**(1S), 369–375 (2021)
27. Vadivel, A., Thangaraja, P., Sundar, C.J.: Neutrosophic e-open maps, neutrosophic e-closed maps and neutrosophic e-homeomorphisms in neutrosophic topological spaces. AIP Confer. Proc. **2364**, 020016 (2021)
28. Vadivel, A., Sundar, C.J., Kirubadevi, K., Tamilselvan, S.: More on neutrosophic nano open sets. Int. J. Neutrosoph. Sci. **18**(4), 204–222 (2022)
29. Vadivel, A., Sundar, C.J., Saraswathi, K., Tamilselvan, S.: Neutrosophic nano M open sets. Int. J. Neutrosoph. Sci. **19**(1), 132–147 (2022)
30. Vadivel, A., Moogambigai, N., Tamilselvan, S., Thangaraja, P.: Application of neutrosophic sets based on neutrosophic score function in material selection. In: 2022 First International Conference on Electrical, Electronics, Information and Communication Technologies, pp. 1–5 (2022)
31. Vijayalakshmi, R., Mookambika, A.P.: Neutrosophic nano semi frontier. Neutrosoph. Sets Syst. **36**, 131–143 (2020)
32. Zadeh, L.A.: Fuzzy sets. Informat. Control. **8**, 338–353 (1965)

Approximation of Functions in a Weighted L^p-Norm by Summability Means of Fourier Series

Soshal Saini and Uaday Singh

Keywords Approximation · Weighted norm · Product means

1 Introduction

Let f be a 2π periodic function belonging to $L^p := L^p[0, 2\pi](p \geq 1)$-space. The trigonometric Fourier series of f is defined as

$$f(x) \sim \frac{a_0}{2} + \sum_{k=1}^{\infty}(a_k \cos kx + b_k \sin kx). \tag{1}$$

The nth partial sums of the Fourier series (1), i.e.,

$$s_n(f; x) := \frac{a_0}{2} + \sum_{\nu=1}^{n}(a_\nu \cos \nu x + b_\nu \sin \nu x), \ \ \forall n \in \mathbb{N} \text{ with } s_0(f; x) = \frac{a_0}{2}, \tag{2}$$

called trigonometric polynomial of degree (or order)$\leq n$.

The conjugate series of the Fourier series of f is defined by $\sum_{k=1}^{\infty}(a_k \sin kx - b_k \cos kx)$, with its nth partial sums

$$\tilde{s}_n(f; x) := \sum_{\nu=1}^{n}(a_\nu \sin \nu x - b_\nu \cos \nu x), \ \ \forall n \in \mathbb{N} \text{ and } \tilde{s}_0(f; x) = 0. \tag{3}$$

S. Saini (✉)
Department of Mathematics, Shaheed Mangal Pandey Government Girls PG College, Meerut, India

U. Singh
Department of Mathematics, Institute of Technology Roorkee, Roorkee, India
e-mail: uadayfma@iitr.ac.in

H.-H. Leung et al. (eds.), *Recent Developments in Algebra and Analysis*,
Trends in Mathematics, https://doi.org/10.1007/978-3-031-37538-5_31

The conjugate of f, denoted by $\tilde{f}$, is defined as

$$\tilde{f}(x) = -\frac{1}{2\pi} \lim_{\epsilon \to 0} \int_{\epsilon}^{\pi} \psi(x,t) \cot(t/2) dt, \tag{4}$$

where $\psi(x,t) = f(x+t) - f(x-t)$, and we also write $\phi(x,t) := f(x+t) + f(x-t) - 2f(x)$ [6].

Let $T \equiv (a_{n,k})$ be a lower triangular matrix. Then the sequence-to-sequence transformation

$$t_n(f;x) = \sum_{k=0}^{n} a_{n,k} s_k(f;x), \ n \in \mathbb{N}_0 = \mathbb{N} \cup \{0\},$$

defines the matrix means of $\{s_n(f;x)\}$. The Fourier series (1) is said to be summable to s by T-means, if $\lim_{n\to\infty} t_n(f;x) = s$, where s is a finite number. A similar definition can be given for the summability of the conjugate Fourier series also.

Let $A \equiv (a_{n,m})$ and $B \equiv (b_{n,m})$ be two infinite lower triangular matrices of real numbers such that

$$A(\text{or } B) = \begin{cases} a_{n,m}(\text{or } b_{n,m}) \geq 0, \ m = 0,1,2,\ldots n, \\ a_{n,m}(\text{or } b_{n,m}) = 0, \ m > n. \end{cases} \tag{5}$$

When we superimpose the $B-$summability on $A-$summability, we get BA means of $\{s_k(f;x)\}$ and $\{\tilde{s}_k(f;x)\}$ defined by

$$t_n^{BA}(f;x) = \sum_{m=0}^{n}\sum_{k=0}^{m} b_{n,m} a_{m,k} s_k(f;x), \quad n = 0,1,2,\ldots .$$

$$\tilde{t}_n^{BA}(f;x) = \sum_{m=0}^{n}\sum_{k=0}^{m} b_{n,m} a_{m,k} \tilde{s}_k(f;x), \quad n = 0,1,2,\ldots .$$

We write

$$(BA)_n(t) = \frac{1}{2\pi} \sum_{m=0}^{n}\sum_{k=0}^{m} b_{n,m} a_{m,k} \frac{\sin(k+1/2)t}{\sin(t/2)}.$$

$$(\tilde{BA})_n(t) = \frac{1}{2\pi} \sum_{m=0}^{n}\sum_{k=0}^{m} b_{n,m} a_{m,k} \frac{\cos(k+1/2)t}{\sin(t/2)}.$$

The weighted L^p-norm of $f \in L^p[0,2\pi]$ with the weighted function $\sin^{\beta p}(x/2)$ is defined by

$$\|f\|_{p,\beta} = \left(\frac{1}{2\pi}\int_0^{2\pi} |f(x)|^p \sin^{\beta p}(x/2)dx\right)^{1/p}, \quad 1 \le p < \infty. \tag{6}$$

The degree of approximation $E_n(f)$ of a function $f \in L^p$-space by a trigonometric polynomial $T_n(x)$ of degree $\le n$ with respect to the weighted norm is given by

$$E_n(f) = \min_{T_n} \| f(x) - T_n(x) \|_{p,\beta} .$$

This method of approximation is called the trigonometric Fourier approximation, and the $T_n(x)$ is called Fourier approximant of f.

$$f \in W(L^p, \xi(t)) \text{ if } \left\|(f(x+t) - f(x)) \sin^\beta\left(\frac{x}{2}\right)\right\|_p = O(\xi(t)),$$

where $\xi(t)$ is a positive integer function, $t > 0$, $\alpha > 0$, $p \ge 1$, and $\beta \ge 0$.

We write $K_1 \ll K_2$ if $\exists$ a positive constant C (it may depend on some parameters), such that $K_1 \le CK_2$.

2 Known Results

Many investigators such as Lal [1], Singh, et al. [4], and Mishra et al. [2] have considered the $C^1.N_p$ means in various directions. Lal [1] have obtained the degree of approximation of $f \in W(L^p, \xi(t))$-class by using the $C^1.N_p$ means of the Fourier series of f under some assumption conditions on the function $\xi(t)$ as follows:

Theorem A [1, Theorem 2] *If f is a 2π periodic function and Lebesgue integrable on $[0, 2\pi]$ and is belonging to $W(L^p, \xi(t))$-class, then its degree of approximation by $C^1.N_p$ means of its Fourier series is given by*

$$\left\|t_n^{CN}(f; x) - f(x)\right\|_p = O\left((n+1)^{\beta+1/p}\xi(1/(n+1))\right).$$

Singh et al. [4] studied the results of Lal [1] further and pointed out some errors [4, Remark 2.4, p.4]. The authors [4] improved the earlier results of Lal [1] by replacing the monotonicity on $\{p_n\}$ by the condition $(n+1)p_n = O(P_n)$ [4, Theorem 3.2, pp. 4–5].

In the sequel, recently, Mishra et al. [2] have obtained the subsequent results for conjugate Fourier series using the $C^1.N_p$ means [2, Theorem 3.1].

Recently, Zhang [5] has pointed out that results of Mishra et al. [2] hold only for the function which is constant almost everywhere under their assumption conditions. The same observations are also true for the results of Singh et al. [4].

We study the above problems further to get better degree of approximation with less assumption conditions on $\xi(t)$ and resolve the issue raised by Zhang [5].

3 Main Results

In this chapter, we extend the earlier results using the more general summability means, the product means, for Fourier series, and its conjugate series also. We note that several authors define the function class $W(L^p, \xi(t))$ with weight function $\sin^{\beta p}(x/2)$ or $\sin^{\beta p}(x)$, but the degree of approximation is measured in ordinary L^p-norm. Also, the function class $W(L^p, \xi(t))$ is a subclass of the weighted $L^p[0, 2\pi]$-space with the weight function $\sin^{\beta p}(x/2)$, so it is relevant to measure the degree of approximation in the weighted norm defined by (6). More precisely, we prove the following:

Theorem 1 *Let f be a 2π-periodic function belonging to $W(L^p, \xi(t))$ with $p \geq 1$, $\beta \geq 0$ and let the entries of the lower triangular matrices $A \equiv (a_{n,k})$ and $B \equiv (b_{n,k})$ satisfy the following conditions:*

$$b_{n,n} \ll \frac{1}{n+1}, \quad n \in \mathbb{N}_0, \tag{7}$$

$$|b_{n,m}a_{m,0} - b_{n,m+1}a_{m+1,1}| \ll \frac{b_{n,m}}{(m+1)^2} \quad \text{for } 0 \leq m \leq n-1 \tag{8}$$

and

$$\sum_{k=0}^{m-1} |(b_{n,m}a_{m,m-k} - b_{n,m+1}a_{m+1,m+1-k}) - (b_{n,m}a_{m,m-k-1} - b_{n,m+1}a_{m+1,m-k})|$$

$$\ll \frac{b_{n,m}}{(m+1)^2} \quad \text{for } 0 \leq m \leq n-1, \tag{9}$$

with $A_{n,n} = \sum_{m=0}^{n} a_{n,m} = B_{n,n} = \sum_{m=0}^{n} b_{n,m} = 1$ for $n = 0, 1, 2, \ldots$. Then the degree of approximation of f by BA means of its Fourier series is given by

$$\left\| t_n^{BA}(f; x) - f(x) \right\|_{p,\beta} = O\left(\xi(\pi/(n+1)) + (n+1)^{1-\sigma} \sum_{m=0}^{n} \frac{b_{n,m}}{m+1} \right) \tag{10}$$

provided that the positive nondecreasing function $\xi(t)$ satisfies the condition:

$$\xi(t)/t^{\sigma} \text{ is non–decreasing function for some } 0 < \sigma < 1. \tag{11}$$

Theorem 2 *Let f be a 2π-periodic function belonging to $W(L^p, \xi(t))$ with $p \geq 1$, $\beta \geq 0$ and let the entries of the lower triangular matrices $A \equiv (a_{n,k})$ and $B \equiv (b_{n,k})$ satisfy the conditions (7), (8) and (9) of Theorem 1 with $A_{n,n} = B_{n,n} = 1$*

for $n = 0, 1, 2, \ldots$. *Then the degree of approximation of* $\tilde{f}$, *conjugate of* f, *by* BA *means of its conjugate Fourier series is given by*

$$\left\| \tilde{t}_n^{BA}(f; x) - \tilde{f}(x) \right\|_{p,\beta} = O\left(\xi(\pi/(n+1)) + (n+1)^{1-\sigma} \sum_{m=0}^{n} \frac{b_{n,m}}{m+1} \right) \tag{12}$$

where $\xi(t)$ *and* σ *are the same as in Theorem 1.*

Remark 1 If the entries of matrix B satisfy one more condition, that is, $\sum_{m=0}^{n} \frac{b_{n,m}}{m+1} = O(1/(n+1))$, then the degree of approximation in our results reduces to $O\left(\xi(\pi/(n+1)) + (n+1)^{-\sigma}\right)$, which is a better approximation.

4 Lemmas

For the proof of our theorems, we need the following lemmas:

Lemma 1 ([3]) *If the conditions (8) and (9) hold, then*

$$|b_{n,r}a_{r,r-l} - b_{n,r+1}a_{r+1,r+1-l}| \ll \frac{b_{n,r}}{(r+1)^2}, \quad \textit{for} \ \ 0 \le l \le r-1 \le n-2. \tag{13}$$

For more details, one can see [3, Lemma 4.1, p.27].

Lemma 2 *If the matrices A and B satisfy the conditions (7)–(9) of Theorem 1, then*

$$|(BA)_n(t)| = O\,(n+1)\,, \quad \textit{for} \ \ 0 < t \le \pi/(n+1). \tag{14}$$

For proof, one can see [3, Lemma 4.2, p.6].

Lemma 3 *If the matrices A and B satisfy the conditions (7)–(9) of Theorem 1, then*

$$|(BA)_n(t)| = O\left(\frac{1}{t^2} \left(\sum_{m=0}^{n} \frac{b_{n,m}}{m+1} + \frac{1}{n+1} \right) \right), \ \textit{for} \ \pi/(n+1) < t \le \pi. \tag{15}$$

For the proof, one can see [3, Lemma 4.3, pp.6–7].

Lemma 4 *If the matrices A and B satisfy the conditions (7)–(9) of Theorem 1, then*

$$|(\tilde{BA})_n(t)| = O\left(\frac{1}{t}\right), \quad \textit{for} \ \ 0 < t \le \pi/(n+1). \tag{16}$$

Proof Using $\frac{1}{\sin(t/2)} = O\left(\frac{\pi}{t}\right)$ for $0 < t \leq \pi/(n+1)$, we have

$$\begin{aligned}|(\tilde{BA})_n(t)| &\leq \frac{1}{2\pi}\sum_{m=0}^{n}\sum_{k=0}^{m} b_{n,m}a_{m,k}\left|\frac{\cos(k+1/2)t}{\sin(t/2)}\right| \\ &= O\left(\frac{1}{t}\sum_{m=0}^{n} b_{n,m}A_{m,m}\right) = O\left(\frac{1}{t}B_{n,n}\right) = O\left(\frac{1}{t}\right),\end{aligned}$$

in view of $A_{n,n} = B_{n,n} = 1$.

Lemma 5 *If the matrices A and B satisfy the conditions (7)–(9) of Theorem 1, then*

$$|(\tilde{BA})_n(t)| = O\left(\frac{1}{t^2}\left(\sum_{m=0}^{n}\frac{b_{n,m}}{m+1} + \frac{1}{n+1}\right)\right), \text{ for } \pi/(n+1) < t \leq \pi. \tag{17}$$

Proof Using $\frac{1}{\sin(t/2)} = O\left(\frac{\pi}{t}\right)$, for $\pi/(n+1) < t \leq \pi$,

$$\begin{aligned}|(\tilde{BA})_n(t)| &= \left|\frac{1}{2\pi}\sum_{m=0}^{n}\sum_{k=0}^{m} b_{n,m}a_{m,k}\frac{\cos(k+1/2)t}{\sin(t/2)}\right| \\ &= O\left(\frac{1}{t}\right)\left|\sum_{m=0}^{n}\sum_{k=0}^{m} b_{n,m}a_{m,k}\cos(k+1/2)t\right|. \end{aligned} \tag{18}$$

Now, using Abel's transformation after changing the order of summation, we have

$$\begin{aligned}&\left|\sum_{m=0}^{n}\sum_{k=0}^{m} b_{n,m}a_{m,k}\cos(k+1/2)t\right| \\ &= \left|\sum_{m=0}^{n}\sum_{k=0}^{m} b_{n,m}a_{m,m-k}\cos(m-k+1/2)t\right| \\ &= \left|\sum_{k=0}^{n}\left[\sum_{m=k}^{n-1}(b_{n,m}a_{m,m-k} - b_{n,m+1}a_{m+1,m+1-k})\sum_{l=m}^{k}\cos(l-k+1/2)t\right.\right. \\ &\qquad \left.\left. + b_{n,n}a_{n,n-k}\sum_{l=k}^{n}\cos(l-k+1/2)t\right]\right| \\ &= O\left(\frac{1}{t}\right)\left(\sum_{m=0}^{n-1}\left[\sum_{k=0}^{m}|b_{n,m}a_{m,m-k} - b_{n,m+1}a_{m+1,m+1-k}|\right] + \sum_{k=0}^{n} b_{n,n}a_{n,n-k}\right)\end{aligned}$$

$$= O\left(\frac{1}{t}\right)\left[\sum_{m=0}^{n-1}\sum_{k=0}^{m-1}|b_{n,m}a_{m,m-k} - b_{n,m+1}a_{m+1,m+1-k}| + b_{n,n}\right.$$

$$\left. + \sum_{m=0}^{n-1}|b_{n,m}a_{m,0} - b_{n,m+1}a_{m+1,1}|\right]$$

$$= O\left(\frac{1}{t}\right)\left[\sum_{m=0}^{n-1} m.\frac{b_{n,m}}{(m+1)^2} + b_{n,n} + \sum_{m=0}^{n-1}\frac{b_{n,m}}{(m+1)^2}\right]$$

$$= O\left(\frac{1}{t}\right)\left[\sum_{m=0}^{n}\frac{b_{n,m}}{(m+1)} + \frac{1}{(n+1)}\right], \tag{19}$$

in view of Lemma 1, conditions (7) and (8), and $A_{n,n} = 1$.

Hence, collecting (18) and (19), we get

$$|(\tilde{BA})_n(t)| = O\left(\frac{1}{t^2}\left(\sum_{m=0}^{n}\frac{b_{n,m}}{m+1} + \frac{1}{n+1}\right)\right).$$

Lemma 6 ([6]) *Let $g(x,t) \in L^p([a,b] \times [c,d])$, for $p \geq 1$. Then*

$$\left\{\int_a^b \left|\int_c^d g(x,t)dt\right|^p dx\right\}^{1/p} \leq \int_c^d \left\{\int_a^b |g(x,t)|^p dx\right\}^{1/p} dt. \tag{20}$$

This is known as generalized form of Minkowski's inequality.

5 Proof of Main Results

Proof of Theorem 1 We have,

$$t_n^{BA}(f;x) - f(x) = \sum_{m=0}^{n}\sum_{k=0}^{m} b_{n,m}a_{m,k}(s_k(f;x) - f(x))$$

$$= \frac{1}{2\pi}\int_0^{\pi}\phi(x,t)\sum_{m=0}^{n}\sum_{k=0}^{m} b_{n,m}a_{m,k}\frac{\sin(k+1/2)t}{\sin(t/2)}dt$$

$$= \int_0^{\pi}\phi(x,t)(BA)_n(t)dt.$$

Similarly, Lemma 5 changes as $|(\tilde{C^1 A})_n(t)| = O\left(\frac{t^{-2}}{n+1}\right)$, and for C^1A-means, Theorem 2 becomes $\left\|\tilde{t}_n^{C^1 A}(f; x) - \tilde{f}(x)\right\|_{p,\beta} = O\left(\xi(\pi/(n+1)) + (n+1)^{-\sigma}\right)$.

Acknowledgments The authors express their sincere gratitude to the reviewers for their valuable suggestions for improving the paper.

References

1. Lal, S.: Approximation of functions belonging to the generalized Lipschitz class by $C^1 \cdot N_p$ summability method of Fourier series. Appl. Math. Comput. **209**(2), 346–350 (2009)
2. Mishra, L.N., Mishra, V.N., Khatri, K.: Deepmala: on the trigonometric approximation of signals belonging to generalized weighted Lipschitz $W(L^r, \xi(t))(r \geq 1)$-class by matrix $(C^1 \cdot N_p)$ operator of conjugate series of its Fourier series. Appl. Math. Comput. **237**, 252–263 (2014)
3. Singh, U.: Soshal: approximation of periodic integrable functions in terms of modulus of continuity, Acta et commentationes Univeritatis Tartuensis de Mathematica **20**(1), 23–34 (2016)
4. Singh, U., Mittal, M.L., Sonker, S.: Trigonometric approximation of signals (functions) belonging to $W(L^r, \xi(t))$ class by matrix $(C^1 \cdot N_p)$ operator. Int. J. Math. Math. Sci., 11 (2012). https://doi.org/10.1155/2012/964101
5. Zhang, R.J.: On the trigonometric approximation of the generalized weighted Lipschitz class. Appl. Math. Comput. **247**, 1139–1140 (2014)
6. Zygmund, A.: Trigonometric Series, 3rd edn. Cambridge University Press, Cambridge (2002)

Third Hankel Determinants $H_3(1)$ and $H_3(2)$ for Bi-starlike Functions

Janani Thambidurai

Keywords Hankel determinant · Analytic functions · Univalent functions · Bi-univalent · Bi-starlike

1 Introduction

Let $\mathcal{A}$ represent the class which has analytic functions,

$$f(z) = z + \sum_{n=2}^{\infty} a_n z^n \text{ in } \Delta = \{z : |z| < 1\}, \text{ open unit disc,} \tag{1}$$

normalized by $f(0) = 0$ and $f'(0) = 1$. Let $\mathcal{S}$ represent all univalent functions in Δ. It is known that every function $f \in \mathcal{S}$ has an inverse f^{-1}, where

$$\begin{aligned} f^{-1}(w) = g(w) = w - a_2 w^2 &+ (2a_2^2 - a_3)w^3 - (5a_2^3 - 5a_2a_3 + a_4)w^4 + (14a_2^4 \\ &-21a_2^2a_3 + 6a_4a_2 + 3a_3^2 - a_5)w^5 + (-42a_2^5 + 84a_2^3a_3 - 28a_4a_2^2 \\ &-28a_2a_3^2 + 7a_5a_2 + 7a_4a_3 - a_6)w^6 + (132a_2^6 - 330a_2^4a_3 + 120a_2^3a_4 \\ &+180a_2^2a_3^2 - 36a_5a_2^2 - 72a_2a_3a_4 + 8a_6a_2 - 12a_3^3 + 8a_5a_3 + 4a_4^2 \\ &-a_7)w^7 + \cdots \end{aligned} \tag{2}$$

Let Σ represent all bi-univalent functions in Δ, where both f and f^{-1} are univalent. Famous bi-starlike functions of order α $[\mathcal{S}^*_\Sigma(\alpha)]$ and bi-convex function of order α $[\mathcal{K}_\Sigma(\alpha)]$ were introduced by Brannan and Taha [3].

J. Thambidurai (✉)
SCOPE, Vellore Institute of Technology, Chennai, TN, India
e-mail: janani.t@vit.ac.in

H.-H. Leung et al. (eds.), *Recent Developments in Algebra and Analysis*, Trends in Mathematics, https://doi.org/10.1007/978-3-031-37538-5_32

A function f is subordinate to g $[f \prec g]$, if there is w, an analytic function defined on Δ with $w(0) = 0$, $|w(z)| < 1$ satisfying $f(z) = g(w(z))$. It is assumed that function ϕ of the form

$$\phi(z) = 1 + B_1 z + B_2 z^2 + B_3 z^3 + \cdots, \quad (B_1 > 0), \phi(0) = 1, \phi'(0) > 0, \tag{3}$$

with positive real part, $\phi(\Delta)$ is symmetric with respect to the real axis.

Various starlike and convex subclasses for which either $\frac{z f'(z)}{f(z)}$ or $1 + \frac{z f''(z)}{f'(z)}$ is subordinate to a function which is more general superordinate were unified by Ma and Minda [12]. Ma-Minda starlike subclass has f satisfying $\frac{z f'(z)}{f(z)} \prec \phi(z)$, the subordination. Ma-Minda convex subclass has f satisfying $1 + \frac{z f''(z)}{f'(z)} \prec \phi(z)$ in a similar manner.

$\mathcal{S}^*_\Sigma(\phi)$ or $\mathcal{K}_\Sigma(\phi)$ represents Ma-Minda bi-starlike or bi-convex, if both f and f^{-1} are Ma-Minda starlike or convex, respectively.

Noonan and Thomas [15] analyzed in 1976 , qth Hankel determinant,

$$H_q(n) = \begin{vmatrix} a_n & a_{n+1} & \cdots & a_{n+q-1} \\ a_{n+1} & a_{n+2} & \cdots & a_{n+q} \\ \vdots & \vdots & \vdots & \vdots \\ a_{n+q-1} & a_{n+q} & \cdots & a_{n+2q-2} \end{vmatrix}, q \geq 1.$$

For $n = 1$, $q = 2$, Fekete and Szegö [6] studied the Hankel determinant, $H_2(1) = \begin{vmatrix} a_1 & a_2 \\ a_2 & a_3 \end{vmatrix}$. With μ real, they analyzed the estimates of $|a_3 - \mu a_2^2|$, $a_1 = 1$. According to their study, if $f \in \mathcal{A}$, then

$$|a_3 - \mu a_2^2| \leq \begin{cases} 4\mu - 3 & \text{if} \quad \mu \geq 1, \\ 1 + 2\, exp(\frac{-2\mu}{1-\mu}) & \text{if} \quad 0 \leq \mu \leq 1, \\ 3 - 4\mu & \text{if} \quad \mu \leq 0. \end{cases}$$

Furthermore, Hummel [7, 8] obtained for convex functions, $|a_3 - \mu a_2^2|$, the sharp estimates, whereas Keogh and Merkes [10] estimated for starlike,convex and close-to-convex, the sharp estimates.

For $q = 3$ and $n = 1, 2$, we know that the Hankel estimates are

$$H_3(1) = \begin{vmatrix} a_1 & a_2 & a_3 \\ a_2 & a_3 & a_4 \\ a_3 & a_4 & a_5 \end{vmatrix},$$

$$H_3(1) = a_3(a_2 a_4 - a_3^2) - a_4(a_4 - a_2 a_3) + a_5(a_3 - a_2^2), \text{ since } a_1 = 1, \tag{4}$$

$$H_3(2) = \begin{vmatrix} a_2 & a_3 & a_4 \\ a_3 & a_4 & a_5 \\ a_4 & a_5 & a_6 \end{vmatrix}.$$

After simplification, we get

$$H_3(2) = a_6(a_2a_4 - a_3^2) + a_5(a_3a_4 - a_2a_5) + a_4(a_3a_5 - a_4^2). \tag{5}$$

Motivated by second Hankel determinant study of various subclasses of bi-univalent function class [1, 4, 5, 13, 14, 16, 20–22] and third Hankel determinant analysis for various univalent functions subclass [2, 11, 18, 19, 23], in the present article, a bi-starlike function subclass of Σ is considered along with its initial Taylor-Maclaurin coefficient estimates $|a_2|, |a_3| \cdots |a_6|$ for functions in subclass of Σ to find the estimates of third Hankel determinants, $H_3(1)$ and $H_3(2)$, of the functions of this class.

The following bi-starlike class definition is considered from the literature for the study.

Definition 1 ([12]) Function f is said to be in $\mathcal{S}^*_\Sigma(\phi)$ if

$$\frac{z\, f'(z)}{f(z)} \prec \phi(z) \tag{6}$$

and

$$\frac{w\, g'(w)}{g(w)} \prec \phi(w) \tag{7}$$

where $z, w \in \Delta$ and g is given in (2).

2 Hankel Estimates of $\mathcal{S}^*_\Sigma(\phi)$

The following lemma and theorem are used to derive our main results.

Lemma 1 ([17]) *Let $\mathcal{P}$ represent all functions with real part > 0 and of the form*

$$h_p(z) = 1 + h_1 z + h_2 z^2 + \cdots \quad for\ z \in \Delta, \tag{8}$$

then $|h_i| \leq 2$ for each i,

Theorem 1 ([9]) *Let f be in $\mathcal{S}^*_\Sigma(\phi)$. Then initial Taylor-Maclaurin coefficient estimates are*

$$\begin{aligned}
|a_2| &\leq B_1\\
|a_3| &\leq B_1^2 + B_1/2\\
|a_4| &\leq 2B_1^3/3 + 5B_1^2/4 + 4B_1/3 + 4|B_2|/3 + |B_3|/3\\
|a_5| &\leq 7B_1/4 + |B_3|(B_1 + 3/4) + 35B_1^2/8 + 5B_1^3/4 + |B_2|(4B_1 + 9/4)\\
|a_6| &\leq 16B_1/5 + 8|B_4|/5 + |B_5|/5 + |B_3|(B_1^2 + 77B_1/24 + 24/5) + 63B_1^2/8\\
&\quad +19B_1^3/4 + 7B_1^4/12 + B_1^5/5 + |B_2|(4B_1^2 + 77B_1/8 + 32/5)
\end{aligned}$$

Theorem 2 *Let f be in $\mathcal{S}^*_\Sigma(\phi)$. Then the Third Hankel estimate, when $n = 1$ is*

$$\begin{aligned}
|H_3(1)| \leq\ & |B_3^2|/9 + |2B_1^3/9 + 5B_1^2/3 + 91B_1/72 + 8B_2/9||B_3| + 16|B_2^2|/9\\
&+|8B_1^3/9 + 20B_1^2/3 + 337B_1/72||B_2| + |B_1^6|/9 + 4|B_1^5|\\
&+859|B_1^4|/144 + 335|B_1^3|/48 + 191|B_1^2|/72
\end{aligned} \tag{9}$$

Proof In order to analyze Third Hankel determinant, we need to find the estimates of the following:

$$V_1 = a_2a_4 - a_3^2 \tag{10}$$

$$V_2 = a_4 - a_2a_3 \tag{11}$$

$$V_3 = a_3 - a_2^2 \tag{12}$$

From Theorem 1, by substituting the coefficient estimates, we can get the following equations easily:

$$\begin{aligned}
V_1 =\ & (B_1p_1((B_1p_3)/12 - (B_1q_3)/12 - (B_1q_1^3)/24 + (B_2q_1^3)/12 - (B_3q_1^3)/24\\
&+(B_1^3p_1^3)/12 + (B_1p_2q_1)/12 - (B_2p_2q_1)/12 + (B_1q_1q_2)/12 - (B_2q_1q_2)/12\\
&+(5B_1^2p_1p_2)/32 - (5B_1^2p_1q_2)/32))/2 - ((B_1^2p_1^2)/4\\
&+(p_2/8 - q_2/8)B_1)^2\\
V_2 =\ & (B_1p_3)/12 - (B_1q_3)/12 - (B_1q_1^3)/24 + (B_2q_1^3)/12\\
&-(B_3q_1^3)/24 + (B_1^3p_1^3)/12\\
&+(B_1p_2q_1)/12 - (B_2p_2q_1)/12 + (B_1q_1q_2)/12 - (B_2q_1q_2)/12\\
&+(5B_1^2p_1p_2)/32\\
&-(5B_1^2p_1q_2)/32 - (B_1p_1((B_1^2p_1^2)/4 + (p_2/8 - q_2/8)B_1))/2\\
V_3 =\ & B_1(p_2/8 - q_2/8)
\end{aligned}$$

After collecting the B_1, B_2, B_3 terms, we get

$$
\begin{aligned}
V_1 = & (-(B_1 p_1 q_1^3)/48)B_3 - (p_1^4 B_1^4)/48 + ((p_1((5p_1p_2)/32 - (5p_1q_2)/32))/2 \\
& -(p_1^2(p_2/8 - q_2/8))/2)B_1^3 \\
& +((p_1(p_3/12 - q_3/12 + (p_2q_1)/12 + (q_1q_2)/12 \\
& -q_1^3/24))/2 - (p_2/8 - q_2/8)^2)B_1^2 \\
& -(B_2p_1((p_2q_1)/12 + (q_1q_2)/12 - q_1^3/12)B_1)/2 \\
V_2 = & -p_1^3B_1^3/24 + (5p_1p_2/32 - 5p_1q_2/32 \\
& -(p_1(p_2/8 - q_2/8))/2)B_1^2 + (p_3/12 - q_3/12 \\
& +p_2q_1/12 + q_1q_2/12 - q_1^3/24)B_1 - B_3q_1^3/24 \\
& -B_2(p_2q_1/12 + q_1q_2/12 - q_1^3/12) \\
V_3 = & (p_2/8 - q_2/8)B_1
\end{aligned}
$$

From the definition of Hankel determinant, Eq. (4), and by using the calculated expression V_1, V_2, V_3, we get the following by using MATLAB:

$$
\begin{aligned}
H_3(1) = & ((p_1^3((5p_1p_2)/32 - (5p_1q_2)/32))/24 - (p_1^4(p_2/8 - q_2/8))/48 \\
& -(p_1^2((p_1^2(p_2/8 - q_2/8))/2 - (p_1((5p_1p_2)/32 - (5p_1q_2)/32))/2))/4 \\
& +(p_1^3((5p_1q_2)/32 - (5p_1p_2)/32 + (p_1(p_2/8 - q_2/8))/2))/12)B_1^5 \\
& -B_1^3(((p_2/8 - q_2/8)^2 - (p_1(p_3/12 - q_3/12 + (p_2q_1)/12 + (q_1q_2)/12 \\
& -q_1^3/24))/2)(p_2/8 - q_2/8) - ((5p_1q_2)/32 - (5p_1p_2)/32 + (p_1(p_2/8 \\
& -q_2/8))/2)(p_3/12 - q_3/12 + (p_2q_1)/12 + (q_1q_2)/12 - q_1^3/24) - (p_2/8 \\
& -q_2/8)((3p_2^2)/128 - (3p_2q_2)/64 + (p_1q_1p_2)/8 \\
& +(3q_2^2)/128 + (p_1q_1q_2)/8 \\
& +p_1p_3/8 - p_1q_3/8 - p_1q_1^3/16) \\
& +(5p_1p_2/32 - 5p_1q_2/32)(p_3/12 - q_3/12 + \\
& +(p_2q_1)/12 + (q_1q_2)/12 - q_1^3/24)) - B_2(B_1^2(((p_2q_1)/12 + (q_1q_2)/12 \\
& -q_1^3/12)((5p_1q_2)/32 - (5p_1p_2)/32 \\
& +(p_1(p_2/8 - q_2/8))/2) - (5p_1p_2/32 \\
& -(5p_1q_2)/32)((p_2q_1)/12 + (q_1q_2)/12 - q_1^3/12) \\
& +(p_2/8 - q_2/8)(p_1p_2q_1/8
\end{aligned}
$$

$$-(p_1q_1^3)/8 + (p_1q_1q_2)/8) + (p_1(p_2/8 - q_2/8)((p_2q_1)/12 + (q_1q_2)/12$$
$$-q_1^3/12))/2) - B_1(2(p_2q_1/12 + q_1q_2/12 - q_1^3/12)(p_3/12 - q_3/12$$
$$+p_2q_1/12 + q_1q_2/12 - q_1^3/24) - (p_2/8 - q_2/8)((p_3q_1)/16 + q_1q_3/16$$
$$+3p_2q_1^2/32 - 3q_1^2q_2/32 - p_2^2/32 + q_2^2/32))$$
$$+(B_1^3p_1^3(p_2q_1/12 + q_1q_2/12$$
$$-q_1^3/12))/12) - B_1^2((p_3/12 - q_3/12 + p_2q_1/12 + q_1q_2/12 - q_1^3/24)^2$$
$$-(p_2/8 - q_2/8)(p_4/16 - q_4/16 + (p_3q_1)/16$$
$$+(q_1q_3)/16 + (3p_2q_1^2)/64$$
$$-(3q_1^2q_2)/64 - p_2^2/32 + q_2^2/32)) + (-p_1^6/576)B_1^6 + (-q_1^6/576)B_3^2$$
$$+(((5p_1p_2)/32 - (5p_1q_2)/32)((5p_1q_2)/32 - (5p_1p_2)/32 + (p_1(p_2/8$$
$$-q_2/8))/2) - ((p_1^2(p_2/8 - q_2/8))/2 - (p_1((5p_1p_2)/32 +$$
$$-(5p_1q_2)/32))/2)(p_2/8 - q_2/8) - (p_1^3(p_3/12 - q_3/12 + (p_2q_1)/12$$
$$+(q_1q_2)/12 - q_1^3/24))/24 - (p_1^2((p_2/8 - q_2/8)^2 - (p_1(p_3/12 - q_3/12$$
$$+(p_2q_1)/12 + (q_1q_2)/12 - q_1^3/24))/2))/4 + ((5p_1^2p_2)/64$$
$$-(5p_1^2q_2)/64)(p_2/8 - q_2/8))B_1^4 + (-((p_2q_1)/12 + (q_1q_2)/12 + \ldots$$

$$+ \ldots -q_1^3/12)^2)B_2^2 + (-(p_1^3q_1^3B_1^3)/288 + ((q_1^3((5p_1p_2)/32 - (5p_1q_2)/32))/24$$
$$-(q_1^3((5p_1q_2)/32$$
$$-(5p_1p_2)/32 + (p_1(p_2/8 - q_2/8))/2))/24 - (p_1q_1^3(p_2/8$$
$$-q_2/8))/12)B_1^2 + ((q_1^3(p_3/12 - q_3/12 + (p_2q_1)/12 + (q_1q_2)/12$$
$$-q_1^3/24))/12 + ((3p_2q_1^2)/64 - (3q_1^2q_2)/64)(p_2/8 - q_2/8))B_1$$
$$-(B_2q_1^3((p_2q_1)/12 + (q_1q_2)/12 - q_1^3/12))/12)B_3$$

After applying modulus on both sides of the equations and applying the Lemma 1, we obtain

$$|H_3(1)| \leq |B_3^2|/9 + |2B_1^3/9 + 5B_1^2/3 + 91B_1/72 + 8B_2/9||B_3| + 16|B_2^2|/9$$
$$+|8B_1^3/9 + 20B_1^2/3 + 337B_1/72||B_2| + |B_1^6|/9 + 4|B_1^5|$$
$$+859|B_1^4|/144 + 335|B_1^3|/48 + 191|B_1^2|/72 \tag{13}$$

Hence the proof.

Theorem 3 *Let f be in $\mathcal{S}^*_\Sigma(\phi)$. Then the Third Hankel estimate, when $n = 2$ is*

$$
\begin{aligned}
|H_3(2)| \le\ & |19B_1^2/60 + 9B_1^3/20 + B_1^4/15 + 4B_1B_2/15 + B_1B_3/15||B_5| + |B_3^3|/27 \\
& +64|B_2^3|/27 + 31|B_1^9|/135 + 179|B_1^8|/45 \\
& +7009|B_1^7|/360 + |B_2^2||320B_1^3/9 \\
& +293B_1^2/6 + 17{,}069B_1/720| + |B_2||392B_1^6/45 + 3503B_1^5/72 \\
& +182{,}257B_1^4/1440 + 17{,}889B_1^3/160 \\
& +12{,}499B_1^2/360| + 27{,}719|B_1^3|/2160 \\
& +26{,}081|B_1^4|/480 + 241{,}837|B_1^5|/2880 + 31{,}609|B_1^6|/576 + |2B_1^3/9 \\
& +239B_1^2/72 + 2057B_1/720 + 4B_2/9||B_3^2| + |B_3||1055B_1^2/72 \\
& +54{,}187B_1^3/1440 + 16B_2^2/9 + 135{,}967B_1^4/4320 \\
& +773B_1^5/72 + 22B_1^6/45 \\
& +B_2(344B_1^3/27 + 1871B_1^2/72 + 6197B_1/360)| + |38B_1^2/15 + 18B_1^3/5 \\
& +8B_1^4/15 + 32B_1B_2/15 + 8B_1B_3/15||B_4|
\end{aligned}
$$

Proof In order to find the estimate of Third Hankel determinant, when $n = 2$, we need to find the estimates of the following:

$$J_1 = a_2a_4 - a_3^2 \tag{14}$$

$$J_2 = a_3a_4 - a_2a_5 \tag{15}$$

$$J_3 = a_3a_5 - a_4^2 \tag{16}$$

From Theorem 1, by substituting the coefficient estimates, we can get the following equations easily, by collecting the B_1, B_2, B_3, B_4, B_5 terms:

$$
\begin{aligned}
J_1 =\ & ((p_1(5p_1p_2/32 - 5p_1q_2/32))/2 - (p_1^2(p_2/8 - q_2/8))/2)B_1^3 + ((p_1(p_3/12 \\
& -q_3/12 + p_2q_1/12 + q_1q_2/12 - q_1^3/24))/2 \\
& -(p_2/8 - q_2/8)^2)B_1^2 - (p_1^4/48)B_1^4 \\
& -(B_1p_1q_1^3/48)B_3 + (-(B_1p_1((p_2q_1)/12 + q_1q_2/12 - q_1^3/12))/2)B_2 \\
J_2 =\ & ((p_1^2(5p_1p_2/32 - 5p_1q_2/32))/4 + (p_1^3(p_2/8 - q_2/8))/12 - (p_1(5p_1^2p_2/64 \\
& -5p_1^2q_2/64))/2)B_1^4 + (p_1^5/48)B_1^5 \\
& +(B_1^2p_1^2q_1^3/48 - B_1((q_1^3(p_2/8 - q_2/8))/24 \\
& +(p_1((3p_2q_1^2)/64 - (3q_1^2q_2)/64))/2))B_3
\end{aligned}
$$

$$
\begin{aligned}
&+B_1^2((p_2/8 - q_2/8)(p_3/12 - q_3/12\\
&+p_2q_1/12 + q_1q_2/12 - q_1^3/24)\\
&-(p_1(p_4/16 - q_4/16 + (p_3q_1)/16 + (q_1q_3)/16\\
&+(3p_2q_1^2)/64 - (3q_1^2q_2)/64 - p_2^2/32 + q_2^2/32))/2) + (((p_1((p_1p_2q_1)/8\\
&-(p_1q_1^3)/8 + (p_1q_1q_2)/8))/2 - (p_1^2((p_2q_1)/12\\
&+(q_1q_2)/12 - q_1^3/12))/4)B_1^2\\
&+((p_1((p_3q_1)/16 + (q_1q_3)/16\\
&+(3p_2q_1^2)/32 - (3q_1^2q_2)/32 - p_2^2/32 + q_2^2/32))/2\\
&-(p_2/8 - q_2/8)((p_2q_1)/12 + (q_1q_2)/12 - q_1^3/12))B_1)B_2 + (((5p_1p_2)/32\\
&-(5p_1q_2)/32)(p_2/8 - q_2/8) - (p_1((3p_2^2)/128 - (3p_2q_2)/64 + (p_1q_1p_2)/8\\
&+(3q_2^2)/128 + (p_1q_1q_2)/8 + (p_1p_3)/8\\
&-(p_1q_3)/8 - (p_1q_1^3)/16))/2 + (p_1^2(p_3/12\\
&-q_3/12 + (p_2q_1)/12 + (q_1q_2)/12 - q_1^3/24))/4)B_1^3
\end{aligned}
$$

$$
\begin{aligned}
J_3 = &((p_1^2((5p_1^2p_2)/64 - (5p_1^2q_2)/64))/4\\
&-(p_1^3((5p_1p_2)/32 - (5p_1q_2)/32))/6)B_1^5\\
&+B_2(((p_1^3((p_2q_1)/12 + (q_1q_2)/12 - q_1^3/12))/6\\
&-(p_1^2((p_1p_2q_1)/8 - (p_1q_1^3)/8\\
&+(p_1q_1q_2)/8))/4)B_1^3 + (2((5p_1p_2)/32\\
&-(5p_1q_2)/32)((p_2q_1)/12 + (q_1q_2)/12\\
&-q_1^3/12) - (p_1^2((p_3q_1)/16 + (q_1q_3)/16 + (3p_2q_1^2)/32\\
&-(3q_1^2q_2)/32 - p_2^2/32\\
&+q_2^2/32))/4 - (p_2/8 - q_2/8)((p_1p_2q_1)/8 - (p_1q_1^3)/8 + (p_1q_1q_2)/8))B_1^2\\
&+(2((p_2q_1)/12 + (q_1q_2)/12 - q_1^3/12)(p_3/12 - q_3/12\\
&+(p_2q_1)/12 + (q_1q_2)/12\\
&-q_1^3/24) - (p_2/8 - q_2/8)((p_3q_1)/16 + (q_1q_3)/16\\
&+(3p_2q_1^2)/32 - (3q_1^2q_2)/32\\
&-p_2^2/32 + q_2^2/32))B_1) + B_1^3((p_1^2(p_4/16 - q_4/16 + (p_3q_1)/16 + (q_1q_3)/16\\
&+(3p_2q_1^2)/64 - (3q_1^2q_2)/64 - p_2^2/32 + q_2^2/32))/4
\end{aligned}
$$

$$
\begin{aligned}
&+(p_2/8 - q_2/8)((3p_2^2)/128 \\
&-(3p_2q_2)/64 + (p_1q_1p_2)/8 + (3q_2^2)/128 \\
&+(p_1q_1q_2)/8 + (p_1p_3)/8 - (p_1q_3)/8 + \\
&-(p_1q_1^3)/16) - 2((5p_1p_2)/32 - (5p_1q_2)/32)(p_3/12 - q_3/12 + (p_2q_1)/12 \\
&+(q_1q_2)/12 - q_1^3/24)) + (-(5p_1^3q_1^3B_1^3)/576 + ((q_1^3((5p_1p_2)/32 \\
&-(5p_1q_2)/32))/12 + (p_1^2((3p_2q_1^2)/64 - (3q_1^2q_2)/64))/4 - (p_1q_1^3(p_2/8 \\
&-q_2/8))/16)B_1^2 + ((q_1^3(p_3/12 - q_3/12 + (p_2q_1)/12 + (q_1q_2)/12 \\
&-q_1^3/24))/12 + ((3p_2q_1^2)/64 - (3q_1^2q_2)/64)(p_2/8 - q_2/8))B_1 \\
&-(B_2q_1^3((p_2q_1)/12 + (q_1q_2)/12 - q_1^3/12))/12)B_3 - B_1^2((p_3/12 - q_3/12 \\
&+(p_2q_1)/12 + (q_1q_2)/12 - q_1^3/24)^2 \\
&-(p_2/8 - q_2/8)(p_4/16 - q_4/16 + p_3q_1/16 \\
&+(q_1q_3)/16 + (3p_2q_1^2)/64 - (3q_1^2q_2)/64 \\
&-p_2^2/32 + q_2^2/32)) + (-p_1^6/144)B_1^6
\end{aligned}
\tag{17}
$$

$$
\begin{aligned}
+ \ldots &+(-q_1^6/576)B_3^2 + ((p_1^2((3p_2^2)/128 - (3p_2q_2)/64 \\
&+(p_1q_1p_2)/8 + (3q_2^2)/128 \\
&+(p_1q_1q_2)/8 + (p_1p_3)/8 - (p_1q_3)/8 - (p_1q_1^3)/16))/4 - ((5p_1p_2)/32 \\
&-(5p_1q_2)/32)^2 - (p_1^3(p_3/12 - q_3/12 + (p_2q_1)/12 \\
&+(q_1q_2)/12 - q_1^3/24))/6 \\
&+((5p_1^2p_2)/64 - (5p_1^2q_2)/64)(p_2/8 - q_2/8))B_1^4 + (-((p_2q_1)/12 \\
&+(q_1q_2)/12 - q_1^3/12)^2)B_2^2
\end{aligned}
\tag{18}
$$

From the definition of Hankel determinant and Eq. (5), we get the below estimate by applying modulus and Lemma 1.

$$
\begin{aligned}
|H_3(2)| \le\ & |19B_1^2/60 + 9B_1^3/20 + B_1^4/15 + 4B_1B_2/15 + B_1B_3/15||B_5| + |B_3^3|/27 \\
&+64|B_2^3|/27 + 31|B_1^9|/135 + 179|B_1^8|/45 \\
&+7009|B_1^7|/360 + |B_2^2||320B_1^3/9 \\
&+293B_1^2/6 + 17{,}069B_1/720| + |B_2||392B_1^6/45 + 3503B_1^5/72 \\
&+182{,}257B_1^4/1440 + 17{,}889B_1^3/160
\end{aligned}
$$

$$
\begin{aligned}
&+12{,}499B_1^2/360| + 27{,}719|B_1^3|/2160\\
&+26{,}081|B_1^4|/480 + 241{,}837|B_1^5|/2880 + 31{,}609|B_1^6|/576 + |2B_1^3/9\\
&+239B_1^2/72 + 2057B_1/720 + 4B_2/9||B_3^2| + |B_3||1055B_1^2/72\\
&+54{,}187B_1^3/1440 + 16B_2^2/9 + 135{,}967B_1^4/4320\\
&+773B_1^5/72 + 22B_1^6/45\\
&+B_2(344B_1^3/27 + 1871B_1^2/72 + 6197B_1/360)| + |38B_1^2/15 + 18B_1^3/5\\
&+8B_1^4/15 + 32B_1B_2/15 + 8B_1B_3/15||B_4|
\end{aligned}
$$

Thus the proof.

Considering the function ϕ to be

$$\phi(z) = \left(\frac{1+z}{1-z}\right)^{\alpha} = 1 + B_1 z + B_2 z^2 + \cdots \quad (0 < \alpha \leq 1), \tag{19}$$

for the class of strongly starlike functions, we have

$$B_1 = 2\alpha, \tag{20}$$

$$B_2 = 2\alpha^2, \tag{21}$$

$$B_3 = (4\alpha^3)/3 + (2\alpha)/3, \tag{22}$$

$$B_4 = 2\alpha(\alpha^3/3 + \alpha/3) + (2\alpha^2)/3, \tag{23}$$

$$B_5 = (2\alpha)/5 + 2\alpha(2\alpha(\alpha^3/15 + \alpha/9) + (2\alpha^2)/9) + (4\alpha^3)/9. \tag{24}$$

Also, if we consider

$$\phi(z) = \frac{1 + (1-2\beta)z}{1-z} \quad (0 \leq \beta < 1), \tag{25}$$

then we have $B_1, B_2, \ldots B_5 = 2(1-\beta)$.

Remark 1 While choosing $\phi(z)$ to be of form (19), one can easily obtain the third Hankel determinants $|H_3(1)|$, $|H_3(2)|$ based on the result discussed in Theorems 2 and 3, respectively, of which $|H_3(2)|$ estimate is obtained newly for the class of strongly starlike functions based on the literature survey.

$$
\begin{aligned}
|H_3(1)| &\leq |784\alpha^6/81 + 4144\alpha^5/27 + 13{,}012\alpha^4/81 + 2165\alpha^3/27 + 1000\alpha^2/81|\\
|H_3(2)| &\leq |2{,}968{,}288\alpha^9/18{,}225 + 1{,}172{,}992\alpha^8/405 + 47{,}160{,}056\alpha^7/6075\\
&\quad +751{,}081\alpha^6/81 + 64{,}224{,}083\alpha^5/12{,}150 + 191{,}206\alpha^4/135\\
&\quad +2{,}639{,}522\alpha^3/18{,}225|
\end{aligned}
$$

Remark 2 One can easily obtain the third Hankel determinants $|H_3(1)|$, $|H_3(2)|$ based on the result discussed in the Theorems 2 and 3, respectively, of which $|H_3(2)|$ estimate is obtained for the first time for the class, where $\phi(z)$ is of the form (25).

$$\begin{aligned}|H_3(1)| \leq\ & |64\beta^6/9 - 512\beta^5/3 + 7739\beta^4/9 - 35{,}957\beta^3/18 + 2479\beta^2 \\ & -28{,}693\beta/18 + 1249/3| \\ |H_3(2)| \leq\ & |18{,}688\beta^8/9 - 15{,}872\beta^9/135 - 722{,}192\beta^7/45 + 642{,}505\beta^6/9 \\ & -3{,}637{,}745\beta^5/18 + 5{,}630{,}791\beta^4/15 - 907{,}991\beta^3/2 + 3{,}429{,}129\beta^2/10 \\ & -6{,}590{,}158\beta/45 + 7{,}274{,}293/270|\end{aligned}$$

References

1. Altinkaya, S., Yalcin, S.: Second hankel determinant for a general subclass of bi-univalent functions. TWMS J. Pure Appl. Math **7**(1), 98–104 (2016)
2. Altinkaya, S., Yalcin, S.: Third Hankel determinant for Bazilevic functions. Adv. Math. **5**, 91–96 (2016)
3. Brannan, D.A., Taha, T.S.: On some classes of bi-univalent functions. Studia Univ. Babeś-Bolyai Math. Some Math. J. **31**(2), 70–77 (1986)
4. Caglar, M., Deniz, E., Srivastava, H.M.: Second Hankel determinant for certain subclasses of bi-univalent functions. Turkish J. Math. **41**(3), 694–706 (2017)
5. Deniz, E., Caglar, M., Orhan, H.: Second Hankel determinant for bi-starlike and bi-convex functions of order ß. Appl. Math. Comput. **271**, 301–307 (2015)
6. Fekete, M., Szegö, G.: Eine Bemerkung über ungerade schlichte Funktionen. J. London. Math. Soc. **8**, 85–89 (1933)
7. Hummel, J.: The coefficient regions of starlike functions. Pacific. J. Math. **7**, 1381–1389 (1957)
8. Hummel, J.: Extremal problems in the class of starlike functions. Proc. Amer. Math. Soc. **11**, 741–749 (1960)
9. Janani, T., Yalcin, S.: Initial seven co-efficient estimates for a subclass of bi-starlike functions. Ukr. Math. J. **73**(11), 1576–1584 (2021)
10. Keogh, F.R., Merkes, E.P.: A coefficient inequality for certain classes of analytic functions. Proc. Amer. Math. Soc. **20**, 8–12 (1969)
11. Kowalczyk, B., Lecko, A., Sim, Y.J.: The sharp bound for the Hankel determinant of the third kind for convex functions. Bull. Austral. Math. Soc. **97**(3), 435–445 (2018)
12. Ma, W.C., Minda, D.: A unified treatment of some special classes of functions. In: Proceedings of the Conference on Complex Analysis, Tianjin, 1992. Conference Proceedings Lecture Notes Anal, vol. 1, pp. 157–169. International Press, Cambridge (1994)
13. Motamednezhad, A., Bulboaca, T., Adegani, E.A., Dibagar, N.: Second Hankel determinant for a subclass of analytic bi-univalent functions defined by subordination. Turkish J. Math. **42**(5), 2798–2808 (2018)
14. Mustafa, N., Murugusundaramoorthy, G., Janani, T.: Second Hankel determinant for a certain subclass of bi-univalent functions. Mediterr. J. Math. **15**(3), 1–17 (2018)
15. Noonan, J.W., Thomas, D.K.: On the second Hankel determinant of areally mean $p-$valent functions. Trans. Amer. Math. Soc. **223**(2), 337–346 (1976)
16. Orhan, H., Magesh, N., Yamini, J.: Bounds for the second Hankel determinant of certain bi-univalent functions. Turkish J. Math. **40**(3), 679–687 (2016)
17. Pommerenke, C.: Univalent Functions. Vandenhoeck & Ruprecht, Göttingen (1975)

18. Shanmugam, G., Stephen, B.A., Babalola, K.O.: Third Hankel determinant for alpha-starlike functions. Gulf J. Math. **2**(2), 107–113 (2014)
19. Shi, L., Srivastava, H.M., Arif, M., Hussain, S., Khan, H.: An investigation of the third Hankel determinant problem for certain subfamilies of univalent functions involving the exponential function. Symmetry **11**(5), 598 (2019)
20. Srivastava, H.M., Altinkaya, S., Yalcin, S.: Hankel determinant for a subclass of bi-univalent functions defined by using a symmetric q-derivative operator. Filomat **32**(2), 503–516 (2018)
21. Srivastava, H.M., Murugusundaramoorthy, G., Bulboaca, T.: The second Hankel determinant for subclasses of Bi-univalent functions associated with a nephroid domain. Revista de la Real Academia de Ciencias Exactas, Físicas y Naturales. Serie A. Matemáticas **116**(4), 1–21 (2022)
22. Zaprawa, P.: On the Fekete-Szego problem for classes of bi-univalent functions. Bull. Belgian Math. Soc.-Simon Stevin **21**(1), 169–178 (2014)
23. Zaprawa, P.: Third Hankel determinants for subclasses of univalent functions. Mediterr. J. Math. **14**(1), 1–10 (2017)

(j,k)th-Proximate Order and (j,k)th-Proximate Type of Entire Function

Banani Dutta and Md. Azizul Hoque

Keywords Proximate order · Proximate type · Proximate (j,k)th order · Proximate (j,k)th type

1 Introduction

In the discussion of growth of an entire function, order and type play a major role. In general, order is the limit superior ratio of $\log^{[2]} M(t,c)$ and $\log t$, where $M(t,c)$ denotes maximum modulus function. In 1946, Shah [7] introduced the concept and existence of proximate order of an entire function. Later C. Ghosh et al. [2] prove the existence of proximate L- order of an entire function. Here we define proximate (j,k)th order, proximate (j,k)th type of an entire function and proved their existence. For the discussion , we need the following definitions.

2 Basic Definitions

Definition 1 ([1]) The (j,k)th order of entire function is

$$\beta_c^{(j,k)} = \limsup_{t\to\infty} \frac{\log^{[j-1]} M_c(t)}{\log^{[k]} t}$$

where j, k are positive integers greater than 1.

Definition 2 ([13]) The relative (j,k)th type of entire function is

B. Dutta
Mathematics Department, Darjeeling Government College, Darjeeling, W.B, India

M. A. Hoque (✉)
Mathematics Department, Sreegopal Banerjee College, Mogra, Dist-Hooghly, W.B, India

H.-H. Leung et al. (eds.), *Recent Developments in Algebra and Analysis*,
Trends in Mathematics, https://doi.org/10.1007/978-3-031-37538-5_33

$$S = \limsup_{t\to\infty} \frac{log^{[j-2]} M_c(t)}{log^{[k-1]} t^{\beta_{(j,k)}}}.$$

In 1923, Valiron [11] generalized Proximate order, and in 1946, Shah [7] proved existence of proximate order, now, we define Proximate (j,k) order as:

Definition 3 (Proximate (j,k) Order) Let ζ be an entire function with finite (j,k)-order $\beta_{(j,k)}$. $\beta_{(j,k)}(r)$ is said to be proximate-(j,k)-order of $\zeta(z)$ if $\beta_{(j,k)}(t)$ satisfies the followings:

(i.) $\beta'_{(j,k)}(t)$ exists for $t > t_0$ except at isolated points at which left-hand and right-hand derivatives exist,
(ii.) $\lim_{t\to\infty} \beta_{(j,k)}(t) = \beta_{(j,k)}$
(iii.) $\lim_{t\to\infty} t\beta'_{(j,k)}(t) \log^{[k]} t = 0$,
(iv.) $\limsup_{t\to\infty} \dfrac{\log^{[j-2]} M(t)}{\log^{[k-1]} t^{\beta_{j,k}}(t)} = 1$,for $j > 2, k > 1$.

3 Main Results

Theorem 1 (Existence of Proximate (j,k) Order) *For an entire function $c(z)$ of finite (j,k) order $\beta_{(j,k)}$, there exists a proximate (j,k) order $\beta_{(j,k)}(t)$ for $j > 2$, $k > 1$.*

Proof Let $\sigma_{j,k}(t) = \frac{\log^{[j-1]} M_c(t)}{\log^{[k]} t}$
Case- I: When $\sigma_{j,k}(t) > \beta_{(j,k)}$ for a sequence of values of $t \to \infty$,
we define $\phi_{j,k}(t) = \max_{x\geq t} \sigma_{j,k}(x)$. Since $\sigma_{j,k}(t)$ is continuous ,

$$\limsup_{t\to\infty} \sigma_{j,k}(t) = \beta_{(j,k)},$$

$\sigma_{j,k}(t) > \beta_{(j,k)}$ for a sequence of values of $t \to \infty$,
So $\phi_{j,k}(t)$ exists and is a decreasing function for t.

Let $t_1 > e^{e^e}$ and $\phi_{j,k}(t_1) = \sigma_{j,k}(t_1)$; such values will exist for a sequence of values of $t \to \infty$.

Let $\phi_{j,k}(t_1) = \beta_{j,k}(t_1)$ and d_1 be the least integer not less than $t_1 + 1$ such that ,

$$\phi_{j,k}(t_1) > \phi_{j,k}(d_1)$$

and also let

$$\beta_{(j,k)}(t) = \beta_{(j,k)}(t_1) = \phi_{j,k}(t_1);\ t_1 < t \leq d_1$$

For $p_1 > d_1$ define

$\beta_{(j,k)}(t) = \beta_{(j,k)}(t_1) - \log^{[k+1]} t + \log^{[k+1]} d_1$ for $d_1 \leq t \leq p_1$
$\beta_{(j,k)}(t) = \phi_{j,k}(t)$ for $t = p_1$ but $\beta_{(j,k)}(t) > \phi_{j,k}(t)$ for $d_1 \leq t < p_1$.

Let t_2 be the least value of t for which $t_2 \geq p_1$ and $\phi_{j,k}(t_2) = \sigma_{j,k}(t_2)$. If $t_2 > p_1$ then let $\beta_{(j,k)}(t) = \phi_{j,k}(t)$ for $p_1 \leq t \leq t_2$, Since $\phi_{j,k}(t)$ is constant for $p_1 \leq t \leq t_2$, therefore $\beta_{(j,k)}(t)$ is constant for $p_1 \leq t \leq t_2$.

Proceeding with similar argument, we can prove that $\beta'_{(j,k)}(t)$ exists in corresponding intervals. Further,
$\beta'_{(j,k)}(t) = 0$ or $-\frac{1}{\log^{[k]} t \log^{[k-1]} t ...... \log t.t}$
$\Rightarrow t \log^{[k]} t\beta'_{(j,k)}(t) = 0$ or $\frac{-1}{\log^{[k-1]} t ...... \log t}$
hence, $\lim_{t\to\infty} t\beta'_{(j,k)}(t) \log^{[k]} t = 0$ since $k > 1$.

Also note that $\beta_{(j,k)}(t) \geq \phi_{j,k}(t) \geq \sigma_{j,k}(t)$ for $t \geq t_1$.

Further ,
$\beta_{(j,k)}(t) = \phi_{j,k}(t)$ for $t_1, t_2, t_3 ....$ and $\beta_{(j,k)}(t)$ is non-increasing and

$$\lim_{t\to\infty} \phi_{(j,k)}(t) = \beta_{(j,k)}.$$

Hence, $\limsup_{t\to\infty} \beta_{(j,k)}(t) = \lim_{t\to\infty} \beta_{(j,k)}(t) = \beta_{(j,k)}$. Again since
$\sigma_{j,k}(t) = \frac{\log^{[j-1]} M_c(t)}{\log^{[k]} t}$
$\Rightarrow \log^{[j-1]} M_c(t) = \sigma_{j,k}(t) \log^{[k]} t$
$= \sigma_{j,k}(t) \log(\log^{[k-1]} t)$
$= \log(\log^{[k-1]} t)^{\sigma_{j,k}(t)}$
$\Rightarrow \log^{[j-2]} M_c(t) = (\log^{[k-1]} t)^{\sigma_{j,k}(t)}$
$= (\log^{[k-1]} t)^{\beta_{j,k}(t)}$,
for infinitely many values of t
and $\log^{[j-2]} M_c(t) < (\log^{[k-1]} t)^{\beta_{j,k}(t)}$ for the remaining t.

Hence, $\limsup_{t\to\infty} \frac{\log^{[j-2]} M_c(t)}{(\log^{[k-1]} t)^{\beta_{j,k}(t)}} = 1.$

Case-II: when $\sigma_{j,k}(t) \leq \beta_{j,k}$ for all large values of t,we consider the two cases.
subcase-I: $\sigma_{j,k}(t) = \beta_{j,k}$,for atleast a sequence of values of $t \to \infty$. Here we take $\beta_{j,k}(t) = \beta_{j,k}$ for all large t.
subcase-II: Let $\sigma_{j,k}(t) < \beta_{j,k}$ for all large t, let $X' > e^{e^e}$ such that $\sigma_{j,k}(t) < \beta_{j,k}$, where $t \geq X'$.We define $\xi_{j,k}(t) = \max_{X' \leq x \leq t} \sigma_{j,k}(x)$.

Therefore $\xi_{j,k}(t)$ is non-decreasing. For $t_1 > X'$, let $\beta_{j,k}(t_1) = \beta_{j,k}$
$\beta_{j,k}(t) = \beta_{j,k} + \log^{[k+1]} t - \log^{[k+1]} t_1$ for $w_1 \leq t \leq t_1$ where $w_1 < t_1$ is such that $\xi_{j,k}(s) = \beta_{j,k}(w_1)$.

If $\xi_{j,k}(w_1) \neq \sigma_{j,k}(w_1)$, we take $\xi_{j,k}(t) = \beta_{j,k}(t)$ for $d_1 \leq t \leq w_1$ where d_1 is the nearest point $(d_1 < w_1)$ at which $\xi_{j,k}(d_1) = \beta_{j,k}(d_1)$.

Therefore $\beta_{j,k}(t)$ is constant for $d_1 \leq t \leq w_1$. If $\xi_{j,k}(w_1) = \sigma_{j,k}(w_1)$, then let $d_1 = w_1$. Choose $t_2 > t_1$ suitable large, and let
$\beta_{j,k}(t_2) = \beta_{j,k}$ and
$\beta_{j,k}(t) = \beta_{j,k} + \log^{[k+1]} t - \log^{[k+1]} t_2$ for $w_2 \leq t \leq t_2$ where $w_2 (< t_2)$ is such that $\xi_{j,k}(w_2) = \beta_{j,k}(w_2)$.

If $\xi_{j,k}(w_2) \neq \sigma_{j,k}(w_2)$, then we take $\xi_{j,k}(t) = \beta_{j,k}(t)$ for $d_2 \leq t \leq w_2$ where d_2 is the nearest point ($d_2 < w_2$) at which $\xi_{j,k}(d_2) = \sigma_{j,k}(d_2)$.

If $\xi_{j,k}(w_2) = \sigma_{j,k}(w_2)$, then let $d_2 = w_2$. For $t < d_2$,let $\beta_{j,k}(t) = \beta_{j,k}(d_2) + \log^{[k+1]} d_2 - \log^{[k+1]} t$ for $p_1 \leq t \leq t_2$,where $p_2(< d_2)$ is the point of intersection of $y = \beta_{j,k}$ with $y = \beta_{j,k}(d_2) + \log^{[k+1]} d_2 - \log^{[k+1]} t$. Let $\beta_{j,k}(t) = \beta_{j,k}$ for $t_1 \leq t \leq p_1$; we can find t_2 so large that $t_1 < p_1$.

Similarly we can prove $\beta'_{j,k}(t)$ exists in adjacent intervals. Further , $\beta'_{(j,k)}(t) = 0$ or $-\frac{1}{\log^{[k]} t \log^{[k-1]} t ...... \log t . t}$. Hence, $\lim_{t\to\infty} t\beta'_{(j,k)}(t) \log^{[k]} t = 0$. Also $\beta_{j,k}(t) \geq \xi_{j,k}(t) \geq \sigma_{j,k}(t)$ for all large t and $\beta_{j,k}(t) = \sigma_{j,k}(t)$ for $t = d_1, d_2, d_3, ...$ Hence, $\lim_{t\to\infty} \beta_{(j,k)}(t) = \beta_{(j,k)}$

$$\textit{and} \limsup_{t\to\infty} \frac{\log^{[j-2]} M_c(t)}{(\log^{[k-1]} t)^{\beta_{j,k}(t)}} = 1.$$

Example Let us consider the entire function $c(z) = e^z$. Let $j = 4, k = 2$

$$\begin{aligned}
\beta_c^{(j,k)} &= \limsup_{t\to\infty} \frac{\log^{[j-1]} M_c(t)}{\log^{[k]} t} \\
&= \limsup_{t\to\infty} \frac{\log^{[3]} e^t}{\log^{[2]} t} \\
&= \limsup_{t\to\infty} \frac{\log^{[2]} t}{\log^{[2]} t} \\
&= 1.
\end{aligned}$$

and $\sigma_{j,k}(t) = \frac{\log^{[j-1]} M_c(t)}{\log^{[k]} t} = \frac{\log^{[3]} e^t}{\log^{[2]} t}$.

Then $\limsup_{t\to\infty} \sigma_{j,k}(t) = \beta_c^{(j,k)}$.

Let $\phi_{j,k}(t) = \max_{x\geq t}\{\sigma_{j,k}(x)\}$.

$\beta_{(j,k)}(t) = \beta_{(j,k)}(t_1) - \log^{[k+1]} t + \log^{[k+1]} d_1 = 1 - \log^{[3]} t + \log^{[3]} d_1$ for $d_1 \leq t \leq p_1$
and $\beta_{(j,k)}(t) = \phi_{j,k}(t)$ for $t = p_1$

Definition 4 (Lower Proximate-(j, k)th-Order) Let $c(z)$ be an entire function of finite lower (j, k)th order $\eta_{(j,k)}$. A function $\eta_{(j,k)}(t)$ is said to be lower proximate-(j, k)th-order of $c(z)$ if $\eta_{(j,k)}(t)$ satisfies the followings:

(i.) $\eta'_{(j,k)}(t)$ exists for $t > t_0$ except at isolated points at which left-hand and right-hand derivatives exist.

(ii.) $\lim_{t\to\infty} \eta_{(j,k)}(t) = \eta_{(j,k)}$

(iii.) $\lim_{t\to\infty} t\eta'_{(j,k)}(t)\log^k t = 0$

(iv.) $\liminf_{t\to\infty} \frac{\log^{[j-2]} M_c(t)}{(\log^{[k-1]} t)^{\eta_{j,k}(t)}} = 1,\ j > 2, k > 1.$

Theorem 2 (Existence of Lower Proximate-(j, k)th-Order) *For an entire function $c(z)$ of finite (j, k)th lower order $\eta_{(j,k)}$ there exists proximate (j, k)th lower order $\eta_{(j,k)}(t)$, for $j > 2, k > 1$.*

Proof Same as Theorem 1.

Definition 5 (Proximate (j, k)th-Type) A function $S_{(j,k)}(t)$ is said to be a proximate-(j, k)th-type of an entire function $c(z)$ of finite (j, k)th-type $S_{(j,k)}$ if $S_{(j,k)}(t)$ satisfies followings:

(i.) $S_{(j,k)}(t)$ is real, differentiable for $t > t_0$ except at isolated points at which left hand and right hand derivatives exist.

(ii.) $\lim_{t\to\infty} S_{(j,k)}(t) = S_{(j,k)}$

(iii.) $\lim_{t\to\infty} tS'_{(j,k)}(t) = 0$

(iv.) $\limsup_{t\to\infty} \frac{\log^{[j-3]} M_c(t)}{\exp[\log^{[k-1]} t^{\beta_{j,k}(t)} S_{j,k}(t)]} = 1$ for $j > 3, k > 1$.

Theorem 3 (Existence of Proximate-(j, k)th-Type) *For an entire function $c(z)$ of finite (j, k)th-order $\beta_{(j,k)}$ and finite (j, k)th-type $S_{(j,k)}$, there exists a (j, k)th-proximate type $S_{(j,k)}(t)$ for $j > 3, k > 1$.*

Proof Let $s_{j,k}(t) = \frac{\log^{[j-2]} M(t)}{\{\log^{[k-1]} t\}^{\beta(j,k)}}$

Case- I: When $s_{j,k}(t) > S_{(j,k)}$ for a sequence of values of $t \to \infty$,

We define $\psi_{j,k}(t) = \max_{x\geq t}\{s_{j,k}(x)\}$

Since $s_{j,k}(t)$ is continuous , $\limsup_{t\to\infty} s_{j,k}(t) = S_{(j,k)}$, $s_{j,k}(t) > S_{(j,k)}$ for a sequence of values of $t \to \infty$, So $\psi_{j,k}(t)$ exists and is a non increasing function for t. Let $t_1 > e^e$and $\psi_{j,k}(t_1) = S_{j,k}(t_1)$ such values will exist for a sequence of values of $t \to \infty$.

Let $\psi_{j,k}(t_1) = S_{j,k}(t_1)$ and d_1 be the smallest integer not less than $t_1 + 1$ such that ,

$\psi_{j,k}(t_1) > \psi_{j,k}(d_1)$ and also let

$$S_{(j,k)}(t) = S_{(j,k)}(t_1) = \psi_{j,k}(t_1);\ t_1 < t \leq d_1$$

For $p_1 > d_1$ define as

$S_{(j,k)}(t) = S_{(j,k)}(t_1) - \log\log t + \log\log d_1$ for $d_1 \leq t \leq p_1$

$S_{(j,k)}(t) = \psi_{j,k}(t)$ for $t = p_1$ but

$S_{(j,k)}(t) > \psi_{j,k}(t)$ for $d_1 \leq t \leq p_1$. Let t_2 be the smallest value of t for which $t_2 \geq p_1$ and $\psi_{j,k}(t_2) = s_{j,k}(t_2)$. If $t_2 > p_1$ then let $S_{(j,k)}(t) = \psi_{j,k}(t)$ for $p_1 \leq t \leq t_2$, since $\psi_{j,k}(t)$ is constant for $p_1 \leq t \leq t_2$, therefore $S_{(j,k)}(t)$ is constant for $p_1 \leq t \leq t_2$.

Similarly we can that $S'_{(j,k)}(t)$ exists in adjacent intervals.

Further, $S'_{(j,k)}(t) = 0$ or $S'_{(j,k)}(t) = -\frac{1}{\log t.t} \Rightarrow tS'_{(j,k)}(t) = 0$

hence, $\lim_{t\to\infty} tS'_{(j,k)}(t) = 0$. Also note that $S_{(j,k)}(t) \geq \psi_{j,k}(t) \geq s_{j,k}(t)$ for $t \geq t_1$.
Further ,

$S_{(j,k)}(t) = s_{j,k}(t)$ for t_1, t_2, t_3........... and $S_{(j,k)}(t)$ is non-increasing and $\lim_{t\to\infty} \psi_{(j,k)}(t) = S_{(j,k)}$. Hence, $\limsup_{t\to\infty} S_{(j,k)}(t) = \lim_{t\to\infty} S_{(j,k)}(t) = S_{(j,k)}$.

Again since

$s_{j,k}(t) = \frac{\log^{[j-2]} M(t)}{\{\log^{[k-1]} t\}^{\beta_{j,k}}}$

$\Rightarrow \log^{[j-2]} M(t) = s_{j,k}(t)\log^{[k-1]} t^{\beta_{(j,k)}}$ for infinite many values of t

$\Rightarrow \log^{[j-3]} M(t) = \exp[S_{j,k}(t)(\log^{[k-1]} t)^{\beta_{j,k}}]$

$\Rightarrow \frac{\log^{[j-3]} M(t)}{\exp[S_{j,k}(t)(\log^{[k-1]} t)^{\beta_{j,k}}]} = 1$ and for the remaining t,

$$\log^{[j-2]} t < s_{(j,k)}(t)(\log^{[k-1]} t)^{\beta_{(j,k)}}$$

Hence, $\limsup_{t\to\infty} \frac{\log^{[j-3]} M(t)}{\exp[(\log^{[k-1]} t)^{\beta_{(j,k)}} S_{j,k}(t)]} = 1.$

Case-II: when $s_{j,k}(t) \leq S_{j,k}$ for all large values of t, we consider two cases. subcase-I: $s_{j,k}(t) = S_{j,k}$ for at least a sequence of values of $t \to \infty$,

Here we take $S_{j,k}(t) = S_{j,k}$ for all large t.

subcase-II: Let $s_{j,k}(t) < S_{j,k}$ for all large t. Let $X' > e^e$ such that $s_{j,k}(t) < S_{j,k}$, where $t \geq X'$.We define , $\xi_{j,k}(t) = \max_{X'\leq x\leq t} s_{j,k}(x)$.

Therefore $\xi_{j,k}(t)$ is non-decreasing. Considering $t_1 > X'$ and let $S_{j,k}(t_1) = S_{j,k}$ and

$S_{j,k}(t) = S_{j,k} + \log\log t - \log\log t_1$ for $w_1 \leq t \leq t_1$ where $w_1 < t_1$ is such that $\xi_{(j,k)}(w_1) = S_{j,k}(w_1)$. If $\xi_{(j,k)}(w_1) \neq s_{(j,k)}(w_1)$, we take $\xi_{(j,k)}(t) = S_{(j,k)}(t)$ for $d_1 \leq t \leq w_1$ where d_1 is the nearest point $(d_1 < w_1)$ at which $\xi_{(j,k)}(d_1) = s_{(j,k)}(d_1)$. Therefore $S_{j,k}(t)$ is constant for $d_1 \leq t \leq w_1$.

If $\xi_{(j,k)}(w_1) = s_{(j,k)}(w_1)$, then let $d_1 = w_1$, choose $t_2 > t_1$ suitable large d and let

$S_{j,k}(t_2) = \beta_{j,k}$

$S_{j,k}(t) = \beta_{j,k} + \log\log t - \log\log t_2$ for $w_2 \leq t \leq t_2$. where $w_2(< t_2)$ is such that $\xi_{j,k}(w_2) = S_{j,k}(w_2)$.If $\xi_{j,k}(w_2) \neq s_{j,k}(w_2)$ then we take $\xi_{j,k}(t) = S_{j,k}(t)$ for $d_2 \leq t \leq w_2$ where d_2 is the nearest point $(d_2 < w_2)$ at which $\xi_{j,k}(d_2) = s_{j,k}(d_2)$.

If $\xi_{j,k}(w_2) = s_{j,k}(w_2)$, then let $d_2 = w_2$. For $t < d_2$, let $S_{j,k}(t) = S_{j,k}(d_2) + \log\log d_2 - \log\log t$ for $p_1 \leq t \leq d_2$,where $p_2(< d_2)$ is the point of intersection of $y = S_{j,k}$ with $y = S_{j,k}(d_2) + \log\log d_2 - \log\log t$.

Let $S_{j,k}(t) = S_{j,k}$ for $t_1 \leq t \leq p_1$, we can find t_2 so large that $t_1 < p_1$.

Similarly we can prove that $S'_{j,k}(t)$ exists in adjacent intervals.

Further , $S'_{(j,k)}(t) = 0$ or $S'_{(j,k)}(t) = -\frac{1}{\log t.t}$

Hence, $\lim_{t\to\infty} tS'_{(j,k)}(t) = 0$. Also $S_{j,k}(t) \geq \xi_{j,k}(t) \geq s_{j,k}(t)$ for all large t and $S_{j,k}(t) = s_{j,k}(t)$ for $d_1, d_2, d_3, \ldots\ldots$

Hence,

$$\lim_{t\to\infty} S_{(j,k)}(t) = S_{(j,k)} \text{ and } \limsup_{t\to\infty} \frac{\log^{[j-3]} M_c(t)}{\exp[S_{j,k}(t)(\log^{[k-1]} t)^{\beta_{j,k}(t)}]} = 1.$$

Definition 6 (Lower (j, k)th-Proximate Type) Let $c(z)$ be an entire function of finite (j, k)th-lower type $\vartheta_{(j,k)}$. A function $\vartheta_{(j,k)}(t)$ is said to be lower proximate-(j, k)th-type of $c(z)$, if $\vartheta_{(j,k)}(t)$ satisfies the following properties:

(i.) $\vartheta'_{(j,k)}(t)$ exists for $t > t_0$ except at isolated points at which left hand and right hand derivatives exist.

(ii.) $\lim_{t\to\infty} \vartheta_{(j,k)}(t) = \vartheta_{(j,k)}$

(iii.) $\lim_{t\to\infty} t\vartheta'_{(j,k)}(t) = 0$

(iv.) $\liminf_{t\to\infty} \dfrac{\log^{[j-3]} M_c(t)}{\exp[(\log^{[k-1]} t)^{\beta_{j,k}(t)} \vartheta_{j,k}(t)]} = 1$ where $j > 3, k > 1$.

Theorem 4 (Existence of Lower Proximate-(j, k)th-Type) *For every entire function $c(z)$ of finite lower (j, k)th-order $\eta_{(j,k)}$ and finite lower (j, k)th-type $\vartheta_{(j,k)}$, there exists an proximate-(j, k)th-type $\vartheta_{(j,k)}(t)$ for $j > 3, k > 1$.*

Proof Same as previous theorem.

4 Conclusion and Future Scope

Here, we have proved the existence of proximate-(j, k)th-order for an entire function. One may try to prove for proximate -(j, k, L)th-order taken with respect to any other function.

References

1. Banerjee, D., Batabyal, S.: Hyper relative order (p, q) of entire functions. Ana Univerritatii de Vest Timisoara Seria Mathematical Informatica LV **2**, 65–84 (2017)
2. Ghosh, C., Bandyopadhyay, A., Mondal, S.: On the L-proximate order and L-proximate type of an entirefunction. Afrika Matematika **33**, 53 (2022)
3. Ishimura, R., Jin, X.: Infinite order differential equations in the space of entire functions of normal type with respect to a proximate order. North-West. Eur. J. Math. **5**, 69–87 (2019)
4. Jiménez-Garrido J., Sanz J., Schindl G.: A Phragmé n–Lindel öf theorem via proximate orders, and the propagation of asymptotics. J. Geom. Anal. **30**, 3468–3483 (2020)
5. Sánchez Ruiz, L.M., Datta, S.K., Biswas, T., Mondal, G.K.: On some growth properties of entire functions using their maximum moduli focusing (p, q)th relative order. Sci. World J. **2014**, 7 (2014)

6. Sánchez Ruiz, L.M., Datta, S.K., Biswas, T., Ghosh, C.: A note on relative (p, q)th proximate order of entire function. J. Math. Res. **8**(5), 1–11 (2016)
7. Shah, S.M.: On proximate orders of integral functions. Bull. Am. Math. Soc. **52**, 326–328 (1946)
8. Singh, S.K., Baker, G.P.: Slowly changing functions and their applications. Indian J. Math. **19**, 1–6 (1977)
9. Somasundaram, D., Thamizharasi, R.: A note on the entire functions of L-bounded index and L-type. Indian J. Pure Appl. Math. **19**, 284–293 (1988)
10. Srivastava, R.S.L., Juneja, O.P.: On proximate type of entire functions. Compositio Mathematica, tome **18**(1–2), 7–12 (1967)
11. Valiron, G.: Lectures on the General Theory of Integral Functions Toulouse. Edouard Privat, Toulouse (1923)
12. Whittaker, J.M.: The lower order of integral functions. J. Lond. Math. Soc. **8**, 20–27 (1933)
13. Vaish S.K, On the (p,q) order of entire functions. Bull. Math. de la Soc. Sci. Math. de la R.S.de Roumanic Tome **30**(78), nr. 2 (1986)

Fractals via Self-Similar Group of Fisher Contractions

C. Thangaraj and D. Easwaramoorthy

Keywords Fractal analysis · Fisher contraction · Iterated function system · Topological group · Self-similar group · Profinite group

1 Introduction

Mandelbrot introduced the fractal geometry in 1975 [1], and it has since been popularized by many researchers. In the real world, fractals can be defined as objects that seem self-similar (SS) under different magnifications. Mathematically, fractals can be defined as sets with a Hausdorff dimension that strictly exceeds the topological dimension. Then Hutchinson formalized the important concept called iterated function system (IFS) [2]. Consequently, Barnsley created the mathematical theory called the Hutchinson-Barnsley (HB) theory [3–5]. A fractal set is generated through the IFS of Banach contractions and defined in terms of a compact invariant subset of the complete metric space (CMS), [6, 7]. To put it another way, Hutchinson defined an operator, called HB operator, on the hyperspace of nonempty compact sets. It involves the Banach fixed point theorem to define a unique fixed point as a fractal set in CMS with the distinguished dimensional measures [8–14]. Secelean explored the idea of countable iterated function system [15]. Secelean proposed the idea of creating new IFS by combining various contractions into F-contractions. The authors developed the notion of a topological IFS attractor in reference, which generalizes the familiar IFS attractor. That is, every IFS attractor is also a topological IFS attractor, but the converse is not true [16, 17].

C. Thangaraj · D. Easwaramoorthy (✉)
Department of Mathematics, School of Advanced Sciences, Vellore Institute of Technology, Vellore, Tamil Nadu, India

H.-H. Leung et al. (eds.), *Recent Developments in Algebra and Analysis*, Trends in Mathematics, https://doi.org/10.1007/978-3-031-37538-5_34

The concept of self-similarity lends itself well to performing intensive research on fractals. To completely comprehend the group structure, one must first fully comprehend self-similar sets [18]. A compact fractal space has a natural propensity to be thought of as being unable to accommodate an infinite number of motions. It clearly depicts the set of notions relevant to the modification of a mathematical pattern when the measured scale changes [19]. A rearrangement is a cyclic group, a continuous parameter group, or a cyclic group whose inverse is typically not invertible and is made up of specialized transformations to partially solve a mathematical problem. As a result, it is a semi-group [19, 20].

A subclass of the non-cyclic renormalization group is self-similar groups (SSG). SSG can also be described in other ways. One approach is to define mealy-type automata-generated groups. This creates SSG that are generally feasible for computer science [19]. A second approach to define SSG is the recursive functional structure of compact topological groups (CTG). Thus, it is close to many problems in dynamical systems.

The most fundamental attribute of fractals is self-similarity. To thoroughly examine the self-similarity sets, one must first comprehend their group structure. Based on the group structure, S. Kocak and M. Saltan et al. describe the self-similar property of fractal sets [21, 22]. The Koch curve, Cantor set, and Sierpinski triangle are typical examples of fractals. The aforementioned sets are employed with IFS with the Banach contraction to get the attractor. In addition, we invoked HB theory to obtain the fractals, which are constructed as self-similar sets.

In 2010, Sahu et al. presented the Kannan iterated function system (KIFS) and used these contractions for creating the fractal sets, which was based on the iterated function system [3, 23]. In 2015, Uthayakumar et al. introduced Kannan contraction's strong self-similar group (SSSG) and SSG and described the KIFS associating with SSSG and profinite groups [24, 25].

So far, the research study on the generation of fractals has not been discussed through the strong self-similar group by using Fisher contractions. It has motivated us to explore the notion of SSSG of Fisher contractions to generate the fractal set and also several consequences of SSSG to describe the FIFS relationship between SSSG and profinite groups. In this study, SSSG of Fisher contractions is proved as an attractor of FIFS, in which every Fisher contraction is a group homomorphism.

The remaining chapter of this study is structured as follows. Section 2 explains the basic facts that are necessary for this study. Also, Sect. 3 discusses Fisher fractals, profinite groups, and self-similarity in the sense of IFS. The self-similar and strong self-similar groups are discussed in Sect. 4 in the sense of FIFS of CTG to generate the fractals as the main focus of this study. Finally, the results are concluded in Sect. 5.

2 Basic Facts

2.1 Metric Fractals

Definition 1 ([2, 3, 26, 27]) If (H, ρ) is a metric space, and $\mathscr{K}_0(H)$ is a nonempty collection of all compact subsets of H. Define $\rho(a, B) = \inf_{b \in B} \rho(a, b)$ and $\rho(A, B) = \sup_{a \in A} \rho(a, B)$ for all $a \in H$ and $A, B \in \mathscr{K}_0(H)$. The Hausdorff metric (H_ρ) is a function $H_\rho : \mathscr{K}_0(H) \times \mathscr{K}_0(H) \to \mathbb{R}$ defined by $H_\rho(A, B) = \max\{\rho(A, B), \rho(B, A)\}$. Then the pair $(\mathscr{K}_0(H), H_\rho)$ is said to be Hausdorff metric space.

Theorem 1 ([2, 3, 26, 27]) *If (H, ρ) is complete, then $(\mathscr{K}_0(H), H_\rho)$ is also complete.*

Definition 2 ([2, 3, 26, 27]) If (H, ρ) is a metric space. Then the self map $T : H \to H$ is known as contraction, if $\exists \quad \alpha \in [0, 1)$ such that $\rho(T(a), T(b)) \leq \alpha\rho(a, b) \quad \forall a, b \in H$. Here α is known as the contraction ratio of T.

Definition 3 ([2, 3, 26, 28]) If (H, ρ) is a metric space and $T_n : H \to H, n = 1, 2, ...N_0 (N_0 \in \mathbb{N})$ are N_0- contraction mappings associating the contraction ratio $\alpha_n, n = 1, 2, ...N_0$. Then $\{H, T_n; n = 1, 2, ..., N_0\}$ the system is said to be the hyperbolic iterated function system (IFS) with the contraction ratio $\alpha = \max_{n=1}^{N_0} \alpha_n$.

Definition 4 ([2, 3, 26, 28]) Let (H, ρ) be a metric space and $\{H, T_n; n = 1, 2, ..., N_0, N_0 \in \mathbb{N}\}$ be IFS of contractions. Then the Hutchinson-Barnsley (HB) operator is a function $\mathscr{F} : \mathscr{K}_0(H) \to \mathscr{K}_0(H)$ defined by $\mathscr{F}(B) = \bigcup_{n=1}^{N_0} T_n(B) \quad \forall B \in \mathscr{K}_0(H)$.

Theorem 2 ([2, 3, 26, 28]) *Let (H, ρ) be a metric space. Let $\{H, T_n; n = 1, 2, ..., N_0, N_0 \in \mathbb{N}\}$ be IFS of contractions. Then, the HB-Operator $\mathscr{F}$ is a Banach contraction mapping on $(\mathscr{K}_0(H), H_\rho)$.*

Theorem 3 ([27, 29, 30]) *If (H, ρ) is CMS, and an IFS of Banach contractions is $\{H, T_n; n = 1, 2, ..., N_0, N_0 \in \mathbb{N}\}$. Then $\mathscr{F}$ has a unique compact invariant set $A_\infty \in \mathscr{K}_0(H)$.*

Definition 5 (([27, 29, 30]) Metric Fractals) The invariant set $A_\infty \in \mathscr{K}_0(H)$ that exists in Theorem 3 is called the unique fixed point of $\mathscr{F}$ (or) attractor (fractals) of IFS of Banach contraction.

3 Fisher Fractals

This section discusses about Fisher IFS and Fisher fractals through the HB-theory generated by Fisher contractions and by using Fisher fixed point theorem. In 1976, Fisher introduced a mapping and proved the fixed point theorem for the Fisher contraction, which are described as follows.

Definition 6 ([31]) Let (H, ρ) be CMS. The function $T : H \to H$ is known as Fisher contractions if there exists real numbers α, β; $0 < \alpha, \beta < \frac{1}{2}$ such that $\rho(T(a), T(b)) \leq \alpha[\rho(a, T(a)) + \rho(b, T(B))] + \beta\rho(a, b)$, for all $a, b \in H$. Here α, β are F-contractivity factors of Fisher contractions T.

If $\alpha = 0$, then the Fisher contraction is reduced to a usual contraction, but the converse is not always true. Similarly, if $\beta = 0$, then the Fisher contraction is reduced to a Kannan contractions, but the converse is not always true.

Theorem 4 *If (H, ρ) is CMS and the mapping $T : (H, \rho) \to (H, \rho)$ is Fisher contraction, then T has a unique fixed point.*

On the basis of IFS provided by Barnsley [3], Sahu [23], Uthayakumar et al. [24], initiated the invariant set of KIFS, now we can introduce invariant set of FIFS as follows:

Definition 7 A FIFS composed of CMS (H, ρ) along with a finite number of Fisher contractions $T_n : H \to H, n = 1, 2, 3, ..., N_0 \quad (N_0 \in \mathbb{N})$ with F-contractivity factors $\alpha_n, \beta_n, n = 1, 2, 3, ..., N_0$.

Definition 8 If (H, ρ) is a metric space. Let $\{H, T_n; n = 1, 2, ..., N_0, N_0 \in \mathbb{N}\}$ be a Fisher IFS that consists of finite number of Fisher contractions. Then the HB-operator of the Fisher IFS of Fisher contractions is a mapping $\mathscr{F} : \mathscr{K}_0(H) \to \mathscr{K}_0(H)$ defined as $\mathscr{F}(B) = \bigcup_{n=1}^{N_0} T_n(B) \quad \forall B \in \mathscr{K}_0(H)$.

Lemma 1 *If (H, ρ) is CMS and $(\mathscr{K}_0(H), H_\rho)$ is the associating Hausdorff metric space. Let $T : H \longrightarrow H$ be a continuous Fisher contractions on (H, ρ) with F-contractivity factors α, β. Then $\mathscr{F} : \mathscr{K}_0(H) \longrightarrow \mathscr{K}_0(H)$ constructed by $T(B) = \{T(x) : x \in B\} \quad \forall B \in \mathscr{K}_0(H)$ is a Fisher mapping on $(\mathscr{K}_0(H), H_\rho)$ with $\mathscr{F}$-contractions ratios α, β.*

Lemma 2 *If (H, ρ) is CMS and $(\mathscr{K}_0(H), H_\rho)$ is the appropriate Hausdorff metric space. Let $T_n : H \longrightarrow H$, $n = 1, 2, 3, ..., N_o$ $(N_o \in \mathbb{N})$ be continuous Fisher contractions on (H, ρ) with F-contractivity factors $\alpha_n, \beta_n, n = 1, 2, 3, ..., N_o$. Then the HB operator $\mathscr{F} : \mathscr{K}_0(H) \longrightarrow \mathscr{K}_0(H)$ of the Fisher IFS of continuous Fisher contractions is also a Fisher contraction on $(\mathscr{K}_0(H), H_\rho)$ with F-contractivity factors $\alpha = \max_{n=1}^{N_o} \alpha_n$ and $\beta = \max_{n=1}^{N_o} \beta_n$.*

Theorem 5 *If (H, ρ) is CMS and $\{H; T_n, n = 1, 2, 3, ..., N_o; N_o \in \mathbb{N}\}$ be FIFS of continuous Fisher contractions with F-contractivity factors α, β. Then, there exists a unique fixed point $A_\infty \in \mathscr{K}_0(H)$ is the HB operator for Fisher IFS or, likewise, $\mathscr{F}$ has unique compact invariant set $A_\infty \in \mathscr{K}_0(H)$. Here, $A_\infty \in \mathscr{K}_0(H)$ obeys $A_\infty = \mathscr{F}(A_\infty) = \bigcup_{n=1}^{N_o} T_n(A_\infty)$, and is provided that $A_\infty = lim_{n\to\infty}\mathscr{F}^{o(n)}(B)$ for any $B \in \mathscr{K}_o(X)$.*

Theorem 6 *If $\{H; T_0, T_1, T_2, ..., T_n, n = 1, 2, 3, ..., N_o; N_o \in \mathbb{N}\}$ is FIFS with the unique fixed point B. If the Fisher contractions mappings $T_0, T_1, ..., T_n$ are one-to-one on B and $T_i(B) \cap T_j(B) = \emptyset \quad \forall, \quad 0 \leq i, j \leq n$* and *$i \neq j$ then B is totally disconnected.*

3.1 Strong Self Similar and Profinite Groups for FIFS

Definition 9 ([21, 22]) Assume that (A, ρ) is CTG with the metric ρ being translation-invariant. Group A is denoted as self-similar if it has a proper subgroup B with a finite index and a onto homomorphism $\chi : A \to B$ (contractions) with respect to ρ.

Definition 10 ([22]) Assume that (A, ρ) is CTG with the metric ρ being translation-invariant. If a group has a proper subgroup B of the finite index and a group isomorphism $\chi : A \to B$ (contractions) with regard to ρ, then A is said to be strong self-similar.

Definition 11 ([22]) Let A be a topological group is known as profinite if it is topologically isomorphic to a finite discrete topological group's inverse limit. Equivalently, a topological group that is totally disconnected, Hausdorff, and compact is known as a profinite group.

4 FIFS of Compact Topological Groups, Self-Similar and Profinite Groups

This section defines SSG and their properties, including SSSG and Fisher contractions. Additionally, a few SSSG and profinite group characteristics are also depicted as well.

Definition 12 If (A, ρ) is CTG with the metric ρ being translation-invariant. If A has a proper subgroup B with a finite index and a onto homomorphism $\chi : A \to B$ (i.e., a Fisher contraction) with regard to ρ, then A is said to be self-similar.

Definition 13 Assume that (A, ρ) is CTG and that its metric is translation-invariant. If a group isomorphism $\chi : A \to B$ where B has an proper subgroup of an finite index of A and (i.e., Fisher contractions) with regard to ρ, then A is said to be strong self-similar.

Theorem 7 *If A is the SSSG of Fisher contractions, then A is FIFS the attractor.*

Proof Assume that (A, ρ) is CTG and that its metric is translation-invariant. If a group has an proper subgroup B of the finite index and a group isomorphism $\chi : A \to B$, that is, Fisher contractions with regard to ρ.

Take $[A : B] = n$ and also take $a_0 = e$ is an identity of A, $i \neq j \quad \forall i, j \in [0, n]$, there exist cosets of B in A such that $(B * a_i) \cap (B * a_j) = \emptyset$ and $A = B \cup (B * a_1) \cup (B * a_2) \cup ... \cup (B * a_{n-1})$. Define $\chi_i : A \to A$ by $\chi_i(g) = \chi_i(g) * a_i, \quad 0 \leq i \leq n-1$.

Here, to claim $\chi_i(g)$ is Fisher contractions $\forall i$. From that $\chi_i(A) = B * a_i$ as a result of χ_0 is onto (surjective). After that χ_0 is Fisher contractions with the contractivity factors α, β and ρ is translation invariant metric, we obtain that

$$\begin{aligned}\rho(\chi_i(g), \chi_i(h)) &= \rho(\chi_0(g) * a_i, \chi_0(h) * a_i) \\ &= \rho(\chi_0(g), \chi_0(h)) \\ &\leq \alpha[\rho(g, \chi_0(g)) + \rho(h, \chi_0(h))] + \beta\rho(g, h),\end{aligned}$$

for each $g, h \in A$. For these reasons, χ_i is Fisher mappings and the corresponding contraction ratios α_i and β_i for $1 \leq i \leq n-1$ and

$$\begin{aligned}A &= B \cup (B * a_1) \cup (B * a_2) \cup \ldots \cup (B * a_{n-1}) \\ &= \chi_0(A) \cup \chi_1(A) \cup \chi_2(A) \cup \ldots \cup \chi_{n-1}(A) \\ A &= \bigcup_{n=0}^{n-1} \chi_n(A).\end{aligned}$$

Thus, A is the attractor of the FIFS $\{A; \chi_0, \chi_1, \ldots, \chi_{n-1}\}$.

Theorem 8 *Allow the topological groups $(A, *, \rho)$ and $(A', *', \rho')$ to be compact. If $T : A \longrightarrow A'$ is both an group isomorphism and isometry map and A is SSSG of Fisher contractions, then prove that A' is also SSSG of Fisher contractions.*

Proof T is onto and T is isometry, hence $\exists\, a, b, c \in A$ such that $T(a) = a'$, $T(b) = b'$ and $T(c) = c'$ for all $a', b', c' \in A'$. ρ is translation-invariant metric, we compute

$$\begin{aligned}\rho'\left(a' *' c', b' *' c'\right) &= \rho'\left(T(a) *' T(c), T(b) *' T(c)\right) \\ &= \rho'(T(x), T(y)) \\ &= \rho'\left(a', b'\right).\end{aligned}$$

A SSSG of Fisher mapping is A, and B is a subgroup of the finite index set A; $\chi : A \longrightarrow B$ is a group. Let us take $T(B) = B'$. And T is a group asymmetry. Similarly B' is a subgroup of A' with a finite index.

Define $T_{|B} : B \longrightarrow B'$ by $T_{|B}(a) = T(a)$ for all $a \in B \subseteq A$. Now we prove that $\chi' = T_{|B} \circ \chi \circ T^{-1} : A' \longrightarrow B'$; this two are satisfying Fisher contractions mapping and a group isomorphism condition. Also, the T, $T_{|B}$ and χ satisfy the (group) isomorphisms condition; naturally χ' satisfies the group isomorphism. χ is a Fisher map, and the contractions ratio α and β and T, $T_{|B}$ are isometries; we get

$$\begin{aligned}\rho'\left(\chi'\left(g'\right), \chi'\left(h'\right)\right) =& \rho'\left(T_{|B} \circ \chi \circ T^{-1}\left(g'\right), T_{|B} \circ \chi \circ T^{-1}\left(h'\right)\right) \\ \leq& \alpha\left[\rho'\left(T^{-1}\left(g'\right), \chi\left(T^{-1}\left(g'\right)\right)\right)\right. \\ & \left. +\rho'\left(T^{-1}\left(h'\right), \chi\left(T^{-1}\left(h'\right)\right)\right)\right] + \beta\rho(T^{-1}(g'), T^{-1}(h')) \\ =& \alpha\left[\rho'\left(g', \chi'\left(g'\right)\right) + \rho'\left(h', \chi'\left(h'\right)\right)\right] + \beta\rho'(g', h'),\end{aligned}$$

$\forall g', h' \in A'$. This implies, χ' is Fisher contractions function on A'.

Theorem 9 *If $A_1, A_2, \ldots, A_n$ are SSSG of Fisher Mapping then $A_1 \times A_2 \times \ldots \times A_n$ is SSSG of Fisher Mapping.*

Proof Since $(A_1, *_1, \rho_1), (A_2, *_2, \rho_2), \ldots, (A_n, *_n, \rho_n)$ are CTG, $A_1 \times A_2 \times \ldots \times A_n$ is CTG. In addition, there exist subgroups $B_1, B_2, \ldots, B_n$ of $A_1, A_2, \ldots, A_n$, respectively; $\chi_i : A_i \longrightarrow B_i$ are Fisher contractions with corresponding contractions given that these groupings exhibit substantial Fisher contractions self-similarity. Define the map $\chi : A_1 \times A_2 \times \ldots \times A_n \longrightarrow B_1 \times B_2 \times \ldots \times B_n$ by $\chi(g_1, g_2, \ldots, g_n) = (\chi_1(g_1), \chi_2(g_2), \ldots, \chi_n(g_n))$.

It is obvious that $B_1 \times B_2 \times \ldots \times B_n$ is a subgroup of $A_1 \times A_2 \times \ldots \times A_n$ and $[A_1 \times A_2 \times \ldots \times A_n : B_1 \times B_2 \times \ldots \times B_n] = m_1 m_2 \ldots m_n$. Since $\chi_1, \chi_2, \ldots, \chi_n$ are group homomorphisms, we compute

$$\begin{aligned}
\chi(g * h) =& \chi((g_1, g_2, \ldots, g_n) * (h_1, h_2, \ldots, h_n)) \\
=& \chi((g_1), (g_2), \ldots, (g_n)) * \chi((h_1), (h_2), \ldots, (h_n)) \\
=& \chi(g) * \chi(h).
\end{aligned}$$

It becomes clear that χ is one to one and onto (bijective) because of the definitions $\chi_1, \chi_2, \ldots, \chi_n$. Hence χ is a group homomorphism. Let $\alpha = \max\{\alpha_1, \alpha_2, \ldots \alpha_n\}$ and $\beta = \max\{\beta_1, \beta_2, \ldots \beta_n\}$, for $1 \leq i \leq n$. Next, to derive that

$$\begin{aligned}
\rho(\chi(g), \chi(h)) =& \rho(\chi(g_1, g_2, \ldots, g_n) * \chi(h_1, h_2, \ldots, h_n)) \\
\chi(g * h) =& \chi((g_1, g_2, \ldots, g_n) * (h_1, h_2, \ldots, h_n)) \\
=& \rho((\chi_1(g_1), \ldots, \chi_n(g_n)) * (\chi_1(h_1), \ldots, \chi_n(h_n))) \\
=& \alpha[\rho(g, \chi(g) + \rho(h, \chi(h))] + \beta\rho(g, h).
\end{aligned}$$

In light of the fact that χ is a Fisher contraction with the contractions ratios α, β. Therefore, $A_1 \times A_2 \times \ldots \times A_n$ is SSSG.

Theorem 10 *A self-similar group of continuous Fisher contractions is a disconnected set.*

Proof Assume that A is SSG of (Fisher) contractions. This implies that A is a topological group. A is the attractor of the FIFS $\{\chi_0, \ldots, \chi_{n-1}\}$ by using Theorem 7 and $\forall\, i$ such that $1 \leq i \leq n-1$, the mappings

$$\chi_i : A \longrightarrow \chi_i(A)$$

are Fisher contractions. Furthermore, we have

$G = \chi_0(A) \cup \chi_1(A) \cup \ldots \cup \chi_{n-1}(A)$

$\Phi = \chi_i(A) \cap \chi_j(A)$,

for all $1 \leq i, j \leq n-1$ and $i \neq j$. It is commonly known that every compact subspace of a Hausdorff space is closed and that the image of a compact set under a continuous map is compact. Therefore, for $1 \leq i \leq n-1$ $\chi_i(A)$ is closed set. As a result,

$$A = \chi_0(A) \cup [\chi_1(A) \cup \ldots \cup \chi_{n-1}(A)]$$
$$\Phi = \chi_0(A) \cap [\chi_1(A) \cup \ldots \cup \chi_{n-1}(A)]) ,$$

we obtain that $\{\chi_0(A), [\chi_1(A) \cup \ldots \cup \chi_{n-1}(A)]\}$ is closed separation of A. Therefore, A is proved (disconnected set).

Theorem 11 *SSSG with continuous Fisher contractions are totally disconnected groups.*

Proof Assume that A is SSSG of continuous Fisher contractions. Theorem 7 clearly shows A is the invariant set (attractor) of a FIFS $\{\chi_0, \ldots, \chi_{n-1}\}$. Since $\chi_0 : A \longrightarrow B$ is one to one, we get

$$\chi_i(g) = \chi_i(h)$$
$$\chi_0(g) * a_i = \chi_0(h) * a_i$$
$$\chi_0(g) = \chi_0(h)$$
$$g = h,$$

$\forall g, h \in A$. From this χ_i is injective for $1 \leq i \leq n-1$. Likewise, $\chi_i(A) \cap \chi_j(A) = \emptyset$, $i \neq j$ and for every $i, j \in 1 \leq i, j \leq n-1$. For this reason, A is proven (totally disconnected by using Theorem 6).

The relationship between a profinite group and a strong Fisher contractions SSG is given by Theorem 12.

Theorem 12 *A profinite group is SSSG of Fisher contractions.*

Proof Let A be SSSG Fisher contractions. By Definition 13, A is CTG. It is known that every metric space is a Hausdorff space, the set A is Hausdorff. Besides, Theorem 11 clears that A is a totally disconnected set. Hence, A is totally disconnected, Hausdorff, and compact. As a result, we have the traits of profinite groups. This demonstrates that SSSG of Fisher contractions is a profinite group.

If all the contractivity factors $\alpha_n = 0$ $(n = 1, 2, \cdots, N_0)$, then the Fisher IFS becomes a standard IFS; and if all the contractivity factors $\beta_n = 0$ $(n = 1, 2, \cdots, N_0)$, then the Fisher IFS becomes a Kannan IFS; the converse of both cases is not always true. Hence, the method of constructing the Fisher fractal is a generalized case of the method of constructing the existing fractals through the classical IFS [2, 5, 6] and Kannan IFS (K-IFS) [23].

The importance of this research work is to generate fractal sets in SSSG through the iterated function system of Fisher contractions. It is demonstrated with the idea of constructing a new type of fractals in SSSG through interesting theorems and results. It is believed that the proposed research work will lead to a new path for developing the strong self-similar groups and their consequences based on Fisher contractions.

5 Conclusion

In this context, we have introduced the concepts of SSG and SSSG with Fisher contractions. We have proved SSSG of Fisher contractions as an attractor of a FIFS. A new type of fractal has been constructed as a self-similar group and strong self-similar group by the Fisher iterated function system on the compact topological group. Furthermore, the relations between the profinite group and the strong version of self-similar group with Fisher contractions have been proved mathematically.

References

1. Mandelbrot, B.B.: The Fractal Geometry of Nature, vol. 8, p. 406. WH Freeman and Company. New York (1983)
2. Hutchinson, J.E.: Fractals and self similarity. Indiana Univ. Math. J **30**, 713–747 (1981)
3. Barnsley, M.F.: Fractals Everywhere, 2nd edn. Academic, Cambridge (1993)
4. Barnsley, M.F., Vince, A.: Developments in fractal geometry. Bull. Math. Sci. **3**, 299–348 (2013)
5. Barnsley, M.F., Wilson, D.C., Lesniak, V.: Some Recent Progress Concerning Topology of Fractals. Springer, Berlin (2014)
6. Falconer, K.: Fractal Geometry, Mathematical Foundations and Applications. Wiley, England (2003)
7. Banerjee, S., Easwaramoorthy, D., Gowrisankar, A.: Fractal Functions, Dimensions and Signal Analysis, Understanding Complex Systems, Springer: Complexity. Springer, Cham (2021)
8. Prasad, B., Katiyar, K.: Multi fuzzy fractal theorems in fuzzy metric spaces. Fuzzy Informat. Eng. **9**(2), 225–236 (2017)
9. Prasad, B., Katiyar, K.: The attractors of Fuzzy Super Iterated Function Systems. Indian J. Sci. Technol. **10**(28), 90453 (2017)
10. Singh, S.L., Prasad, B., Kumar, A.: Fractals via iterated functions and multifunctions. Chaos, Solitons Fractals. **39**, 1224–1231 (2009)
11. Selmi, B.: Slices of Hewitt-Stromberg measures and co-dimensions formula. Analysis **42**(1), 23–39 (2022)
12. Selmi, B.: The relative multifractal analysis, review and examples. Acta Sci. Math. **86**(3–4), 635–666 (2020)
13. Douzi, Z., Selmi, B.: On the mutual singularity of Hewitt–Stromberg measures for which the multifractal functions do not necessarily coincide. Ricerche di Matematica **72**, 1–32 (2021)
14. Selmi, B.: A review on multifractal analysis of Hewitt-Stromberg measures. The J. Geometr. Analys. **32**(1), 1–44 (2022)
15. Secelean, N.A.: Iterated function systems consisting of F-contractions. Fixed Point Theory Appl. **277**, 13 (2013)
16. Pasupathi, R., Chand, A.K.B., Navascués, M.A.: Cyclic iterated function systems. J. Fixed Point Theory Appl. **22**(3), 1–17 (2020)

17. Pasupathi, R., Chand, A.K.B., Navascués, M.A., Sebastián, M.V.: Cyclic generalized iterated function systems. Comput. Math. Methods **3**(6), e1202 (2021)
18. Moran, M.: The group of isometries of a self-similar set. J. Math. Analys. Appl. **392**(1), 89–98 (2012)
19. Nekrashevych, V.: Self-similar groups. Bull. Amer. Math. Soc. **44**(3), 505–512 (2007)
20. Shirkov, D.V.: The evolution of Bogolyubov's renormalization group. Ukrainian J. Phys. **45**, 409–424 (2000)
21. Demir, B., Saltan, M.: A self-similar group in the sense of iterated function system, Far East J. Math. Sci., **60**(1), 83–99 (2012)
22. Saltan, M., Demir, B.: Self-similar groups in the sense of an iterated function system and their properties. J. Math. Analy. Appl. **408**(2), 694–704 (2013)
23. Sahu, D.R., Chakraborty, A., Dubey, R.P.: K-iterated function system. Fractals **18**(01), 139–144 (2010)
24. Uthayakumar, R., Gowrisankar, A.: Generation of fractals via self-similar group of Kannan iterated function system. Appl. Math. Informat. Sci. **9**(6), 3245 (2015)
25. Uthayakumar, R., Gowrisankar, A.: Attractor and self-similar group of generalized fuzzy contraction mapping in fuzzy metric space. Cogent Math. **2**(1), 1024579 (2015)
26. Easwaramoorthy, D., Uthayakumar, R.: Analysis on fractals in fuzzy metric spaces. Fractals **19**(3), 379–386 (2011)
27. Uthayakumar, R., Easwaramoorthy, D.: Hutchinson-Barnsley operator in fuzzy metric spaces. Int. J. Math. Comput. Sci. World Acad. Sci. Eng. Technol. **5**(8), 1418–1422 (2011)
28. Easwaramoorthy, D., Uthayakumar, R.: Intuitionistic fuzzy fractals on complete and compact spaces. In: Control, Computation and Information Systems, Communications in Computer and Information Science, vol. 140, pp. 89–96 (2011)
29. Easwaramoorthy, D., Uthayakumar, R.: Multivalued intuitionistic fuzzy fractals. Global J. Pure Appl. Math. **11**(4), 2519–2535 (2015)
30. Gowrisankar, A., Easwaramoorthy, D.: Local countable iterated function systems. Adv. Algebra Analy. Trends Math. **1**, 169–175 (2018)
31. Fisher, B.: A fixed point theorem for compact metric spaces. Publ. Math. Debrecen **25**, 193–194 (1978)

On Alternating Direction Implicit Solutions of 2D Kelvin–Helmholtz Instability Problem

Aziz Takhirov

1 Introduction

The goal of this chapter is to extend a highly parallelizable alternating direction implicit algorithm for solving the unsteady incompressible Navier–Stokes equations of [3] to more general boundary conditions and validate it on two-dimensional Kelvin–Helmholtz mixing layer problem [5]. A projection based similar scheme developed earlier in [1] has been thoroughly studied and validated for two-dimensional and three-dimensional lid-driven cavity problems. However, the validation of these schemes for high Reynolds number flows has not performed yet, to the best of our knowledge, and this is a first step in that direction.

Turbulence is inherently a three-dimensional phenomenon. As such, there are many good numerical benchmark problems for Re $\gg$ 1 incompressible flows in three dimensional. On the other hand, first studying two-dimensional examples is preferred because of faster implementation and significantly shorter computing times compared to three-dimensional simulations.

The most popular numerical benchmark for a high Reynolds number flow in two-dimensions seems to be the Kelvin–Helmholtz instability or mixing layer problem given in [5] and thoroughly studied in [7] (referred to as DNS henceforth). The velocity field is initiated with a perturbation noise, and through the action of the nonlinear term, small vortices arise, which then merge into a larger and larger vortices until finally one vortex is formed. Dozens of numerical studies have been dedicated for the Kelvin–Helmholtz instability problem. Although the existing results seem to be qualitatively correct, there are often a considerable quantitative differences, as discussed in more detail in the numerical section. Our experience

A. Takhirov (✉)
University of Sharjah, Sharjah, UAE
e-mail: takhirov@sharjah.ac.ae

H.-H. Leung et al. (eds.), *Recent Developments in Algebra and Analysis*,
Trends in Mathematics, https://doi.org/10.1007/978-3-031-37538-5_35

with simulating this problem also showed that slight changes in discretization parameters can yield substantially different results.

This chapter is organized as follows. Next we recall the setting of Kelvin–Helmholtz problem. In Sect. 2, we recall a few notations, present the time-discrete numerical scheme. In the section that follows, we prove a simple but an important lemma that allows us to state the stability of our scheme in the context of the Kelvin–Helmholtz problem. Section 4 discusses the numerical results, and Sect. 5 concludes the manuscript.

1.1 Kelvin–Helmholtz Instability Problem

The flow of Kelvin–Helmholtz instability problem is given by incompressible Navier–Stokes system in $\Omega = [0, 1]^2$:

$$\begin{aligned} \frac{\partial \mathbf{u}}{\partial t} + \left(\mathbf{u} \cdot \vec{\nabla}\right) \mathbf{u} + \nabla p + \nu \vec{\Delta} \mathbf{u} &= \mathbf{f} \text{ in } \Omega \times (0, T_f] \\ \nabla \cdot \mathbf{u} &= 0 \text{ in } \Omega \times (0, T_f] \end{aligned} \tag{1}$$

with periodic boundary conditions on the x boundary and no penetration $\mathbf{u} \cdot \mathbf{n} = 0$ and free-slip boundary conditions $(-\nu \nabla \mathbf{u} \cdot \mathbf{n}) \times \mathbf{n} = 0$ on the y boundary.

The initial velocity is given by

$$\begin{aligned} u_0 &= \begin{bmatrix} u_\infty \tanh\left((2y-1)/\alpha_0\right) \\ 0 \end{bmatrix} + 10^{-3} \begin{bmatrix} \partial_y \psi(x, y) \\ -\partial_x \psi(x, y) \end{bmatrix}, \\ \psi(x, y) &= \exp\left(-784\,(y - 0.5)^2 \right) [\cos(8\pi x) + \cos(20\pi x)], \end{aligned} \tag{2}$$

where the reference velocity is $u_\infty = 1$.

2 Notations and Numerical Scheme

2.1 Notations

To simplify the discussion, we only consider time-discretized numerical schemes.

Define the functional space for appropriate for the Kelvin–Helmholtz problem:

$$\begin{aligned} H^1_\# := \Big\{ \mathbf{u} \in H^1(\Omega) : \mathbf{u}|_{x=0} = \mathbf{u}|_{x=1} \text{ and } u_2|_{y=0} = u_2|_{y=1} = 0, \\ \partial_y u_1|_{y=0} = \partial_y u_1|_{y=1} = 0 \Big\}. \end{aligned}$$

The scalar product of $L^2(\Omega)$ and $L^2_{\int=0}(\Omega)$ is denoted by $(\cdot, \cdot)$.

The splitting error operator of the scheme is defined as

$$Sv := \partial_{xx}\partial_{yy}v$$

with its domain

$$D(S) := \left\{\mathbf{u} \in H^1_{\#}(\Omega) : S\mathbf{u} \in L^2(\Omega)\right\}.$$

2.2 Direction Splitting Scheme

Next we present the direction splitting scheme for the Navier–Stokes equations (1). Our numerical approximation is constructed via the artificial compressibility regularization:

$$\begin{aligned} \partial_t \mathbf{u} + \left(\mathbf{u} \cdot \overrightarrow{\nabla}\right)\mathbf{u} + \nabla p - \nu \overrightarrow{\Delta} \mathbf{u} &= \mathbf{0} \\ \tau \partial_t p + \chi \nabla \cdot \mathbf{u} &= 0, \end{aligned} \tag{3}$$

where $\chi = \mathcal{O}(1)$. The resulting approximation $(\mathbf{u}, p)$ has first-order temporal accuracy (see [8]), and higher-order schemes can be constructed using the bootstrapping approach of [2–4].

Denoting $NL := \left(\mathbf{u} \cdot \overrightarrow{\nabla}\right)\mathbf{u}$, the Douglass–Gunn factorized Euler scheme takes the form:

Algorithm Given initial data $\mathbf{u}^0$ and properly initialized initial pressure field p^0, solve for $\left(\mathbf{u}^{n+1}, p^{n+1}\right)$:

$$\begin{aligned} [\mathrm{I} - (\chi + \nu)\tau\partial_{xx}]\left[\mathrm{I} - \nu\tau\partial_{yy}\right]\frac{u_1^{n+1} - u_1^n}{\tau} &= \nu\Delta u_1^n + \chi\partial_{xx}u_1^n + \chi\partial_{xy}u_2^n \\ &\quad + f_1^{n+1} - \partial_x p^n - NL_1^n \end{aligned} \tag{4}$$

$$\begin{aligned} [\mathrm{I} - \nu\tau\partial_{xx}]\left[\mathrm{I} - (\chi + \nu)\tau\partial_{yy}\right]\frac{u_2^{n+1} - u_2^n}{\tau} &= \nu\Delta u_2^n + \chi\partial_{yx}u_1^{n+1} + \chi\partial_{yy}u_2^n \\ &\quad + f_2^{n+1} - \partial_y p^n - NL_2^n \end{aligned} \tag{5}$$

$$p^{n+1} - p^n + \chi\nabla \cdot \mathbf{u}^{n+1} = 0. \tag{6}$$

The algorithm can be rewritten in a more recognizable Crank–Nicolson form:

$$\frac{\mathbf{u}^{n+1}-\mathbf{u}^n}{\tau} + NL^n - \nu\Delta\mathbf{u}^{n+1} - \chi\nabla\left(\nabla\cdot\tilde{\mathbf{u}}^{n+1}\right) - \left(0, \partial_{yy}\left(\mathbf{u}^{n+1}-\mathbf{u}^n\right)\right)^T$$
$$+\nu(\chi+\nu)\tau^2 S\left[\frac{\mathbf{u}^{n+1}-\mathbf{u}^n}{\tau}\right] + \nabla p^n = \mathbf{f}^{n+1} \tag{7}$$

where $\tilde{\mathbf{u}}^{n+1} := \left(u_1^{n+1}, u_2^n\right)$.

3 Stability of the Unsteady Stokes Approximation

The key result in establishing the stability of the scheme is the following coercivity result for the operator S:

Lemma 1 *The bilinear form $D(S) \times D(S) \ni (\mathbf{u}, \mathbf{v}) \to (\mathbf{u}, S\mathbf{v}) \in R$ is symmetric positive and*

$$(\mathbf{u}, S\mathbf{u}) = \|\partial_{xy}\mathbf{u}\|^2, \quad \forall\, \mathbf{u} \in D(S). \tag{8}$$

Proof Using Fubini–Tonelli theorem and integration by parts gives

$$\begin{aligned}
(\mathbf{u}, S\mathbf{u}) &= (u_1, \partial_{xx}\partial_{yy}u_1) + (u_2, \partial_{xx}\partial_{yy}u_2) \\
&= \underbrace{\int_{y=0}^{y=1} u_1\partial_{xyy}u_1\Big|_{x=0}^{x=1} dy}_{=0 \text{ by periodicity}} - (\partial_x u_1, \partial_{xyy}u_1) \\
&\quad + \underbrace{\int_{y=0}^{y=1} u_2\partial_{xyy}u_2\Big|_{x=0}^{x=1} dy}_{=0 \text{ by periodicity}} - (\partial_x u_2, \partial_{xyy}u_2) \\
&= \underbrace{-\int_{x=0}^{x=1} \partial_x u_1\partial_{xy}u_1\Big|_{y=0}^{y=1} dx}_{=0 \text{ by free-slip}} + (\partial_{xy}u_1, \partial_{xy}u_1) \\
&\quad \underbrace{-\int_{x=0}^{x=1} \partial_x u_2\partial_{xy}u_2\Big|_{y=0}^{y=1} dx}_{=0 \text{ by free-slip}} + (\partial_{xy}u_2, \partial_{xy}u_2) \\
&= \|\partial_{xy}\mathbf{u}\|^2.
\end{aligned}$$

Setting $\mathbf{f} = \mathbf{0}$ and neglecting the nonlinear term, coercivity of the splitting error operator S indicates stability of the scheme with respect to the initial data. The proof directly follows that of [3][Theorem 4.1] and shall be omitted.

4 Numerical Example

Herein we report the results of our simulations for Reynolds number $\mathrm{Re} = 10^4$ case. A non-dimensional time unit $\bar{t} = \alpha_0/u_\infty$ is used for reporting the results. In what follows, we present the simulation results corresponding to $\tau = 2.5e - 5$ and 256×256 uniform grid. The time interval considered is $\bar{t} \in [0, 400]$. In Fig. 2, the plots produced by our Algorithm are labelled as "ADI," and those of [7] are labelled as "DNS."

The expected behavior is as follows: the nonlinearity amplifies the initial perturbations, and four vortices are formed around the horizontal line $y = 0.5$. After a while, those four vortices merge into larger two vortices. Later on, those two vortices coalesce into a single vortex centered at the origin (merging of the two vortices with vortices from neighboring cells has been reported in [6] at $\mathrm{Re} = 100$).

The vorticity contours obtained with our scheme are reported in Fig. 1. The reliable prediction of the pairing time of the last two vortices is very difficult, and a wide spectrum of merging times has been reported. This discrepancy has been attributed to the high sensitivity of this problem to inherent perturbations that occur in any numerical simulation [7]. Even on 256×256 uniform grid with 8-th degree DG scheme and a timestep of $\tau = 3.6 \times 10^{-5}$, mesh convergence has been found only for the time interval $\bar{t} < 200$.

4.1 Kinetic Energy Evolution

The top left graph in Fig. 2 is the plot of the kinetic energy $\frac{\|u\|^2}{2}$ vs. $\bar{t}$. Kinetic energy is supposed to monotonically decrease with time, and that is observed in both graphs. Energy of the scheme seems to decay faster, due to the scheme being more diffusive.

4.2 Enstrophy Evolution

The top right figure in Fig. 2 is the plot of the enstrophy $\frac{\|\nabla \times u\|^2}{2}$ vs. $\bar{t}$. Enstrophy is also a monotonically decreasing quantity for the Navier–Stokes system. Looking at the graphs, we see the expected behavior in both cases.

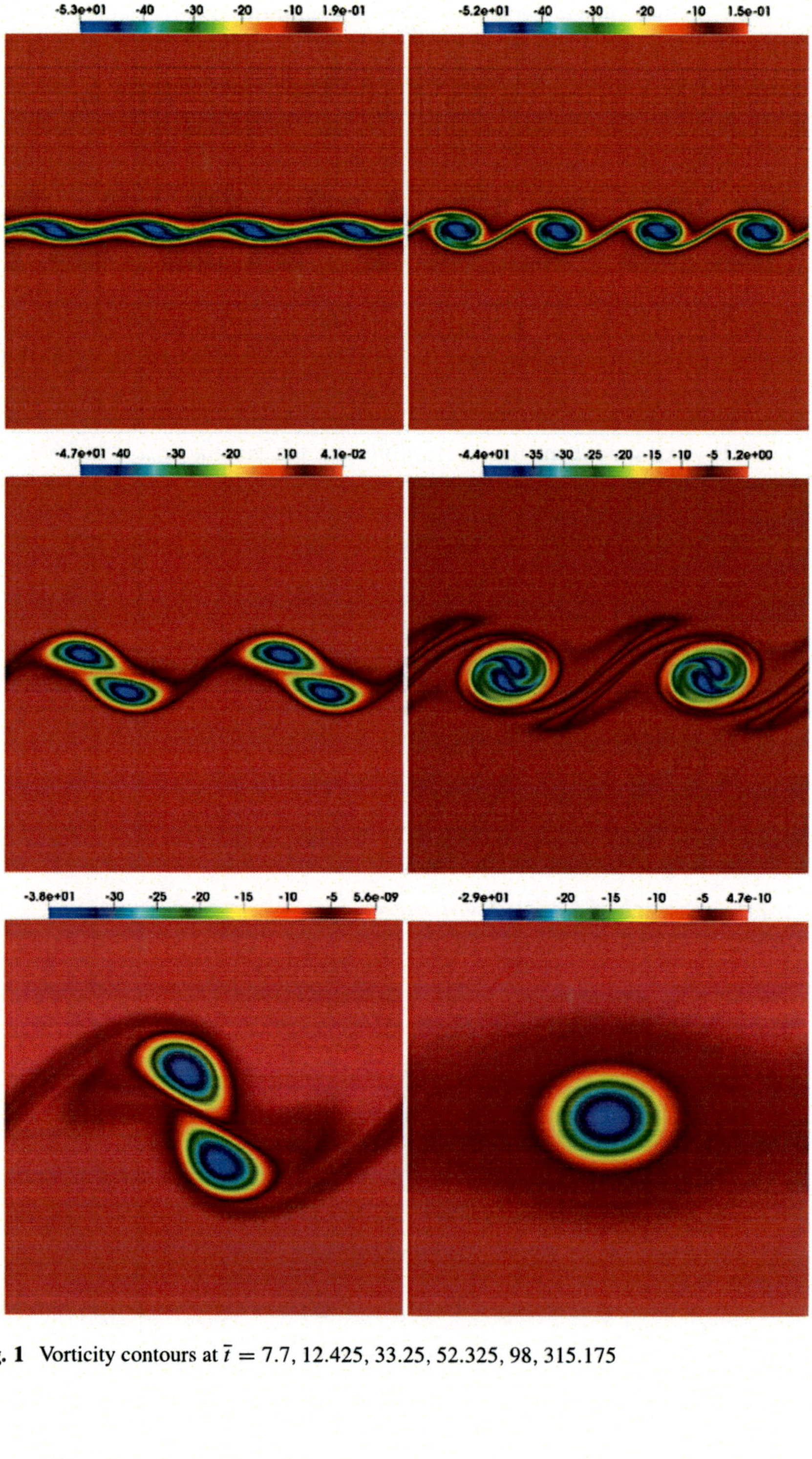

Fig. 1 Vorticity contours at $\bar{t} = 7.7, 12.425, 33.25, 52.325, 98, 315.175$

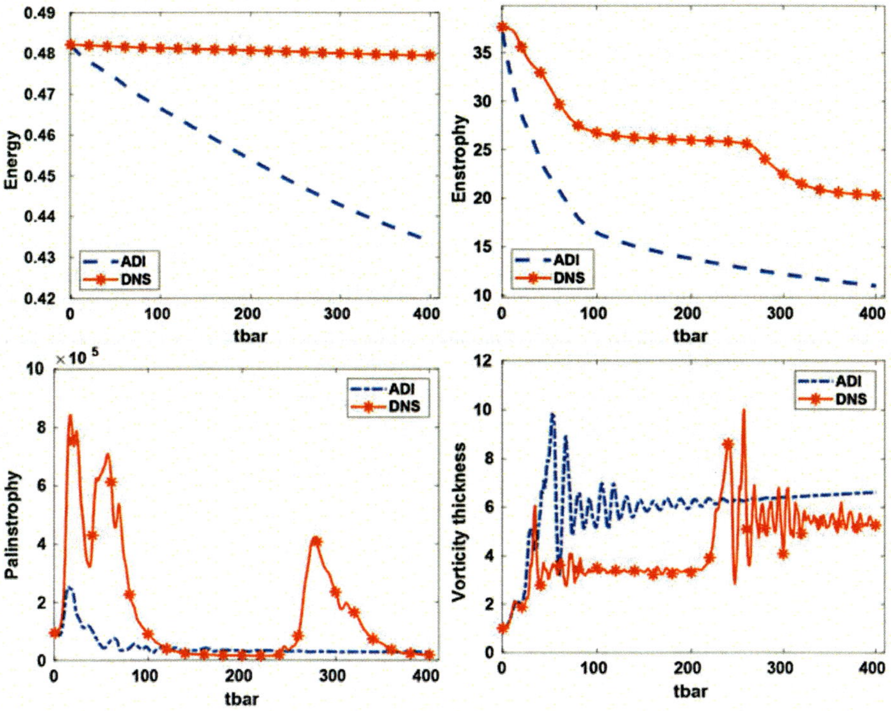

Fig. 2 Top row: kinetic energy and enstrophy. Bottow row: palinstrophy and the relative vorticity thickness

4.3 Palinstrophy Evolution

The bottom left figure in Fig. 2 is the plot of the palinstrophy $\frac{\| \nabla (\nabla \times u) \|^2}{2}$ vs. $\bar{t}$. In the current formulation of Navier–Stokes system, there is no control over the accuracy of palinstrophy, and thus it is the most difficult quantity to predict. As it can be seen, the ADI result follows the overall trend of DNS.

4.4 Vorticity Thickness Evolution

The bottom-right figure in Fig. 2 is the plot of the relative vorticity thickness

$$\delta(t) = \frac{2u_\infty}{\sup\limits_{y \in [0,1]} |< \omega > (t, y)|}, \qquad < \omega > (t, y) = \int_0^1 \omega(t, x, y) dx$$

against $\bar{t}$. Oscillations in the graph of $\alpha(t)$ indicate ellipsoidal vortices, while its smoothness means they are of circular shape. The current plot of $\delta(t)$ vs. t indicates that merging occurs at $\bar{t} \simeq 100$, much earlier than the expected result $\bar{t} \simeq 250$.

5 Conclusions

We described the extension of the massively parallel ADI scheme for solving the Navier–Stokes system (1) subject to more general boundary conditions. We discussed the stability of the unsteady Stokes approximation in Lemma 1. Numerical testing of our scheme on the Kelvin–Helmholtz benchmark problem gives qualitatively accurate results even on under-resolved mesh. Quantitative comparison has overall marginal accuracy, and fully resolved numerical simulation of the problem will be carried out in the future.

References

1. Guermond, J.L., Minev, P.D.: Start-up flow in a three-dimensional lid-driven cavity by means of a massively parallel direction splitting algorithm. Int. J. Numer. Methods Fluids **68**, 856–871 (2012)
2. Guermond, J.-L., Minev, P.D.: High-order time stepping for the incompressible Navier-Stokes equations. SIAM J. Sci. Comput. **37**, A2656–A2681 (2015)
3. Guermond, J.-L., Minev, P.D.: High-order time stepping for the Navier-Stokes equations with minimal computational complexity. J. Comput. Appl. Math. **310**, 92–103 (2017)
4. Guermond, J.-L., Minev, P.D.: High-order adaptive time stepping for the incompressible navier–stokes equations. SIAM J. Sci. Comput. **41**, A770–A788 (2019)
5. Lesieur, P.L.R.P.C.M., Staquet, C.: The mixing layer and its coherence examined from the point of view of two-dimensional turbulence. J. Fluid Mech. **192**, 511–534 (1988)
6. Rebholz, L.G., Vargun, D., Xiao, M.: Enabling convergence of the iterated penalty Picard iteration with $O(1)$ penalty parameter for incompressible Navier-Stokes via Anderson acceleration. Comput. Methods Appl. Mech. Eng. **387**, Paper No. 114178, 17 (2021)
7. Schroeder, P.W., John, V., Lederer, P.L., Lehrenfeld, C., Lube, G., Schöberl, J.: On reference solutions and the sensitivity of the 2D Kelvin-Helmholtz instability problem. Comput. Math. Appl. **77**, 1010–1028 (2019)
8. Shen, J.: On error estimates of the penalty method for unsteady Navier-Stokes equations. SIAM J. Numer. Anal. **32**, 386–403 (1995)

On Hyper-relative (m, s) Order of Entire Functions in Light of Central Index

Md. Azizul Hoque and Banani Dutta

Keywords Central index · Order · (m,s) order

1 Introduction

Considering an entire function $c(z)$ defined on the set of complex numbers. To characterize the growth we need, maximum modulus function on $|z| = r$ is introduced as $M(r, c) = \max_{|z|=r} |c(z)|$.

All the preliminary definitions regarding this discussion are available in [5, 6]. Here we will develop the idea of hyper-relative (m,s) order using the function ψ in light of central index.

1.1 Definitions

The function ψ defined in [5] satisfies the following relations:

(i) $\lim_{r\to\infty} \dfrac{\log^{[m]} r}{\log^{[s]} \psi(r)} = 0$

(ii) $\lim_{r\to\infty} \dfrac{\log^{[m]} \psi(\alpha r)}{\log^{[s]} \psi(r)} = 1$ where $\alpha > 1, m, s \in Z^+, m > s$.

For example, we can consider the function z^n (where, $n > 2$, an integer).

Definition 1 ([5]) Let c(z) , d(z) be two entire functions, then the relative (m,s)$\psi-$ order of c w.r.t. d is

M. A. Hoque
Department of Mathematics, Sreegopal Banerjee College, Mogra, Hooghly, West Bengal, India

B. Dutta (✉)
Department of Mathematics, Darjeeling Government College, Darjeeling, West Bengal, India

H.-H. Leung et al. (eds.), *Recent Developments in Algebra and Analysis*,
Trends in Mathematics, https://doi.org/10.1007/978-3-031-37538-5_36

$$\rho_{d,\psi}^{(m,s)}(c) = \limsup_{r\to\infty} \frac{\log^{[m-1]} \nu_d{}^{-1}\nu(r,c)}{\log^{[s]} \psi(r)}$$

and $\lambda_{d,\psi}^{(m,s)}(c) = \liminf_{r\to\infty} \frac{\log^{[m-1]} \nu_d^{-1}\nu(r,c)}{\log^{[s]} \psi(r)}$, where m,s $\in Z^+$, $m > s > 1$.

Definition 2 ([5]) Let c(z) , d(z) be two entire functions, then the hyper-relative (m,s)$\psi-$ order of c w.r.t. d is

$$\bar{\rho}_{d,\psi}^{(m,s)}(c) = \limsup_{r\to\infty} \frac{\log^{[m]} \nu_d^{-1}\nu(r,c)}{\log^{[s]} \psi(r)}$$

and $\bar{\lambda}_{d,\psi}^{(m,s)}(c) = \liminf_{r\to\infty} \frac{\log^{[m]} \nu_d^{-1}\nu(r,c)}{\log^{[s]} \psi(r)}$, where m,s $\in Z^+$, $m > s > 1$.

1.2 Lemma

The following lemmas will be needed in the discussion. Here we are modifying the lemmas given in [1] by using central index.

Lemma 1 *We know that for an entire function c(z) which satisfies property (A),* $[M_c(r)]^2 \leq [M_c(r^\sigma)]$, $\sigma > 1$ *,holds for all large r.*

We also have $\nu(r,c) \leq M(r,c) \leq \frac{R}{R-r}\nu(R,c), \, for 0 \leq r < R$.............(1) *Now putting R=2r in inequality (1) ,*

$$\nu(r,c) \leq M(r,c) \leq 2\nu(2r,c)$$

Using this we get

$$[\nu_c(r)]^n \leq [2\nu_c(2r^\sigma)].$$

Lemma 2 *For an entire function c(z) and* α^*, β^* *be such that* $\alpha^* > 1, 0 < \beta^* < \alpha^*, s^* > 1$ *,then*

$$M_c(\alpha^* r) > \beta^* M_c(r).$$

we get

$$2\nu_c(2\alpha^* r) > \beta^* \nu_c(r).$$

Lemma 3 *There exists* $K(s^*, c) > 0$*,*$s^* > 1$ *such that* $[M_c(r)]^{s^*} \leq K[M_c(r^{s^*})]$ *, for* $r > 0$*. Using inequality (1) we get*

$$[\nu_c(r)]^{s^*} \leq 2K[\nu_c((2r)^{s^*})].$$

Lemma 4 *If c is transcendental, then*

$$\lim_{r\to\infty} \frac{\nu_c(2^{s*}r^{s*})}{r^n \nu_c(r)} = \infty = \lim_{r\to\infty} \frac{\nu_c(2^\lambda r^\lambda)}{r^n \nu_c(r^\mu)}$$

where $0 < \mu < \lambda$.

Lemma 5 *Let c and d be two entire functions ,* $d(0) = 0$ *, then for sufficiently large values of r*

$$M_c(r) \geq M_c(\frac{1}{16} M_d(\frac{r}{2})).$$

Again we have

$$M_{c\circ d}(r) \leq M_c(M_d(r)).$$

Using inequality (1) we get ,

$$2\nu_{c\circ d}(r) \geq \nu_c(\frac{1}{16} \nu_d(\frac{r}{2})).$$

and

$$\nu_{c\circ d}(r) \leq 2\nu_c(2\nu_d(4r)).$$

2 Main Results

Theorem 1 *Considering four entire functions* f_1, f_2 *and* g, h *with* $0 < \lambda_{h,\psi} \leq \rho_{h,\psi} < \infty$, *then for* $m > 2$

$$\bar{\rho}_{g,\psi}^{(m,s)}(f_1 \pm f_2) \leq \max[\rho_{g\circ h,\psi}^{(m,s)}(f_1), \rho_{g\circ h,\psi}^{(m,s)}(f_2)]$$

equality holds when $\rho_{g\circ h,\psi}^{(m,s)}(f_1) \neq \rho_{g\circ h,\psi}^{(m,s)}(f_2)$ *, where* ψ *is defined earlier.*

Proof We consider the theorem for $f_1 + f_2$. Let $f = f_1 + f_2$ and

$$\rho_{g\circ h,\psi}^{(m,s)}(f_1) \leq \rho_{g\circ h,\psi}^{(m,s)}(f_2).$$

Let $\epsilon > 0$ be arbitrary, from the definition of $\rho_{g\circ h,\psi}^{(m,s)}(f_1)$, for all larger values of r, we have

$$\nu(r, f_1) \le \nu_{goh}[\exp^{[m-1]}[(\rho^{(m,s)}_{goh,\psi}(f_1) + \epsilon)\log^{[s]}\psi(r)]]$$

$$\le \nu_{goh}[\exp^{[m-1]}[(\rho^{(m,s)}_{goh,\psi}(f_2) + \epsilon)\log^{[s]}\psi(r)]]$$

and

$$\nu(r, f_2) \le \nu_{goh}[\exp^{[m-1]}[(\rho^{(m,s)}_{goh,\psi}(f_2) + \epsilon)\log^{[s]}\psi(r)]]$$

Now,

$$\nu(r, f) \le \nu(r, f_1) + \nu(r, f_2)$$

$$< 2\nu_{goh}[\exp^{[m-1]}[(\rho^{(m,s)}_{goh,\psi}(f_2) + \epsilon)\log^{[s]}\psi(r)]]$$

$$< 2\nu_{goh}[2.3\exp^{[m-1]}[(\rho^{(m,s)}_{goh,\psi}(f_2) + \epsilon)\log^{[s]}\psi(r)]], \textit{by Lemma } 2$$

$$\Rightarrow \nu(r, f)$$
$$< 2.2\nu_g[2\nu_h[4.2.3\exp^{[m-1]}[(\rho^{(m,s)}_{goh,\psi}(f_2) + \epsilon)\log^{[s]}\psi(r)]]], \textit{by Lemma } 5.$$

$$= 4\nu_g[2\nu_h[24\exp^{[m-1]}[(\rho^{(m,s)}_{goh,\psi}(f_2) + \epsilon)\log^{[s]}\psi(r)]]]$$

$$\Rightarrow \frac{1}{4}\nu_f(r) \le \nu_g[2\nu_h[24\exp^{[m-1]}[(\rho^{(m,s)}_{goh,\psi}(f_2) + \epsilon)\log^{[s]}\psi(r)]]]$$

$$\Rightarrow \nu_g^{-1}\frac{1}{4}\nu_f(r) \le 2\nu_h[24\exp^{[m-1]}[(\rho^{(m,s)}_{goh,\psi}(f_2) + \epsilon)\log^{[s]}\psi(r)]]$$

$$\Rightarrow \log \nu_g^{-1}\nu_f(r) + O(1) < 2\nu_h[24\exp^{[m-1]}[(\rho^{(m,s)}_{goh,\psi}(f_2) + \epsilon)\log^{[s]}\psi(r)]] + O(1)$$

$$< [24\exp^{[m-1]}[(\rho^{(m,s)}_{goh,\psi}(f_2) + \epsilon)\log^{[s]}\psi(r)]^{\rho_{h,\psi}+\epsilon}$$

$$\Rightarrow \log\log \nu_g^{-1}\nu_f(r) + O(1)$$
$$< (\rho_{h,\psi} + \epsilon)\log[24\exp^{[m-1]}[(\rho^{(m,s)}_{goh,\psi}(f_2) + \epsilon)\log^{[s]}\psi(r)]$$

$$\Rightarrow \log\log \nu_g^{-1}\nu_f(r) + O(1)$$
$$< (\rho_{h,\psi} + \epsilon)[\exp^{[m-2]}[(\rho^{(m,s)}_{goh,\psi}(f_2) + \epsilon)\log^{[s]}\psi(r)]] + O(1)$$

$$\Rightarrow \log^{[m]} \nu_g^{-1} \nu_f(r) < (\rho_{goh,\psi}^{(m,s)}(f_2) + \epsilon) \log^{[s]} \psi(r) + O(1)$$

$$\Rightarrow \frac{\log^{[m]} \nu_g^{-1} \nu_f(r)}{\log^{[s]} \psi(r)} \leq (\rho_{goh,\psi}^{(m,s)}(f_2) + \epsilon) + O(1)$$

$\Rightarrow \bar{\rho}_{g,\psi}^{(m,s)}(f) \leq \rho_{goh,\psi}^{(m,s)}(f_2)$, as $\epsilon > 0$ is arbitrary.
So,

$$\bar{\rho}_{g,\psi}^{(m,s)}(f_1 \pm f_2) \leq \max[\rho_{goh,\psi}^{(m,s)}(f_1), \rho_{goh,\psi}^{(m,s)}(f_2)]$$

Next let $\rho_{goh,\psi}^{(m,s)}(f_1) < \rho_{goh,\psi}^{(m,s)}(f_2)$.

Then from the definition of $\rho_{goh,\psi}^{(m,s)}(f_1)$, for all large values of r
$\nu(r, f_1) < \nu_{goh}[\exp^{[m-1]}[(\rho_{goh,\psi}^{(m,s)}(f_1 + \epsilon)\log^{[s]} \psi(r)]]$ ....(4)

Then from the definition of $\rho_{goh,\psi}^{(m,s)}(f_2)$, there exists a sequence $r_n \to \infty$ such that
$\nu_{r_n, f_2} > \nu_{goh}[\exp^{[m-1]}[(\rho_{goh,\psi}^{(m,s)}(f_2) - \epsilon)\log^{[s]} \psi(r_n)]]$ ....(5)

From Lemma (4) we get

$$\lim_{r\to\infty} \frac{\nu_{goh}[\exp^{[m-1]}[(\rho_{goh,\psi}^{(m,s)}(f_2) - \epsilon)\log^{[s]} \psi(r)]]}{\nu_{goh}[\exp^{[m-1]}[(\rho_{goh,\psi}^{(m,s)}(f_1) + \epsilon)\log^{[s]} \psi(r)]]} < \frac{\nu(r, f_2)}{\nu(r, f_1)} \to \infty.$$

Then for all large r,

$$\nu_{goh}[\exp^{[m-1]}[(\rho_{goh,\psi}^{(m,s)}(f_2) - \epsilon)\log^{[s]} \psi(r)]]$$
$$> 2.\nu_{goh}[\exp^{[m-1]}[(\rho_{goh,\psi}^{(m,s)}(f_1) + \epsilon)\log^{[s]} \psi(r)]]$$

$$\Rightarrow \nu(r_n, f_2) > \nu_{goh}[\exp^{[m-1]}[(\rho_{goh,\psi}^{(m,s)}(f_2) - \epsilon)\log^{[s]} \psi(r)]]$$
$$> 2.\nu_{goh}[\exp^{[m-1]}[(\rho_{goh,\psi}^{(m,s)}(f_1) + \epsilon)\log^{[s]} \psi(r)]].$$

For a sequence of r tending to infinity, by using (4)
$\nu(r_n, f_2) > 2\nu(r_n, f_1)$ for $n \in N$

$$\Rightarrow \nu(r_n, f) \geq \nu(r_n, f_2) - \nu(r_n, f_1) > \nu(r_n, f_2) - \frac{1}{2}\nu(r_n, f_2) = \frac{1}{2}\nu(r_n, f_2)$$

$$> \frac{1}{2}\nu_{goh}[\exp^{[m-1]}[(\rho_{goh,\psi}^{(m,s)}(f_2) - \epsilon)\log^{[s]} \psi(r_n)]], using (5).$$

Then using Lemma 2 we get,

$$\nu(f, r_n) > \nu_{goh}[\frac{1}{3}\exp^{[m-1]}[(\rho^{(m,s)}_{goh,\psi}(f_2) - \epsilon)\log^{[s]}\psi(r_n)]]$$

$$> \frac{1}{2}\nu_g[\frac{1}{16}\nu_h[\frac{\exp^{[m-1]}[(\rho^{(m,s)}_{goh,\psi}(f_2) - \epsilon)\log^{[s]}\psi(r_n)}{6}]], \text{ by Lemma 5.}$$

$$\Rightarrow \nu(f, r_n) \geq \nu_g[\frac{1}{16}\nu_h[\frac{\exp^{[m-1]}[(\rho^{(m,s)}_{goh,\psi}(f_2) - \epsilon)\log^{[s]}\psi(r_n)}{6}]]$$

$$\Rightarrow \nu_g^{-1}2\nu(f, r_n) > [\frac{1}{16}\nu_h[\frac{\exp^{[m-1]}[(\rho^{(m,s)}_{goh,\psi}(f_2) - \epsilon)\log^{[s]}\psi(r_n)}{6}]]$$

$$\Rightarrow \log\nu_g^{-1}\nu(f, r_n) + O(1)$$

$$> [\log\nu_h[\frac{\exp^{[m-1]}[(\rho^{(m,s)}_{goh,\psi}(f_2) - \epsilon)\log^{[s]}\psi(r_n)}{6}]] + O(1)$$

$$> [\frac{\exp^{[m-1]}[(\rho^{(m,s)}_{goh,\psi}(f_2) - \epsilon)\log^{[s]}\psi(r_n)}{6}]^{\lambda_{h,\psi}-\epsilon} + O(1)$$

$$\Rightarrow \log\log\nu_g^{-1}\nu(f, r_n) > (\lambda_{h,\psi} - \epsilon)[\exp^{[m-2]}[(\rho^{(m,s)}_{goh,\psi}(f_2) - \epsilon)\log^{[s]}\psi(r_n)] + O(1)$$

$$\Rightarrow \log^{[m]}\nu_g^{-1}\nu(f, r_n) > (\rho^{(m,s)}_{goh,\psi}(f_2) - \epsilon)\log^{[s]}\psi(r_n) + O(1)$$

$$\Rightarrow \limsup_{r\to\infty}(\frac{\log^{[m]}\nu_g^{-1}\nu(f, r_n)}{\log^{[s]}\psi(r_n}) \geq (\rho^{(m,s)}_{goh,\psi}(f_2) - \epsilon)$$

i.e. $\bar{\rho}^{(m,s)}_{g,\psi}(f) \geq \rho^{(m,s)}_{goh,\psi}(f_2)$, since $\epsilon > 0$ is arbitrary.
So,

$$\bar{\rho}^{(m,s)}_{g,\psi}(f_1 \pm f_2) \geq \rho^{(m,s)}_{goh,\psi}(f_2) = \max[\rho^{(m,s)}_{goh,\psi}(f_1), \rho^{(m,s)}_{goh,\psi}(f_2)]$$

So we have

$$\bar{\rho}^{(m,s)}_{g,\psi}(f_1 \pm f_2) = \max[\rho^{(m,s)}_{goh,\psi}(f_1), \rho^{(m,s)}_{goh,\psi}(f_2)]$$

This proves the theorem.

Theorem 2 *Let S be a polynomial, and c, g, h are entire functions with* $0 < \lambda_{h,\psi} \leq \rho_{h,\psi} < \infty$, *where c is transcendental and* ψ *is defined earlier. Then for* $m > 2$,

$$\bar{\rho}^{(m,s)}_{g,\psi}(S.c) = \rho^{(m,s)}_{goh,\psi}(c).$$

Proof Let the degree of $S(z)$ be m. Then there exists α with $0 < \alpha < 1$, and $n(> m) \in Z^+$, $2\alpha < |S(z)| < r^n$ holds on $|z| = r$ for all large r. Using Lemma (2)

$$2\nu_c(2\frac{1}{\alpha}\alpha r) > \frac{1}{2\alpha}\nu_c(\alpha r)$$

$$\Rightarrow \nu_c(\alpha r) < 4\alpha\nu_c(2r).$$

Let $k(z) = S(z)c(z)$, then for all large r and $p > 1$

$$\nu_c(\alpha r) < 4\alpha\nu_c(2r) \le 2\nu_k(r) \le r^n\nu_c(r) < \nu_c(2^p r^p), \; by \; Lemma \; 5$$

Let $\epsilon > 0$ be arbitrary and for all large r,

$$2\nu_k(r) < \nu_c(2^p r^p)$$

$$< \nu_{goh}[\exp^{[m-1]}[(\rho^{(m,s)}_{goh,\psi}(c) + \epsilon)\log^{[s]}\psi(2^p r^p)]]$$

$$< 2\nu_g[2\nu_h[4\exp^{[m-1]}[(\rho^{(m,s)}_{goh,\psi}(c) + \epsilon)\log^{[s]}\psi(2^p r^p)]]]$$

$$\Rightarrow \nu_g^{-1}\nu_k(r) < [2\nu_h[4\exp^{[m-1]}[(\rho^{(m,s)}_{goh,\psi}(c) + \epsilon)\log^{[s]}\psi(2^p r^p)]]]$$

$$\Rightarrow \log\nu_g^{-1}\nu_k(r) < \log\nu_h[4\exp^{[m-1]}[(\rho^{(m,s)}_{goh,\psi}(c) + \epsilon)\log^{[s]}\psi(2^p r^p)]] + O(1)$$

$$< [4\exp^{[m-1]}[(\rho^{(m,s)}_{goh,\psi}(c) + \epsilon)\log^{[s]}\psi(2^p r^p)]]^{(}\rho_{h,\psi} + \epsilon) + O(1)$$

$$\Rightarrow \log\log\nu_g^{-1}\nu_k(r) < (\rho_{h,\psi} + \epsilon)\exp^{[m-2]}[(\rho^{(m,s)}_{goh,\psi}(c) + \epsilon)\log^{[s]}\psi(2^p r^p)] + O(1)$$

$$\Rightarrow \log^{[m]}\nu_g^{-1}\nu_k(r) < [(\rho^{(m,s)}_{goh,\psi}(c) + \epsilon)\log^{[s]}\psi(2^p r^p)] + O(1)$$

$$\Rightarrow \frac{\log^{[m]}\nu_g^{-1}\nu_k(r)}{\log^{[s]}\psi(r)} < (\rho^{(m,s)}_{goh,\psi}(c) + \epsilon)\frac{\log^{[s]}\psi(2^p r^p)}{\log^{[s]}\psi(r)} + O(1)$$

$$\Rightarrow \bar{\rho}^{(m,s)}_{g,\psi}(S.c) \le \rho^{(m,s)}_{goh,\psi}(c) + \epsilon.$$

Since $\epsilon > 0$ is arbitrary,

$$\bar{\rho}^{(m,s)}_{g,\psi}(S.c) \le \rho^{(m,s)}_{goh,\psi}(c).$$

Next for a sequence of values of $r_n \to \infty$

$$2\nu_k(r_n) > \nu_c(\alpha r_n)$$

$$> \nu_{goh}[\exp^{[m-1]}[(\rho^{(m,s)}_{goh,\psi}(c) - \epsilon)\log^{[s]} \psi(\alpha r_n)]]$$

$$> \frac{1}{2}\nu_g[\frac{1}{16}\nu_h[\frac{\exp^{[m-1]}[(\rho^{(m,s)}_{goh,\psi}(c) - \epsilon)\log^{[s]} \psi(\alpha r_n)}{2}]], \textit{ by Lemma 5.}$$

$$\Rightarrow \nu_g^{-1}\nu_k(r_n) + O(1) > [\log \nu_h[\frac{\exp^{[m-1]}[(\rho^{(m,s)}_{goh,\psi}(c) - \epsilon)\log^{[s]} \psi(\alpha r_n)}{2}]] + O(1)$$

$$\Rightarrow \log \nu_g^{-1}\nu_k(r_n) > [\frac{1}{16}\nu_h[\frac{\exp^{[m-1]}[(\rho^{(m,s)}_{goh,\psi}(c) - \epsilon)\log^{[s]} \psi(\alpha r_n)}{2}]]$$

$$> [\frac{\exp^{[m-1]}[(\rho^{(m,s)}_{goh,\psi}(c) - \epsilon)\log^{[s]} \psi(\alpha r_n)}{2}]^{\lambda_{h,\psi}-\epsilon} + O(1)$$

$$\Rightarrow \log\log \nu_g^{-1}\nu(k, r_n) > (\lambda_{h,\psi} - \epsilon)[\exp^{[m-2]}[(\rho^{(m,s)}_{goh,\psi}(c) - \epsilon)\log^{[s]} \psi(\alpha r_n)] + O(1)$$

$$\Rightarrow \log^{[m]} \nu_g^{-1}\nu(k, r_n) > (\rho^{(m,s)}_{goh,\psi}(c) - \epsilon)\log^{[s]} \psi(\alpha r_n) + O(1)$$

$$\Rightarrow \limsup_{r\to\infty}(\frac{\log^{[m]} \nu_g^{-1}\nu(k, r_n)}{\log^{[s]} \psi(r_n}) > (\rho^{(m,s)}_{goh,\psi}(c) - \epsilon). \limsup_{r\to\infty} \frac{\log^{[s]} \psi(\alpha r_n)}{\log^{[s]} \psi(r_n)}$$

i.e. $\bar{\rho}^{(m,s)}_{g,\psi}(k) \geq (\rho^{(m,s)}_{goh,\psi}(c) - \epsilon).1$

So,

$$\bar{\rho}^{(m,s)}_{g,\psi}(k) \geq \rho^{(m,s)}_{goh,\psi}(c).$$

So we have $\bar{\rho}^{(m,s)}_{g,\psi}(S.c) = \rho^{(m,s)}_{goh,\psi}(c)$.

This proves the theorem.

Theorem 3 *Considering* $n(> 1) \in Z^+$ *and* c, g, h *are entire functions,* $0 < \lambda_{h,\psi} \leq \rho_{h,\psi} < \infty$, ψ *is defined earlier. Then for* $m > 2$,

$$\bar{\rho}^{(m,s)}_{g,\psi}(c^n) = \rho^{(m,s)}_{goh,\psi}(c).$$

Proof From Lemmas 2, 3 we obtain

$$[\nu_c(r)]^n \leq 2K^*\nu_c(2^n r^n)$$

$< 2\nu_c(Gr^n)$, where $G = 2^{n+1}(2K^* + 1)$

$$< 2\nu_{goh}[\exp^{[m-1]}[(\rho^{(m,s)}_{goh,\psi}(c) + \epsilon)\log^{[s]} \psi(Gr^n)]]$$

$$< 4\nu_g[2\nu_h[4\exp^{[m-1]}[(\rho^{(m,s)}_{goh,\psi}(c)+\epsilon)\log^{[s]}\psi(Gr^n)]]],$$

where $K^* = K^*(n, c) > 0, n > 1, m > 1$.

So,

$$\Rightarrow \nu_g^{-1}\frac{1}{4}[\nu_c(r)]^n < [2\nu_h[4\exp^{[m-1]}[(\rho^{(m,s)}_{goh,\psi}(c)+\epsilon)\log^{[s]}\psi(Gr^n)]]]$$

$$\Rightarrow \log\nu_g^{-1}[\nu_c(r)]^n < \log\nu_h[4\exp^{[m-1]}[(\rho^{(m,s)}_{goh,\psi}(c)+\epsilon)\log^{[s]}\psi(Gr^n)]] + O(1)$$

$$< [4\exp^{[m-1]}[(\rho^{(m,s)}_{goh,\psi}(c)+\epsilon)\log^{[s]}\psi(Gr^n)]]^{(}\rho_{h,\psi}+\epsilon) + O(1)$$

$$\Rightarrow \log\log\nu_g^{-1}[\nu_c(r)]^n < (\rho_{h,\psi}+\epsilon)\exp^{[m-2]}[(\rho^{(m,s)}_{goh,\psi}(c)+\epsilon)\log^{[s]}\psi(Gr^n)]+O(1)$$

$$\Rightarrow \log^{[p]}\nu_g^{-1}[\nu_c(r)]^n < [(\rho^{(m,s)}_{goh,\psi}(c)+\epsilon)\log^{[s]}\psi(Gr^n)] + O(1)$$

$$\Rightarrow \limsup_{r\to\infty}\frac{\log^{[m]}\nu_g^{-1}[\nu_c(r)]^n}{\log^{[s]}\psi(r)} \le \limsup_{r\to\infty}\frac{(\rho^{(m,s)}_{goh,\psi}(c)+\epsilon)\log^{[s]}\psi(Gr^n)}{\log^{[s]}\psi(Gr^n)}$$

$$.\limsup_{r\to\infty}\frac{\log^{[s]}\psi(Gr^n)}{\log^{[s]}\psi(r)}$$

$$\Rightarrow \bar{\rho}^{(m,s)}_{g,\psi}(c^n) \le (\rho^{(m,s)}_{goh,\psi}(c)+\epsilon).1$$

we get,

$$\bar{\rho}^{(m,s)}_{g,\psi}(c^n) \le \rho^{(m,s)}_{goh,\psi}(c).$$

For a sequence of $r = r_n$

$$[\nu_c(r_n)]^n > \nu_c(r_n)$$

$$> \nu_{goh}[\exp^{[m-1]}[(\rho^{(m,s)}_{goh,\psi}(c)-\epsilon)\log^{[s]}\psi(r_n)]]$$

$$> \frac{1}{2}\nu_g[\frac{1}{16}\nu_h[\frac{\exp^{[m-1]}[(\rho^{(m,s)}_{goh,\psi}(c)-\epsilon)\log^{[s]}\psi(r_n)}{2}]], \textit{by Lemma } 5.$$

$$\Rightarrow \nu_g^{-1}2[\nu_c(r_n)]^n > [\frac{1}{16}\nu_h[\frac{\exp^{[m-1]}[(\rho^{(m,s)}_{goh,\psi}(c)-\epsilon)\log^{[s]}\psi(r_n)}{2}]]$$

$$\Rightarrow \log \nu_g^{-1}[\nu_c(r_n)]^n + O(1)$$

$$> \log \nu_h[\frac{\exp^{[m-1]}[(\rho_{g\circ h,\psi}^{(m,s)}(c) - \epsilon)\log^{[s]} \psi(r_n)}{2}]] + O(1)$$

$$> [\frac{\exp^{[m-1]}[(\rho_{g\circ h,\psi}^{(m,s)}(c) - \epsilon)\log^{[s]} \psi(r_n)}{2}]^{\lambda_{h,\psi}-\epsilon} + O(1)$$

$$\Rightarrow \log\log \nu_g^{-1}[\nu_c(r_n)]^n > (\lambda_{h,\psi} - \epsilon)[\exp^{[m-2]}[(\rho_{g\circ h,\psi}^{(m,s)}(c) - \epsilon)\log^{[s]} \psi(r_n)] + O(1)$$

$$\Rightarrow \log^{[m]} \nu_g^{-1}[\nu_c(r_n)]^n > (\rho_{g\circ h,\psi}^{(m,s)}(c) - \epsilon)\log^{[s]} \psi(r_n)) + O(1)$$

$$i.e. \bar{\rho}_{g,\psi}^{(m,s)}(c^n) > (\rho_{g\circ h,\psi}^{(m,s)}(c) - \epsilon),$$

So,

$$\bar{\rho}_{g,\psi}^{(m,s)}(c^n) \geq \rho_{g\circ h,\psi}^{(m,s)}(c).$$

So we have

$$\bar{\rho}_{g,\psi}^{(m,s)}(c^n) = \rho_{g\circ h,\psi}^{(m,s)}(c).$$

This proves the theorem.

Conclusion and Future Scope

Here we have established few results on hyper-relative (m,s) order of entire function. One may try to prove on other growth factors of composition of entire functions.

References

1. Banerjee, D., Batabyal, S.: Hyper relative order (p,q) of entire functions. Analele Universitatii de Vest, Timisoara, Seria Mathematica-Informatica LV **2**, 65–84 (2017)
2. Bergweiler, W., Jank, G., Volkmann, L.: Wachstumsverhalten ZusammengesetzerFunktionen. Results Math. **7**, 35–53 (1984)
3. Bernal, L.: Orden relativ de crecimiento de funciones enteras. Collect Mth. **39**, 209–229 (1988)
4. Datta, S.K., Dutta, B.: On the growth analysis of complex linear differential equations with entire and meromorphic coefficients. Int. J. Eng. Sci. Math. **7**(5), 44–55 (2018)
5. Datta, S.K., Dutta, B.: On the central index of composite entire functions. Int. J. Eng. Sci. Math. **7**(5), 90–101 (2018)
6. Lahiri, B.K., Banerjee, D.: Generalised relativ order of entire functions. Proc. Nat. Acad. Sci. India **72**(A) (IV), 351–371 (2002)

7. Lahiri, B.K., Banerjee, D.: Entire functions of relative order (p, q). Soochow J. Math. **31**(4), 497–513 (2005)
8. Levin, B.J.: Distribution of Zeros of Entire Functions. American Mathematical Society, Providence (1980)
9. Lu, F., Qi, J.: A note on the hyper order of entire functions. Bull. Korean Math. Soc. **50**(4), 1209–1219 (2013)

Application of Subordination and q-Differentiation to Classes of Regular Functions Associated with Certain Special Functions and Bell Numbers

Ezekiel Abiodun Oyekan and Ayotunde Olajide Lasode

Keywords Regular function · Bell numbers · Coefficient inequality · Error function · Struve function · Fekete-Szegö inequality · q-Differentiation

1 Introduction

In this study, we let $\mathcal{A}$ represent the set of regular functions. We also let $\mathcal{S}$ represent the set of regular functions that are also schlicht (i.e., univalent) in the domain $\Xi = \{z \in \mathbb{C}$ such that $|z| < 1\}$, so that functions in $\mathcal{S}$ can be expressed in the series

$$F(z) = z + \sum_{y=2}^{\infty} A_y z^y, \quad z \in \Xi \tag{1}$$

which satisfies the following conditions: $F(0) = 0$ and $F'(0) - 1 = 0$. Some known subsets of $\mathcal{S}$ include $\mathcal{C}$ and $\mathcal{S}^{\star}$, known to contain functions $F(z) \in \mathcal{S}$ that, respectively, satisfy the following conditions: $\mathcal{R}e\left(1 + (zF''/F')\right) > 0$, and $\mathcal{R}e\left(zF'/F\right) > 0$ for $z \in \Xi$. They are, respectively, known as the sets of convex and starlike functions. The essence of these two subclasses of $\mathcal{S}$ is so evident in many literature through several classes defined via them and the numerous results arising therefrom.

Suppose there exists a function $s(z)$ ($|s(z)| = |z| < 1$) regular in Ξ, where $F(z) = J(s(z))$, $z \in \Xi$. Then F is said to be subordinate to J usually represented as $F(z) \prec J(z)$, $z \in \Xi$ for $F, J \in \mathcal{A}$. Indeed, if $J \in \mathcal{S}$, then $F \prec J \iff F(0) = J(0)$ and $F(\Xi) \subset J(\Xi)$. Function $s(z)$ is known as Schwarz function. Also, for

E. A. Oyekan (✉)
Department of Mathematical Sciences, Olusegun Agagu University of Science and Technology, Okitipupa, Nigeria
e-mail: ea.oyekan@oaustech.edu.ng

A. O. Lasode
Department of Mathematics, University of Ilorin, Ilorin, Nigeira

H.-H. Leung et al. (eds.), *Recent Developments in Algebra and Analysis*, Trends in Mathematics, https://doi.org/10.1007/978-3-031-37538-5_37

$F(z)$ in (1) and $J(z) = z + \sum_{y=2}^{\infty} b_y z^y \in \mathcal{A}$, the convolution of F and J denoted by $(F \star J)(z)$ is defined by $(F \star J)(z) = z + \sum_{y=2}^{\infty} A_y b_y z^y$, $z \in \varXi$.

Bell [3, 4] introduced the numbers: 1, 1, 2, 5, 15, 52, 203, 877, 4140, ... usually represented by β_y, $y \in \mathbb{N} \cup \{0\}$. The Bell numbers are generated from the number of possible partitions of a set. Some properties of Bell numbers can be found in [2, 5, 15]. In particular, Kumar et al. [8] investigated the function

$$\mathcal{B}(z) = e^{e^z - 1} = \sum_{y=0}^{\infty} \beta_y \frac{z^y}{y!} = 1 + z + z^2 + \frac{5}{6}z^3 + \frac{5}{8}z^4 + \cdots, \quad z \in \varXi. \tag{2}$$

It was also demonstrated in [8] that function $\mathcal{B}(z)$ is starlike with respect to 1.

The Struve functions

$$T_p(z) = z + \sum_{y=0}^{\infty} \frac{(-1)^y}{\Gamma(y + \frac{3}{2})\Gamma(p + y + \frac{3}{2})} \left(\frac{z}{2}\right)^{2y+p+1}, \quad z \in \mathbb{C} \tag{3}$$

and

$$W_p^{b,c}(z) = z + \sum_{y=0}^{\infty} \frac{(-1)^y c^y}{\Gamma(y + \frac{3}{2})\Gamma(p + y + \frac{b+2}{2})} \left(\frac{z}{2}\right)^{2y+p+1}, \quad z \in \mathbb{C} \tag{4}$$

where $p, b, c \in \mathbb{C}$ are particular solutions of certain second-order non-homogeneous differential equations. The aforementioned functions $T_p(z)$ and $W_p^{b,c}(z)$ are, respectively, called Struve and generalized Struve functions. Let

$$U_p^{b,c}(z) = 2^p \sqrt{\pi}\, \Gamma\left(p + \frac{b+2}{2}\right) z^{\frac{-p-1}{2}} W_p^{b,c}(\sqrt{z}) \tag{5}$$

so that by utilizing the Pochhammer symbol

$$(t)_j = \frac{\Gamma(t + j)}{\Gamma(t)} = t(t + 1)\ldots(t + j - 1),$$

one can write (5) as

$$U_p^{b,c}(z) = \sum_{y=0}^{\infty} \frac{(-\frac{c}{4})^y}{(\frac{3}{2})_y (t)_y} z^y = b_0 + b_1 z + b_2 z^2 + \cdots \tag{6}$$

for $t = p + \frac{b+2}{2} \neq 0, -1, -2, \cdots$, $b_y = \frac{(-1)^y c^y \Gamma(\frac{3}{2})\Gamma(t)}{4^y \Gamma(y+\frac{3}{2})\Gamma(y+t)}$, $y \geqq 0$ and $b_0 = 1$. We note that function $U_p^{b,c}(z)$ in (6) is regular in $\mathbb{C}$, and it is also clear that $U_p^{b,c}(0) = 1$. (For more details, see [11, 14, 17, 19]). Let

$$V_p^{b,c}(z) = zU_p^{b,c}(z) = z + \sum_{y=2}^{\infty} \frac{(-\frac{c}{4})^{y-1}}{(\frac{3}{2})_{y-1}(t)_{y-1}} z^y. \tag{7}$$

$V_p^{b,c}$ is the well-known generalized Struve function. Indeed, the univalence, starlike, and convex properties of function (7) were investigated by Orhan and Yagmur [11]. Also, using the principle of convolution, and in view of function (1) and (7), Raza and Yagmur [17] defined and investigated the function

$$\begin{aligned} G(z) = (F \star V_p^{b,c})(z) &= z + \sum_{y=2}^{\infty} \frac{(-\frac{c}{4})^{y-1}}{(\frac{3}{2})_{y-1}(t)_{y-1}} a_y z^y \\ &= z - \frac{c}{6t_1} a_2 z^2 + \frac{c^2}{20t_2} a_3 z^3 - \frac{c^3}{56t_3} a_4 z^4 + \frac{c^4}{144t_4} a_5 z^5 - \cdots \end{aligned} \tag{8}$$

where $(t)_1 = t_1$, $(t)_2 = t_2, \ldots$ and $(t)_y = \frac{\Gamma(t+y)}{\Gamma(t)} = t_y \ \forall y \in \mathbb{N}$. Next, let

$$\mathcal{G}(p, b, c; F) := \left\{ G(z) : G(z) = z + \sum_{y=2}^{\infty} \frac{(-\frac{c}{4})^{y-1}}{(\frac{3}{2})_{y-1}(t)_{y-1}} a_y \, z^y, \ z \in \Xi \right\}. \tag{9}$$

Another special function of interest in this work is the error function erF. It got its name as a result of its importance in the study of errors from numerical computations. The error function finds its usefulness in error theory in probability theory and in mathematical physics where it can be expressed as a special case of the Whittaker function. Some practical applications include such study areas like the theory of optics, combustion, and quantum mechanics. The error function $erF(z)$ was reported by Abramowitz and Stegun [1] as

$$erF(z) = \frac{2}{\sqrt{\pi}} \int_0^z \exp(-\eta^2) d\eta = \frac{2}{\sqrt{\pi}} \sum_{y=0}^{\infty} \frac{(-1)^{y-1} z^{y+1}}{(2y+1)!}, \quad z \in \mathbb{C} \tag{10}$$

whose properties are also investigated in [2, 5, 6]. Ramachandran et al. [16] reported a modified version of (10) as

$$ErF(z) = z + \sum_{y=2}^{\infty} \frac{(-1)^{y-1}}{(2y-1)(y-1)!} z^y \tag{11}$$

and used it to define a certain set of regular-univalent functions. Some properties of the set were thereafter obtained. Indeed, the relevance of the aforementioned special functions cannot be underrated especially in areas of numerical computations in mathematical physics and probability.

The q-differentiation of F in (1) is defined by

$$\left.\begin{aligned} \mathcal{D}_qF(z) &= \begin{cases} F'(z) & \text{for } z = 0 \\ \frac{F(z)-F(qz)}{(1-q)z} & \text{for } z \neq 0 \end{cases} \\ \mathcal{D}_q^2F(z) &= \mathcal{D}_q(\mathcal{D}_qF(z)) \end{aligned}\right\} \tag{12}$$

where

$$q \in (0, 1), \quad [y]_q = \frac{1-q^y}{1-q} = 1 + q + q^2 + \cdots + q^{y-1}, \quad \text{and} \quad \lim_{q \uparrow 1}[y]_q = y. \tag{13}$$

It was introduced by Jackson [7] (for more details, see [9, 10]).

In view of (8) and (11), we define the function

$$H(z) = (G \star ErF)(z) = z + \sum_{y=2}^{\infty} \frac{(-1)^{y-1}(-\frac{c}{4})^{y-1}}{(2y-1)(y-1)!(\frac{3}{2})_{y-1}(t)_{y-1}} a_y z^y \tag{14}$$

so that with (12) we obtain

$$\mathcal{D}_qH(z) = 1 + \sum_{y=2}^{\infty} \frac{(-1)^{y-1}(-\frac{c}{4})^{y-1}[y]_q}{(2y-1)(y-1)!(\frac{3}{2})_{y-1}(t)_{y-1}} a_y z^{y-1}. \tag{15}$$

Henceforth, we shall let $b \in \mathbb{C} \setminus \{0\}$, $t \neq 0, -1, -2, \cdots$, $c \in \mathbb{R}^+$, $q \in (0, 1)$, and the functions $\mathcal{B}(z)$ and $H(z)$ are as defined in (2) and (14), respectively, unless otherwise explicitly declared.

Definition 1 We say that function $H \in \mathcal{A}$ is a member of the set $\phi\mathcal{S}_q^\star(b, H, \mathcal{B})$ if the condition

$$1 + \frac{1}{b}\Big(\frac{z\mathcal{D}_qH(z)}{H(z)} - 1\Big) \prec \mathcal{B}(z), \quad z \in \Xi \tag{16}$$

is satisfied and that $H \in \mathcal{A}$ is a member of the set $\phi\mathcal{C}_q(b, H, \mathcal{B})$ if the condition

$$1 + \frac{1}{b}\Big(\frac{z\mathcal{D}_qH(z)}{\mathcal{D}_qH(z)}\Big) \prec \mathcal{B}(z), \quad z \in \Xi \tag{17}$$

is satisfied.

2 Needed Lemmas

Let the regular function $s(z) = \sum_{y=1}^{\infty} w_y z^y \in \Omega$ where Ω represents the set of Schwarz functions defined in Ξ. Then the following lemmas hold.

Lemma 1 **([18])** *Let* $s(z) \in \Omega$, *then* $|w_y| \leqq 1$ $(y \in \mathbb{N})$. *Equality occurs for functions* $s(z) = e^{i\vartheta} z^y$ $(\vartheta \in [0, 2\pi))$.

Lemma 2 **([18])** *Let* $s(z) \in \Omega$, *then for* $\gamma \in \mathbb{C}$, $|w_2 + \gamma w_1^2| \leqq \max\{1, |\gamma|\}$. *Equality occurs for function* $s(z) = z^2$.

Lemma 3 **([14])** *Let* $y \in \mathbb{N}$, $t = p + \frac{b+2}{2} \neq 0, -1, -2.\cdots$ *and* $p, b \in \mathbb{C}$. *If* $(t)_1 = t_1$, $(t)_2 = t_2, \ldots, (t)_y = \frac{\Gamma(t+y)}{\Gamma(t)} \equiv t(t+1)\cdots(t+y-1)$; *then it is clear that:*

(i) *If* $p = -1$ *and* $b = 2$, *then* $t_1 = 1$,
(ii) *If* $p = -1$ *and* $b = 2$, *then* $t_2 = 2$,
(iii) *If* $p = -1$ *and* $b = 2$, *then* $t_3 = 6$.

3 Main Results

Motivated by the works in [2, 12, 13], we shall now present our main results as follows.

Theorem 1 *Let* $H \in \phi S_q^{\star}(b, H, \mathcal{B})$, *then*

$$|a_2| \leqq \frac{18|b|t_1}{cq}, \tag{18}$$

$$|a_3| \leqq \frac{200|b|t_2}{c^2(1+q)q} \max\left\{1, \left|\frac{b+q}{q}\right|\right\}, \tag{19}$$

$$|a_4| \leqq \frac{2,352|b|t_3}{c^3(1+q+q^2)q} \max\left\{1, \left|\sigma\left[\frac{\gamma}{\sigma} + \left(\frac{q+b}{q}\right) + \left(1 + \frac{2}{\sigma}\right)\right]\right|\right\} \tag{20}$$

and

$$|a_3 - \mu a_2^2| \leqq \frac{200|b|t_2}{c^2(1+q)q} \max\left\{1, \left|\frac{50t_2q^2 + 50t_2qb + 81\mu t_1^2 q(1+q)b}{50t_2q^2}\right|\right\}, \tag{21}$$

where $\mu \in \mathbb{C}$, $\sigma = \frac{([2]_q-1)+([3]_q-1)}{([2]_q-1)([3]_q-1)} b$ *and* $\gamma = \frac{5}{6} - \frac{b^2}{([2]_q-1))^2}$.

Proof Suppose $H \in \phi S_q^\star(b, H, \mathcal{B})$; then the principle of subordination allows us to write (16) as

$$1+\frac{1}{b}\Big(\frac{z\mathcal{D}_q H(z)}{H(z)}-1\Big)=\mathcal{B}(s(z)) \tag{22}$$

where function $s(z)$ is given by (18) so that (22) simplifies to

$$[z\mathcal{D}_q H(z)-H(z)]H^{-1}(z)=b[\mathcal{B}(s(z))-1]. \tag{23}$$

Using (14) and (15) in the LHS of (23) we obtain

$$\begin{aligned} z\mathcal{D}_q H(z)-H(z) =& z\Big(1+\sum_{y=2}^{\infty}\frac{(-1)^{y-1}(-\frac{c}{4})^{y-1}[y]_q}{(2y-1)(y-1)!(\frac{3}{2})_{y-1}(t)_{y-1}}a_y z^{y-1}\Big)\\ &-\Big(z+\sum_{y=2}^{\infty}\frac{(-1)^{y-1}(-\frac{c}{4})^{y-1}}{(2y-1)(y-1)!(\frac{3}{2})_{y-1}(t)_{y-1}}a_y z^{y}\Big)\\ =&\sum_{y=2}^{\infty}\frac{([y]_q-1)\,(-1)^{y-1}(-\frac{c}{4})^{y-1}}{(2y-1)(y-1)!(\frac{3}{2})_{y-1}(t)_{y-1}}a_y z^{y}. \end{aligned}$$

or

$$\begin{aligned} z\mathcal{D}_q H(z)-H(z) = ([2]_q-1)\frac{c}{18t_1}a_2z^2+([3]_q-1)\frac{c^2}{200t_2}a_3z^3\\ +([4]_q-1)\frac{c^3}{2352t_3}a_4z^4+\cdots. \end{aligned} \tag{24}$$

Using the binomial expansion theorem, we obtain

$$\begin{aligned} H^{-1}(z) =& z^{-1}+(-1)z^{-2}\Big(\sum_{y=2}^{\infty}\frac{(-1)^{y-1}(-\frac{c}{4})^{y-1}}{(2y-1)(y-1)!(\frac{3}{2})_{y-1}(t)_{y-1}}a_y z^{y}\Big)\\ &+\frac{(-1)(-2)}{2!}z^{-3}\Big(\sum_{y=2}^{\infty}\frac{(-1)^{y-1}(-\frac{c}{4})^{y-1}}{(2y-1)(y-1)!(\frac{3}{2})_{y-1}(t)_{y-1}}a_y z^{y}\Big)^2+\cdots\\ =&\frac{1}{z}-\frac{c}{18t_1}a_2-\frac{c^2}{200t_2}a_3z-\frac{c^3}{2352t_3}a_4z^2-\cdots\\ &+\frac{c^2}{324t_1^2}a_2^2z+\frac{c^3}{3600t_1t_2}a_2a_3z^2+\frac{c^4}{42{,}336t_1t_3}a_2a_4z^3+\cdots \end{aligned}$$

$$+\frac{c^3}{3600t_1t_2}a_2a_3z^2+\frac{c^4}{40{,}000t_2^2}a_3^2z^3+\frac{c^5}{470{,}400t_2t_3}a_3a_4z^4+\cdots$$

$$+\frac{c^4}{42{,}336t_1t_3}a_2a_4z^3+\frac{c^5}{470{,}400t_2t_3}a_3a_4z^4+\cdots$$

so that we obtain a simplified function

$$H^{-1}(z)=\frac{1}{z}-\frac{c}{18t_1}a_2+\Big(\frac{c^2}{324t_1^2}a_2^2-\frac{c^2}{200t_2}a_3\Big)z$$
$$+\Big(\frac{c^3}{1800t_1t_2}a_2a_3-\frac{c^3}{2352t_3}a_4\Big)z^2+\cdots. \tag{25}$$

Now the product of (24) and (25) yields

$$[z\mathcal{D}_qH(z)-H(z)]H^{-1}(z)=([2]_q-1)\frac{c}{18t_1}a_2z$$
$$+\left(([3]_q-1)\frac{c^2}{200t_2}a_3-([2]_q-1)\frac{c^2}{324t_1^2}a_2^2\right)z^2$$
$$+\left(([4]_q-1)\frac{c^3}{2352t_3}a_4+([2]_q-1)\frac{c^3}{5832t_1^3}a_2^3\right.$$
$$\left.-\Big[([2]_q-1)+([3]_q-1)\Big]\frac{c^3}{3600t_1t_2}a_2a_3\right)z^3+\cdots. \tag{26}$$

Expanding the RHS of (23) gives

$$b[\mathcal{B}(s(z))-1]=bw_1z+b(w_2+w_1^2)z^2+b(w_3+2w_1w_2+\frac{5}{6}w_1^3)z^3+\cdots. \tag{27}$$

The comparison of coefficients in (26) and (27) shows that

$$([2]_q-1)\frac{c}{18t_1}a_2=bw_1, \tag{28}$$

$$([3]_q-1)\frac{c^2}{200t_2}a_3-([2]_q-1)\frac{c^2}{324t_1^2}a_2^2=b(w_2+w_1^2) \tag{29}$$

and

$$([4]_q-1)\frac{c^3}{2352t_3}a_4+([2]_q-1)\frac{c^3}{5832t_1^3}a_2^3-\Big[([2]_q-1)+([3]_q-1)\Big]\frac{c^3}{3600t_1t_2}a_2a_3$$
$$= b(w_3+2w_1w_2+\frac{5}{6}w_1^3). \qquad (30)$$

Simple computation shows that from (28), we will obtain

$$a_2=\frac{18t_1bw_1}{c([2]_q-1)} \qquad (31)$$

and by the application of triangle inequality, (13), Lemmas 1 and 3, we obtain inequality (18). Also, by using (31) in (29) and simplifying, we obtain

$$a_3=\left[\frac{200t_2b}{c^2q(1+q)}\right]\left[w_2+\left(\frac{[2]_q-1+b}{[2]_q-1}\right)w_1^2\right] \qquad (32)$$

and by applying triangle inequality, (13) and Lemmas 2 and 3 for $\gamma=\frac{([2]_q-1)+b}{1-[2]_q}$, we obtain inequality (19). Lastly, putting (31) and (32) into (30), using $\sigma=\frac{([2]_q-1)+([3]_q-1)}{([2]_q-1)([3]_q-1)}b$ and $\gamma=\frac{5}{6}-\frac{b^2}{([2]_q-1)^2}$ and simplifying, we obtain

$$a_4=\left[\frac{2352t_3b}{c^3([4]_q-1)}\right]\left[w_3+\sigma\left(\frac{\gamma}{\sigma}+\left(\frac{[2]_q-1+b}{[2]_q-1}\right)\right)w_1^3+\sigma\left(\frac{2}{\sigma}+1\right)w_1w_2\right]$$

and by applying triangle inequality, (13), Lemmas 1, 2 and 3, we obtain inequality (20).

Next, by using (31) and (32), and for $\mu\in\mathbb{C}$, we obtain

$$a_3-\mu a_2^2=\left[\frac{200t_2b}{c^2([3]_q-1)}\right]\Bigg[w_2+\left(\frac{50t_2([2]_q-1)^2+50bt_2([2]_q-1)+81\mu t_1^2b([3]_q-1)}{50t_2([2]_q-1)^2}\right)w_1^2\Bigg]$$

and by applying triangle inequality, (13), Lemmas 2 and 3, we obtain inequality (21).

Theorem 2 *Let $H\in\phi\mathcal{C}_q(b,H,\mathcal{B})$, then*

$$|a_2|\leqq\frac{18|b|t_1}{c(1+q)}, \qquad (33)$$

$$|a_3|\leqq\frac{200|b|t_2}{c^2(1+q)q}\max\left\{1,\left|1+b\right|\right\}, \qquad (34)$$

$$|a_3 - \mu a_2^2| \leqq \frac{200|b|t_2}{c^2(1+q)q} \max\left\{1, \left|\frac{50t_2(1+q)^2(1+b) - 81\mu t_1^2 q(1+q)b}{50t_2(1+q)^2}\right|\right\} \tag{35}$$

where $\mu \in \mathbb{C}$.

Proof Suppose that $H \in \phi\mathcal{C}_q(b, H, \mathcal{B})$, then the principle of subordination allows us to write (17) as

$$1 + \frac{1}{b}\left(\frac{z\mathcal{D}_q H(z)}{\mathcal{D}_q H(z)}\right) = \mathcal{B}(s(z))$$

or

$$[z\mathcal{D}_q H(z)][\mathcal{D}_q H(z)]^{-1} = b[\mathcal{B}(s(z)) - 1]. \tag{36}$$

From (15),

$$z\mathcal{D}_q H(z) = z + \sum_{y=2}^{\infty} \frac{(-1)^{y-1}(-\frac{c}{4})^{y-1}[y]_q}{(2y-1)(y-1)!(\frac{3}{2})_{y-1}(t)_{y-1}} a_y z^y$$

so that further simplification gives

$$z\mathcal{D}_q H(z) = z + [2]_q \frac{c}{18t_1} a_2 z^2 + [3]_q \frac{c^2}{200t_2} a_3 z^3 + [4]_q \frac{c^3}{2352t_3} a_4 z^4 + [5]_q \frac{c^4}{3110t_4} a_5 z^5 + \cdots. \tag{37}$$

By binomial expansion,

$$[\mathcal{D}_q H(z)]^{-1} = 1 - \left(\sum_{y=2}^{\infty} \frac{(-1)^{y-1}(-\frac{c}{4})^{y-1}[y]_q}{(2y-1)(y-1)!(\frac{3}{2})_{y-1}(t)_{y-1}} a_y z^{y-1}\right) + \left(\sum_{y=2}^{\infty} \frac{(-1)^{y-1}(-\frac{c}{4})^{y-1}[y]_q}{(2y-1)(y-1)!(\frac{3}{2})_{y-1}(t)_{y-1}} a_y z^{y-1}\right)^2 + \cdots \tag{38}$$

which simplifies to

$$[\mathcal{D}_q H(z)]^{-1} = 1 - [2]_q \frac{c}{18t_1} a_2 z - [3]_q \frac{c^2}{200t_2} a_3 z^2 - [4]_q \frac{c^3}{2352t_3} a_4 z^3$$

$$-[5]_q\frac{c^4}{3110t_4}a_5z^4-\cdots+[2]_q^2\frac{c^2}{324t_1^2}a_2^2z^2+[2]_q[3]_q\frac{c^3}{1800t_1t_2}a_2a_3z^3. \quad (39)$$

Putting (37) and (39) into LHS of (36) simplifies to

$$[z\mathcal{D}_qH(z)][\mathcal{D}_qH(z)]^{-1}=z-[2]_q\frac{c}{18t_1}a_2z^2+\left([3]_q\frac{c^2}{200t_2}a_3-[2]_q^2\frac{c^2}{324t_1^2}a_2^2\right)z^3+\cdots \quad (40)$$

and equating the coefficients in (27) and (40) gives

$$-[2]_q\frac{c}{18t_1}a_2=bw_1 \quad (41)$$

$$[3]_q\frac{c^2}{200t_2}a_3-[2]_q^2\frac{c^2}{324t_1^2}a_2^2=b(w_2+w_1^2). \quad (42)$$

From (41), we obtain

$$a_2=\frac{-18t_1bw_1}{c(1+q)} \quad (43)$$

and the application of triangle inequality, (13), and Lemmas 1 and 3 gives the inequality in (33). To get bound for a_3 we put (43) into (42) to obtain

$$a_3=\left[\frac{200t_2b}{c^2[3]_q}\right]\left[w_2+(1+b)w_1^2\right] \quad (44)$$

and for $\gamma=1+b$ and the application of triangle inequality, (13) and Lemmas 1 and 3 give the inequality in (34). Next, considering (43) and (44), then we obtain

$$a_3-\mu a_2^2=\left[\frac{200t_2b}{c^2[3]_q}\right]\left[w_2+\left(\frac{50t_2[2]_q^2(1+b)-81\mu t_1^2b[3]_q}{50t_2[2]_q^2}\right)w_1^2\right]$$

and the application of triangle inequality, (13) and Lemmas 1 and 3 gives the inequality in (35).

By making use of Lemma 3 and setting $c=2$ in Theorem 1, we obtain

Corollary 1 *Let $q\in(0,1)$, $b\in\mathbb{C}\setminus\{0\}$, $t\neq 0,-1,-2,\cdots$ and $\mathcal{B}(z)$ is defined in (2). Then for $H\in\phi\mathcal{S}_q^{\star}(b,H,\mathcal{B})$, we obtain*

$$|a_2|\leqq\frac{9|b|t_1}{q},$$

$$|a_3| \leqq \frac{50|b|t_2}{(1+q)q} \max\left\{1, \left|\frac{b+q}{q}\right|\right\}$$

$$|a_4| \leqq \frac{294|b|t_3}{(1+q+q^2)q} \max\left\{1, \left|\sigma\left[\frac{\gamma}{\sigma} + \left(\frac{q+b}{q}\right) + \left(1+\frac{2}{\sigma}\right)\right]\right|\right\}$$

and

$$|a_3 - \mu a_2^2| \leqq \frac{50|b|t_2}{(1+q)q} \max\left\{1, \left|\frac{100t_2q^2 + 100qb + 81\mu q(1+q)b}{100q^2}\right|\right\}.$$

where $\mu \in \mathbb{C}$, $\sigma = \frac{([2]_q-1)+([3]_q-1)}{([2]_q-1)([3]_q-1)} b$ *and* $\gamma = \frac{5}{6} - \frac{b^2}{([2]_q-1))^2}$.

By making use of Lemma 3 and setting $c = 2$ in Theorem 2, we obtain

Corollary 2 *Let* $q \in (0, 1)$, $b \in \mathbb{C} \setminus \{0\}$, $t \neq 0, -1, -2, \cdots$ *and* $\mathcal{B}(z)$ *is defined in* (2). *Then for* $H \in \phi\mathcal{C}_q(b, H, \mathcal{B})$, *we obtain*

$$|a_2| \leqq \frac{9|b|t_1}{(1+q)},$$

$$|a_3| \leqq \frac{50|b|t_2}{(1+q)q} \max\left\{1, \left|1+b\right|\right\},$$

$$|a_3 - \mu a_2^2| \leqq \frac{50|b|t_2}{(1+q)q} \max\left\{1, \left|\frac{100(1+q)^2(1+b) - 81\mu q(1+q)b}{100(1+q)^2}\right|\right\}.$$

References

1. Abramowitz, M., Stegun, I.A. (eds.): Handbook of Mathematical Functions with Formulas, Graphs, and Mathematical Tables. Dover Publications, New York (1965)
2. Altinkaya, S., Olatunji, S.O.: Generalized distribution for analytic function classes associated with error functions and Bell numbers. Bull. de la Sociedad Math. Mexicana. **26**, 377–384 (2020). https://doi.org/10.1007/s40590-019-00265-z
3. Bell, E.T.: Exponential polynomials. Ann. Math. **35**, 258–277 (1934). https://doi.org/10.2307/1968431
4. Bell, E.T.: The iterated exponential integers. Ann. Math. **39**, 539–557 (1938). https://doi.org/10.2307/1968633
5. Canifield, E.R.: Engel's inequality for Bell-numbers. J. Combinat. Theo. Ser. **72**, 184–187 (1995). https://doi.org/10.1016/0097-3165(95)90033-0
6. Elbert, A., Laforgia, A.: The zeros of the complementary error function. Numer. Algorithms. **49**(1), 153–157 (2008). https://doi.org/10.1007/s11075-008-9186-7
7. Jackson, F.H.: On q-functions and a certain difference operator. Trans. Roy. Soc. Edinb. **46**(2), 253–281 (1908). https://doi.org/10.1017/S0080456800002751
8. Kumar, V., Cho, N.E., Ravichandran, V., Srivastava, H.M.: Sharp coefficient bounds for starlike functions associated with the Bell numbers. Math. Slovaca. **69**, 1053–1064 (2019). https://doi.org/10.1515/ms-2017-0289

9. Lasode, A.O.: Estimates for a generalized class of analytic and bi-univalent functions involving two q-operators. Earthline J. Math. Sci. **10**(2), 211–225 (2022). https://doi.org/10.34198/ejms.10222.211225
10. Lasode, A.O., Opoola, T.O.: Some investigations on a class of analytic and univalent functions involving q-differentiation. Eur. J. Math. Anal. **2**(12), 1–9 (2022). https://doi.org/10.28924/ada/ma.2.12
11. Orhan, H., Yagmur, N.: Geometric properties of generalized Struve functions. In: International Congress in Honour of Professor H.M. Srivastava, Bursa, Turkey, pp. 23–26 (2012)
12. Oyekan, E.A.: Certain geometric properties of functions involving Galué-type Struve functions. Ann. Math. Comput. Sci. **8**, 43–53 (2022). annalsmcs.org
13. Oyekan, E.A., Aderibola, S.O.: New results on the Chebyshev polynomial bounds for classes of univalent functions. Asia Pac. J. Math. **7**(24), 12–22 (2020). https://doi.org/10.28924/APJM/7-24
14. Oyekan, E.A., Opoola, T.O.: Hankel determinant for a subclass of analytic functions associated with generalized Struve function of order p bounded by conical regions. Palestine J. Math. **11**(2), 395–405 (2022). ppu.edu
15. Qi, F.: Some inequalities for Bell-numbers. Proc. Sci. Math. Sci. **127**, 551–564 (2017). https://doi.org/10.1007/s12044-017-0355-2
16. Ramachandran, C., Dhanalakshmi, K., Vanitha, L.: Hankel determinant for a subclass of analytic functions associated with error functions bounded by conical regions. Intern. J. Math. Anal. **11**(12), 571–581 (2017). https://doi.org/10.12988/ijma.2017.7468
17. Raza, M., Yagmur, N.: Some properties of a class of analytic functions defined by geometric Struve functions. Turkish J. Math. **39**(6), 1–14 (2015). https://doi.org/10.3906/mat-1501-48
18. Thomas, D.K., Tuneski, N., Vasudevarao, A.: Univalent Functions: A Primer. Walter de Gruyter, Berlin (2018). https://doi.org/10.1515/9783110560961-001
19. Yagmur, N., Orhan, H.: Hardy space of generalized Struve functions. Complex Var. Elliptic Equ. **59**(7), 1–8 (2014). https://doi.org/10.1080/17476933.2013.799148